THE INTERNATIONAL SERIES OF MONOGRAPHS ON PHYSICS

GENERAL EDITORS

R. K. ADAIR   R. J. ELLIOTT
W. MARSHALL   D. H. WILKINSON
M. REES   H. EHRENREICH

THEORY OF NEUTRON SCATTERING FROM CONDENSED MATTER

Volume 2: Polarization Effects and Magnetic Scattering

STEPHEN W. LOVESEY

Rutherford Appleton Laboratory

CLARENDON PRESS · OXFORD

This book has been printed digitally and produced in a standard specification in order to ensure its continuing availability

OXFORD
UNIVERSITY PRESS

Great Clarendon Street, Oxford OX2 6DP

Oxford University Press is a department of the University of Oxford.
It furthers the University's objective of excellence in research, scholarship, and education by publishing worldwide in

Oxford New York

Auckland Bangkok Buenos Aires Cape Town Chennai
Dar es Salaam Delhi Hong Kong Istanbul Karachi Kolkata
Kuala Lumpur Madrid Melbourne Mexico City Mumbai Nairobi
São Paulo Shanghai Taipei Tokyo Toronto

Oxford is a registered trade mark of Oxford University Press
in the UK and in certain other countries

Published in the United States
by Oxford University Press Inc., New York

© Stephen W. Lovesey, 1984

The moral rights of the author have been asserted
Database right Oxford University Press (maker)

Reprinted 2003

All rights reserved. No part of this publication may be reproduced, stored in a retrieval system, or transmitted, in any form or by any means, without the prior permission in writing of Oxford University Press, or as expressly permitted by law, or under terms agreed with the appropriate reprographics rights organization. Enquiries concerning reproduction outside the scope of the above should be sent to the Rights Department, Oxford University Press, at the address above

You must not circulate this book in any other binding or cover and you must impose this same condition on any acquirer

ISBN 0-19-852029-8

ACKNOWLEDGEMENTS

I am indebted to the following for permission to use diagrams as a basis for figures in the text: M. F. Collins, J. M. Hastings, W. C. Koehler, H. A. Mook, R. M. Moon, R. Nathans, T. Riste, C. G. Shull, F. A. Wedgwood, The Royal Society, American Institute of Physics, *The Physical Review* and *Physical Review Letters*, and Institute of Physics (London).

CONTENTS

CONTENTS OF VOLUME 1

IMPORTANT SYMBOLS

$A = \{(i+1)b^{(+)} + ib^{(-)}\}/(2i+1)$;
$= \bar{\hat{b}}$ coherent scattering amplitude; § 1.6
$\hat{a}^+$, $\hat{a}$ Bose creation and annihilation operators
Å (angstrom) $= 10^{-8}$ cm

$B = 2\{b^{(+)} - b^{(-)}\}/(2i+1)$; § 1.6
$b^{(+)}$ scattering length for $i+\frac{1}{2}$ state } for bound nuclei
$b^{(-)}$ scattering length for $i-\frac{1}{2}$ state } for bound nuclei
$\hat{b}$ scattering amplitude operator; §§ 1.6, 10.2
$\bar{b}$ coherent scattering length for bound nuclei; Table 1.1

c concentration of impurities
$\hat{c}^+$, $\hat{c}$ Fermi creation and annihilation operators

$\mathbf{d}$ position vector
$\mathrm{d}\sigma/\mathrm{d}\Omega$ differential cross-section
$\mathrm{d}^2\sigma/\mathrm{d}\Omega\,\mathrm{d}E'$ partial differential cross-section; eqns (1.23), (3.4), (7.9)

e magnitude of charge of single electron
E, E' initial and final energy of neutron, respectively
eV electron volt

$F(\boldsymbol{\kappa})$ form factor
$F_{\mathrm{N}}(\boldsymbol{\tau})$ unit-cell structure factor
$F_{\mathrm{M}}(\boldsymbol{\tau})$ magnetic unit-cell structure factor

g gyromagnetic ratio

H magnetic field
$\mathscr{H}$ quantum mechanical Hamiltonian
$\hbar$ Planck's constant/2π

$\hat{\mathbf{i}}$ nuclear spin operator of magnitude i
Im imaginary part

$\hat{\mathbf{J}} = \hat{\mathbf{L}} + \hat{\mathbf{S}}$ total angular momentum operator of magnitude J
$J(\mathbf{n})$ Heisenberg exchange parameter § 9.2
$\mathscr{J}(\mathbf{q})$ Fourier transform of $J(\mathbf{n})$
$j_K(x)$ spherical Bessel function of order K

$\mathbf{k}$, $\mathbf{k}'$ initial and final wave vector of neutron, respectively
k_B Boltzmann's constant

l cell indices
$\mathbf{l}$ position vector; § 2.1

m neutron mass
m_e electron mass
M mass of nucleus, effective electron mass
meV (millielectron volt) $= 10^{-3}$ eV
$M^\alpha = -g\mu_B\langle\hat{S}^\alpha\rangle$ magnetic moment

N number of unit cells in crystal or particles
$n(\omega) = [\exp(\hbar\omega\beta) - 1]^{-1}$ Bose factor

$\hat{\mathbf{p}} = -i\hbar\boldsymbol{\nabla}$ momentum operator
p_f Fermi wave vector
p_σ incident neutron spin probability
p_λ probability distribution for initial target states
$\mathbf{P}$, $\mathbf{P}'$ initial and final polarization; Chapter 10

$\mathbf{q}$ wave vector
$\hat{\mathbf{Q}}_\perp = \tilde{\boldsymbol{\kappa}} \times (\hat{\mathbf{Q}} \times \tilde{\boldsymbol{\kappa}})$ magnetic interaction operator; eqns (7.10), (11.7)

$r_0 = (\gamma e^2/m_e c^2) = -0.54 \times 10^{-12}$ cm; Chapter 7
$\mathbf{R}_{ld} = \mathbf{l} + \mathbf{d}$ position vector of nucleus in rigid lattice
$R_{AB}(t)$ relaxation function; Table 8.1
$R_{AB}(\omega)$ Fourier transform of relaxation function
$R_{\mathbf{q}}(t)$ spin relaxation function; § 8.3
Re real part

$\hat{\mathbf{s}}$ electron spin operator
$\hat{\mathbf{S}}$ spin operator of magnitude S
$S(\boldsymbol{\kappa}, \omega)$ response function; eqns (7.11b), (7.91) and Table 8.1

T absolute temperature $\beta = (k_B T)^{-1}$
t time variable

$\hat{\mathbf{u}}(l, t)$ displacement operator

V volume of target sample
v_0 volume of unit cell

$\hat{V}(\boldsymbol{\kappa})$ Fourier transform of neutron-matter interaction potential multiplied by $(m/2\pi\hbar^2)$

$W(\boldsymbol{\kappa})$ exponent in Debye–Waller factor; §§ 4.3, 8.1

$Y_{jj'}(\boldsymbol{\kappa}, t)$ correlation function; eqns (3.6), (10.146)

$Z = \mathrm{Tr}\exp(-\beta\hat{\mathcal{H}})$ partition function

α (with β) Cartesian component index
$\hat{\boldsymbol{\alpha}}$ neutron-matter interaction operator; § 10.2

β (with α) Cartesian component index
$\beta = (k_{\mathrm{B}}T)^{-1}$
$\hat{\beta}$ neutron-matter interaction operator; § 10.2

γ_q eqn (9.17)

$\delta(x)$ Dirac delta function; eqns (1.19), (1.20), (1.87)
$\delta_{l,m}$ Kronecker delta function

ϵ_{f} Fermi energy

$\hat{\boldsymbol{\eta}}_j(\boldsymbol{\kappa})$ unit vector in direction of moment at site $\mathbf{R}_j$; § 10.5

$\theta(t)$ unit step function
θ scattering angle $\mathbf{k}\cdot\mathbf{k}' = kk'\cos\theta$

$\boldsymbol{\kappa} = \mathbf{k} - \mathbf{k}'$ neutron wave vector change; Fig. 10.6

μ_{B} Bohr magneton
$\mu_{\mathrm{N}} = (m_{\mathrm{e}}\mu_{\mathrm{B}}/m)$

$\boldsymbol{\rho}$ vector to nearest neighbour atoms

$\frac{1}{2}\hat{\boldsymbol{\sigma}}$ neutron spin operator, $\hat{\boldsymbol{\sigma}}$ Pauli matrix
σ, σ' initial and final neutron spin quantum numbers in cross-section
σ_{c} single (bound) nucleus coherent cross-section
$\sigma_{\mathrm{i}} = \sigma - \sigma_{\mathrm{c}}$ single (bound) nucleus incoherent cross-section
σ total single (bound) nucleus cross-section; Table 1.1

$\boldsymbol{\tau} = t_1\boldsymbol{\tau}_1 + t_2\boldsymbol{\tau}_2 + t_3\boldsymbol{\tau}_3$ reciprocal lattice vectors; § 2.1

$\varphi(\mathbf{r})$ Bloch wave function

$\phi(t)$ linear response function; Table 8.1
$\phi_{\mathbf{q}}(t)$ wave vector dependent linear response function

χ isothermal susceptibility $= -\chi[0]$
$\chi[\omega] = \chi'[\omega] + i\chi''[\omega]$ generalized susceptibility; Table 8.1

$\omega_{\mathbf{q}}$ magnon frequency

$\hbar\omega = \hbar^2(k^2 - k'^2)/2m$ neutron energy change

Ω solid angle

∂_t derivative with respect to t

7
PRINCIPAL FEATURES OF MAGNETIC SCATTERING

Let us express the magnetic moment of a neutron in terms of a Pauli (spin) matrix $\hat{\boldsymbol{\sigma}}$; the spin operator is then $(\hat{\boldsymbol{\sigma}}/2)$. Denoting the nuclear Bohr magneton by μ_N, the magnetic moment operator for a neutron is

$$\hat{\boldsymbol{\mu}} = \gamma\mu_N\hat{\boldsymbol{\sigma}} \tag{7.1}$$

where the gyromagnetic ratio $\gamma = -1.91$. The interaction of a neutron with a magnetic field $\mathbf{H}$ is described by the potential

$$-\hat{\boldsymbol{\mu}} \cdot \mathbf{H} = -\gamma\mu_N\hat{\boldsymbol{\sigma}} \cdot \mathbf{H}. \tag{7.2}$$

When the magnetic field is generated by electrons in the target sample, we find that a natural unit for the strength of the interaction (7.2) is the classical electron radius, $(e^2/m_e c^2) = 0.282 \cdot 10^{-12}$ cm. The quantity

$$r_0 = (\gamma e^2/m_e c^2) = -0.54 \cdot 10^{-12}\ \text{cm} \tag{7.3}$$

is then a useful unit for magnetic scattering lengths. From the magnitude of r_0 it follows that magnetic and nuclear scattering cross-sections are of a similar size. The cross-section for scattering by the magnetic field generated by the nuclei in the target sample is usually insignificant since the corresponding scattering length $= (m_e r_0/m)$.

In this chapter we derive an expression for the cross-section for scattering by unpaired electrons, starting from the Born approximation (1.23). The interaction operator in the latter is taken to be eqn (7.2). Our first task is therefore to obtain an expression for the magnetic field generated by the electrons, and to evaluate the matrix element of (7.2) with respect to the initial and final neutron states.

7.1. Cross-section for electrons

For magnetic salts and rare-earth metals it is reasonable to consider the electrons as well localized about the ions, whereas the unpaired electrons in semiconductors and 3d transition metals are quite mobile. We are therefore led to consider two principal models for electrons in materials which are usually referred to as the atomic, and the itinerant- or band-electron models. For some materials, e.g. actinides, neither of these (extreme) models are really appropriate. We consider the features of scattering from atomic and band electrons after deriving a general expression for the partial differential cross-section for scattering unpolarized neutrons by electrons.

The magnetic field due to a single electron moving with velocity $\mathbf{v}_e$ is (Landau and Lifschitz 1981),

$$\mathbf{H} = \operatorname{curl}\left\{\frac{\boldsymbol{\mu}_e \times \mathbf{R}}{|\mathbf{R}|^3}\right\} + \frac{(-e)}{c}\frac{\mathbf{v}_e \times \mathbf{R}}{|\mathbf{R}|^3} \tag{7.4}$$

where $\mathbf{R}$ is the distance from the electron to the point at which the field is measured. The magnetic moment operator of an electron $\hat{\boldsymbol{\mu}}_e$ is

$$\hat{\boldsymbol{\mu}}_e = -2\mu_B \hat{\mathbf{s}},$$

so that the interaction potential between the neutron and the electron becomes

$$-\gamma\mu_N \hat{\boldsymbol{\sigma}} \cdot \mathbf{H} = \gamma\mu_N\left\{2\mu_B \hat{\boldsymbol{\sigma}} \cdot \operatorname{curl}\left(\frac{\hat{\mathbf{s}} \times \mathbf{R}}{|\mathbf{R}|^3}\right) - \frac{e}{2m_e c}\left(\hat{\mathbf{p}}_e \cdot \frac{\hat{\boldsymbol{\sigma}} \times \mathbf{R}}{|\mathbf{R}|^3} + \frac{\hat{\boldsymbol{\sigma}} \times \mathbf{R}}{|\mathbf{R}|^3} \cdot \hat{\mathbf{p}}_e\right)\right\} \tag{7.5}$$

with

$$\hat{\mathbf{p}}_e = -i\hbar \boldsymbol{\nabla}_e.$$

Note that the second term has been written as one-half of the sum of $\hat{\mathbf{p}}_e \cdot (\hat{\boldsymbol{\sigma}} \times \mathbf{R}/|\mathbf{R}|^3)$ and $(\hat{\boldsymbol{\sigma}} \times \mathbf{R}/|\mathbf{R}|^3) \cdot \hat{\mathbf{p}}_e$ to take proper account of the operator nature of the momentum of the electron.

The first term on the right-hand side of (7.5) is classically the dipole–dipole interaction between the neutron and the electron while the second term is the interaction of the neutron with the electron that arises from the translational motion of the latter.

From eqn (1.23), the cross-section for the scattering of neutrons through the interaction potential given in eqn (7.5) is

$$\frac{d^2\sigma}{d\Omega\, dE'} = (2\gamma\mu_N\mu_B)^2 \frac{k'}{k} \sum_{\lambda\lambda'\sigma} p_\lambda p_\sigma \left|\left\langle \mathbf{k}'\lambda'\sigma' \left| \sum_i \hat{\boldsymbol{\sigma}} \cdot \operatorname{curl}\left(\frac{\hat{\mathbf{s}}_i \times \mathbf{R}}{|\mathbf{R}|^3}\right) - \frac{1}{2\hbar}\left(\hat{\mathbf{p}}_i \cdot \frac{\hat{\boldsymbol{\sigma}} \times \mathbf{R}}{|\mathbf{R}|^3} + \frac{\hat{\boldsymbol{\sigma}} \times \mathbf{R}}{|\mathbf{R}|^3} \cdot \hat{\mathbf{p}}_i\right)\right| \mathbf{k}\lambda\sigma \right\rangle\right|^2 \delta(\hbar\omega + E_\lambda - E_{\lambda'}). \tag{7.6}$$

In (7.6), $|\lambda\rangle$ and $|\lambda'\rangle$ are the initial and final states of the target and E_λ and $E_{\lambda'}$ the corresponding energies. The operators $\hat{\mathbf{s}}_i$ and $\hat{\mathbf{p}}_i$ are the spin and linear momentum of the ith electron.

The integration over the neutron coordinates in the matrix element of the interaction potential required in eqn (7.6) is quite straightforward if we make use of the well-known identities

$$\mathbf{R}/|\mathbf{R}|^3 = -\boldsymbol{\nabla}(1/|\mathbf{R}|) \tag{7.7}$$

and

$$\frac{1}{|\mathbf{R}|}=\frac{1}{2\pi^2}\int \mathrm{d}\mathbf{q}\frac{1}{q^2}\exp(\mathrm{i}\mathbf{q}\cdot\mathbf{R}).$$

Thus, for example,

$$\begin{aligned}\operatorname{curl}(\hat{\mathbf{s}}\times\mathbf{R}/|\mathbf{R}|^3) &= -\operatorname{curl}\left(\hat{\mathbf{s}}\times\boldsymbol{\nabla}\left(\frac{1}{|\mathbf{R}|}\right)\right)\\ &= -\boldsymbol{\nabla}\times(\hat{\mathbf{s}}\times\boldsymbol{\nabla})\left(\frac{1}{|\mathbf{R}|}\right)\\ &= -\frac{1}{2\pi^2}\int \mathrm{d}\mathbf{q}\frac{1}{q^2}\{\boldsymbol{\nabla}\times\hat{\mathbf{s}}\times\boldsymbol{\nabla})\}\exp(\mathrm{i}\mathbf{q}\cdot\mathbf{R})\\ &= \frac{1}{2\pi^2}\int \mathrm{d}\mathbf{q}\frac{1}{q^2}\{\mathbf{q}\times(\hat{\mathbf{s}}\times\mathbf{q})\}\exp(\mathrm{i}\mathbf{q}\cdot\mathbf{R}).\end{aligned}$$

Using the definition of the scattering vector

$$\boldsymbol{\kappa}=\mathbf{k}-\mathbf{k}',$$

we find,

$$\left(\frac{2\pi\hbar^2}{m}\right)\left\langle\mathbf{k}'\middle|\,\hat{\boldsymbol{\sigma}}\cdot\operatorname{curl}\left(\frac{\hat{\mathbf{s}}_i\times\mathbf{R}}{|\mathbf{R}|^3}\right)\middle|\mathbf{k}\right\rangle = 4\pi\exp(\mathrm{i}\boldsymbol{\kappa}\cdot\mathbf{r}_i)\hat{\boldsymbol{\sigma}}\cdot\{\tilde{\boldsymbol{\kappa}}\times(\hat{\mathbf{s}}_i\times\tilde{\boldsymbol{\kappa}})\} \tag{7.8a}$$

and

$$\left(\frac{2\pi\hbar^2}{m}\right)\left\langle\mathbf{k}'\middle|\,\hat{\mathbf{p}}_i\cdot\frac{\hat{\boldsymbol{\sigma}}\times\mathbf{R}}{|\mathbf{R}|^3}\middle|\mathbf{k}\right\rangle = -\frac{4\pi\mathrm{i}}{|\boldsymbol{\kappa}|}\exp(\mathrm{i}\boldsymbol{\kappa}\cdot\mathbf{r}_i)\hat{\boldsymbol{\sigma}}\cdot(\tilde{\boldsymbol{\kappa}}\times\hat{\mathbf{p}}_i) \tag{7.8b}$$

where $\mathbf{r}_i$ is the position vector of the ith electron and $\tilde{\boldsymbol{\kappa}}=\boldsymbol{\kappa}/|\boldsymbol{\kappa}|$. The results (7.8a), (7.8b) together with the fact that $\tilde{\boldsymbol{\kappa}}\times\hat{\mathbf{p}}_i$ commutes with $\exp(\mathrm{i}\boldsymbol{\kappa}\cdot\mathbf{r}_i)$ enable us to write the neutron cross-section in the form

$$\frac{\mathrm{d}^2\sigma}{\mathrm{d}\Omega\,\mathrm{d}E'}=\left(\frac{m}{2\pi\hbar^2}\right)^2(2\gamma\mu_{\mathrm{N}}\mu_{\mathrm{B}})^2(4\pi)^2\frac{k'}{k}\sum_{\lambda\lambda'\sigma\sigma'}p_\lambda p_\sigma \times\langle\lambda\sigma|\,(\hat{\boldsymbol{\sigma}}\cdot\hat{\mathbf{Q}}_\perp)^+\,|\lambda'\sigma'\rangle\langle\lambda'\sigma'|\,\hat{\boldsymbol{\sigma}}\cdot\hat{\mathbf{Q}}_\perp\,|\lambda\sigma\rangle\,\delta(\hbar\omega+E_\lambda-E_{\lambda'}), \tag{7.9}$$

where the operator $\hat{\mathbf{Q}}_\perp$ is defined by

$$\hat{\mathbf{Q}}_\perp=\sum_i\exp(\mathrm{i}\boldsymbol{\kappa}\cdot\mathbf{r}_i)\left\{\tilde{\boldsymbol{\kappa}}\times(\hat{\mathbf{s}}_i\times\tilde{\boldsymbol{\kappa}})-\frac{\mathrm{i}}{\hbar\,|\boldsymbol{\kappa}|}\tilde{\boldsymbol{\kappa}}\times\hat{\mathbf{p}}_i\right\}. \tag{7.10}$$

The two terms in $\hat{\mathbf{Q}}_\perp$ are usually referred to as the spin and orbital contributions to the magnetic interaction. For unpolarized neutrons

$$\sum_\sigma p_\sigma\langle\sigma|\,\hat{\sigma}_\alpha\hat{\sigma}_\beta\,|\sigma\rangle=\delta_{\alpha,\beta}.$$

Also,

$$\frac{m}{2\pi\hbar^2}\cdot 2\gamma\mu_{\mathrm{N}}\mu_{\mathrm{B}}\cdot 4\pi=\frac{m}{2\pi\hbar^2}\cdot 2\gamma\frac{e\hbar}{2m_{\mathrm{p}}c}\cdot\frac{e\hbar}{2m_{\mathrm{e}}c}\cdot 4\pi=\gamma e^2/m_{\mathrm{e}}c^2=r_0.$$

Thus we have, finally,

$$\frac{\mathrm{d}^2\sigma}{\mathrm{d}\Omega\,\mathrm{d}E'}=r_0^2\frac{k'}{k}S(\boldsymbol{\kappa},\omega) \tag{7.11a}$$

where the response function $S(\boldsymbol{\kappa},\omega)$ is

$$S(\boldsymbol{\kappa},\omega)=\sum_{\lambda\lambda'}p_\lambda\langle\lambda|\,\hat{\mathbf{Q}}_\perp^+\,|\lambda'\rangle\cdot\langle\lambda'|\,\hat{\mathbf{Q}}_\perp\,|\lambda\rangle\,\delta(\hbar\omega+E_\lambda-E_{\lambda'})$$
$$=\sum_{\alpha\beta}(\delta_{\alpha\beta}-\tilde{\kappa}_\alpha\tilde{\kappa}_\beta)\sum_{\lambda\lambda'}p_\lambda\langle\lambda|\,\hat{Q}_\alpha^+\,|\lambda'\rangle\langle\lambda'|\,\hat{Q}_\beta\,|\lambda\rangle\,\delta(\hbar\omega+E_\lambda-E_{\lambda'}). \tag{7.11b}$$

In the second form $\hat{\mathbf{Q}}$ is any operator related to the operator $\hat{\mathbf{Q}}_\perp$ through

$$\hat{\mathbf{Q}}_\perp=\tilde{\boldsymbol{\kappa}}\times(\hat{\mathbf{Q}}\times\tilde{\boldsymbol{\kappa}}) \tag{7.12a}$$

and use has been made of the identity

$$\hat{\mathbf{Q}}_\perp^+\cdot\hat{\mathbf{Q}}_\perp=\sum_{\alpha\beta}(\delta_{\alpha\beta}-\tilde{\kappa}_\alpha\tilde{\kappa}_\beta)\hat{Q}_\alpha^+\hat{Q}_\beta. \tag{7.12b}$$

A useful expression for $\hat{\mathbf{Q}}$ is given in (8.19). In § 11.1 we demonstrate the relation of $\hat{\mathbf{Q}}_\perp$ to the total magnetic moment operator, namely eqn (11.7).

The response function can be expressed in terms of a correlation function formed with $\hat{\mathbf{Q}}_\perp$, or alternatively $\hat{\mathbf{Q}}$. Such a representation is used in § 7.6 to calculate the response of an electron fluid and a general discussion of correlation functions for magnetic scattering is given in Chapter 8. The reader should be alert to the fact that it is often convenient to extract factors in (7.11b) which arise largely from the wave functions for the unpaired electrons; the factors include an atomic form factor for localized electrons. In view of this, we do not have a general and universal definition of the response function for magnetic neutron scattering.

7.2. Spin-only scattering

In this and the following two sections we focus our attention on target samples in which the unpaired electrons possess wave functions localized about the sites of a crystal lattice, i.e. about sites defined by the vectors $\mathbf{R}_{ld}=\mathbf{l}+\mathbf{d}$. (The scattering cross-section when the ions are not stationary, as is the case here, but vibrate about the equilibrium positions $\mathbf{R}_{ld}$, is discussed in § 8.1.) In other words we assume the magnetic properties of the crystal to be well described by the atomic Hartree–Fock picture, each unpaired electron being associated with an atomic orbital of the form

$$f(r)\,Y^{l}_{m}(\tilde{\mathbf{r}})$$

(l = orbital quantum number, m = magnetic quantum number), where $f(r)$ is the radial part of the wave function and $Y^{l}_{m}(\tilde{\mathbf{r}})$ a spherical harmonic or some linear combination as dictated by the crystal field.

In many cases of interest the total orbital angular momentum of the magnetic ions is either zero (e.g. the half-filled shell configurations Mn^{2+}, Fe^{3+}, and Gd^{3+}) or quenched by the crystalline field. Under these circumstances the second part of $\hat{\mathbf{Q}}_{\perp}$ is zero. It follows that in this particular instance we can take

$$\hat{\mathbf{Q}} = \sum_{i} \exp(i\boldsymbol{\kappa}\cdot\mathbf{r}_i)\hat{\mathbf{s}}_i. \tag{7.13}$$

The sum in (7.13) runs over all sites in the crystal (i.e. all $\mathbf{R}_{ld}$) and all the unpaired electrons associated with the ions.

Each cell in the crystal is identical, by definition, so that we can rewrite (7.13) as

$$\sum_{l} \exp(i\boldsymbol{\kappa}\cdot\mathbf{l}) \sum_{i}{}_{\text{cell}} \exp(i\boldsymbol{\kappa}\cdot\mathbf{r}_i)\hat{\mathbf{s}}_i$$

where now the second sum is over all the unpaired electrons in a single cell, as indicated by the suffix cell.

Within the cell the ions are situated at the sites defined by the vectors $\mathbf{d}$ so that if we denote the position of the νth electron of the ion at the site $\mathbf{d}$ relative to this site position by $\mathbf{r}_\nu$, we can write

$$\sum_{i}{}_{\text{cell}} \exp(i\boldsymbol{\kappa}\cdot\mathbf{r}_i)\hat{\mathbf{s}}_i = \sum_{d} \exp(i\boldsymbol{\kappa}\cdot\mathbf{d}) \sum_{\nu(d)} \exp(i\boldsymbol{\kappa}\cdot\mathbf{r}_\nu)\hat{\mathbf{s}}_\nu.$$

Thus (7.13) is

$$\hat{\mathbf{Q}} = \sum_{l,d} \exp(i\boldsymbol{\kappa}\cdot\mathbf{R}_{ld}) \sum_{\nu(d)} \exp(i\boldsymbol{\kappa}\cdot\mathbf{r}_\nu)\hat{\mathbf{s}}_\nu. \tag{7.14}$$

The unpaired electrons of each ion will couple together to give a ground state with a certain total spin $\hat{\mathbf{S}}_{ld}$, and the neutrons will not usually have enough energy to break down this coupling. $\hat{\mathbf{S}}_{ld}$ being the only vector associated with $\hat{\mathbf{Q}}$, the matrix elements of the latter must be proportional to those of $\hat{\mathbf{S}}_{ld}$ (Wigner–Eckart theorem; Elliott and Dawber 1979). We can therefore write

$$\langle\lambda'|\,\hat{\mathbf{Q}}\,|\lambda\rangle = \sum_{l,d} \exp(i\boldsymbol{\kappa}\cdot\mathbf{R}_{ld})F_d(\boldsymbol{\kappa})\langle\lambda'|\,\hat{\mathbf{S}}_{ld}\,|\lambda\rangle \tag{7.15}$$

where $F_d(\boldsymbol{\kappa})$ is the form factor defined as the Fourier transform of the normalized spin density associated with the ion at the dth site in the unit cell, i.e.

$$F_d(\boldsymbol{\kappa}) = \int \mathrm{d}\mathbf{r} \exp(\mathrm{i}\boldsymbol{\kappa} \cdot \mathbf{r}) s_d(\mathbf{r}) \tag{7.16}$$

where $s_d(\mathbf{r})$ is the normalized spin density. Note that, by definition,

$$F_d(0) = 1. \tag{7.17}$$

If we insert (7.15) into (7.11b), the partial differential cross-section for the magnetic scattering by ions with only spin angular momentum is

$$\frac{\mathrm{d}^2\sigma}{\mathrm{d}\Omega\,\mathrm{d}E'} = r_0^2 \frac{k'}{k} \sum_{\alpha,\beta} (\delta_{\alpha\beta} - \tilde{\kappa}_\alpha \tilde{\kappa}_\beta) \sum_{\lambda,\lambda'} p_\lambda \sum_{l,d} \sum_{l',d'} F_d^*(\boldsymbol{\kappa}) F_{d'}(\boldsymbol{\kappa}) \exp\{\mathrm{i}\boldsymbol{\kappa} \cdot (\mathbf{R}_{l'd'} - \mathbf{R}_{ld})\}$$
$$\times \langle\lambda| \hat{S}_{ld}^\alpha |\lambda'\rangle\langle\lambda'| \hat{S}_{l'd'}^\beta |\lambda\rangle\, \delta(\hbar\omega + E_\lambda - E_{\lambda'}). \tag{7.18}$$

7.3. Scattering by ions with spin and orbital angular momentum

In general a magnetic ion in a crystal possesses both spin and orbital angular momentum and so it is necessary to include both terms in eqn (7.10). The calculation of the matrix elements of $\hat{\mathbf{Q}}_\perp$ is then no longer as simple a task as for ths spin-only case considered in the preceding section. A detailed discussion is given in Chapter 11. Here we merely quote certain results obtained there, for the sake of completeness.

If the mean radius of the wave function of the unpaired electrons is much less than $|\boldsymbol{\kappa}|^{-1}$, as is often the case, then $\hat{\mathbf{Q}}$ has the form (dipole approximation)

$$\hat{\mathbf{Q}} \simeq \hat{\mathbf{Q}}^{(\mathrm{D})} = \sum_{l,d} \exp(\mathrm{i}\boldsymbol{\kappa} \cdot \mathbf{R}_{ld})\{\bar{\jmath}_0 \hat{\mathbf{S}}_{ld} + \tfrac{1}{2}(\bar{\jmath}_0 + \bar{\jmath}_2)\hat{\mathbf{L}}_{ld}\} \tag{7.19}$$

where $\hat{\mathbf{L}}_{ld}$ is the angular momentum operator for the ion at the site $\mathbf{R}_{ld}$. The coefficient $\bar{\jmath}_K(\kappa)$, usually referred to as a radial integral, is given by

$$\bar{\jmath}_K(\kappa) = \int_0^\infty \mathrm{d}r\, r^2 j_K(\kappa r)\, |f(r)|^2, \tag{7.20}$$

where j_K is a spherical Bessel function of order K. It has been assumed in (7.19) that all the ions in the crystal are identical, otherwise the radial integrals would depend upon d. Moreover, the radial wave function $f(r)$ is assumed to be the same for all magnetic electrons. In the limit $|\boldsymbol{\kappa}| \to 0$,

$$\bar{\jmath}_0 \hat{\mathbf{S}} + \tfrac{1}{2}(\bar{\jmath}_0 + \bar{\jmath}_2)\hat{\mathbf{L}} \to \tfrac{1}{2}(\hat{\mathbf{L}} + 2\hat{\mathbf{S}}), \tag{7.21}$$

so that the *scattering amplitude in the forward direction is proportional to the total magnetic moment per ion.*

Some magnetic ions that would have no orbital angular momentum because of quenching by the crystalline field are found in some cases to have gyromagnetic ratios g that differ from the spin-only value $g = 2$, i.e. they have a finite orbital angular momentum. This arises because of the action of spin–orbit coupling, which, by mixing excited states into the ground state, aligns some orbital moment parallel or antiparallel to the spin. For these ions, by definition of the g value within the ground state manifold, $\hat{\mathbf{L}}$ may be replaced by $(g-2)\hat{\mathbf{S}}$. Hence,

$$\begin{aligned}\hat{\mathbf{Q}}^{(\mathrm{D})} &= \sum_{l,d} \exp(\mathrm{i}\boldsymbol{\kappa}\cdot\mathbf{R}_{ld})\{\bar{j}_0 + \tfrac{1}{2}(g-2)(\bar{j}_0 + \bar{j}_2)\}\hat{\mathbf{S}}_{ld} \\ &= \tfrac{1}{2}gF(\boldsymbol{\kappa})\sum_{l,d}\exp(\mathrm{i}\boldsymbol{\kappa}\cdot\mathbf{R}_{ld})\hat{\mathbf{S}}_{ld}\end{aligned} \tag{7.22}$$

where $F(\boldsymbol{\kappa})$ is the form factor of the ion and is given by

$$F(\boldsymbol{\kappa}) = \bar{j}_0 + \left(\frac{g-2}{g}\right)\bar{j}_2. \tag{7.23}$$

Examples where the orbital moment is unquenched are the rare-earth ions. In these ions the spin–orbit coupling combines $\hat{\mathbf{S}}$ and $\hat{\mathbf{L}}$ to give various $\hat{\mathbf{J}}$ values. Within the states of given $\mathbf{J}$ we write

$$2\hat{\mathbf{S}} = g_S\hat{\mathbf{J}},$$

$$\hat{\mathbf{L}} = g_L\hat{\mathbf{J}},$$

and

$$g\hat{\mathbf{J}} = \hat{\mathbf{L}} + 2\hat{\mathbf{S}}.$$

To calculate g_S, g_L, and g we take the scalar product of the defining equations with $\hat{\mathbf{J}}$ and use the relation

$$2\hat{\mathbf{L}}\cdot\hat{\mathbf{S}} = \hat{\mathbf{J}}^2 - \hat{\mathbf{L}}^2 - \hat{\mathbf{S}}^2.$$

We then find

$$\begin{aligned} g_S &= \frac{J(J+1) - L(L+1) + S(S+1)}{J(J+1)}, \\ g_L &= \frac{J(J+1) + L(L+1) - S(S+1)}{2J(J+1)},\end{aligned} \tag{7.24}$$

and finally g, the Landé splitting factor for the ion

$$g = 1 + \frac{J(J+1) - L(L+1) + S(S+1)}{2J(J+1)}.$$

With these definitions,

$$\hat{\mathbf{Q}}^{(\mathrm{D})} = \tfrac{1}{2}gF(\boldsymbol{\kappa})\sum_{l,d}\exp(\mathrm{i}\boldsymbol{\kappa}\cdot\mathbf{R}_{ld})\hat{\mathbf{J}}_{ld} \tag{7.25}$$

where

$$F(\boldsymbol{\kappa}) = \bar{j}_0 \frac{g_S}{g} + (\bar{j}_0 + \bar{j}_2) \frac{g_L}{g} \tag{7.26}$$

is the form factor of the ion.

We emphasize that the expression for $\hat{\mathbf{Q}}$ given by eqn (7.19) is only approximate; a detailed analysis of the effects of orbital angular momentum on the cross-section for magnetic scattering is given in Chapter 11.

The results (7.22) and (7.25) allow us to write, within the dipole approximation and for identical magnetic ions,

$$\frac{d^2\sigma}{d\Omega\, dE'} = r_0^2 \frac{k'}{k} \{\tfrac{1}{2} g F(\boldsymbol{\kappa})\}^2 \sum_{\alpha\beta} (\delta_{\alpha\beta} - \tilde{\kappa}_\alpha \tilde{\kappa}_\beta) \sum_{\lambda,\lambda'} p_\lambda$$
$$\times \sum_{l,d} \sum_{l',d'} \exp\{i\boldsymbol{\kappa} \cdot (\mathbf{R}_{l'd'} - \mathbf{R}_{ld})\} \langle \lambda | \hat{S}^\alpha_{ld} | \lambda' \rangle \langle \lambda' | \hat{S}^\beta_{l'd'} | \lambda \rangle\, \delta(\hbar\omega + E_\lambda - E_{\lambda'}), \tag{7.27}$$

where the spin operator $\hat{\mathbf{S}}$ may be the actual spin, or the total angular momentum $\hat{\mathbf{J}}$, or some effective spin operator in the case of scattering by ions with partially quenched orbital angular momentum. The form factor $F(\boldsymbol{\kappa})$ in (7.27) is

$$F(\boldsymbol{\kappa}) = \bar{j}_0 \frac{g_S}{g} + (\bar{j}_0 + \bar{j}_2) \left(\frac{g - g_S}{g} \right), \tag{7.28}$$

where $\mu_B g_S S$ is the spin contribution and $\mu_B (g - g_S) S$ is the orbital moment contribution to the total magnetic moment of an ion $\mu_B g S$.

7.4. Paramagnets

We now illustrate the evaluation of the cross-section (7.27) by considering two simple examples (a) a perfect paramagnet and (b) a paramagnet in a magnetic field. In both examples we assume the ions to form a Bravais lattice.

7.4.1. *Perfect paramagnets*

In a perfect paramagnet the energy is independent of spin orientation and so the sum over λ' in (7.27) can be done immediately by closure and recognizing that the scattering must be elastic. We get

$$\frac{d\sigma}{d\Omega} = r_0^2 \{\tfrac{1}{2} g F(\boldsymbol{\kappa})\}^2 \sum_{\alpha\beta} (\delta_{\alpha\beta} - \tilde{\kappa}_\alpha \tilde{\kappa}_\beta) \sum_{l,l'} \exp\{i\boldsymbol{\kappa} \cdot (\mathbf{l}' - \mathbf{l})\} \sum_\lambda p_\lambda \langle \lambda | \hat{S}^\alpha_l \hat{S}^\beta_{l'} | \lambda \rangle. \tag{7.29}$$

Now

$$\sum_{\lambda} p_{\lambda}\langle\lambda|\,\hat{S}_l^{\alpha}\hat{S}_{l'}^{\beta}\,|\lambda\rangle=\langle\hat{S}_l^{\alpha}\hat{S}_{l'}^{\beta}\rangle, \tag{7.30}$$

where, as usual, the angular brackets denote a thermal average at a temperature $T=1/k_{\rm B}\beta$. For a paramagnet there is no spatial correlation between spins, i.e.

$$\langle\hat{S}_l^{\alpha}\hat{S}_{l'}^{\beta}\rangle=\langle\hat{S}_l^{\alpha}\rangle\langle\hat{S}_{l'}^{\beta}\rangle, \quad \text{if} \quad l\neq l'. \tag{7.31}$$

By definition $\langle\hat{\mathbf{S}}\rangle=0$ so that there is no contribution to the cross-section from terms with $l\neq l'$.

For $l=l'$ we have

$$\begin{aligned}\langle\hat{S}_l^{\alpha}\hat{S}_l^{\beta}\rangle&=\delta_{\alpha\beta}\langle(\hat{S}_l^{\alpha})^2\rangle\\&=\tfrac{1}{3}\,\delta_{\alpha\beta}\langle(\hat{\mathbf{S}})^2\rangle=\tfrac{1}{3}\,\delta_{\alpha\beta}\,S(S+1).\end{aligned} \tag{7.32}$$

Thus, with this result the cross-section (7.29) becomes

$$\begin{aligned}\frac{d\sigma}{d\Omega}&=r_0^2\{\tfrac{1}{2}gF(\boldsymbol{\kappa})\}^2\sum_{\alpha\beta}(\delta_{\alpha\beta}-\tilde{\kappa}_{\alpha}\tilde{\kappa}_{\beta})N\,\tfrac{1}{3}\,\delta_{\alpha\beta}S(S+1)\\&=r_0^2\{\tfrac{1}{2}gF(\boldsymbol{\kappa})\}^2N\tfrac{2}{3}S(S+1).\end{aligned} \tag{7.33}$$

This formula shows that the cross-section is large for large spin values S, just as we would expect. There is no coherent scattering because the paramagnetic ions are randomly orientated, and the dependence on the scattering vector $\boldsymbol{\kappa}$ comes only through the form factor $F(\boldsymbol{\kappa})$.

7.4.2. *Paramagnets in a magnetic field*

A perfect paramagnet at low enough temperatures will be appreciably polarized if a magnetic field **H** is applied. In this case, unless the neutron wavelength is exceptionally long, the energy changes on reversing a spin are small compared to the energy of an incident neutron, and we may therefore ignore such energy changes and take the scattering processes to be purely elastic.† Thus we can still use the cross-section (7.29), but the thermal average must be evaluated with care. The calculation is an example of the static approximation discussed in § 1.8.

Clearly,

$$\langle\hat{S}_l^{x}\hat{S}_{l'}^{y}\rangle=\langle\hat{S}_l^{x}\hat{S}_{l'}^{z}\rangle=0, \text{ etc.,}$$

and, with **H** parallel to the z-axis,

$$\langle\hat{S}_l^{x}\hat{S}_{l'}^{x}\rangle=\langle\hat{S}_l^{y}\hat{S}_{l'}^{y}\rangle=\delta_{l,l'}\langle(\hat{S}^{x})^2\rangle. \tag{7.34}$$

Also,

$$\langle\hat{S}_l^{z}\hat{S}_{l'}^{z}\rangle=\langle\hat{S}^{z}\rangle^2+\delta_{l,l'}\{\langle(\hat{S}^{z})^2\rangle-\langle\hat{S}^{z}\rangle^2\}. \tag{7.35}$$

† A useful result is $2\mu_{\rm B}H=0.116\cdot 10^{-4}$ meV with the field H is gauss.

Thus,

$$\sum_{\alpha,\beta} (\delta_{\alpha,\beta} - \tilde{\kappa}_\alpha \tilde{\kappa}_\beta)\langle \hat{S}_l^\alpha \hat{S}_{l'}^\beta \rangle = (1+\tilde{\kappa}_z^2)\langle \hat{S}_l^x \hat{S}_{l'}^x \rangle + (1-\tilde{\kappa}_z^2)\langle \hat{S}_l^z \hat{S}_{l'}^z \rangle \qquad (7.36)$$

and so, using (2.9),

$$\sum_{l,l'} \exp\{\mathrm{i}\boldsymbol{\kappa}\cdot(\mathbf{l}'-\mathbf{l})\} \sum_{\alpha,\beta} (\delta_{\alpha\beta} - \tilde{\kappa}_\alpha \tilde{\kappa}_\beta)\langle \hat{S}_l^\alpha \hat{S}_{l'}^\beta \rangle$$
$$= (1-\tilde{\kappa}_z^2)\langle \hat{S}^z \rangle^2 \left| \sum_l \exp(\mathrm{i}\boldsymbol{\kappa}\cdot\mathbf{l}) \right|^2$$
$$+ N[(1+\tilde{\kappa}_z^2)\langle (\hat{S}^x)^2 \rangle + (1-\tilde{\kappa}_z^2)\{\langle (\hat{S}^z)^2 \rangle - \langle \hat{S}^z \rangle^2\}]$$
$$= (1-\tilde{\kappa}_z^2)\langle \hat{S}^z \rangle^2 N \frac{(2\pi)^3}{v_0} \sum_{\tau} \delta(\boldsymbol{\kappa}-\boldsymbol{\tau})$$
$$+ N[\tfrac{1}{2}S(S+1) + \tfrac{1}{2}\langle (\hat{S}^z)^2 \rangle - \langle \hat{S}^z \rangle^2 + \tilde{\kappa}_z^2\{\tfrac{1}{2}S(S+1) - \tfrac{3}{2}\langle (\hat{S}^z)^2 \rangle + \langle \hat{S}^z \rangle^2\}]. \qquad (7.37)$$

where $\boldsymbol{\tau}$ is a reciprocal lattice vector.
In obtaining the second line of (7.37) we used

$$2\langle (\hat{S}^x)^2 \rangle = S(S+1) - \langle (\hat{S}^z)^2 \rangle.$$

The structure of (7.37) is to be noted. The first term, which gives rise to coherent, Bragg scattering, is proportional to the square of the average value of the z-component of spin; the second term, which gives diffuse scattering, is proportional to the square of the fluctuations of the transverse and the z-component of the spin.

The average values of $\hat{S}^z$ and $(\hat{S}^z)^2$ are readily calculated. The partition function Z is given by

$$Z = \sum_{m=-S}^{S} \exp(g\mu_B H\beta m).$$

Some simple algebra gives the result

$$Z(u) = \sinh u(S+\tfrac{1}{2})/\sinh(\tfrac{1}{2}u) \qquad (7.38)$$

where we have defined $u = g\mu_B H\beta$, and

$$\langle \hat{S}^z \rangle = \frac{1}{Z(u)} \sum_{m=-S}^{S} m e^{mu} = \frac{1}{Z(u)} \frac{\mathrm{d}Z(u)}{\mathrm{d}u},$$

which from (7.38) leads to the result

$$\langle \hat{S}^z \rangle = \tfrac{1}{2}[(2S+1)\coth\{\tfrac{1}{2}u(2S+1)\} - \coth(\tfrac{1}{2}u)]. \qquad (7.39)$$

When $u = g\mu_B H\beta \ll 1$ we have

$$\langle \hat{S}^z \rangle \simeq \tfrac{1}{3} g\mu_B H\beta S(S+1). \tag{7.40}$$

Furthermore,

$$\langle (\hat{S}^z)^2 \rangle = \frac{1}{Z(u)} \sum_{m=-S}^{S} m^2 e^{mu} = \frac{1}{Z(u)} \frac{d^2 Z(u)}{du^2}$$

$$= \frac{1}{4}\left[(2S+1)^2 + \frac{\cosh u + 3}{\cosh u - 1} - 2(2S+1)\coth\{\tfrac{1}{2}u(2S+1)\}\coth(\tfrac{1}{2}u)\right]. \tag{7.41}$$

When $u \ll 1$,

$$\langle (\hat{S}^z)^2 \rangle \simeq \tfrac{1}{3} S(S+1)\{1 + \tfrac{1}{30} u^2 (2S-1)(2S+3)\}. \tag{7.42}$$

In the limit of $H = 0$ (i.e. $u = 0$), the cross-section given by (7.37) becomes identical to that for the scattering from a perfect paramagnet, (7.33), as, of course, it must.

7.5. Band electrons (Harrison 1980; Morgan *et al.* 1986)

In the preceding sections we have taken the electrons responsible for the magnetic field in (7.5) to possess wave functions that are strongly localized about the ions in the target crystal. While this is quite reasonable for ionic crystals and rare-earth metals, it would seem a doubtful assumption for transition metals, for instance, whose magnetic properties are mainly determined by their 3d electrons, which are known to give a large contribution to the low-temperature specific heat. Thus we have also to consider the calculation of the cross-section for the scattering of neutrons from a solid that is more appropriately described in terms of band theory rather than the atomic theory considered hitherto.

The band theory of a solid is constructed by envisaging the solid as being built by taking free atoms and placing them at the sites of a perfect crystal-lattice structure. The mutual interaction of the atoms causes the energy levels of the constituent atoms to broaden into bands of energy states, the width of these bands being determined by the degree of perturbation to which the levels are subject on forming the solid. Thus, for instance, the 3s electrons of sodium form a wide band because their corresponding wave function in the free atom has a large mean radius as compared, say, to the mean radius of the wave function of the 4f electrons in the rare earths, which are deeply buried in the atom and form a very narrow band in the solid, centred about the energy of the 4f level in the free atom. Each electron state of the free atom becomes a band in the solid but we can discard all but the bands coming from the outermost shells of the free atoms, since the core states play no part in the dynamical behaviour of the solid but go to form the field of ion potentials in which the remaining electrons move. Of course, it must also be remembered that the magnetic interaction of the neutrons which the solid involves only the unpaired electron spins, so that here also we do not need to consider the closed shells of electrons in the ion cores.

It is true, of course, that there is an unpairing of electron spin in the ion cores due to configurational interaction within the ion, the unpaired s-electron spin density giving rise to the major part of the magnetic field at the nucleus of the ion. However, this unpaired spin gives a negligible contribution to the scattering amplitude because of its smallness.

The spin and orbital contributions to the cross-section for scattering by band electrons are very different, particularly at small scattering vectors. We set aside cross-terms that might be induced by spin–orbit coupling, for example, so that the cross-section is taken to be the sum of a spin and an orbital contribution derived, respectively, from the first and second terms in eqn (7.10). We show in § 7.6 that in the limit $\kappa \to 0$ the spin and orbital cross-sections for a degenerate perfect electron fluid are proportional to κ^{-1} and κ^{-3}; the corresponding integrated intensities are proportional to κ (spin) and κ^{-1} (orbital). Consequently, orbital scattering dominates at small scattering vectors.

More generally, the κ^{-1} and κ^{-3} behaviour for $\kappa \to 0$ is obtained from intraband transitions, whereas for interband transitions the cross-sections are proportional to κ^2 and κ^{-2} for small scattering vectors. It follows that, for band electrons, the dominant term in the cross-section at small scattering vectors arises from intraband transitions in the orbital contribution, and this term is proportional to κ^{-3} for $\kappa \to 0$. A different result is obtained from a model in which the electrons are assumed to occupy a very narrow band derived from a single atomic symmetry. For example, an extreme tight-binding model for a transition metal based on d symmetry bands, and no overlap of wave functions on different atomic sites, possesses an orbital cross-section that is independent of the scattering vector in the limit $\kappa \to 0$. In this instance the cross-section is dominated by the spin contribution at small scattering vectors. This result is misleading because in any real material there will be some hybridization of the atomic states and some overlap of atomic wave functions on different sites, and the orbital cross-section will probably dominate at sufficiently small scattering vectors. The limiting form of the orbital contribution to the cross-section depends on diamagnetic screening and is discussed in § 7.6.

Let there be $\mathcal{N}$ itinerant electrons involved in determining the magnetic properties of the solid. We denote the different bands to which these belong by the index λ, e.g. for d electrons in a Bravais lattice there are $2l+1=2\times2+1=5$ bands to be taken into account. We then need the eigenstates φ of the Hamiltonian

$$\mathcal{H}=\sum_{i=1}^{\mathcal{N}}\frac{1}{2m_e}\hat{\mathbf{p}}_i^2+\hat{V}(\mathbf{r}_i) \tag{7.43}$$

where $\hat{V}(\mathbf{r})$ is the periodic potential due to the ion cores. The periodicity of $\hat{V}(\mathbf{r})$ means that

$$\hat{V}(\mathbf{r})=\hat{V}(\mathbf{r}+\mathbf{l}), \tag{7.44}$$

where $\mathbf{l}$ is any lattice vector (cf. § 2.1). The eigenfunctions of (7.43) with $\hat{V}$ satisfying (7.44) are Bloch functions,†

$$\varphi_{\mathbf{k}\lambda}(\mathbf{r})=\exp(\mathrm{i}\mathbf{k}\cdot\mathbf{r})u_{\mathbf{k}\lambda}(\mathbf{r}) \tag{7.45a}$$

where

$$u_{\mathbf{k}\lambda}(\mathbf{r}+\mathbf{l})=u_{\mathbf{k}\lambda}(\mathbf{r}) \tag{7.45b}$$

and

$$\left\{\frac{1}{2m_e}\hat{\mathbf{p}}^2+\hat{V}(\mathbf{r})\right\}\varphi_{\mathbf{k}\lambda}(\mathbf{r})=\mathscr{E}_\lambda(\mathbf{k})\varphi_{\mathbf{k}\lambda}(\mathbf{r}). \tag{7.46}$$

All the functions $\varphi_{\mathbf{k}\lambda}(\mathbf{r})$ with different λ or $\mathbf{k}$ are orthogonal. In particular, the orthogonality of $\varphi_{\mathbf{k}\lambda}(\mathbf{r})$ with different λ and the same $\mathbf{k}$ implies that the functions $u_{\mathbf{k}\lambda}(\mathbf{r})$ are orthogonal.

The N values of $\mathbf{k}$ (N being the number of unit cells) are determined, as usual, by applying periodic boundary conditions to the solid and have a density $V/(2\pi)^3$ in reciprocal space. The total number of one-particle eigenstates is therefore $2\lambda N$, the factor two coming from the two possible spin states of an electron. We shall use the index σ to denote spin states, σ taking the values $+1$ and -1, which we often represent by $\uparrow$ and $\downarrow$ respectively. The inclusion of the spin state of the electrons means that the Bloch function becomes a two-component spinor $\varphi_{\mathbf{k}\lambda\sigma}$, but we often drop the spin index for brevity where this is unlikely to cause confusion. One last point concerning the Bloch states (7.45) is that since we shall always assume the non-relativistic Hamiltonian to be invariant under space inversion, we have

$$\varphi_{\mathbf{k}\lambda\sigma}(\mathbf{r})=\varphi_{-\mathbf{k}\lambda\sigma}(-\mathbf{r}). \tag{7.47}$$

Invariance under time reversal means that the band energies $\mathscr{E}_\lambda(\mathbf{k})$ obey

$$\mathscr{E}_\lambda(\mathbf{k})=\mathscr{E}_\lambda(-\mathbf{k}). \tag{7.48}$$

† We denote the eigenvalues of (7.43) by the wave vector $\mathbf{k}$ in accord with current usage. This should cause no confusion with the wave vector associated with incident neutron.

We can express the operator $\hat{\mathbf{Q}}_\perp$ (eqn (7.10)) in terms of Bloch states, or for that matter any complete set of states, with the aid of second quantization formalism (Harrison 1980; Fetter and Walecka 1971). This method of describing many-particle systems is based upon the concept that if ψ_ν ($\nu = 1, \ldots, \mathcal{N}$) are any complete set of orthogonal normalized wave functions, then operators $\hat{c}_\nu^+$ and $\hat{c}_\nu$ can be defined to create and annihilate particles in the state ψ_ν. For particles that obey Fermi–Dirac statistics, $\hat{c}_\nu^+$ and $\hat{c}_\mu$ satisfy the following *anticommutation* rule

$$\hat{c}_\nu \hat{c}_\mu^+ + \hat{c}_\mu^+ \hat{c}_\nu = [\hat{c}_\nu, \hat{c}_\mu^+]_+ = \delta_{\nu,\mu}, \tag{7.49}$$

all other anticommutators being zero. We also note that the particle-number operator for the state ψ_ν, namely $\hat{n}_\nu = \hat{c}_\nu^+ \hat{c}_\nu$, obeys the relation

$$\hat{n}_\nu \hat{n}_\nu = \hat{n}_\nu \tag{7.50}$$

as a consequence of the condition

$$(\hat{c}_\nu^+)^2 = (\hat{c}_\nu)^2 = 0, \tag{7.51}$$

which is itself a direct consequence of the Pauli exclusion principle.

In order to express the Hamiltonian describing the particles, in our case electrons, in terms of the operators $\hat{c}$ and $\hat{c}^+$, we require prescriptions for operators that are the sum of one-particle operators, two-particle operators, etc. These are as follows: If $\hat{f}_i^{(1)}$ acts only on the coordinates of the ith particle (by coordinates we mean both spin and space coordinates) and

$$\hat{F}^{(1)} = \sum_i \hat{f}_i^{(1)},$$

then

$$\hat{F}^{(1)} = \sum_{\nu,\mu} f_{\nu\mu}^{(1)} \hat{c}_\nu^+ \hat{c}_\mu \tag{7.52}$$

where

$$f_{\nu\mu}^{(1)} = \langle \psi_\nu | \hat{f}^{(1)} | \psi_\mu \rangle.$$

Similarly, if $\hat{f}_{ij}^{(2)}$ is an operator acting on particles i and j at once and

$$\hat{F}^{(2)} = \sum_{i>j} \hat{f}_{ij}^{(2)},$$

then

$$\hat{F}^{(2)} = \tfrac{1}{2} \sum_{\mu,\nu,\mu',\nu'} (f^{(2)})_{\mu'\nu'}^{\mu\nu} \hat{c}_\mu^+ \hat{c}_\nu^+ \hat{c}_{\nu'} \hat{c}_{\mu'} \tag{7.53}$$

where

$$(f^{(2)})_{\mu'\nu'}^{\mu\nu} = \langle \psi_\mu, \psi_\nu | \hat{f}^{(2)} | \psi_{\mu'}, \psi_{\nu'} \rangle.$$

The procedure can be extended to the sum of three-particle operators, etc., but we shall not have occasion to use them.

As an example of an operator that is the sum of single-particle operators, we can take the Hamiltonian (7.43). From (7.52) it follows that this can be written as

$$\hat{\mathcal{H}} = \sum_{\mu,\nu} \left\langle \psi_\mu \left| \frac{1}{2m_e}\hat{\mathbf{p}}^2 + \hat{V}(\mathbf{r}) \right| \psi_\nu \right\rangle \hat{c}_\mu^+ \hat{c}_\nu. \tag{7.54}$$

If we now choose the states ψ_ν to be Bloch states $\varphi_{\mathbf{k}\lambda\sigma}$, (eqn (7.45a)), then it follows from (7.46) that (7.54) reduces to the diagonal form

$$\hat{\mathcal{H}} = \sum_{\mathbf{k}\lambda\sigma} \mathscr{E}_\lambda(\mathbf{k}) \hat{c}_{\mathbf{k}\lambda\sigma}^+ \hat{c}_{\mathbf{k}\lambda\sigma}. \tag{7.55}$$

A useful concept in second quantization is that of a particle field operator. If ξ denotes the coordinates of a particle, then the particle field operator $\hat{\psi}(\xi)$ is defined to be

$$\hat{\psi}(\xi) = \sum_\nu \psi_\nu(\xi)\hat{c}_\nu$$

and

$$\hat{\psi}^+(\xi) = \sum_\nu \psi_\nu^*(\xi)\hat{c}_\nu^+. \tag{7.56}$$

Note the analogy between these expressions and the expansion of a wave function in terms of a complete set of eigenfunctions. Also, by what has been said concerning the creation and annihilation operators $\hat{c}_\nu^+$ and $\hat{c}_\nu$ it is evident that the operator $\hat{\psi}$ decreases the *total* number of particles in the system by one, while $\hat{\psi}^+$ increases it by one.

It follows from the anticommutation relations for $\hat{c}$ and $\hat{c}^+$ that $\hat{\psi}$ and $\hat{\psi}^+$ obey

$$\hat{\psi}(\xi)\hat{\psi}^+(\xi') + \hat{\psi}^+(\xi')\hat{\psi}(\xi) = [\hat{\psi}(\xi), \hat{\psi}^+(\xi')]_+ = \delta(\xi - \xi'), \tag{7.57}$$

and all other anticommutators are zero. In (7.57), $\delta(\xi-\xi')$ is shorthand for $\delta(x-x')\,\delta(y-y')\,\delta(z-z')\,\delta_{\sigma,\sigma'}$.

The microscopic particle density $\hat{\rho}(\mathbf{r})$ introduced in Chapter 3 is given by

$$\hat{\rho}(\mathbf{r}) = \hat{\psi}^+(\mathbf{r})\hat{\psi}(\mathbf{r}). \tag{7.58}$$

Let us now proceed to express $\hat{\mathbf{Q}}_\perp$, eqn (7.10), in terms of operators referring to Bloch states, $\varphi_{\mathbf{k}\lambda\sigma}$.

Consider first the spin part of (7.10). Clearly we require an expression for

$$\sum_i \exp(\mathrm{i}\boldsymbol{\kappa}\cdot\mathbf{r}_i)\hat{\mathbf{s}}_i.$$

The microscopic spin density $\hat{\mathbf{s}}(\mathbf{r})$ is defined as

$$\hat{\mathbf{s}}(\mathbf{r}) = \sum_i \delta(\mathbf{r}-\mathbf{r}_i)\hat{\mathbf{s}}_i. \tag{7.59}$$

If we now write

$$\hat{s}(\mathbf{r}) = \frac{1}{V}\sum_{\mathbf{q}} \exp(i\mathbf{q}\cdot\mathbf{r})\hat{s}(\mathbf{q}), \tag{7.60}$$

with

$$\hat{s}(\mathbf{q}) = \int d\mathbf{r}\, \exp(-i\mathbf{q}\cdot\mathbf{r})\hat{s}(\mathbf{r}), \tag{7.61}$$

then

$$\sum_i \exp(i\boldsymbol{\kappa}\cdot\mathbf{r}_i)\hat{\mathbf{s}}_i = \int d\mathbf{r}\, \exp(i\boldsymbol{\kappa}\cdot\mathbf{r})\sum_i \delta(\mathbf{r}-\mathbf{r}_i)\hat{\mathbf{s}}_i$$
$$= \int d\mathbf{r}\, \exp(i\boldsymbol{\kappa}\cdot\mathbf{r})\hat{s}(\mathbf{r}) = \hat{s}(-\boldsymbol{\kappa}). \tag{7.62}$$

By analogy with (7.58), the spin density $\hat{\mathbf{s}}(\mathbf{r})$ is given in the second quantized formalism by

$$\hat{s}(\mathbf{r}) = \hat{\psi}^{+}(\mathbf{r})\hat{\mathbf{s}}\,\hat{\psi}(\mathbf{r}), \tag{7.63}$$

which in terms of Bloch states reads

$$\hat{s}(\mathbf{r}) = \sum_{\substack{\mathbf{k},\lambda \\ \mathbf{k}',\lambda'}} \sum_{\sigma,\sigma'} \varphi^{*}_{\mathbf{k}\lambda\sigma}(\mathbf{r})\hat{c}^{+}_{\mathbf{k}\lambda\sigma}\hat{\mathbf{s}}\,\varphi_{\mathbf{k}'\lambda'\sigma'}(\mathbf{r})\hat{c}_{\mathbf{k}'\lambda'\sigma'}. \tag{7.64}$$

We now separate the spin spinor in the bloch function $\varphi_{\mathbf{k}\lambda\sigma}$ from the spatial eigenfunction, i.e. write

$$\varphi_{\mathbf{k}\lambda\sigma} = \varphi_{\mathbf{k}\lambda}\chi_{\sigma}, \tag{7.65}$$

so that $\hat{s}(\mathbf{q})$ is

$$\hat{s}(\mathbf{q}) = \int d\mathbf{r}\, \exp(-i\mathbf{q}\cdot\mathbf{r})\sum_{\substack{\mathbf{k}\lambda \\ \mathbf{k}'\lambda'}} \varphi^{*}_{\mathbf{k}\lambda}(\mathbf{r})\varphi_{\mathbf{k}'\lambda'}(\mathbf{r})\sum_{\sigma,\sigma'} \hat{c}^{+}_{\mathbf{k}\lambda\sigma}\chi^{+}_{\sigma}\hat{\mathbf{s}}\chi_{\sigma'}\hat{c}_{\mathbf{k}'\lambda'\sigma'}. \tag{7.66}$$

Let us evaluate the spin part of (7.66). We take

$$\chi_{\uparrow} = \begin{pmatrix}1\\0\end{pmatrix} \quad \text{and} \quad \chi_{\downarrow} = \begin{pmatrix}0\\1\end{pmatrix}$$

and write

$$\hat{\mathbf{s}} = \tfrac{1}{2}(\hat{\sigma}_x, \hat{\sigma}_y, \hat{\sigma}_z),$$

where $\hat{\sigma}_\alpha$ are Pauli matrices. Explicitly

$$\hat{\sigma}_x = \begin{pmatrix}0 & 1\\1 & 0\end{pmatrix}, \quad \hat{\sigma}_y = \begin{pmatrix}0 & -i\\i & 0\end{pmatrix}, \quad \hat{\sigma}_z = \begin{pmatrix}1 & 0\\0 & -1\end{pmatrix},$$

from which the following relations are readily derived

$$\hat{\sigma}_z\chi_\uparrow = \chi_\uparrow \qquad \hat{\sigma}_z\chi_\downarrow = -\chi_\downarrow$$
$$\hat{\sigma}_y\chi_\uparrow = \mathrm{i}\chi_\downarrow \qquad \hat{\sigma}_y\chi_\downarrow = -\mathrm{i}\chi_\uparrow$$
$$\hat{\sigma}_x\chi_\uparrow = \chi_\downarrow \qquad \hat{\sigma}_x\chi_\downarrow = \chi_\uparrow.$$

It is then a trivial matter to show that

$$\sum_{\sigma,\sigma'} \hat{c}^+_{\mathbf{k}\sigma}\chi^+_\sigma\hat{\mathbf{s}}\chi_{\sigma'}\hat{c}_{\mathbf{k}'\sigma'}$$
$$= \left\{\tfrac{1}{2}(\hat{c}^+_{\mathbf{k}\uparrow}\hat{c}_{\mathbf{k}'\downarrow} + \hat{c}^+_{\mathbf{k}\downarrow}\hat{c}_{\mathbf{k}'\uparrow}), \frac{1}{2\mathrm{i}}(\hat{c}^+_{\mathbf{k}\uparrow}\hat{c}_{\mathbf{k}'\downarrow} - \hat{c}^+_{\mathbf{k}\downarrow}\hat{c}_{\mathbf{k}'\uparrow}), \tfrac{1}{2}(\hat{c}^+_{\mathbf{k}\uparrow}\hat{c}_{\mathbf{k}'\uparrow} - \hat{c}^+_{\mathbf{k}\downarrow}\hat{c}_{\mathbf{k}'\downarrow})\right\},$$

and hence (7.66) becomes

$$\hat{s}(\mathbf{q}) = \int \mathrm{d}\mathbf{r}\,\exp(-\mathrm{i}\mathbf{q}\cdot\mathbf{r}) \sum_{\substack{\mathbf{k}\lambda \\ \mathbf{k}'\lambda'}} \varphi^*_{\mathbf{k}\lambda}(\mathbf{r})\varphi_{\mathbf{k}'\lambda'}(\mathbf{r})$$
$$\times \left\{\tfrac{1}{2}(\hat{c}^+_{\mathbf{k}\lambda\uparrow}\hat{c}_{\mathbf{k}'\lambda'\downarrow} + \hat{c}^+_{\mathbf{k}\lambda\downarrow}\hat{c}_{\mathbf{k}'\lambda'\uparrow}), \frac{1}{2\mathrm{i}}(\hat{c}^+_{\mathbf{k}\lambda\uparrow}\hat{c}_{\mathbf{k}'\lambda'\downarrow} - \hat{c}^+_{\mathbf{k}\lambda\downarrow}\hat{c}_{\mathbf{k}'\lambda'\uparrow}), \tfrac{1}{2}(\hat{c}^+_{\mathbf{k}\lambda\uparrow}\hat{c}_{\mathbf{k}'\lambda'\uparrow} - \hat{c}^+_{\mathbf{k}\lambda\downarrow}\hat{c}_{\mathbf{k}'\lambda'\downarrow})\right\}. \tag{7.67}$$

Since the spin angular momentum operators $\hat{S}^x$ and $\hat{S}^y$, when expressed in terms of raising and lowering operators $\hat{S}^+$ and $\hat{S}^-$, read

$$\hat{S}^x = \tfrac{1}{2}(\hat{S}^+ + \hat{S}^-) \quad \text{and} \quad \hat{S}^y = \frac{1}{2\mathrm{i}}(\hat{S}^+ - \hat{S}^-), \tag{7.68}$$

(7.67) implies that the equivalent electron-spin raising and lowering operators (or spin-flip operators) are

$$\hat{c}^+_\uparrow\hat{c}_\downarrow \quad \text{and} \quad \hat{c}^+_\downarrow\hat{c}_\uparrow,$$

respectively, while the operator for its z-component of spin is

$$\tfrac{1}{2}(\hat{c}^+_\uparrow\hat{c}_\uparrow - \hat{c}^+_\downarrow\hat{c}_\downarrow).$$

We now consider the orbital interaction operator in (7.10). The required operator is proportional to

$$\sum_i \exp(\mathrm{i}\boldsymbol{\kappa}\cdot\mathbf{r}_i)(\boldsymbol{\kappa}\times\hat{\mathbf{p}}_i) = \int \mathrm{d}\mathbf{r}\,\exp(\mathrm{i}\boldsymbol{\kappa}\cdot\mathbf{r})\sum_{\mathbf{k}\lambda}\sum_{\mathbf{k}'\lambda'} \varphi^*_{\mathbf{k}\lambda}(\mathbf{r})(\boldsymbol{\kappa}\times\hat{\mathbf{p}})\varphi_{\mathbf{k}'\lambda'}(\mathbf{r})$$
$$\times \sum_{\sigma\sigma'} \delta_{\sigma,\sigma'}\hat{c}^+_{\mathbf{k}\lambda\sigma}\hat{c}_{\mathbf{k}'\lambda'\sigma'} \tag{7.69}$$

where the second form follows directly from (7.52) and is the analogue of eqn (7.66) for the Fourier transform of the spin density. In evaluating the matrix element in (7.69), it is often useful to exploit the identity

$$\int \mathrm{d}\mathbf{r}\, \exp(\mathrm{i}\boldsymbol{\kappa}\cdot\mathbf{r})\varphi^*_{\mathbf{k}\lambda}(\mathbf{r})(\boldsymbol{\kappa}\times\hat{\mathbf{p}})\varphi_{\mathbf{k}'\lambda'}(\mathbf{r}) = -\int \mathrm{d}\mathbf{r}\, \exp(\mathrm{i}\boldsymbol{\kappa}\cdot\mathbf{r})\varphi_{\mathbf{k}'\lambda'}(\mathbf{r})(\boldsymbol{\kappa}\times\hat{\mathbf{p}})\varphi^*_{\mathbf{k}\lambda}(\mathbf{r}), \quad (7.70)$$

which follows from the result

$$\int \mathrm{d}\mathbf{r}\, \boldsymbol{\nabla}\{\varphi^*_{\mathbf{k}\lambda}(\mathbf{r})\varphi_{\mathbf{k}'\lambda'}(\mathbf{r})\exp(\mathrm{i}\boldsymbol{\kappa}\cdot\mathbf{r})\} = 0.$$

In the limit $\kappa \to 0$ the matrix element is simply that of the velocity operator. We find

$$\sum_i \exp(\mathrm{i}\boldsymbol{\kappa}\cdot\mathbf{r}_i)(\boldsymbol{\kappa}\times\hat{\mathbf{p}}_i) = (m_e/\hbar)\sum_{\mathbf{k}\sigma}\Big\{\sum_{\lambda}\hat{c}^+_{\mathbf{k}\lambda\sigma}\hat{c}_{\mathbf{k}\lambda\sigma}(\boldsymbol{\kappa}\times\boldsymbol{\nabla}_{\mathbf{k}})\mathscr{E}_\lambda(\mathbf{k}) + \sum_{\lambda\lambda'}\hat{c}^+_{\mathbf{k}\lambda\sigma}\hat{c}_{\mathbf{k}\lambda'\sigma}[\mathscr{E}_{\lambda'}(\mathbf{k}) - \mathscr{E}_\lambda(\mathbf{k})]\int \mathrm{d}\mathbf{r}\, u_{\mathbf{k}\lambda'}(\mathbf{r})(\boldsymbol{\kappa}\times\boldsymbol{\nabla}_{\mathbf{k}})u_{\mathbf{k}\lambda}(\mathbf{r})\Big\}, \qquad \boldsymbol{\kappa}\to 0.$$

The second contribution to (7.71) is nonzero for $\lambda \neq \lambda'$. For a spherical band with an effective mass M_λ,

$$\mathscr{E}_\lambda(\mathbf{k}) = (\hbar^2 k^2/2M_\lambda),$$

so that the various contributions in (7.71) are proportional to the mass ratio (m_e/M_λ). This ratio is very small for narrow bands, e.g. for Fe we find $(m_e/M) \sim 0.1$, and it vanishes in the extreme tight-binding limit.

Let us pause in our discussion of the formalism to obtain an estimate of the size of the neutron cross-section for inelastic scattering from paramagnetic electrons. We may as a crude estimate for orientation expect the cross-section to be of the order of the spin response averaged over $\boldsymbol{\kappa}$. Starting with (7.66), or more directly from (7.103), the response per electron in this instance is found to be

$$\int_{-\infty}^{\infty} f(u)\{1 - f(\omega + u)\}G(u)G(u+\omega)\,\mathrm{d}u,$$

where $f(\omega)$ is the Fermi occupation function and $G(\omega)$ is the density of electronic states at energy $\hbar\omega$. Observe that our expression has the form of a joint density of states (Harrison 1980, p. 103).

For simple metals the Fermi energy ϵ_f is typically a few eV, and hence at room temperature the Fermi occupation function approximates to a step function at ϵ_f. For this case the expression simplifies to

$$\int_0^{\omega} G(u + \epsilon_f - \omega)G(u + \epsilon_f)\,\mathrm{d}u, \qquad \hbar\omega < \epsilon_f$$

or,

$$\int_0^{\epsilon_f} G(u)G(u+\omega)\,\mathrm{d}u, \qquad \hbar\omega > \epsilon_f.$$

In the limit $\hbar\omega \ll \epsilon_f$ we obtain the result $\omega G^2(\epsilon_f)$ as an estimate of the cross section for scattering from band electrons when averaged over the neutron wave vector. For a 'top hat' electron density of states of total width Δ the estimate is zero except within the interval $0 < \hbar\omega < \Delta$ where it is a pyramid of height $(1/2\Delta)$, i.e. it has the value $(\hbar\omega/\Delta^2)$ for $0 < \hbar\omega < (\Delta/2)$. From these results we conclude that scattering from band electrons is particularly strong for materials with a narrow band width/high density of states.

Pronounced structure in the electronic density of states is found in the cross-section. This is evident from the foregoing estimate of the cross-section, and borne out in detail in band structure calculations (Morgan *et al.* 1986). Within the framework of the one-electron model of metals there are strong similarities between the magnetic neutron and Thomson (charge) photon cross-sections (Schülke *et al.* 1986). However, the magnetic and charge response functions are not affected in the same way by electron correlations, e.g. the plasmon excitation is not engaged in the magnetic neutron response unless a field is applied (Lovesey and Trohidou 1986) or the electrons order magnetically, cf. § 9.3.

To proceed further in our task of obtaining an expression for the neutron cross-section in terms of the formalism of band theory, we introduce the tight-binding representation (Callaway 1974). This consists of expanding Bloch functions in terms of a set of wave functions that are strongly localized about the ion-core potentials, these functions being taken to resemble atomic wave functions closely. This approximation scheme is based upon the argument that if the radial part of the wave function of the unpaired electrons in the free atom is sufficiently small for them to suffer only a small distortion on forming the solid, then the wave function of these electrons in the solid must surely be well represented by some linear combination of the corresponding atomic wave functions, i.e. tight-binding theory is expected to yield a good description of electrons in narrow energy bands. The form of the linear combination is dictated by the requirement that the resultant wave function should satisfy Bloch's theorem, namely

$$\varphi_{\mathbf{k}}(\mathbf{r}+\mathbf{l}) = \exp(i\mathbf{k}\cdot\mathbf{l})\varphi_{\mathbf{k}}(\mathbf{r}).$$

For a single band the combination

$$\sum_{\mathbf{l}} \exp(i\mathbf{k}\cdot\mathbf{l})\phi(\mathbf{r}-\mathbf{l})$$

satisfies this requirement, where ϕ are the localized or tight-binding wave functions. In order that we may introduce operators that create and annihilate electrons in the states ϕ, as we shall in Chapter 9, we must assume the wave functions ϕ to be orthogonal, i.e. to satisfy

$$\int d\mathbf{r}\phi^*(\mathbf{r})\phi(\mathbf{r}-\mathbf{l}) = 0 \quad (\mathbf{l} \neq 0).$$

If the tight-binding functions are taken to be atomic orbitals then, clearly, this is *not* true. However, for our purposes we shall merely assume that a set of orthogonal wave functions ϕ exist, and that these can, if need be, be approximated to by replacing them in the formalism by atomic orbitals, this being the equivalent to the usual tight-binding theory. In the limit of zero bandwidth this replacement is, of course, an exact one. (In the theory of the transport properties of Bloch electrons in external fields the set of functions ϕ that we have defined are called Wannier functions.)

In the general case of several bands and also more than one atom per unit cell we have to replace the expansion coefficients $\exp(i\mathbf{k}\cdot\mathbf{l})$ by more complicated ones. We shall verify that the required (exact) expansion in this general case is

$$\varphi_{\mathbf{k}\lambda}(\mathbf{r})=\frac{1}{\sqrt{N}}\sum_{l,d,\Lambda}\exp(i\mathbf{k}\cdot\mathbf{l})a_{\lambda}^{\Lambda}(\mathbf{d};\mathbf{k})\phi_{\Lambda}(\mathbf{r}-\mathbf{R}_{ld}), \tag{7.72}$$

where Λ specifies the various wave functions that contribute to a given band. (We draw attention to the similarity between the following argument and that given in § 4.1 for the expansion of the displacement operator u for a general lattice structure.) For d electrons there would be five different types of wave function, so $\Lambda=1,2,\ldots,5$. The coefficients $a_{\lambda}^{\Lambda}(\mathbf{d};\mathbf{k})$ in (7.72) will be shown to be solutions of an eigenfunction equation (see (7.78)). The wave functions ϕ_{Λ} satisfy the orthonormality condition

$$\int d\mathbf{r}\,\phi_{\Lambda}^{*}(\mathbf{r}-\mathbf{R}_{ld})\phi_{\Lambda'}(\mathbf{r}-\mathbf{R}_{l'd'})=\delta_{l,l'}\,\delta_{d,d'}\,\delta_{\Lambda,\Lambda'}. \tag{7.73}$$

The operator that creates an electron in the state ϕ_{Λ} localized about $\mathbf{R}_{ld}$ is given in terms of the operators for the Bloch states $\varphi_{\mathbf{k}\lambda}$ by

$$\hat{c}_{ld\Lambda\sigma}^{+}=\frac{1}{\sqrt{N}}\sum_{\mathbf{k}\lambda}\exp(-i\mathbf{k}\cdot\mathbf{l})\{a_{\lambda}^{\Lambda}(\mathbf{d};\mathbf{k})\}^{*}\hat{c}_{\mathbf{k}\lambda\sigma}^{+}, \tag{7.74}$$

which, because (7.74) is a canonical transformation, satisfy

$$[\hat{c}_{ld\Lambda\sigma},\hat{c}_{l'd'\Lambda'\sigma'}^{+}]_{+}=\delta_{l,l'}\,\delta_{d,d'}\,\delta_{\Lambda,\Lambda'}\,\delta_{\sigma,\sigma'} \tag{7.75}$$

all other anticommutators being zero.

Finally, in terms of the operators $\hat{c}_{ld\Lambda\sigma}^{+}$ and $\hat{c}_{ld\Lambda\sigma}$ the Hamiltonian (7.55) is, from (7.52),

$$\hat{\mathcal{H}}=\sum_{\substack{l,l'\\d,d'}}\sum_{\Lambda,\Lambda'}\sum_{\sigma}T_{\Lambda\Lambda'}^{dd'}(\mathbf{l}-\mathbf{l}')\hat{c}_{ld\Lambda\sigma}^{+}\hat{c}_{l'd'\Lambda'\sigma} \tag{7.76}$$

where the matrix elements $T_{\Lambda\Lambda'}^{dd'}(\mathbf{l}-\mathbf{l}')$ are†

$$T_{\Lambda\Lambda'}^{dd'}(\mathbf{l}-\mathbf{l}')=\int d\mathbf{r}\,\phi_{\Lambda}^{*}(\mathbf{r}-\mathbf{R}_{ld})\left\{\frac{1}{2m_{e}}\hat{\mathbf{p}}^{2}+\hat{V}(\mathbf{r})\right\}\phi_{\Lambda'}(\mathbf{r}-\mathbf{R}_{l'd'}). \tag{7.77}$$

† $\hat{V}$ will be spin-dependent in a magnetically ordered system and thus $T_{\Lambda\Lambda'}^{dd'}$ and a_{λ}^{Λ}, from eqn (7.78), will also depend on spin, i.e. $\lambda\rightarrow\lambda,\sigma$.

The eigenvalue equation for the transformation coefficients is

$$\sum_{l'd'\Lambda'} T^{dd'}_{\Lambda\Lambda'}(\mathbf{l}-\mathbf{l}')a^{\Lambda'}_{\lambda}(\mathbf{d}';\mathbf{k})\exp(i\mathbf{k}\cdot\mathbf{l}') = \mathscr{E}_{\lambda}(\mathbf{k})a^{\Lambda}_{\lambda}(\mathbf{d};\mathbf{k})\exp(i\mathbf{k}\cdot\mathbf{l}). \tag{7.78}$$

We now review some of the algebra for the transformation from the Bloch states $\varphi_{\mathbf{k}\lambda}$ to the states $\phi_{\Lambda}(\mathbf{r}-\mathbf{R}_{ld})$.

First we observe that $T^{dd'}_{\Lambda\Lambda'}(\mathbf{l}-\mathbf{l}')$ has the property

$$T^{dd'}_{\Lambda\Lambda'}(\mathbf{l}-\mathbf{l}') = T^{d'd}_{\Lambda'\Lambda}(\mathbf{l}'-\mathbf{l}), \tag{7.79}$$

hence†

$$a^{\Lambda}_{\lambda}(\mathbf{d};\mathbf{k}) = \{a^{\Lambda}_{\lambda}(\mathbf{d};-\mathbf{k})\}^{*}, \tag{7.80}$$

as is readily seen on using (7.79) in (7.78). To prove (7.79), note that because $T^{dd'}_{\Lambda\Lambda'}(\mathbf{l}-\mathbf{l}')$ is the matrix element of a Hermitian operator (the single-particle Hamiltonian) it must satisfy

$$T^{dd'}_{\Lambda\Lambda'}(\mathbf{l}-\mathbf{l}') = \{T^{d'd}_{\Lambda'\Lambda}(\mathbf{l}'-\mathbf{l})\}^{*}$$

and further, since the wave functions ϕ represent stationary states they can always be taken to be purely real; hence (7.79) follows.

Also, the coefficients a^{Λ}_{λ} must satisfy the relations

$$\sum_{d,\Lambda} \{a^{\Lambda}_{\lambda}(\mathbf{d};\mathbf{k})\}^{*}a^{\Lambda}_{\lambda'}(\mathbf{d};\mathbf{k}) = \delta_{\lambda,\lambda'} \tag{7.81a}$$

and

$$\sum_{\lambda} \{a^{\Lambda'}_{\lambda}(\mathbf{d}';\mathbf{k})\}^{*}a^{\Lambda}_{\lambda}(\mathbf{d};\mathbf{k}) = \delta_{d,d'}\,\delta_{\Lambda,\Lambda'}. \tag{7.81b}$$

The inverse transformation to (7.72) is obtained by multiplying both sides by $\{\exp(i\mathbf{k}\cdot\mathbf{l}')a^{\Lambda'}_{\lambda}(\mathbf{d}';\mathbf{k})\}^{*}$ and summing over $\mathbf{k}$ and λ, and

$$\phi_{\Lambda}(\mathbf{r}-\mathbf{R}_{ld}) = \frac{1}{\sqrt{N}}\sum_{\mathbf{k},\lambda} \exp(-i\mathbf{k}\cdot\mathbf{l})\{a^{\Lambda}_{\lambda}(\mathbf{d};\mathbf{k})\}^{*}\varphi_{\mathbf{k}\lambda}(\mathbf{r}). \tag{7.82}$$

Similarly,

$$\hat{c}^{+}_{\mathbf{k}\lambda\sigma} = \frac{1}{\sqrt{N}}\sum_{ld\Lambda} \exp(i\mathbf{k}\cdot\mathbf{l})a^{\Lambda}_{\lambda}(\mathbf{d};\mathbf{k})\hat{c}^{+}_{ld\Lambda\sigma}. \tag{7.83}$$

Consider the spatial part of $\hat{\mathfrak{z}}(\mathbf{q})$, eqn (7.66), namely

$$\int d\mathbf{r}\,\exp(-i\mathbf{q}\cdot\mathbf{r})\sum_{\substack{\mathbf{k},\lambda\\ \mathbf{k}',\lambda'}} \varphi^{*}_{\mathbf{k}\lambda}(\mathbf{r})\varphi_{\mathbf{k}'\lambda'}(\mathbf{r}) \tag{7.84}$$

† The atomic wave functions in (7.72) are often written as linear combinations that transform as the irreducible representation of the continuous group l in the crystal symmetry. In this instance the coefficients (7.80) are purely real.

and transform to the localized functions ϕ. From (7.72) this is

$$\frac{1}{N}\int \mathrm{d}\mathbf{r}\exp(-\mathrm{i}\mathbf{q}\cdot\mathbf{r})\sum_{\substack{\mathbf{k},\lambda\\ \mathbf{k}',\lambda'}}\sum_{l,d,\Lambda}\sum_{l',d',\Lambda'}\exp(-\mathrm{i}\mathbf{k}\cdot\mathbf{l})\{a_\lambda^\Lambda(\mathbf{d};\mathbf{k})\}^*\phi_\Lambda^*(\mathbf{r}-\mathbf{R}_{ld})$$
$$\times\exp(\mathrm{i}\mathbf{k}'\cdot\mathbf{l}')a_{\lambda'}^{\Lambda'}(\mathbf{d}';\mathbf{k}')\phi_{\Lambda'}(\mathbf{r}-\mathbf{R}_{l'd'}),$$

which, on changing the integration variable $\mathbf{r}\rightarrow\mathbf{r}+\mathbf{R}_{ld}$, becomes

$$\frac{1}{N}\sum_{l,d,\Lambda}\sum_{l',d',\Lambda'}\int \mathrm{d}\mathbf{r}\sum_{\substack{\mathbf{k},\lambda\\ \mathbf{k}',\lambda'}}\exp(-\mathrm{i}\mathbf{q}\cdot\mathbf{r}-\mathrm{i}\mathbf{q}\cdot\mathbf{R}_{ld}-\mathrm{i}\mathbf{k}\cdot\mathbf{l}+\mathrm{i}\mathbf{k}'\cdot\mathbf{l}')$$
$$\times a_{\lambda'}^{\Lambda'}(\mathbf{d}';\mathbf{k}')\{a_\lambda^\Lambda(\mathbf{d};\mathbf{k})\}^*\phi_\Lambda^*(\mathbf{r})\phi_{\Lambda'}(\mathbf{r}+\mathbf{R}_{ld}-\mathbf{R}_{l'd'})$$
$$=\frac{1}{N}\sum_{l,l'}\sum_{d,\Lambda}\sum_{d',\Lambda'}\int \mathrm{d}\mathbf{r}\sum_{\substack{\mathbf{k},\lambda\\ \mathbf{k}',\lambda'}}\exp(-\mathrm{i}\mathbf{q}\cdot\mathbf{r}-\mathrm{i}\mathbf{q}\cdot\mathbf{d})\exp\{-\mathrm{i}\mathbf{l}'\cdot(\mathbf{k}-\mathbf{k}'+\mathbf{q})\}$$
$$\times\exp\{-\mathrm{i}\mathbf{l}\cdot(\mathbf{k}+\mathbf{q})\}a_{\lambda'}^{\Lambda'}(\mathbf{d}';\mathbf{k}')\{a_\lambda^\Lambda(\mathbf{d};\mathbf{k})\}^*\phi_\Lambda^*(\mathbf{r})\phi_{\Lambda'}(\mathbf{r}+\mathbf{l}+\mathbf{d}-\mathbf{d}'). \quad (7.85)$$

The sum over $\mathbf{l}'$ in (7.85) gives $N\delta_{\mathbf{k}',\mathbf{k}+\mathbf{q}}$ so that the whole expression reduces to

$$\sum_{l}\sum_{d,\Lambda}\sum_{d',\Lambda'}\sum_{\substack{\mathbf{k},\lambda\\ \mathbf{k}',\lambda'}}\delta_{\mathbf{k}',\mathbf{k}+\mathbf{q}}\int \mathrm{d}\mathbf{r}\exp(-\mathrm{i}\mathbf{q}\cdot\mathbf{r})\exp(-\mathrm{i}\mathbf{q}\cdot\mathbf{d}-\mathrm{i}\mathbf{k}'\cdot\mathbf{l})$$
$$\times a_{\lambda'}^{\Lambda'}(\mathbf{d}';\mathbf{k}')\{a_\lambda^\Lambda(\mathbf{d};\mathbf{k})\}^*\phi_\Lambda^*(\mathbf{r})\phi_{\Lambda'}(\mathbf{r}+\mathbf{l}+\mathbf{d}-\mathbf{d}'). \quad (7.86)$$

If $\mathbf{q}=0$, the orthonormality condition (7.73) gives for (7.86)

$$\sum_{l}\sum_{d,\Lambda}\sum_{d',\Lambda'}\sum_{\substack{\mathbf{k},\lambda\\ \mathbf{k}',\lambda'}}\delta_{\mathbf{k},\mathbf{k}'}\,\delta_{l,0}\,\delta_{d,d'}\,\delta_{\Lambda,\Lambda'}a_{\lambda'}^{\Lambda}(\mathbf{d};\mathbf{k})\{a_\lambda^\Lambda(\mathbf{d};\mathbf{k})\}^*=\sum_{\substack{\mathbf{k},\lambda\\ \mathbf{k}',\lambda'}}\delta_{\mathbf{k},\mathbf{k}'}\delta_{\lambda,\lambda'},$$

which is seen to agree with the result obtained from (7.84) for $\mathbf{q}=0$, when it is recalled that the Bloch functions are orthogonal. This result means that for $\boldsymbol{\kappa}=0$ the scattering amplitude is proportional to the total spin density within a unit cell.

Eqn (7.86) is an *exact* expression for (7.84) in terms of the localized functions ϕ, its utility being that we can use the fact that the functions ϕ are well localized to obtain a meaningful approximation for $\hat{\mathscr{s}}(\mathbf{q})$ that is simpler than that given by (7.66). We therefore introduce the assumption that the overlap of the ϕ's on different sites can be ignored in (7.86) and thus set $\mathbf{l}=0$ and $\mathbf{d}=\mathbf{d}'$. In this approximation, the expression for $\hat{\mathscr{s}}(\mathbf{q})$ becomes

$$\hat{\mathscr{s}}(\mathbf{q})=\sum_{\Lambda,\Lambda'}\int \mathrm{d}\mathbf{r}\exp(-\mathrm{i}\mathbf{q}\cdot\mathbf{r})\phi_\Lambda^*(\mathbf{r})\phi_{\Lambda'}(\mathbf{r})\sum_{\mathbf{k},\lambda,\lambda'}\sum_{d}\exp(-\mathrm{i}\mathbf{q}\cdot\mathbf{d})$$
$$\times a_{\lambda'}^{\Lambda'}(\mathbf{d};\mathbf{k}+\mathbf{q})\{a_\lambda^\Lambda(\mathbf{d};\mathbf{k})\}^*\sum_{\sigma,\sigma'}\hat{c}_{\mathbf{k}\lambda\sigma}^{+}\chi_\sigma^{+}\hat{\mathbf{s}}\,\chi_{\sigma'}\hat{c}_{\mathbf{k}+\mathbf{q}\lambda'\sigma'}. \quad (7.87)$$

Note that in the case of a *single* band and a *Bravais lattice* the expansion coefficients equal to unity, when the expression (7.87) for $\hat{\mathfrak{s}}$ reduces to the simple form

$$\hat{\mathfrak{s}}(\mathbf{q}) = \int \mathrm{d}\mathbf{r}\, \exp(-\mathrm{i}\mathbf{q}\cdot\mathbf{r})\, |\phi(\mathbf{r})|^2 \sum_{\mathbf{k}} \sum_{\sigma,\sigma'} \hat{c}^+_{\mathbf{k}\sigma} \chi^+_{\sigma}\, \hat{\mathbf{s}}\, \chi_{\sigma'} \hat{c}_{\mathbf{k}+\mathbf{q}\sigma'}. \tag{7.88}$$

For this particular case we observe that, in the light of the discussion following (7.67), the spin part of the cross-section for the scattering by itinerant electrons has a structure similar to that for spin-only scattering by localized electrons.

It is not appropriate for us to give a lengthy discussion here but we draw attention to one difference, which may be observable between form factors for the itinerant-electron model and for the localized model. It is a characteristic approximation in the itinerant theory, as outlined here, to assume all integrals involving an overlap between different functions, say $\phi(\mathbf{r})$ and $\phi(\mathbf{r}-\mathbf{l})$, to be zero. In this approximation the effective form factor is

$$F(\boldsymbol{\kappa}) = \int \mathrm{d}\mathbf{r}\, \exp(\mathrm{i}\boldsymbol{\kappa}\cdot\mathbf{r})\, |\phi(\mathbf{r})|^2.$$

However, in salts, the atomic form factor is defined as (see (7.16))

$$F_d(\boldsymbol{\kappa}) = \int \mathrm{d}\mathbf{r}\, \exp(\mathrm{i}\boldsymbol{\kappa}\cdot\mathbf{r})\mathfrak{s}_d(\mathbf{r})$$

where $\mathfrak{s}_d(\mathbf{r})$ is the normalized density on atom d. If we denote the atomic wave function as $A(\mathbf{r})$, we have

$$F_d(\boldsymbol{\kappa}) = \int \mathrm{d}\mathbf{r}\, \exp(\mathrm{i}\boldsymbol{\kappa}\cdot\mathbf{r})\, |A(\mathbf{r})|^2.$$

These formulae are not quite the same because $\phi(\mathbf{r})$ is a Wannier function and differs a little from $A(\mathbf{r})$. In general

$$\phi(\mathbf{r}) = \sum_n W_n A(\mathbf{r}-\mathbf{n}),$$

where W_n are the Wannier expansion coefficients with W_0 almost, but not quite, equal to unity. Neglecting all overlap effects,

$$\sum_n W_n^2 = 1$$

and

$$F(\boldsymbol{\kappa}) = F_d(\boldsymbol{\kappa}) \sum_n |W_n|^2 \exp(\mathrm{i}\boldsymbol{\kappa}\cdot\mathbf{n}).$$

For purposes of illustration let us suppose that the Wannier function spreads only to the r nearest neighbours labelled with vectors $\boldsymbol{\rho}$. Then

$$\phi(\mathbf{r}) \simeq (1 - rW_1^2)^{1/2} A(\mathbf{r}) + W_1 \sum_{\rho} A(\mathbf{r} - \boldsymbol{\rho})$$

and, neglecting all overlap terms,

$$F(\boldsymbol{\kappa}) \simeq F_d(\boldsymbol{\kappa}) \left\{ 1 - rW_1^2 + W_1^2 \sum_{\rho} \exp(i\boldsymbol{\kappa} \cdot \boldsymbol{\rho}) \right\}. \tag{7.89}$$

In general, therefore, we might expect $F(\boldsymbol{\kappa})$ to fall off more rapidly with $\boldsymbol{\kappa}$ than $F_d(\boldsymbol{\kappa})$. Note, however, that, for any reciprocal lattice vector $\boldsymbol{\tau}$,

$$F(\boldsymbol{\tau}) \equiv F_d(\boldsymbol{\tau}).$$

Therefore the difference between these functions cannot be observed in ferromagnets by Bragg scattering. In principle, however, a difference in form factor can exist and in some ways the problem is analogous to covalency effects in salts (to be discussed in Chapter 12).

The foregoing discussion is greatly oversimplified compared to the situation we expect to find in a real metal. We use it for illustrative purposes only, and to draw attention to the fact that the form factor for inelastic scattering in a metal need not be a smooth interpolation of the factor measured at reciprocal lattice points in Bragg scattering experiments (Steinsvoll, et al. 1981).

7.6. Perfect electron fluid

The cross-section for magnetic scattering from a perfect electron fluid is readily calculated from (7.11a) using the second quantization formalism introduced in the preceding section. The result is appropriate for a target sample with nearly free electrons which form a spherical band to a good approximation. For such a sample, the N electrons are described by plane waves and their energy dispersion

$$\mathscr{E}(\mathbf{k}) = (\hbar^2 k^2 / 2M) \tag{7.90}$$

where M is an effective mass. The latter can be very different from the electron mass, e.g. in some semiconductors $(M/m_e) \ll 1$. The free-electron Hamiltonian is of the same form as (7.55) without the band index.

An interacting electron fluid supports a collective density oscillation that is called a plasmon (March and Tosi 1984). Neutrons are not scattered by plasmons in a homogeneous fluid because fluctuations in the spin and particle densities are uncorrelated. Thus, the response for magnetic neutron scattering contains only contributions from the particle–hole continuum, and simple treatments of the Coulomb interaction between electrons give results for the response spectrum which are the same as for a perfect electron fluid. An external magnetic field induces a correlation between fluctuations in the spin and particle densities, and also shifts the plasmon frequency. The cross-section for scattering from the plasmon is significant at small scattering vectors. The orbital interaction gives an amplitude for exciting the plasmon which is proportional to (H^2/κ^2) where H is the magnetic field, and $\kappa \to 0$ (Lovesey 1979, Lovesey and Trohidou 1986). The cross-section for a realistic model of electrons in a semiconductor subject to a magnetic field is reported in Elliott and Kleppmann (1975). In the remainder of this section we calculate the magnetic response of a perfect electron fluid in the absence of an external field.

The calculation we describe is similar to the calculation of the cross-section for nuclear scattering from a perfect Fermi fluid given in § 3.6.1. The magnetic interaction operator (eqn (7.10)) is the sum of two terms which represent the spin and orbital interactions, whereas the nuclear interaction is the sum of spin and particle density interactions. We find that the neutron cross-section contains information on the electron Fermi surface (Dieterich 1973).

We begin by expressing the cross-section (7.11a) in terms of a correlation function. A general discussion of correlation functions in magnetic scattering is presented in Chapter 8. For the moment we need the result

$$S(\kappa, \omega) = \frac{1}{N} \sum_{\lambda\lambda'} p_\lambda \langle\lambda|\, \hat{\mathbf{Q}}_\perp^+ \,|\lambda'\rangle \cdot \langle\lambda'|\, \hat{\mathbf{Q}}_\perp \,|\lambda\rangle\, \delta(\hbar\omega + E_\lambda - E_{\lambda'})$$
$$= \frac{1}{2\pi\hbar N} \int_{-\infty}^{\infty} \mathrm{d}t \exp(-\mathrm{i}\omega t) \langle \hat{\mathbf{Q}}_\perp^+ \cdot \hat{\mathbf{Q}}_\perp(t)\rangle. \qquad (7.91)$$

Here the angular bracket denotes the thermal average of the enclosed quantity, $\hat{\mathbf{Q}}_\perp(t)$ is the Heisenberg operator formed from (7.10), and $\hat{\mathbf{Q}}_\perp \equiv \hat{\mathbf{Q}}_\perp(0)$. The reduction of the response function $S(\kappa, \omega)$ to the time Fourier transform of a correlation function hinges on the use of an integral representation of the delta function, namely,

$$\delta(x) = \frac{1}{2\pi} \int_{-\infty}^{\infty} \mathrm{d}t \exp(-\mathrm{i}xt). \qquad (7.92)$$

In (7.91) the integration variable t has the dimension of time, and the response function has the dimension $(\text{energy})^{-1}$. Given the definition (7.91) of $S(\kappa, \omega)$, the partial differential cross-section (7.11a) reduces to

$$\frac{\mathrm{d}^2\sigma}{\mathrm{d}\Omega\, \mathrm{d}E'} = N r_0^2 \frac{k'}{k} S(\kappa, \omega). \qquad (7.93)$$

This expression differs from (7.11a) because we have included a normalization factor $(1/N)$ in the definition of the response function for an electron fluid.

The representation of $\hat{\mathbf{Q}}_\perp$ in second quantization formalism follows from (7.52); specific results for the spin and orbital interactions are given in (7.66) and (7.69), respectively. For a single band of free electrons, the wave functions are

$$\varphi_{\mathbf{k}}(\mathbf{r}) = V^{-1/2} \exp(\mathrm{i}\mathbf{k} \cdot \mathbf{r}) \tag{7.94}$$

where V is the sample volume. The wave vectors $\mathbf{k}$ are defined by imposing cyclic boundary conditions on the target sample, and have a density $V/(2\pi)^3$. We then find, using (7.92),

$$\begin{aligned}\int \mathrm{d}\mathbf{r}\varphi^*_{\mathbf{k}}(\mathbf{r})\varphi_{\mathbf{k}'}(\mathbf{r}) &= \frac{(2\pi)^3}{V}\,\delta(\mathbf{k}-\mathbf{k}')\\ &\equiv \delta_{\mathbf{k},\mathbf{k}'}\end{aligned} \tag{7.95}$$

where the Kronecker delta function is unity for $\mathbf{k} = \mathbf{k}'$, and zero otherwise.
From (7.66) and (7.69) we obtain

$$\hat{\mathbf{Q}}_\perp = \int \mathrm{d}\mathbf{r} \exp(\mathrm{i}\boldsymbol{\kappa} \cdot \mathbf{r}) \sum_{\mathbf{k}\mathbf{k}'} \varphi^*_{\mathbf{k}}(\mathbf{r}) \sum_{\sigma\sigma'} \left\{ \varphi_{\mathbf{k}'}(\mathbf{r}) \chi^+_\sigma \hat{\mathbf{s}}_\perp \chi_{\sigma'} - \frac{\mathrm{i}}{\hbar\kappa^2}\, \delta_{\sigma,\sigma'} (\boldsymbol{\kappa} \times \hat{\mathbf{p}}) \varphi_{\mathbf{k}'}(\mathbf{r}) \right\} \hat{c}^+_{\mathbf{k}\sigma} \hat{c}_{\mathbf{k}'\sigma'}. \tag{7.96}$$

Note that, since the orbital operator is independent of the spin and the spinors are orthogonal, the second term in (7.96) is nonzero only for $\sigma = \sigma'$. For ease of notation we have defined the spin operator $\hat{\mathbf{s}}_\perp$ in terms of $\hat{\mathbf{s}}$ as in (7.12a). From (7.95),

$$\int \mathrm{d}\mathbf{r} \exp(\mathrm{i}\boldsymbol{\kappa} \cdot \mathbf{r}) \varphi^*_{\mathbf{k}}(\mathbf{r}) \varphi_{\mathbf{k}'}(\mathbf{r}) = \delta_{\mathbf{k},\mathbf{k}'+\boldsymbol{\kappa}} \tag{7.97}$$

and, since

$$\hat{\mathbf{p}}\varphi_{\mathbf{k}'}(\mathbf{r}) = -\mathrm{i}\hbar\nabla\varphi_{\mathbf{k}'}(\mathbf{r}) = \hbar\mathbf{k}'\varphi_{\mathbf{k}'}(\mathbf{r}),$$

we obtain for (7.96) the result, valid for free electrons,

$$\hat{\mathbf{Q}}_\perp = \sum_{\mathbf{k}} \left\{ \sum_{\sigma\sigma'} \chi^+_\sigma \hat{\mathbf{s}}_\perp \chi_{\sigma'} \hat{c}^+_{\boldsymbol{\kappa}+\mathbf{k}\sigma} \hat{c}_{\mathbf{k}\sigma'} - \frac{\mathrm{i}}{\kappa^2} (\boldsymbol{\kappa} \times \mathbf{k}) \sum_\sigma \hat{c}^+_{\boldsymbol{\kappa}+\mathbf{k}\sigma} \hat{c}_{\mathbf{k}\sigma} \right\}. \tag{7.98}$$

The Heisenberg operator $\hat{\mathbf{Q}}_\perp(t)$ is readily formed from (7.98) by using

$$\hat{c}_{\mathbf{k}\sigma}(t) = \exp(-\mathrm{i}t\mathscr{E}(\mathbf{k})/\hbar)\hat{c}_{\mathbf{k}\sigma}. \tag{7.99}$$

The result (7.99) can be verified using the equation-of-motion for Heisenberg operators, noting that the Hamiltonian of a free electron gas is quadratic in the Fermi operators $\hat{c}$, $\hat{c}^+$ (eqn (7.55)). Thus,

$$\mathrm{i}\hbar\,\partial_t \hat{c}_{\mathbf{k}\sigma}(t) = [\hat{c}_{\mathbf{k}\sigma}(t), \mathscr{H}] = \mathscr{E}(\mathbf{k})\hat{c}_{\mathbf{k}\sigma}(t).$$

The construction of the correlation function in (7.91) from (7.98) is straightforward. The cross-term between the spin and orbital interactions vanishes by symmetry and, for free electrons,

$$\langle \hat{c}^{+}_{\mathbf{k}\sigma}\hat{c}_{\boldsymbol{\kappa}+\mathbf{k}\sigma}\hat{c}^{+}_{\boldsymbol{\kappa}+\mathbf{k}'\sigma'}\hat{c}_{\mathbf{k}'\sigma'}\rangle = \delta_{\mathbf{k},\mathbf{k}'}\,\delta_{\sigma,\sigma'}\,f_{\mathbf{k}}(1-f_{\boldsymbol{\kappa}+\mathbf{k}}); \qquad \boldsymbol{\kappa}\neq 0. \qquad (7.100)$$

Here, $f_{\mathbf{k}}$ is the Fermi occupation function

$$f_{\mathbf{k}} = \langle \hat{c}^{+}_{\mathbf{k}\sigma}\hat{c}_{\mathbf{k}\sigma}\rangle \qquad (7.101)$$

which is independent of the spin state, in the absence of an external magnetic field. Note that the result (7.100) is valid for $\boldsymbol{\kappa}\neq 0$. The term in $S(\kappa,\omega)$ with $\boldsymbol{\kappa}=0$ corresponds to scattering in the forward direction which is not of interest.

The last step in the calculation of $S(\kappa,\omega)$ is to evaluate the spin matrix elements. Given that the occupation functions are independent of the spin states of the electrons, the spin dependence of the correlation function reduces to one term

$$\begin{aligned}\sum_{\sigma\sigma'} \chi^{+}_{\sigma}\hat{\mathbf{s}}_{\perp}\chi_{\sigma'}\cdot\chi^{+}_{\sigma'}\hat{\mathbf{s}}_{\perp}\chi_{\sigma} &= \sum_{\sigma}\chi^{+}_{\sigma}\hat{\mathbf{s}}_{\perp}\cdot\hat{\mathbf{s}}_{\perp}\chi_{\sigma}\\ &= \tfrac{2}{3}\sum_{\sigma}\chi^{+}_{\sigma}\hat{\mathbf{s}}^{2}\chi_{\sigma}\\ &= \tfrac{4}{3}s(s+1) = 1.\end{aligned}$$

The first equality follows from closure, the second follows from the use of (7.12b) for an isotropic system, and, finally, $\hat{\mathbf{s}}^2 = s(s+1)$ with $s=\tfrac{1}{2}$.

Assembling the results we obtain the correlation function in the response function (7.91)

$$\langle \hat{\mathbf{Q}}^{+}_{\perp}\cdot\hat{\mathbf{Q}}_{\perp}(t)\rangle = \sum_{\mathbf{k}}\left\{1+\frac{2}{\kappa^2}(\tilde{\boldsymbol{\kappa}}\times\mathbf{k})^2\right\}f_{\mathbf{k}}(1-f_{\boldsymbol{\kappa}+\mathbf{k}})\exp\{it[\mathscr{E}(\boldsymbol{\kappa}+\mathbf{k})-\mathscr{E}(\mathbf{k})]/\hbar\}. \qquad (7.102)$$

Using (7.92) the corresponding response function is

$$\begin{aligned}S(\kappa,\omega) &= \frac{1}{N}\sum_{\mathbf{k}}\left\{1+\frac{2}{\kappa^2}(\tilde{\boldsymbol{\kappa}}\times\mathbf{k})^2\right\}f_{\mathbf{k}}(1-f_{\boldsymbol{\kappa}+\mathbf{k}})\,\delta[\hbar\omega+\mathscr{E}(\mathbf{k})-\mathscr{E}(\boldsymbol{\kappa}+\mathbf{k})]\\ &= \{1+n(\omega)\}\frac{1}{N}\sum_{\mathbf{k}}\left\{1+\frac{2}{\kappa^2}(\tilde{\boldsymbol{\kappa}}\times\mathbf{k})^2\right\}f_{\mathbf{k}} \qquad (7.103)\\ &\quad\times\{\delta[\hbar\omega+\mathscr{E}(\mathbf{k})-\mathscr{E}(\boldsymbol{\kappa}+\mathbf{k})]-\delta[-\hbar\omega+\mathscr{E}(\mathbf{k})-\mathscr{E}(\boldsymbol{\kappa}+\mathbf{k})]\}.\end{aligned}$$

Here we have introduced the detailed balance factor

$$\{1+n(\omega)\} = \{1-\exp(-\hbar\omega\beta)\}^{-1} \qquad (7.104)$$

with $\beta = 1/k_{\mathrm{B}}T$. The reduction of $S(\kappa,\omega)$ to the final form exploits the condition of detailed balance

$$S(\kappa, \omega) = \exp(\hbar\omega\beta) S(\kappa, -\omega) \tag{7.105}$$

together with a shift in the wave vector $\mathbf{k} \to \mathbf{k} - \boldsymbol{\kappa}$. The latter operation does not change the orbital matrix element in (7.103) because $\mathbf{k}$ occurs in a vector product, $(\boldsymbol{\kappa} \times \mathbf{k})$.

An explicit expression for $S(\kappa, \omega)$ is readily obtained in the limit $T \to 0$ because the Fermi function approximates to a step function

$$\left.\begin{aligned} f_{\mathbf{k}} &= 1, \quad & |\mathbf{k}| < p_{\mathrm{f}} \\ &= 0, \quad & |\mathbf{k}| > p_{\mathrm{f}} \end{aligned}\right\} T \to 0 \tag{7.106}$$

where the Fermi wave vector p_{f} is obtained from the total number of electrons

$$N = \sum_{\mathbf{k}\sigma} f_{\mathbf{k}} = 2 \frac{V}{(2\pi)^3} \int_{k < p_{\mathrm{f}}} \mathrm{d}\mathbf{k} = \{2V/(2\pi)^3\} \frac{4\pi}{3} p_{\mathrm{f}}^3. \tag{7.107}$$

The integral over the delta function in (7.103) is accomplished using a coordinate system in which $\tilde{\boldsymbol{\kappa}}$ defines the z-axis, say, and $k^2 = k_z^2 + \rho^2$. We find, for example,

$$\begin{aligned} &\frac{1}{N} \sum_{\mathbf{k}} \left\{1 + \frac{2}{\kappa^2} (\tilde{\boldsymbol{\kappa}} \times \mathbf{k})^2\right\} f_{\mathbf{k}}\, \delta[\hbar\omega + \mathscr{E}(\mathbf{k}) - \mathscr{E}(\boldsymbol{\kappa} + \mathbf{k})] \\ &\quad = (M/\hbar^2\kappa)(3/8p_{\mathrm{f}}^3) \left\{(p_{\mathrm{f}}^2 - Q^2) + \frac{1}{\kappa^2} (p_{\mathrm{f}}^2 - Q^2)^2\right\}; \quad T \to 0, \quad \text{and} \quad |Q| < p_{\mathrm{f}} \end{aligned} \tag{7.108}$$

where the wave vector

$$Q = (M/\hbar^2\kappa)(\hbar\omega - \hbar^2\kappa^2/2M). \tag{7.109}$$

The condition for a nonzero result, $|Q| < p_{\mathrm{f}}$, arises from the delta function.

In determining the range of ω and κ for which the condition $|Q| < p_{\mathrm{f}}$ is satisfied, we must bear in mind that the response function $S(\kappa, \omega)$ is zero for $\omega < 0$ in the limit of absolute zero. This arises from the fact that for $T \to 0$ there are no thermally excited states to participate in the response, and it is shown explicitly in the behaviour of the detailed balance factor in (7.103). We find that the response is finite within a range of values for the energy and wave vector transfer. The restrictions on ω and κ stem from the fact that electrons are subject to the Pauli exclusion principle. The electrons which participate in the response are only those which are excited across the Fermi surface.

We present the final result for $S(\kappa, \omega)$ in terms of reduced wave vector and energy variables. The Fermi energy $\epsilon_{\mathrm{f}} = (\hbar^2 p_{\mathrm{f}}^2/2M)$ and our variables are

$$x = \hbar\omega/\epsilon_{\mathrm{f}}, \quad \text{and} \quad y = \kappa/p_{\mathrm{f}}. \tag{7.110}$$

For $0 < y < 2$, the magnetic response of a degenerate electron fluid is

$$S(\kappa, \omega) = (3x/16y\epsilon_{\mathrm{f}})\left\{1 + \frac{2}{y^2}\left[1 - \frac{1}{4y^2}(x^2 + y^4)\right]\right\}; \quad 0 \leq x \leq (2y - y^2) \tag{7.111}$$

and

$$S(\kappa, \omega) = (3/16y\epsilon_{\mathrm{f}})\theta\{1 + \theta/y^2\}; \qquad (2y - y^2) \leq x \leq (2y + y^2),$$

where θ is a function of x and y, namely,

$$\theta = 1 - \left(\frac{x - y^2}{2y}\right)^2.$$

For $y > 2$, the range of x for a nonzero response is

$$(y^2 - 2y) \leq x \leq (y^2 + 2y),$$

and the result for $S(\kappa, \omega)$ is the same as the second expression in (7.111). The Pauli exclusion principle therefore confines the response of a degenerate electron fluid to a continuum bounded above by $(y^2 + 2y)$, and below by $x = 0$ for $0 < y < 2$ and $(y^2 - 2y)$ for $y > 2$. The terms in $S(\kappa, \omega)$ proportional to $(1/y^3)$ arise from the orbital interaction.

The integrated intensity for magnetic scattering from a degenerate electron fluid is obtained from (7.111) and the definition

$$\begin{aligned} S(\kappa) &= \hbar \int_0^\infty \mathrm{d}\omega\, S(\kappa, \omega) \\ &= (3y/16)\left(\frac{5}{3} + \frac{2}{y^2} - \frac{17y^2}{120}\right); \qquad 0 < y \leq 2 \\ &= \frac{1}{2} + \frac{2}{5y^2}; \qquad y \geq 2. \end{aligned} \tag{7.112}$$

The spin contribution to the intensity (7.112) is

$$(3y/8)(1 - y^2/12); \qquad 0 < y \leq 2$$

and

$$\tfrac{1}{2}; \qquad y \geq 2.$$

We conclude that the divergence of $S(\kappa)$ for $y \to 0$ is caused by the orbital contribution, as might be expected.

The calculation of the orbital scattering cross-section for an interacting electron fluid is quite subtle, because it must incorporate the Coulomb interaction, and the effect of diamagnetic screening (Sasaki and Obata 1980). The latter is most important for small scattering vectors, $y \ll 1$, for it eliminates the divergent behaviour found in the preceding calculation for a perfect electron fluid. Indeed, diamagnetic screening constrains the orbital cross-section to a maximum value determined by the Pauli paramagnetic susceptibility $= (\mu_0^2 p_f^3 / 2\pi^2 \epsilon_f)$, where μ_0 is the magnetic moment of an electron.

7.7. Sum rule (Lovesey and Trohidou 1986)

An expression for the first frequency moment or sum rule of the magnetic scattering response (7.91) is obtained directly from (3.60) by replacing the particle density operator by $\hat{\mathbf{Q}}_\perp$ defined in (7.10). Evaluated for a one component plasma, or jellium, model of electrons with charge $-e$ and an effective mass M we obtain

$$\int_{-\infty}^{\infty} \mathrm{d}\omega\, \omega S(\kappa, \omega) = (\kappa^2/4M) + (2\langle \mathrm{KE} \rangle / 3\hbar^2) + (\rho_0/2\kappa^4) \int \mathrm{d}\mathbf{r} \{g(r) - 1\}\{\cos(\boldsymbol{\kappa} \cdot \mathbf{r}) - 1\}(\boldsymbol{\kappa} \cdot \boldsymbol{\nabla})^2 (e^2/r) \quad (7.113)$$

Here, ρ_0 is the electron density, $\langle \mathrm{KE} \rangle$ is the mean kinetic energy of an electron, and $g(r)$ is the pair distribution function defined as in (5.33). It is interesting to observe that, the structure of the right-hand side is similar to the expression (5.289) for the third frequency moment of the response for nuclear scattering from a quantum fluid since $\hat{\mathbf{Q}}_\perp$ contains the momentum density. Even for a perfect electron plasma $g(r)$ differs from unity because of exchange induced correlations discussed in § 3.6.1.

REFERENCES

Callaway, J. (1974). *Quantum theory of the solid state*, Part B. Academic Press, New York.

Dieterich, W. (1973). *Solid State Commun.* **12,** 1191.

Elliott, J. P. and Dawber, P. G. (1979). *Symmetry in physics*, Vols. 1 and 2. The Macmillan Press, London.

Elliott, R. J. and Kleppmann, W. G. (1975). *J. Phys.* **C8,** 2737.

Fetter, A. L. and Walecka, J. D. (1971). *Quantum theory of many-particle systems*. McGraw-Hill, New York.

Harrison, W. A. (1980). *Electronic Structure and the Properties of Solids*. W. H. Freeman, San Francisco.

Landau, L. D. and Lifshitz, E. M. (1981). *Electrodynamics of continuous media*. Pergamon Press, Oxford.

Lovesey, S. W. (1979). *Z. Phys.* **B32,** 189.

—— and Trohidou, K. N. (1986). *Z. Phys.* **B62,** 207.

—— and Trohidou, K. N. (1986). *Il Nuovo Cimento* **8D,** 39.

March, N. H. and Tosi, M. P. (1984). *Coulomb liquids*. Academic Press, London.

Morgan, T., Blackman, J. A., and Cooke, J. F. (1986). *Phys. Rev.* **B33,** 7154.

Sasaki, K. and Obata, Y. (1980). *Prog. Theor. Phys. Suppl.* **69,** 406.

Schülke, W., Nagasawa, H., Maurikis, S., and Lanzki, P. (1986). *Phys. Rev.* **B33,** 6744.

Steinsvoll, O., Moon, R. M., Koehler, W. C., and Windsor, C. G. (1981). *Phys. Rev.* **B24,** 4031.

8

MAGNETIC CORRELATION AND RESPONSE FUNCTIONS

The magnetic cross-section (7.11) can be reduced to an auto-correlation function formed with the operator that describes the neutron–matter interaction. The steps involved in the reduction for nuclear scattering are described in Chapter 3, and an example of the use of a correlation function in magnetic scattering is included in § 7.6.

Here we present a general discussion of correlation functions in magnetic neutron scattering. We will also review the relation between the partial differential cross-section and the linear response function derived in Appendix B. The linear approximation employed in the latter is equivalent to the Born approximation for scattering. Consequently, the cross-section and the linear response function are related, in the guise of the fluctuation-dissipation theorem.

Various techniques and functions are employed to calculate the linear response function, or the generalized susceptibility. Some of the more important functions are listed in Table B.1, and reproduced here for convenience (Table 8.1). The definitions are given in terms of an operator $\hat{B}$, and is Hermitian conjugate $\hat{B}^{+}$. The wave-vector dependence of the operators and functions is suppressed for ease of notation. For a sample that is spatially isotropic, e.g. a fluid or a single Bravais lattice where the scattering particles occupy centres of inversion symmetry, the response functions do not depend on the direction of the scattering vector $\boldsymbol{\kappa}$. Hence, the dependence on $\boldsymbol{\kappa}$ can be suppressed without the loss of generality, and all the interrelationships in Table 8.1 are completely correct as they stand. When the condition of spatial isotropy does not hold and the response functions depend on the direction of $\boldsymbol{\kappa}$, the interrelationships in Table 8.1 are modified; the most general forms are presented in §§ 8.3 and 8.4. The dependence of the response functions on the direction of $\boldsymbol{\kappa}$ is seen in the condition of detailed balance (§ 8.3) which relates the scattering processes described by ω, $\boldsymbol{\kappa}$, and $-\omega$, $-\boldsymbol{\kappa}$. However, we always adhere to the definitions of the response functions, given in Table 8.1, apart from trivial multiplicative factors and the addition of Cartesian component labels where appropriate.

In the theory of magnetic scattering it is often convenient to use spin raising and lowering operators $\hat{S}^{\pm}$ and in this instance some care must be exercised so as not to make invalid assumptions about the analytic properties of the associated response functions. The analogous situation in nuclear scattering is encountered when the displacement correlation

Table 8.1
Summary of principal response functions

Linear response of a system close to thermal equilibrium	Change in $\langle\hat{A}\rangle$ at time t due to a perturbation $= -\hat{B}h(t)$ $\Big\} = \int_{-\infty}^{t} \mathrm{d}t'\phi_{AB}(t-t')h(t')$
	$\phi_{AB}(t) = \frac{\mathrm{i}}{\hbar}\langle[\hat{A}(t), \hat{B}]\rangle$
Generalized susceptibility	$\chi_{AB}[\omega] = -\lim_{\epsilon\to 0^+}\int_0^\infty \mathrm{d}t \exp(-\mathrm{i}\omega t - \epsilon t)\phi_{AB}(t) = \chi'_{AB}[\omega] + \mathrm{i}\chi''_{AB}[\omega]$
Relaxation function	$R_{AB}(t) = \int_0^\beta \mathrm{d}\mu\langle\{\hat{B} - \langle\hat{B}\rangle\}\{\hat{A}(t+\mathrm{i}\hbar\mu) - \langle\hat{A}\rangle\}\rangle = R_{BA}(-t)$
	or $\phi_{AB}(t) = -\partial_t R_{AB}(t)$
	$R_{AB}(\omega) = \frac{1}{2\pi}\int_{-\infty}^{\infty} \mathrm{d}t \exp(-\mathrm{i}\omega t)R_{AB}(t) = R_{BA}(-\omega)$
Green function	$G(t) = -\mathrm{i}\theta(t)\langle[\hat{B}(t), \hat{B}^+]\rangle = -\theta(t)\hbar\phi_{BB^+}(t)$
	$G(\omega) = \int_{-\infty}^{\infty} \mathrm{d}t \exp(\mathrm{i}\omega t)G(t) \equiv \langle\langle\hat{B}; \hat{B}^+\rangle\rangle = G'(\omega) + \mathrm{i}G''(\omega)$
Static isothermal susceptibility	$\chi_{AB} = R_{AB}(t=0) = -\chi'_{AB}[0]$
	$S(\omega) = \frac{1}{2\pi\hbar}\int_{-\infty}^{\infty} \mathrm{d}t \exp(-\mathrm{i}\omega t)\langle\hat{B}\hat{B}^+(t)\rangle$
	$= \frac{1}{2\pi\hbar}\int_{-\infty}^{\infty} \mathrm{d}t \exp(\mathrm{i}\omega t)\langle\hat{B}(t)\hat{B}^+\rangle = \{1+n(\omega)\}\frac{1}{\pi}\chi''_{B^+B}[\omega]$
	$= \{1+n(\omega)\}\omega R_{B^+B}(\omega) = -\{1+n(\omega)\}\frac{1}{\pi\hbar}G''(\omega)$
where the detailed balance factor is	$\{1+n(\omega)\} = \{1-\exp(-\hbar\omega\beta)\}^{-1} = -n(-\omega)$

function is calculated in terms of Bose creation and annihilation operators (cf. § 4.7.1).

Our presentation here is more or less self-contained. Notwithstanding, the reader will probably benefit from reading Appendix B (Volume 1) if the goal is an appreciation of the ubiquity of linear response theory in neutron scattering.

8.1. Correlation functions: localized model (Lovesey 1986)

In Chapter 7 we found that the neutron cross-section for the scattering by identical magnetic ions situated at the sites $\mathbf{R}_{ld}$ of a crystal is

$$\frac{\mathrm{d}^2\sigma}{\mathrm{d}\Omega\,\mathrm{d}E'} = r_0^2\frac{k'}{k}\{\tfrac{1}{2}gF(\boldsymbol{\kappa})\}^2\sum_{\alpha,\beta}(\delta_{\alpha\beta} - \tilde{\kappa}_\alpha\tilde{\kappa}_\beta)\sum_{\lambda,\lambda'}p_\lambda\sum_{l,d}\sum_{l',d'}\langle\lambda|\exp(-\mathrm{i}\boldsymbol{\kappa}\cdot\mathbf{R}_{ld})\hat{S}^\alpha_{ld}|\lambda'\rangle$$
$$\times\langle\lambda'|\exp(\mathrm{i}\boldsymbol{\kappa}\cdot\mathbf{R}_{l'd'})\hat{S}^\beta_{l'd'}|\lambda\rangle\,\delta(\hbar\omega + E_\lambda - E_{\lambda'}). \quad (8.1)$$

Here the orbital motion of the electrons is partially taken into account via the dipole approximation to the form factor and $r_0 = -0.54 \cdot 10^{-12}$ cm.

To express (8.1) in terms of a correlation function we proceed as in Chapter 3, writing the delta function in terms of an integral representation. Thus, using (7.92), and closure for the states λ',

$$\sum_{\lambda,\lambda'} p_\lambda \langle\lambda| \exp(-i\boldsymbol{\kappa}\cdot\mathbf{R}_{ld})\hat{S}^\alpha_{ld} |\lambda'\rangle \langle\lambda'| \exp(i\boldsymbol{\kappa}\cdot\mathbf{R}_{l'd'})\hat{S}^\beta_{l'd'} |\lambda\rangle\, \delta(\hbar\omega + E_\lambda - E_{\lambda'})$$

$$= \sum_{\lambda,\lambda'} p_\lambda \frac{1}{2\pi\hbar}\int_{-\infty}^{\infty} dt \exp(-i\omega t)\langle\lambda| \exp(-i\boldsymbol{\kappa}\cdot\mathbf{R}_{ld})\hat{S}^\alpha_{ld} |\lambda'\rangle$$

$$\times \langle\lambda'| \exp(i\mathcal{H}t/\hbar) \exp(i\boldsymbol{\kappa}\cdot\mathbf{R}_{l'd'})\exp(-i\mathcal{H}t/\hbar + i\mathcal{H}t/\hbar)\hat{S}^\beta_{l'd'} \exp(-i\mathcal{H}t/\hbar) |\lambda\rangle$$

$$= \frac{1}{2\pi\hbar}\int_{-\infty}^{\infty} dt \exp(-i\omega t)\langle \exp(-i\boldsymbol{\kappa}\cdot\hat{\mathbf{R}}_{ld})\hat{S}^\alpha_{ld} \exp\{i\boldsymbol{\kappa}\cdot\hat{\mathbf{R}}_{l'd'}(t)\}\hat{S}^\beta_{l'd'}(t)\rangle. \quad (8.2)$$

We use the shorthand notation $\hat{S}(0) \equiv \hat{S}$, etc.

It is convenient to rewrite (8.2) in the form

$$\frac{1}{2\pi\hbar}\int_{-\infty}^{\infty} dt \exp(-i\omega t)\langle \exp(-i\boldsymbol{\kappa}\cdot\hat{\mathbf{R}}_{ld})\hat{S}^\alpha_{ld} \int d\mathbf{r}' \exp(i\boldsymbol{\kappa}\cdot\mathbf{r}')\delta\{\mathbf{r}' - \hat{\mathbf{R}}_{l'd'}(t)\}\hat{S}^\beta_{l'd'}(t)\rangle$$

$$= \frac{1}{2\pi\hbar}\int_{-\infty}^{\infty} dt \int d\mathbf{r} \exp(i\boldsymbol{\kappa}\cdot\mathbf{r} - i\omega t)$$

$$\times \int d\mathbf{r}' \langle \delta\{\mathbf{r} + \hat{\mathbf{R}}_{ld} - \mathbf{r}'\}\hat{S}^\alpha_{ld}\delta\{\mathbf{r}' - \hat{\mathbf{R}}_{l'd'}(t)\}\hat{S}^\beta_{l'd'}(t)\rangle \quad (8.3)$$

and hence the partial differential cross-section is

$$\frac{d^2\sigma}{d\Omega\, dE'} = r_0^2 \frac{k'}{k}\{\tfrac{1}{2}gF(\boldsymbol{\kappa})\}^2 \sum_{\alpha,\beta}(\delta_{\alpha\beta} - \tilde{\kappa}_\alpha\tilde{\kappa}_\beta)\frac{N}{2\pi\hbar}\int_{-\infty}^{\infty} dt \int d\mathbf{r} \exp(i\boldsymbol{\kappa}\cdot\mathbf{r} - i\omega t)\Gamma_{\alpha\beta}(\mathbf{r}, t)$$

$$(8.4)$$

where

$$\Gamma_{\alpha\beta}(\mathbf{r}, t) = \frac{1}{N}\sum_{l,d}\sum_{l',d'}\int d\mathbf{r}' \langle \delta\{\mathbf{r} + \hat{\mathbf{R}}_{ld} - \mathbf{r}'\}\hat{S}^\alpha_{ld}\, \delta\{\mathbf{r}' - \hat{\mathbf{R}}_{l'd'}(t)\}\hat{S}^\beta_{l'd'}(t)\rangle.$$

$$(8.5)$$

In (8.4), N is the number of unit cells in the crystal.

Now to a good approximation the motion of an ion is independent of its spin orientation because the spin-dependent forces are small. Hence

(8.5) becomes

$$\Gamma_{\alpha\beta}(\mathbf{r}, t) = \frac{1}{N} \sum_{l,d} \sum_{l',d'} \langle \hat{S}^{\alpha}_{ld} \hat{S}^{\beta}_{l'd'}(t) \rangle \int \mathrm{d}\mathbf{r}' \, \langle \delta\{\mathbf{r} + \hat{\mathbf{R}}_{ld} - \mathbf{r}'\} \, \delta\{\mathbf{r}' - \hat{\mathbf{R}}_{l'd'}(t)\} \rangle. \tag{8.6}$$

Let us now take the particular case of a Bravais lattice and write for (8.6)

$$\Gamma_{\alpha\beta}(\mathbf{r}, t) = \sum_{l} \gamma_{\alpha\beta}(l, t) G_l(\mathbf{r}, t)$$

where

$$\gamma_{\alpha\beta}(l, t) = \frac{1}{N} \sum_{m} \langle \hat{S}^{\alpha}_{m} \hat{S}^{\beta}_{m+l}(t) \rangle = \gamma_{\alpha\beta}(l, \infty) + \gamma'_{\alpha\beta}(l, t) \tag{8.7}$$

and

$$G_l(\mathbf{r}, t) = \int \mathrm{d}\mathbf{r}' \, \langle \delta\{\mathbf{r} + \hat{\mathbf{R}}_0 - \mathbf{r}'\} \delta\{\mathbf{r}' - \hat{\mathbf{R}}_l(t)\} \rangle$$

$$= G_l(\mathbf{r}, \infty) + G'_l(\mathbf{r}, t).$$

Therefore,

$$\Gamma_{\alpha\beta}(\mathbf{r}, t) = \sum_{l} \gamma_{\alpha\beta}(l, \infty) G_l(\mathbf{r}, \infty) + \sum_{l} \gamma_{\alpha\beta}(l, \infty) G'_l(\mathbf{r}, t)$$

$$+ \sum_{l} \gamma'_{\alpha\beta}(l, t) G_l(\mathbf{r}, \infty) + \sum_{l} \gamma'_{\alpha\beta}(l, t) G'_l(\mathbf{r}, t). \tag{8.8}$$

There is a contribution to the cross-section (8.4) from each of the four terms in (8.8). The first of these gives elastic magnetic scattering; the second scattering that is elastic in the spin system but inelastic in the phonon system (magnetovibrational scattering); the third term gives inelastic magnetic scattering; the fourth term gives scattering inelastic in both the spin and phonon systems.

We will evaluate the correlation function $G_l(\mathbf{r}, t)$ for a lattice in which the atoms make small-amplitude, harmonic oscillations about equilbirium positions defined by lattice vectors $\mathbf{l}$. The displacement from the site $\mathbf{l}$ is denoted by $\mathbf{u}(l)$, and $\mathbf{R}_l = \mathbf{l} + \mathbf{u}(l)$.

First, we express the delta functions in $G_l(\mathbf{r}, t)$ in terms of an integral over wave vectors by using (7.92), and find

$$G_l(\mathbf{r}, t) = \frac{1}{(2\pi)^3} \int \mathrm{d}\mathbf{q} \exp(-i\mathbf{q} \cdot \mathbf{r}) \langle \exp(-i\mathbf{q} \cdot \hat{\mathbf{R}}_0) \exp\{i\mathbf{q} \cdot \hat{\mathbf{R}}_l(t)\} \rangle$$

$$= \frac{1}{(2\pi)^3} \int \mathrm{d}\mathbf{q} \exp\{i\mathbf{q} \cdot (\mathbf{l} - \mathbf{r})\} \langle \exp\{-i\mathbf{q} \cdot \hat{\mathbf{u}}(0)\} \exp\{i\mathbf{q} \cdot \hat{\mathbf{u}}(l, t)\} \rangle.$$

where $\hat{\mathbf{u}}(0) \equiv \hat{\mathbf{u}}(0, 0)$. From this result we deduce that

$$G_l(\mathbf{r}, \infty) = \frac{1}{(2\pi)^3} \int \mathrm{d}\mathbf{q} \exp\{\mathrm{i}\mathbf{q} \cdot (\mathbf{l} - \mathbf{r}) - 2W(\mathbf{q})\} \tag{8.9}$$

where the Debye–Waller factor (cf. § 4.3)

$$\begin{aligned} \exp\{-W(\mathbf{q})\} &= \langle \exp(\mathrm{i}\mathbf{q} \cdot \hat{\mathbf{u}}) \rangle \\ &= \exp\{-\tfrac{1}{2}\langle (\mathbf{q} \cdot \hat{\mathbf{u}})^2 \rangle\}. \end{aligned} \tag{8.10}$$

The final form is valid for harmonic vibrations. More generally we have

$$\langle \exp\{-\mathrm{i}\mathbf{q} \cdot \hat{\mathbf{u}}(0)\} \exp\{\mathrm{i}\mathbf{q} \cdot \hat{\mathbf{u}}(l, t)\} \rangle = \exp\{-2W(\mathbf{q})\} \exp\langle \mathbf{q} \cdot \hat{\mathbf{u}}(0) \mathbf{q} \cdot \hat{\mathbf{u}}(l, t) \rangle.$$

Thus,

$$G_l'(\mathbf{r}, t) = \frac{1}{(2\pi)^3} \int \mathrm{d}\mathbf{q} \exp\{\mathrm{i}\mathbf{q} \cdot (\mathbf{l} - \mathbf{r}) - 2W(\mathbf{q})\} \{\exp\langle \mathbf{q} \cdot \hat{\mathbf{u}}(0) \mathbf{q} \cdot \hat{\mathbf{u}}(l, t) \rangle - 1\}. \tag{8.11}$$

Explicit expressions for the displacement correlation functions in (8.10) and (8.11) are given in Chapter 4.

Using the result (8.9) in the first and third terms in (8.8), the corresponding cross-sections are given by

$$\left(\frac{\mathrm{d}\sigma}{\mathrm{d}\Omega}\right)_{\mathrm{el}} = r_0^2 \{\tfrac{1}{2} g F(\boldsymbol{\kappa})\}^2 N \exp\{-2W(\boldsymbol{\kappa})\} \sum_{\alpha,\beta} (\delta_{\alpha\beta} - \tilde{\kappa}_\alpha \tilde{\kappa}_\beta) \sum_l \exp(\mathrm{i}\boldsymbol{\kappa} \cdot \mathbf{l}) \gamma_{\alpha\beta}(l, \infty) \tag{8.12}$$

and the inelastic magnetic scattering

$$\left(\frac{\mathrm{d}^2\sigma}{\mathrm{d}\Omega\, \mathrm{d}E'}\right)_{\mathrm{inel}} = r_0^2 \frac{k'}{k} \{\tfrac{1}{2} g F(\boldsymbol{\kappa})\}^2 N \exp\{-2W(\boldsymbol{\kappa})\} \sum_{\alpha,\beta} (\delta_{\alpha\beta} - \tilde{\kappa}_\alpha \tilde{\kappa}_\beta)$$
$$\times \sum_l \exp(\mathrm{i}\boldsymbol{\kappa} \cdot \mathbf{l}) \frac{1}{2\pi\hbar} \int_{-\infty}^{\infty} \mathrm{d}t \exp(-\mathrm{i}\omega t) \gamma'_{\alpha\beta}(l, t). \tag{8.13}$$

The magnetovibrational cross-section is easily rearranged into the form, using (8.11),

$$\left(\frac{\mathrm{d}^2\sigma}{\mathrm{d}\Omega\, \mathrm{d}E'}\right)_{\mathrm{m.v.}} = r_0^2 \frac{k'}{k} \{\tfrac{1}{2} g F(\boldsymbol{\kappa})\}^2 \exp\{-2W(\boldsymbol{\kappa})\} \sum_{\alpha,\beta} (\delta_{\alpha\beta} - \tilde{\kappa}_\alpha \tilde{\kappa}_\beta) \frac{1}{2\pi\hbar} \int_{-\infty}^{\infty} \mathrm{d}t$$
$$\times \exp(-\mathrm{i}\omega t) \sum_{l,l'} \gamma_{\alpha\beta}(l' - l, \infty) \exp\{\mathrm{i}\boldsymbol{\kappa} \cdot (\mathbf{l}' - \mathbf{l})\} \{\exp\langle \boldsymbol{\kappa} \cdot \hat{\mathbf{u}}(l) \boldsymbol{\kappa} \cdot \hat{\mathbf{u}}(l', t) \rangle - 1\}, \tag{8.14}$$

and similarly the cross-section inelastic in both the magnetic and vibrational system is

$$\frac{d^2\sigma}{d\Omega\, dE'} = r_0^2 \frac{k'}{k} \{\tfrac{1}{2}gF(\boldsymbol{\kappa})\}^2 \exp\{-2W(\boldsymbol{\kappa})\} \sum_{\alpha,\beta} (\delta_{\alpha\beta} - \tilde{\kappa}_\alpha \tilde{\kappa}_\beta) \frac{1}{2\pi\hbar} \int_{-\infty}^{\infty} dt$$

$$\times \exp(-i\omega t) \sum_{l,l'} \gamma'_{\alpha\beta}(l'-l, t) \exp\{i\boldsymbol{\kappa} \cdot (\mathbf{l}' - \mathbf{l})\}\{\exp\langle \boldsymbol{\kappa} \cdot \hat{\mathbf{u}}(l) \boldsymbol{\kappa} \cdot \hat{\mathbf{u}}(l', t)\rangle - 1\}. \tag{8.15}$$

For a ferromagnet in which the spins are aligned in the z-direction,

$$\gamma_{\alpha\beta}(l'-l, \infty) = \delta_{\alpha\beta}\, \delta_{\alpha z} \langle \hat{S}^z \rangle^2 \tag{8.16}$$

and the effective scattering length in (8.14) is

$$\bar{b} \to r_0 \tfrac{1}{2} gF(\boldsymbol{\kappa})(1 - \tilde{\kappa}_z^2) \langle \hat{S}^z \rangle.$$

In other words, the magnetovibrational scattering is exactly like the coherent inelastic nuclear scattering, but with the difference that the neutron is interacting with the magnetic moment of the atom through magnetic forces rather than with the nucleus of the atom through nuclear forces (Steinsvoll, et al. 1981).

The cross-section (8.15), which is inelastic in both the magnetic and phonon systems, is not usually very interesting or important, because it is generally of a diffuse character with no distinct features in momentum or energy conservation. One special case is worth noting, however: for perfect paramagnet

$$\gamma'_{\alpha\beta}(l'-l, t) = \delta_{\alpha\beta} \delta_{l',l} \tfrac{1}{3} S(S+1). \tag{8.17a}$$

More precisely, we can use (8.17a) as a reasonable assumption when the exchange energies in the target are negligible compared to the neutron energy. Notice that (8.17a) is *not* applicable to a magnetic system with substantial exchange effects but which is at high enough temperature to be in the paramagnetic phase. So (8.15) becomes, for a perfect paramagnet,

$$\frac{d^2\sigma}{d\Omega\, dE'} = r_0^2 \frac{k'}{k} \{\tfrac{1}{2}gF(\boldsymbol{\kappa})\}^2 \exp\{-2W(\boldsymbol{\kappa})\} \tfrac{2}{3} S(S+1)$$

$$\times \frac{N}{2\pi\hbar} \int_{-\infty}^{\infty} dt \exp(-i\omega t)\{\exp\langle \boldsymbol{\kappa} \cdot \hat{\mathbf{u}} \boldsymbol{\kappa} \cdot \hat{\mathbf{u}}(t)\rangle - 1\}, \tag{8.17b}$$

which is simply the nuclear incoherent inelastic scattering with the substitution

$$\frac{\sigma_i}{4\pi} \to r_0^2 \{\tfrac{1}{2} gF(\boldsymbol{\kappa})\}^2 \tfrac{2}{3} S(S+1).$$

8.2. Correlation functions: itinerant model

A general form of the magnetic partial differential cross-section for unpolarized neutrons is given by (7.11) where the neutron–matter interaction is described by an operator $\hat{\mathbf{Q}}_{\perp}$, or alternatively, $\hat{\mathbf{Q}}$. To express the response function in terms of a correlation function we follow the steps in (8.2); the result is

$$S(\boldsymbol{\kappa}, \omega) = \frac{1}{2\pi\hbar} \int_{-\infty}^{\infty} \mathrm{d}t \exp(-\mathrm{i}\omega t) \langle \hat{\mathbf{Q}}_{\perp}^{+} \cdot \hat{\mathbf{Q}}_{\perp}(t) \rangle$$

$$= \sum_{\alpha\beta} (\delta_{\alpha\beta} - \tilde{\kappa}_{\alpha}\tilde{\kappa}_{\beta}) \frac{1}{2\pi\hbar} \int_{-\infty}^{\infty} \mathrm{d}t \exp(-\mathrm{i}\omega t) \langle \hat{Q}_{\alpha}^{+} \hat{Q}_{\beta}(t) \rangle \quad (8.18)$$

where the second form follows from (7.12b).

It is straightforward to verify that a suitable expression for $\hat{\mathbf{Q}}$ is

$$\hat{\mathbf{Q}} = \sum_{i} \exp(\mathrm{i}\boldsymbol{\kappa} \cdot \mathbf{r}_i) \left\{ \hat{\mathbf{s}}_i - \frac{\mathrm{i}}{\hbar\kappa^2} (\boldsymbol{\kappa} \times \hat{\mathbf{p}}_i) \right\} \quad (8.19)$$

where the sum includes all unpaired electrons and $\hat{\mathbf{s}}$ and $\hat{\mathbf{p}}$ are, respectively, the electron-spin and linear-momentum operators. In discussing the properties of itinerant-electron models it is usually an advantage to employ the second quantization formalism introduced in § 7.5. Bloch states are specified by a wave vector $\mathbf{k}$, band index λ, and spin variables σ. For $\hat{\mathbf{Q}}$ we obtain

$$\hat{\mathbf{Q}} = \int \mathrm{d}\mathbf{r} \exp(\mathrm{i}\boldsymbol{\kappa} \cdot \mathbf{r}) \sum_{\mathbf{k}\lambda\sigma} \sum_{\mathbf{k}'\lambda'\sigma'} \varphi^{*}_{\mathbf{k}\lambda\sigma}(\mathbf{r}) \left\{ \hat{\mathbf{s}} - \frac{\mathrm{i}}{\hbar\kappa^2} (\boldsymbol{\kappa} \times \hat{\mathbf{p}}) \right\} \varphi_{\mathbf{k}'\lambda'\sigma'}(\mathbf{r}) \hat{c}^{+}_{\mathbf{k}\lambda\sigma} \hat{c}_{\mathbf{k}'\lambda'\sigma'}. \quad (8.20)$$

The development of this result for nearly free electrons is easily deduced from the corresponding development of $\hat{\mathbf{Q}}_{\perp}$ for an electron fluid given in § 7.6. A general development is not possible since we must first specify the form of the wave functions, $\varphi_{\mathbf{k}\lambda\sigma}(\mathbf{r})$, and these are largely model-dependent. The matrix element of the orbital interaction satisfies the identity (7.70), and for $\boldsymbol{\kappa} \to 0$ we have the result (7.71).

When we come to consider the dynamical behaviour of electrons in narrow energy bends, we shall simplify the analysis by choosing to examine a model based upon a single band. It is, therefore, pertinent to consider the cross-section for the scattering by electrons in a single, narrow, energy band. In this instance the orbital interaction is negligible, for small $\boldsymbol{\kappa}$, and we find that

$$\int \mathrm{d}\mathbf{r} \exp(-\mathrm{i}\mathbf{q} \cdot \mathbf{r}) \sum_{\mathbf{k},\lambda} \sum_{\mathbf{k}',\lambda'} \phi^{*}_{\mathbf{k}\lambda}(\mathbf{r}) \phi_{\mathbf{k}'\lambda'}(\mathbf{r})$$

can be replaced by

$$\int d\mathbf{r} \exp(-i\mathbf{q} \cdot \mathbf{r}) \, |\phi(\mathbf{r})|^2 \sum_{\mathbf{k},\mathbf{k}'} \delta_{\mathbf{k}',\mathbf{k}+\mathbf{q}}$$

in the approximation in which the overlap of the wave functions $\phi(\mathbf{r})$ centred about the lattice points of a Bravais crystal is neglected. Thus in this scheme

$$\begin{aligned}\hat{\mathbf{s}}(\mathbf{q}) = \int d\mathbf{r} \exp(-i\mathbf{q} \cdot \mathbf{r}) \, |\phi(\mathbf{r})|^2 \\ \times \sum_{\mathbf{k}} \Big(\tfrac{1}{2}[\hat{c}^+_{\mathbf{k}\uparrow} \hat{c}_{\mathbf{k}+\mathbf{q}\downarrow} + \text{c.c.}], \frac{1}{2i}[\hat{c}^+_{\mathbf{k}\uparrow} \hat{c}_{\mathbf{k}+\mathbf{q}\downarrow} - \text{c.c.}], \\ \tfrac{1}{2}[\hat{c}^+_{\mathbf{k}\uparrow} \hat{c}_{\mathbf{k}+\mathbf{q}\uparrow} - \hat{c}^+_{\mathbf{k}\downarrow} \hat{c}_{\mathbf{k}+\mathbf{q}\downarrow}] \Big), \quad (8.21)\end{aligned}$$

where c.c. denotes the complex conjugate of the preceding operator, and we shall in future denote

$$F(\boldsymbol{\kappa}) = \int d\mathbf{r} \exp(i\boldsymbol{\kappa} \cdot \mathbf{r}) \, |\phi(\mathbf{r})|^2,$$

for this is the form factor appropriate to the model.

8.3. Response functions: Bravais lattice

In this section we turn our attention to representing the inelastic magnetic cross-section (8.13) in terms of functions that are of more direct significance in statistical mechanics. We shall be careful to use a notation which in § 8.4 is easily generalized to non-Bravais lattices.

First let us represent $\gamma_{\alpha\beta}$ in terms of the operators $\hat{S}^{\alpha}_{\mathbf{q}}$ defined by

$$\hat{S}^{\alpha}_{l} = \frac{1}{N} \sum_{\mathbf{q}} \hat{S}^{\alpha}_{\mathbf{q}} \exp(i\mathbf{q} \cdot \mathbf{l}), \quad (8.22)$$

i.e.

$$\hat{S}^{\alpha}_{\mathbf{q}} = \sum_{l} \exp(-i\mathbf{q} \cdot \mathbf{l}) \hat{S}^{\alpha}_{l}.$$

Then

$$\begin{aligned}\sum_{l} \exp(i\boldsymbol{\kappa} \cdot \mathbf{l}) \gamma_{\alpha\beta}(l, t) &= \frac{1}{N} \sum_{lm} \exp(i\boldsymbol{\kappa} \cdot \mathbf{l}) \langle \hat{S}^{\alpha}_{m} \hat{S}^{\beta}_{m+l}(t) \rangle \\ &= \frac{1}{N} \langle \hat{S}^{\alpha}_{\boldsymbol{\kappa}} \hat{S}^{\beta}_{-\boldsymbol{\kappa}}(t) \rangle. \quad (8.23)\end{aligned}$$

Taking the *sum* of (8.12) and (8.13), the result (8.25) leads to

$$\frac{d^2\sigma}{d\Omega\, dE'} = r_0^2 \frac{k'}{k} \{\tfrac{1}{2} g F(\boldsymbol{\kappa})\}^2 \exp\{-2W(\boldsymbol{\kappa})\}$$

$$\times \sum_{\alpha\beta} (\delta_{\alpha\beta} - \tilde{\kappa}_\alpha \tilde{\kappa}_\beta) \frac{1}{2\pi\hbar} \int_{-\infty}^{\infty} dt \exp(-i\omega t) \langle \hat{S}^\alpha_{\boldsymbol{\kappa}} \hat{S}^\beta_{-\boldsymbol{\kappa}}(t) \rangle. \quad (8.24)$$

Note that it can be proved that (8.24) satisfies the condition of detailed balance

$$\left(\frac{d^2\sigma}{d\Omega\, dE'}\right)_{\boldsymbol{\kappa},\omega} = \exp(\hbar\omega\beta) \left(\frac{d^2\sigma}{d\Omega\, dE'}\right)_{-\boldsymbol{\kappa},-\omega}.$$

The correlation function in (8.24) can in many simple cases be calculated directly from a theory of the magnetic system. But it is related to quantities of more direct physical interpretation and it is therefore important to bring out these relations clearly—especially for discussions of problems where no strictly rigorous theory can be constructed.

The linear response of a magnetic system to a magnetic field **H** that varies in both space and time can be deduced from the results listed in Table 8.1. We find that the magnetization is given by

$$M_l^\alpha(t) = -g\mu_B \langle \hat{S}_l^\alpha \rangle + \sum_{l'} \int_{-\infty}^{t} dt' \sum_\beta \phi^{\alpha\beta}(l - l', t - t') H_{l'}^\beta(t')$$

where the linear response function

$$\phi^{\alpha\beta}(l, t) = (g\mu_B)^2 \frac{i}{\hbar} \langle [\hat{S}_l^\alpha(t), \hat{S}_0^\beta] \rangle.$$

If the time dependence of **H** is of the form $\exp(i\omega t)$, then

$$M_l^\alpha(t) = -g\mu_B \langle \hat{S}_l^\alpha \rangle - \exp(i\omega t) \sum_{l'} \sum_\beta \chi^{\alpha\beta}[l - l', \omega] H_{l'}^\beta$$

where

$$\chi^{\alpha\beta}[l, \omega] = -\int_0^\infty dt \exp(-i\omega t) \phi^{\alpha\beta}(l, t)$$

is a generalized susceptibility tensor. Because the operators $\hat{S}_l^\alpha(t)$ are Hermitian, $\phi^{\alpha\beta}(l, t)$ is real.

However, it is much more convenient to work in terms of the Fourier transforms of these operators, which are *not* Hermitian. Thus

$$\{\hat{S}^\alpha_{\mathbf{q}}(t)\}^+ = \hat{S}^\alpha_{-\mathbf{q}}(t) \quad \text{for} \quad \alpha = x, y, z.$$

For a magnetic field varying in space like

$$H_{\mathbf{q}}^{\beta}(t')\exp(\mathrm{i}\mathbf{q}\cdot\mathbf{l}),$$

we get

$$M_{\mathbf{q}}^{\alpha}(t) = -g\mu_{\mathrm{B}}\langle\hat{S}_{\mathbf{q}}^{\alpha}\rangle + \int_{-\infty}^{t} \mathrm{d}t' \sum_{\beta} \phi_{\mathbf{q}}^{\alpha\beta}(t-t')H_{\mathbf{q}}^{\beta}(t') \tag{8.25}$$

where

$$\phi_{\mathbf{q}}^{\alpha\beta}(t) = \frac{(g\mu_{\mathrm{B}})^2}{N}\frac{\mathrm{i}}{\hbar}\langle[\hat{S}_{\mathbf{q}}^{\alpha}(t), \hat{S}_{-\mathbf{q}}^{\beta}]\rangle.$$

If the time dependence of $H_{\mathbf{q}}^{\beta}(t')$ is of the form $\exp(\mathrm{i}\omega t')$, then

$$M_{\mathbf{q}}^{\alpha}(t) = -g\mu_{\mathrm{B}}\langle\hat{S}_{\mathbf{q}}^{\alpha}\rangle - \exp(\mathrm{i}\omega t)\sum_{\beta}\chi_{\mathbf{q}}^{\alpha\beta}[\omega]H_{\mathbf{q}}^{\beta} \tag{8.26}$$

where

$$\chi_{\mathbf{q}}^{\alpha\beta}[\omega] = -\int_{0}^{\infty}\mathrm{d}t\,\exp(-\mathrm{i}\omega t)\phi_{\mathbf{q}}^{\alpha\beta}(t)$$

is a generalized susceptibility tensor.

We choose to define the relaxation function through

$$\partial_t\mathscr{R}_{\mathbf{q}}^{\alpha\beta}(t) = \frac{-N}{(g\mu_{\mathrm{B}})^2}\phi_{\mathbf{q}}^{\alpha\beta}(t). \tag{8.27}$$

We note that this differs by a simple factor from the function defined in Table 8.1.

A useful alternative expression for the relaxation function is

$$\mathscr{R}_{\mathbf{q}}^{\alpha\beta}(t) = \int_{0}^{\beta}\mathrm{d}\mu\langle\hat{S}_{-\mathbf{q}}^{\beta}\hat{S}_{\mathbf{q}}^{\alpha}(t+\mathrm{i}\hbar\mu)\rangle - \beta\langle\hat{S}_{-\mathbf{q}}^{\beta}\rangle\langle\hat{S}_{\mathbf{q}}^{\alpha}\rangle. \tag{8.28}$$

This expression is best verified by showing that it satisfies (8.27) with the linear response function given in (8.25). The constant of integration in (8.28) is determined from the requirement $\mathscr{R}_{\mathbf{q}}^{\alpha\beta}(\infty) = 0$. The latter follows because, by definition, the relaxation function is the change in the magnetization induced by a discontinuous field and, for a bulk sample, the change will vanish after a sufficiently long time elapse. We also anticipate that

$$\langle\hat{S}_{-\mathbf{q}}^{\beta}\hat{S}_{\mathbf{q}}^{\alpha}(\infty)\rangle = \langle\hat{S}_{-\mathbf{q}}^{\beta}\rangle\langle\hat{S}_{\mathbf{q}}^{\alpha}\rangle$$

and thus we arrive at the form of the integration constant given in (8.28).

Let us now establish the relation between the relaxation function and the correlation function in the cross-section (8.24). Since the latter contains the time Fourier transform of a spin correlation function we consider the function

$$\mathcal{R}_{\mathbf{q}}^{\alpha\beta}(\omega)=\frac{1}{2\pi}\int_{-\infty}^{\infty}\mathrm{d}t\,\exp(-\mathrm{i}\omega t)\mathcal{R}_{\mathbf{q}}^{\alpha\beta}(t). \tag{8.29}$$

Using the identity

$$\langle\hat{S}_{\mathbf{q}}^{\alpha}(t)\hat{S}_{-\mathbf{q}}^{\beta}\rangle=\langle\hat{S}_{-\mathbf{q}}^{\beta}\hat{S}_{\mathbf{q}}^{\alpha}(t+i\hbar\beta)\rangle, \tag{8.30}$$

we readily obtain from (8.28) the desired result

$$\frac{1}{2\pi\hbar}\int_{-\infty}^{\infty}\mathrm{d}t\,\exp(-\mathrm{i}\omega t)\langle\hat{S}_{\mathbf{q}}^{\alpha}\hat{S}_{-\mathbf{q}}^{\beta}(t)\rangle$$
$$=\omega\{1+n(\omega)\}\mathcal{R}_{\mathbf{q}}^{\alpha\beta}(-\omega)+\delta(\hbar\omega)\langle\hat{S}_{\mathbf{q}}^{\alpha}\rangle\langle\hat{S}_{-\mathbf{q}}^{\beta}\rangle. \tag{8.31}$$

The detailed balance factor $\{1+n(\omega)\}$ is given in Table 8.1. The second term in (8.31) represents purely elastic scattering. Note that $\mathcal{R}_{\mathbf{q}}^{\alpha\beta}(\omega)$ has the dimensions (frequency $\times$ energy)$^{-1}$.

In subsequent developments we make use of two identities for correlation functions. For arbitrary operators $\hat{A}$ and $\hat{B}$ (Lovesey 1980)

$$\langle\hat{A}\hat{B}\rangle^{*}=\langle\hat{B}^{+}\hat{A}^{+}\rangle \tag{8.32}$$

and

$$\langle\hat{A}(t)\hat{B}\rangle=\langle\hat{A}\hat{B}(-t)\rangle. \tag{8.33}$$

The second identity follows from the assumed stationarity of the sample which implies that

$$\langle\hat{A}(t+t_0)\hat{B}(t_0)\rangle=\langle\hat{A}(t)\hat{B}\rangle$$

for any t_0; taking $t_0=-t$, we obtain (8.33). Using these two identities it is straightforward to show that

$$\phi_{\mathbf{q}}^{\alpha\beta}(t)^{*}=\phi_{-\mathbf{q}}^{\alpha\beta}(t)=-\phi_{\mathbf{q}}^{\beta\alpha}(-t), \tag{8.34}$$

and hence

$$\begin{aligned}\chi_{\mathbf{q}}^{\alpha\beta}[\omega]^{*}&=\chi_{-\mathbf{q}}^{\alpha\beta}[-\omega],\\ \mathcal{R}_{\mathbf{q}}^{\alpha\beta}(t)^{*}&=\mathcal{R}_{-\mathbf{q}}^{\alpha\beta}(t)=\mathcal{R}_{\mathbf{q}}^{\beta\alpha}(-t),\\ \mathcal{R}_{\mathbf{q}}^{\alpha\beta}(\omega)^{*}&=\mathcal{R}_{-\mathbf{q}}^{\alpha\beta}(-\omega)=\mathcal{R}_{\mathbf{q}}^{\beta\alpha}(\omega).\end{aligned} \tag{8.35}$$

Combining (8.24) and (8.31), the inelastic cross-section is

$$\left(\frac{\mathrm{d}^2\sigma}{\mathrm{d}\Omega\,\mathrm{d}E'}\right)_{\mathrm{inel}} = r_0^2\frac{k'}{k}\{\tfrac{1}{2}gF(\boldsymbol{\kappa})\}^2\exp\{-2W(\boldsymbol{\kappa})\}\omega\{1+n(\omega)\}$$

$$\times\sum_{\alpha\beta}(\delta_{\alpha\beta}-\tilde{\kappa}_\alpha\tilde{\kappa}_\beta)\mathscr{R}_{\boldsymbol{\kappa}}^{\alpha\beta}(-\omega)$$

$$= r_0^2\frac{k'}{k}\{\tfrac{1}{2}gF(\boldsymbol{\kappa})\}^2\exp\{-2W(\boldsymbol{\kappa})\}\omega\{1+n(\omega)\}$$

$$\times\sum_{\alpha\beta}(\delta_{\alpha\beta}-\tilde{\kappa}_\alpha\tilde{\kappa}_\beta)\tfrac{1}{2}\{\mathscr{R}_{\boldsymbol{\kappa}}^{\alpha\beta}(-\omega)+\mathscr{R}_{\boldsymbol{\kappa}}^{\beta\alpha}(-\omega)\}. \qquad (8.36)$$

Notice that

$$\mathscr{R}_{-\boldsymbol{\kappa}}^{\alpha\beta}(-\omega)+\mathscr{R}_{-\boldsymbol{\kappa}}^{\beta\alpha}(-\omega)=\mathscr{R}_{\boldsymbol{\kappa}}^{\alpha\beta}(\omega)+\mathscr{R}_{\boldsymbol{\kappa}}^{\beta\alpha}(\omega);$$

hence from (8.36) we recover the detailed balance condition.

We see that, in general, $\mathscr{R}_{\mathbf{q}}^{\alpha\beta}(t)$ and $\mathscr{R}_{\mathbf{q}}^{\alpha\beta}(\omega)$ are complex, but the symmetrized form

$$\overline{\mathscr{R}_{\mathbf{q}}^{\alpha\beta}}(\omega)=\tfrac{1}{2}\{\mathscr{R}_{\mathbf{q}}^{\alpha\beta}(\omega)+\mathscr{R}_{\mathbf{q}}^{\beta\alpha}(\omega)\}, \qquad (8.37)$$

which is all that is required in the cross-section formula (8.36), is always real. Furthermore we notice that the symmetrical form

$$\overline{\mathscr{R}_{\mathbf{q}}^{\alpha\beta}}(t)=\tfrac{1}{2}\{\mathscr{R}_{\mathbf{q}}^{\alpha\beta}(t)+\mathscr{R}_{\mathbf{q}}^{\beta\alpha}(t)\}$$

is, in general, complex with a real part that is even in t and an imaginary part that is odd in t. For many simple lattices, and all Bravais lattices, each magnetic atom is a centre of inversion symmetry, and then $\mathbf{q}$ and $-\mathbf{q}$ must be equivalent: in this case $\mathscr{R}_{\mathbf{q}}^{\alpha\beta}(t)$ is itself real, $\overline{\mathscr{R}_{\mathbf{q}}^{\alpha\beta}}(t)$ is real and even in t, and $\mathscr{R}_{\mathbf{q}}^{\alpha\beta}(\omega)$ is real and even in ω.

Now we can prove that

$$\mathscr{R}_{\boldsymbol{\kappa}}^{\alpha\beta}(\omega)=\frac{N}{2\pi\mathrm{i}\omega(g\mu_{\mathrm{B}})^2}\{\chi_{\boldsymbol{\kappa}}^{\alpha\beta}[\omega]-\chi_{\boldsymbol{\kappa}}^{\beta\alpha}[\omega]^*\}; \qquad (8.38)$$

hence

$$\mathscr{R}_{\boldsymbol{\kappa}}^{\alpha\beta}(\omega)+\mathscr{R}_{\boldsymbol{\kappa}}^{\beta\alpha}(\omega)=\frac{N}{\pi\omega(g\mu_{\mathrm{B}})^2}\,\mathrm{Im}\{\chi_{\boldsymbol{\kappa}}^{\alpha\beta}[\omega]+\chi_{\boldsymbol{\kappa}}^{\beta\alpha}[\omega]\}$$

$$=\frac{-N}{\pi\omega(g\mu_{\mathrm{B}})^2}\,\mathrm{Im}\{\chi_{-\boldsymbol{\kappa}}^{\alpha\beta}[-\omega]+\chi_{-\boldsymbol{\kappa}}^{\beta\alpha}[-\omega]\}, \qquad (8.39)$$

and an alternative form for the cross-section (8.36) is therefore

$$\left(\frac{\mathrm{d}^2\sigma}{\mathrm{d}\Omega\,\mathrm{d}E'}\right)_{\text{inel}} = r_0^2 \frac{k'}{k}\{\tfrac{1}{2}gF(\boldsymbol{\kappa})\}^2 \exp\{-2W(\boldsymbol{\kappa})\}\{1+n(\omega)\}$$

$$\times \frac{-N}{\pi(g\mu_\mathrm{B})^2}\sum_{\alpha\beta}(\delta_{\alpha\beta}-\tilde{\kappa}_\alpha\tilde{\kappa}_\beta)\mathrm{Im}\{\chi_{\boldsymbol{\kappa}}^{\alpha\beta}[-\omega]\}. \qquad (8.40)$$

From (8.26) and (8.27) we obtain

$$\chi_{\mathbf{q}}^{\alpha\beta}[\omega] = -\frac{(g\mu_\mathrm{B})^2}{N}\mathcal{R}_{\mathbf{q}}^{\alpha\beta}(t=0) + \frac{(g\mu_\mathrm{B})^2}{N}\mathrm{i}\omega\int_0^\infty \mathrm{d}t\,\exp(-\mathrm{i}\omega t)\mathcal{R}_{\mathbf{q}}^{\alpha\beta}(t). \qquad (8.41)$$

If we take the limit $\omega = 0$, we see that $\chi_{\mathbf{q}}^{\alpha\beta}[0]$ is proportional to the *isothermal susceptibility* $\chi_{\mathbf{q}}^{\alpha\beta}$

$$\chi_{\mathbf{q}}^{\alpha\beta} = -\chi_{\mathbf{q}}^{\alpha\beta}[\omega=0] = \frac{(g\mu_\mathrm{B})^2}{N}\mathcal{R}_{\mathbf{q}}^{\alpha\beta}(t=0). \qquad (8.42)$$

We now define the *spectral-weight function*

$$F_{\mathbf{q}}^{\alpha\beta}(\omega) = \frac{1}{2\pi}\int_{-\infty}^{\infty}\mathrm{d}t\,\exp(-\mathrm{i}\omega t)\left\{\frac{\mathcal{R}_{\mathbf{q}}^{\alpha\beta}(t)}{\mathcal{R}_{\mathbf{q}}^{\alpha\beta}(t=0)}\right\}$$

$$= \mathcal{R}_{\mathbf{q}}^{\alpha\beta}(\omega)/\mathcal{R}_{\mathbf{q}}^{\alpha\beta}(t=0). \qquad (8.43)$$

Then

$$\mathcal{R}_{\mathbf{q}}^{\alpha\beta}(\omega) = \frac{N}{(g\mu_\mathrm{B})^2}\chi_{\mathbf{q}}^{\alpha\beta}F_{\mathbf{q}}^{\alpha\beta}(\omega), \qquad (8.44)$$

and (8.36) can be rewritten as

$$\left(\frac{\mathrm{d}^2\sigma}{\mathrm{d}\Omega\,\mathrm{d}E'}\right)_{\text{inel}} = r_0^2 \frac{k'}{k}\{\tfrac{1}{2}gF(\boldsymbol{\kappa})\}^2 \exp\{-2W(\boldsymbol{\kappa})\}\omega\{1+n(\omega)\}$$

$$\times \frac{N}{(g\mu_\mathrm{B})^2}\sum_{\alpha\beta}(\delta_{\alpha\beta}-\tilde{\kappa}_\alpha\tilde{\kappa}_\beta)\chi_{\boldsymbol{\kappa}}^{\alpha\beta}F_{\boldsymbol{\kappa}}^{\alpha\beta}(-\omega). \qquad (8.45)$$

From (8.43) we notice that

$$\int_{-\infty}^{\infty}\mathrm{d}\omega\,F_{\boldsymbol{\kappa}}^{\alpha\beta}(\omega) = 1. \qquad (8.46)$$

For a Bravais lattice, $F_{\boldsymbol{\kappa}}^{\alpha\beta}(\omega)$ is an even function of ω.

Our purpose in quoting so many alternative forms for the cross-section is that the quantities introduced, particularly $\chi_{\mathbf{q}}^{\alpha\beta}[\omega]$, $\mathcal{R}_{\mathbf{q}}^{\alpha\beta}(t)$, and $\mathcal{R}_{\mathbf{q}}^{\alpha\beta}(\omega)$, have a well-defined physical interpretation, and are more amenable to calculation than are the correlation functions themselves. The isothermal susceptibility $\chi_{\mathbf{q}}^{\alpha\beta}$ is of considerable significance in statistical mechanics and (8.42) shows how it is related to the functions we have defined. From (8.45) and (8.46) together we note that

$$\int_{-\infty}^{\infty} \mathrm{d}\omega \frac{k}{k'} \frac{1-\exp(-\hbar\omega\beta)}{\omega} \left(\frac{\mathrm{d}^2\sigma}{\mathrm{d}\Omega\,\mathrm{d}E'}\right)_{\text{inel}}$$
$$= r_0^2\{\tfrac{1}{2}gF(\boldsymbol{\kappa})\}^2 \exp\{-2W(\boldsymbol{\kappa})\} \frac{N}{(g\mu_{\mathrm{B}})^2} \sum_{\alpha\beta} (\delta_{\alpha\beta} - \tilde{\kappa}_\alpha \tilde{\kappa}_\beta) \chi_{\boldsymbol{\kappa}}^{\alpha\beta}, \quad (8.47)$$

which means that the isothermal susceptibility can be extracted from the neutron cross-section directly and without a knowledge of the spectral weight $F_{\boldsymbol{\kappa}}^{\alpha\beta}(\omega)$ (Als-Nielsen 1976).

In many theoretical studies it is convenient to work in terms of a Green function. We therefore now examine how the cross-sections are related to such functions. The Green function is defined as (Lovesey 1980)

$$G_{\boldsymbol{\kappa}}^{\alpha\beta}(t) = -\mathrm{i}\theta(t)\langle[\hat{S}_{\boldsymbol{\kappa}}^{\alpha}(t), \hat{S}_{-\boldsymbol{\kappa}}^{\beta}]\rangle \quad (8.48)$$

and satisfies the equation-of-motion†

$$\mathrm{i}\hbar\partial_t G_{\boldsymbol{\kappa}}^{\alpha\beta}(t) = \hbar\delta(t)\langle[\hat{S}_{\boldsymbol{\kappa}}^{\alpha}, \hat{S}_{-\boldsymbol{\kappa}}^{\beta}]\rangle - \mathrm{i}\theta(t)\langle[[\hat{S}_{\boldsymbol{\kappa}}^{\alpha}(t), \mathscr{H}], \hat{S}_{-\boldsymbol{\kappa}}^{\beta}]\rangle. \quad (8.49)$$

Then

$$G_{\boldsymbol{\kappa}}^{\alpha\beta}(\omega) = \int_{-\infty}^{\infty} \mathrm{d}t \exp(\mathrm{i}\omega t) G_{\boldsymbol{\kappa}}^{\alpha\beta}(t)$$
$$= -\mathrm{i}\int_{0}^{\infty} \mathrm{d}t \exp(\mathrm{i}\omega t)\langle[\hat{S}_{\boldsymbol{\kappa}}^{\alpha}(t), \hat{S}_{-\boldsymbol{\kappa}}^{\beta}]\rangle$$
$$= \frac{N\hbar}{(g\mu_{\mathrm{B}})^2} \chi_{\boldsymbol{\kappa}}^{\alpha\beta}[-\omega]. \quad (8.50)$$

Hence from (8.40)

$$\left(\frac{\mathrm{d}^2\sigma}{\mathrm{d}\Omega\,\mathrm{d}E'}\right)_{\text{inel}} = r_0^2 \frac{k'}{k} \{\tfrac{1}{2}gF(\boldsymbol{\kappa})\}^2 \exp\{-2W(\boldsymbol{\kappa})\}\{1+n(\omega)\}$$
$$\times \frac{-1}{\pi\hbar} \sum_{\alpha\beta} (\delta_{\alpha\beta} - \tilde{\kappa}_\alpha \tilde{\kappa}_\beta) \operatorname{Im}\{G_{\boldsymbol{\kappa}}^{\alpha\beta}(\omega)\}. \quad (8.51)$$

So far the results we have given in this section have been quite general. However, when the total z-component of spin is a constant of the motion, as it is for a simple Heisenberg model, there is a considerable simplification in the formulae. We now examine this point.

† Here we use the identity $\partial_t\theta(t) = \delta(t)$.

If $\hat{S}^z_{\text{tot}}$ is a constant of motion, i.e. it commutes with the Hamiltonian

$$\sum_l [\hat{S}^z_l, \mathcal{H}] = 0, \tag{8.52}$$

then the raising and lowering operators

$$\hat{S}^+_l = \hat{S}^x_l + \mathrm{i}\hat{S}^y_l, \qquad \hat{S}^-_l = \hat{S}^x_l - \mathrm{i}\hat{S}^y_l \tag{8.53}$$

change the total z-component by plus or minus one unit. Hence

$$\langle \hat{S}^+_l \hat{S}^+_{l'}(t) \rangle = \langle \hat{S}^-_l \hat{S}^-_{l'}(t) \rangle = 0.$$

Because of this, terms in the cross-section like

$$-\tilde{\kappa}_x\tilde{\kappa}_y \langle \hat{S}^x_{\boldsymbol{\kappa}} \hat{S}^y_{-\boldsymbol{\kappa}}(t) + \hat{S}^y_{\boldsymbol{\kappa}} \hat{S}^x_{-\boldsymbol{\kappa}}(t) \rangle$$

vanish, as also do cross-terms like

$$-\tilde{\kappa}_x\tilde{\kappa}_z \langle \hat{S}^x_{\boldsymbol{\kappa}} \hat{S}^z_{-\boldsymbol{\kappa}}(t) \rangle.$$

Also the x- and y-axes are equivalent, so

$$\langle \hat{S}^x_{\boldsymbol{\kappa}} \hat{S}^x_{-\boldsymbol{\kappa}}(t) \rangle = \langle \hat{S}^y_{\boldsymbol{\kappa}} \hat{S}^y_{-\boldsymbol{\kappa}}(t) \rangle.$$

Hence (8.24) simplifies to

$$\begin{aligned}\frac{\mathrm{d}^2\sigma}{\mathrm{d}\Omega\,\mathrm{d}E'} &= r_0^2 \frac{k'}{k} \{\tfrac{1}{2} g F(\boldsymbol{\kappa})\}^2 \exp\{-2W(\boldsymbol{\kappa})\} \\ &\quad\times \left\{ (1-\tilde{\kappa}_z^2) \frac{1}{2\pi\hbar} \int_{-\infty}^{\infty} \mathrm{d}t \exp(-\mathrm{i}\omega t) \langle \hat{S}^z_{\boldsymbol{\kappa}} \hat{S}^z_{-\boldsymbol{\kappa}}(t) \rangle \right. \\ &\quad \left. + (1+\tilde{\kappa}_z^2) \frac{1}{2\pi\hbar} \int_{-\infty}^{\infty} \mathrm{d}t \exp(-\mathrm{i}\omega t) \langle \hat{S}^x_{\boldsymbol{\kappa}} \hat{S}^x_{-\boldsymbol{\kappa}}(t) \rangle \right\}. \end{aligned} \tag{8.54}$$

From (8.25) we find that the cross-terms

$$\phi^{xz}_{\mathbf{q}}(t),\ \phi^{zx}_{\mathbf{q}}(t),\ \phi^{zy}_{\mathbf{q}}(t), \text{ and } \phi^{yz}_{\mathbf{q}}(t) \text{ are zero}$$

and that

$$\phi^{xy}_{\mathbf{q}}(t) + \phi^{yx}_{\mathbf{q}}(t) = 0.$$

The non-zero terms which concern us are then just

$$\phi^{xx}_{\mathbf{q}}(t) = \phi^{yy}_{\mathbf{q}}(t) \quad \text{and} \quad \phi^{zz}_{\mathbf{q}}(t),$$

and similar remarks apply also to $\chi^{\alpha\beta}_{\mathbf{q}}[\omega]$, $\mathcal{R}^{\alpha\beta}_{\mathbf{q}}(t)$, and $\mathcal{R}^{\alpha\beta}_{\mathbf{q}}(\omega)$. The cross-section (8.36) now becomes the sum of two terms, which we denote as longitudinal and transverse respectively, where

$$\left(\frac{\mathrm{d}^2\sigma}{\mathrm{d}\Omega\,\mathrm{d}E'}\right)^{\text{long}}_{\text{inel}} = r_0^2 \frac{k'}{k} \{\tfrac{1}{2} g F(\boldsymbol{\kappa})\}^2 \exp\{-2W(\boldsymbol{\kappa})\} \omega \{1+n(\omega)\}(1-\tilde{\kappa}_z^2) \mathcal{R}^{zz}_{\boldsymbol{\kappa}}(-\omega) \tag{8.55}$$

and

$$\left(\frac{d^2\sigma}{d\Omega\, dE'}\right)^{\text{trans}}_{\text{inel}} = r_0^2 \frac{k'}{k}\{\tfrac{1}{2}gF(\boldsymbol{\kappa})\}^2 \exp\{-2W(\boldsymbol{\kappa})\}\omega\{1+n(\omega)\}(1+\tilde{\kappa}_z^2)\mathcal{R}^{xx}_{\boldsymbol{\kappa}}(-\omega). \tag{8.56}$$

From (8.39), (8.44), and (8.50) we can write down the cross-section in terms of the generalized susceptibility, spectral weight, or retarded Green function by the simple substitution into (8.55) and (8.56) of the relations

$$\mathcal{R}^{zz}_{\boldsymbol{\kappa}}(\omega) = \frac{N}{\pi\omega(g\mu_B)^2}\,\mathrm{Im}\{\chi^{zz}_{\boldsymbol{\kappa}}[\omega]\} \tag{8.57}$$

$$= \frac{N}{(g\mu_B)^2}\,\chi^{zz}_{\boldsymbol{\kappa}} F^{zz}_{\boldsymbol{\kappa}}(\omega) \tag{8.58}$$

$$= \frac{1}{\pi\hbar\omega}\,\mathrm{Im}\{G^{zz}_{\boldsymbol{\kappa}}(-\omega)\} \tag{8.59}$$

and exactly the same equations with superscripts xx replacing zz.

To take full advantage of all these symmetries it is customary in theoretical calculations to work in terms of raising and lowering operators which are Hermitian conjugates of one another. Thus

$$\hat{S}^+_{\mathbf{q}} = \sum_n \exp(-i\mathbf{q}\cdot\mathbf{n})\hat{S}^+_n = \hat{S}^x_{\mathbf{q}} + i\hat{S}^y_{\mathbf{q}},$$

$$\hat{S}^-_{\mathbf{q}} = \sum_n \exp(i\mathbf{q}\cdot\mathbf{n})\hat{S}^-_n = \hat{S}^x_{-\mathbf{q}} - i\hat{S}^y_{-\mathbf{q}}. \tag{8.60}$$

Hence

$$\hat{S}^x_{\mathbf{q}} = \tfrac{1}{2}(\hat{S}^+_{\mathbf{q}} + \hat{S}^-_{-\mathbf{q}}),$$

$$\hat{S}^y_{\mathbf{q}} = \frac{1}{2i}(\hat{S}^+_{\mathbf{q}} - \hat{S}^-_{-\mathbf{q}}). \tag{8.61}$$

In terms of these variables, (8.54) can be rewritten as

$$\frac{d^2\sigma}{d\Omega\, dE'} = r_0^2\frac{k'}{k}\{\tfrac{1}{2}gF(\boldsymbol{\kappa})\}^2\exp\{-2W(\boldsymbol{\kappa})\}$$

$$\times\Big\{(1-\tilde{\kappa}_z^2)\frac{1}{2\pi\hbar}\int_{-\infty}^{\infty} dt\,\exp(-i\omega t)\langle\hat{S}^z_{\boldsymbol{\kappa}}\hat{S}^z_{-\boldsymbol{\kappa}}(t)\rangle$$

$$+\tfrac{1}{4}(1+\tilde{\kappa}_z^2)\frac{1}{2\pi\hbar}\int_{-\infty}^{\infty} dt\,\exp(-i\omega t)\langle\hat{S}^+_{\boldsymbol{\kappa}}\hat{S}^-_{\boldsymbol{\kappa}}(t) + \hat{S}^-_{-\boldsymbol{\kappa}}\hat{S}^+_{-\boldsymbol{\kappa}}(t)\rangle\Big\}. \tag{8.62}$$

Now defining

$$\phi_{\mathbf{q}}^{(0)}(t) = \phi_{\mathbf{q}}^{zz}(t) = \frac{(g\mu_B)^2}{N}\frac{\mathrm{i}}{\hbar}\langle[\hat{S}_{\mathbf{q}}^z(t), \hat{S}_{-\mathbf{q}}^z]\rangle,$$

$$\phi_{\mathbf{q}}^{(+)}(t) = \frac{(g\mu_B)^2}{N}\frac{\mathrm{i}}{\hbar}\langle[\hat{S}_{\mathbf{q}}^+(t), \hat{S}_{\mathbf{q}}^-]\rangle,$$

$$\phi_{\mathbf{q}}^{(-)}(t) = \phi_{\mathbf{q}}^{(+)}(t)^* = \frac{(g\mu_B)^2}{N}\frac{\mathrm{i}}{\hbar}\langle[\hat{S}_{\mathbf{q}}^-(t), \hat{S}_{\mathbf{q}}^+]\rangle, \tag{8.63}$$

and $\chi_{\mathbf{q}}^{(\nu)}[\omega]$ and $\mathscr{R}_{\mathbf{q}}^{(\nu)}(t)$

$$\chi_{\mathbf{q}}^{(\nu)}[\omega] = -\int_0^\infty \mathrm{d}t \exp(-\mathrm{i}\omega t)\phi_{\mathbf{q}}^{(\nu)}(t), \tag{8.64}$$

$$\partial_t \mathscr{R}_{\mathbf{q}}^{(\nu)}(t) = \frac{-N}{(g\mu_B)^2}\phi_{\mathbf{q}}^{(\nu)}(t), \tag{8.65}$$

we find

$$\mathscr{R}_{\mathbf{q}}^{(0)}(t) = \int_0^\beta \mathrm{d}\mu\ \langle\hat{S}_{-\mathbf{q}}^z(-\mathrm{i}\hbar\mu)\hat{S}_{\mathbf{q}}^z(t)\rangle - \beta\langle\hat{S}_{-\mathbf{q}}^z\rangle\langle\hat{S}_{\mathbf{q}}^z\rangle$$

$$\mathscr{R}_{\mathbf{q}}^{(+)}(t) = \int_0^\beta \mathrm{d}\mu\ \langle\hat{S}_{\mathbf{q}}^-(-\mathrm{i}\hbar\mu)\hat{S}_{\mathbf{q}}^+(t)\rangle$$

$$\mathscr{R}_{\mathbf{q}}^{(-)}(t) = \int_0^\beta \mathrm{d}\mu\ \langle\hat{S}_{\mathbf{q}}^+(-\mathrm{i}\hbar\mu)\hat{S}_{\mathbf{q}}^-(t)\rangle. \tag{8.66}$$

and

$$\mathscr{R}_{\mathbf{q}}^{xx}(t) = \tfrac{1}{4}\{\mathscr{R}_{\mathbf{q}}^{(+)}(t) + \mathscr{R}_{-\mathbf{q}}^{(-)}(t)\}. \tag{8.67}$$

Hence

$$\mathscr{R}_{\mathbf{q}}^{xx}(\omega) = \tfrac{1}{4}\{\mathscr{R}_{\mathbf{q}}^{(+)}(\omega) + \mathscr{R}_{-\mathbf{q}}^{(-)}(\omega)\}. \tag{8.68}$$

Now from the definitions (8.63) we have

$$\phi_{\mathbf{q}}^{(0)}(t) = \phi_{-\mathbf{q}}^{(0)}(t)^* = -\phi_{-\mathbf{q}}^{(0)}(-t), \tag{8.69}$$

$$\phi_{\mathbf{q}}^{(\nu)}(t) = \phi_{\mathbf{q}}^{(-\nu)}(t)^* = -\phi_{\mathbf{q}}^{(-\nu)}(-t); \qquad \nu = + \text{ or } - \tag{8.70}$$

and hence

$$\chi_{\mathbf{q}}^{(0)}[\omega]^* = \chi_{-\mathbf{q}}^{(0)}[-\omega],$$

$$\chi_{\mathbf{q}}^{(\nu)}[\omega]^* = \chi_{\mathbf{q}}^{(-\nu)}[-\omega]; \qquad \nu = + \text{ or } - \tag{8.71}$$

$$\mathscr{R}_{\mathbf{q}}^{(0)}(t) = \mathscr{R}_{-\mathbf{q}}^{(0)}(t)^* = \mathscr{R}_{-\mathbf{q}}^{(0)}(-t),$$

$$\mathscr{R}_{\mathbf{q}}^{(\nu)}(t) = \mathscr{R}_{\mathbf{q}}^{(-\nu)}(t)^* = \mathscr{R}_{\mathbf{q}}^{(-\nu)}(-t); \qquad \nu = + \text{ or } - \tag{8.72}$$

$$\mathscr{R}_{\mathbf{q}}^{(0)}(\omega) = \mathscr{R}_{\mathbf{q}}^{(0)}(\omega)^* = \mathscr{R}_{-\mathbf{q}}^{(0)}(-\omega),$$

$$\mathscr{R}_{\mathbf{q}}^{(\nu)}(\omega) = \mathscr{R}_{\mathbf{q}}^{(\nu)}(\omega)^* = \mathscr{R}_{\mathbf{q}}^{(-\nu)}(-\omega); \qquad \nu = + \text{ or } -, \tag{8.73}$$

and

$$\mathrm{Im}\{\chi_{\mathbf{q}}^{(\nu)}[\omega]\} = \frac{\pi(g\mu_B)^2}{N}\,\omega\,\mathcal{R}_{\mathbf{q}}^{(\nu)}(\omega); \qquad \nu = 0,\,+,\,\text{or}\,-. \tag{8.74}$$

Finally, defining Green functions as

$$G_{\mathbf{q}}^{(0)}(t) = -\mathrm{i}\theta(t)\langle[\hat{S}_{\mathbf{q}}^{z}(t),\,\hat{S}_{-\mathbf{q}}^{z}]\rangle$$

and

$$G_{\mathbf{q}}^{(\nu)}(t) = -\mathrm{i}\theta(t)\langle[\hat{S}_{\mathbf{q}}^{\nu}(t),\,\hat{S}_{\mathbf{q}}^{-\nu}]\rangle; \qquad \nu = +\,\text{or}\,-, \tag{8.75}$$

$$\mathrm{Im}\{G_{\mathbf{q}}^{(\nu)}(-\omega)\} = \pi\hbar\omega\,\mathcal{R}_{\mathbf{q}}^{(\nu)}(\omega); \qquad \nu = 0,\,+,\,\text{or}\,-. \tag{8.76}$$

These new variables do not affect the form for the longitudinal cross-section, which remains as (8.55); for ease of reference we repeat the nomenclature

$$\begin{aligned}\mathcal{R}_{\boldsymbol{\kappa}}^{zz}(\omega) &= \mathcal{R}_{\boldsymbol{\kappa}}^{(0)}(\omega)\\ &= \frac{N}{\pi\omega(g\mu_B)^2}\,\mathrm{Im}\{\chi_{\boldsymbol{\kappa}}^{(0)}[\omega]\}\\ &= \frac{1}{\pi\hbar\omega}\,\mathrm{Im}\{G_{\boldsymbol{\kappa}}^{(0)}(-\omega)\},\end{aligned} \tag{8.77}$$

whereas the transverse cross-section (8.56) can now be re-expressed using

$$\begin{aligned}\mathcal{R}_{\boldsymbol{\kappa}}^{xx}(\omega) &= \tfrac{1}{4}\{\mathcal{R}_{\boldsymbol{\kappa}}^{(+)}(\omega) + \mathcal{R}_{-\boldsymbol{\kappa}}^{(-)}(\omega)\}\\ &= \frac{N}{4\pi(g\mu_B)^2\omega}\,\mathrm{Im}\{\chi_{\boldsymbol{\kappa}}^{(+)}[\omega] + \chi_{-\boldsymbol{\kappa}}^{(-)}[\omega]\}\\ &= \frac{1}{4\pi\hbar\omega}\,\mathrm{Im}\{G_{\boldsymbol{\kappa}}^{(+)}(-\omega) + G_{-\boldsymbol{\kappa}}^{(-)}(-\omega)\}.\end{aligned} \tag{8.78a}$$

An alternative form to (8.78a) is given by using (8.73) and (8.74), namely

$$\begin{aligned}\mathcal{R}_{\boldsymbol{\kappa}}^{xx}(\omega) &= \tfrac{1}{4}\{\mathcal{R}_{\boldsymbol{\kappa}}^{(+)}(\omega) + \mathcal{R}_{-\boldsymbol{\kappa}}^{(+)}(-\omega)\}\\ &= \frac{N}{4\pi(g\mu_B)^2\omega}\,\mathrm{Im}\{\chi_{\boldsymbol{\kappa}}^{(+)}[\omega] - \chi_{-\boldsymbol{\kappa}}^{(+)}[-\omega]\}\\ &= \frac{1}{4\pi\hbar\omega}\,\mathrm{Im}\{G_{\boldsymbol{\kappa}}^{(+)}(-\omega) - G_{-\boldsymbol{\kappa}}^{(+)}(\omega)\},\end{aligned} \tag{8.78b}$$

which, when substituted into (8.56) and defining

$$\mathcal{A}(\mathbf{k},\mathbf{k}') = r_0^2\,\frac{k'}{k}\,\{\tfrac{1}{2}gF(\boldsymbol{\kappa})\}^2\exp\{-2W(\boldsymbol{\kappa})\}(1+\tilde{\kappa}_z^2), \tag{8.79}$$

gives

$$\left(\frac{\mathrm{d}^2\sigma}{\mathrm{d}\Omega\,\mathrm{d}E'}\right)^{\text{trans}}_{\text{inel}} = \tfrac{1}{4}\mathscr{A}(\mathbf{k},\mathbf{k}')\omega\{1+n(\omega)\}[\mathscr{R}^{(+)\prime}_{\boldsymbol{\kappa}}(-\omega)+\mathscr{R}^{(+)\prime}_{-\boldsymbol{\kappa}}(\omega)]$$

$$= \mathscr{A}(\mathbf{k},\mathbf{k}')\frac{N}{4\pi(g\mu_{\mathrm{B}})^2}\,\mathrm{Im}[\{1+n(\omega)\}\chi^{(+)\prime}_{-\boldsymbol{\kappa}}[\omega]+n(-\omega)\chi^{(+)\prime}_{\boldsymbol{\kappa}}[-\omega]]$$

$$= -\mathscr{A}(\mathbf{k},\mathbf{k}')\frac{1}{4\pi\hbar}\,\mathrm{Im}[\{1+n(\omega)\}G^{(+)\prime}_{\boldsymbol{\kappa}}(\omega)+n(-\omega)G^{(+)\prime}_{-\boldsymbol{\kappa}}(-\omega)]. \tag{8.80}$$

The generalization of this expression for a polarized incident beam is given by (10.131).

Equation (8.80) brings out the similarity, for simple processes, of creating and annihilating magnons just as phonons could be created or annihilated. To see this similarity unambiguously we can anticipate a result from Chapter 9 which shows for non-interacting spin waves (magnons) that $\mathrm{Im}\{G^{(+)\prime}_{\boldsymbol{\kappa}}(\omega)\}$ is simply proportional to $\delta(\omega-\omega_{\boldsymbol{\kappa}})$, where $\omega_{\boldsymbol{\kappa}}$ is the frequency of a spin wave of wave vector $\boldsymbol{\kappa}$.

Earlier, in connection with (8.37), we remarked that the general formulae simplified when every magnetic atom was a centre of inversion symmetry, so that $\mathbf{q}$ and $-\mathbf{q}$ were equivalent. As we would expect, when we have both this symmetry condition and also the condition that $\hat{S}^z_{\text{tot}}$ be a constant of motion, the formulae become exceptionally simple. We shall now review these formulae when both conditions are satisfied.

First,

$$\phi^{zz}_{\mathbf{q}}(t)^* = \phi^{zz}_{\mathbf{q}}(t) = -\phi^{zz}_{\mathbf{q}}(-t) = \phi^{zz}_{-\mathbf{q}}(t)$$

and

$$\phi^{xx}_{\mathbf{q}}(t)^* = \phi^{xx}_{\mathbf{q}}(t) = -\phi^{xx}_{\mathbf{q}}(-t) = \phi^{xx}_{-\mathbf{q}}(t). \tag{8.81}$$

Hence $\phi^{zz}_{\mathbf{q}}(t)$ and $\phi^{xx}_{\mathbf{q}}(t)$ are real and odd functions of t.

Because $\phi^{zz}_{\mathbf{q}}(t)$ and $\phi^{xx}_{\mathbf{q}}(t)$ are real it follows that the real and imaginary parts of $\chi^{zz}_{\mathbf{q}}[\omega]$ and $\chi^{xx}_{\mathbf{q}}[\omega]$ are related by Kramers–Krönig relations. Furthermore,

$$\chi^{zz}_{\mathbf{q}}[\omega]^* = \chi^{zz}_{\mathbf{q}}[-\omega],$$
$$\chi^{xx}_{\mathbf{q}}[\omega]^* = \chi^{xx}_{\mathbf{q}}[-\omega]. \tag{8.82a}$$

Hence the real parts of $\chi^{zz}_{\mathbf{q}}[\omega]$ and $\chi^{xx}_{\mathbf{q}}[\omega]$ are both even in ω, whereas the imaginary parts are both odd in ω.

From (8.81),

$$\mathscr{R}^{zz}_{\mathbf{q}}(t)^* = \mathscr{R}^{zz}_{\mathbf{q}}(t) = \mathscr{R}^{zz}_{\mathbf{q}}(-t) = \mathscr{R}^{zz}_{-\mathbf{q}}(t),$$

and

$$\mathcal{R}^{xx}_{\mathbf{q}}(t)^* = \mathcal{R}^{xx}_{\mathbf{q}}(t) = \mathcal{R}^{xx}_{\mathbf{q}}(-t) = \mathcal{R}^{xx}_{-\mathbf{q}}(t). \tag{8.82b}$$

Hence both functions are real and even in t. Also, for the Fourier transforms,

$$\begin{aligned}\mathcal{R}^{zz}_{\mathbf{q}}(\omega)^* &= \mathcal{R}^{zz}_{\mathbf{q}}(\omega) = \mathcal{R}^{zz}_{\mathbf{q}}(-\omega) = \mathcal{R}^{zz}_{-\mathbf{q}}(\omega),\\ \mathcal{R}^{xx}_{\mathbf{q}}(\omega)^* &= \mathcal{R}^{xx}_{\mathbf{q}}(\omega) = \mathcal{R}^{xx}_{\mathbf{q}}(-\omega) = \mathcal{R}^{xx}_{-\mathbf{q}}(\omega).\end{aligned} \tag{8.82c}$$

Hence both functions, which are always real, now become even in ω.

It follows immediately from (8.43) that the spectral weights $F^{zz}_{\mathbf{q}}(\omega)$ and $F^{xx}_{\mathbf{q}}(\omega)$, which are always real, are also even in ω.

The functions $\phi^{(0)}_{\mathbf{q}}(t)$, $\chi^{(0)}_{\mathbf{q}}[\omega]$, $\mathcal{R}^{(0)}_{\mathbf{q}}(t)$, $\mathcal{R}^{(0)}_{\mathbf{q}}(\omega)$, and $G^{(0)}_{\mathbf{q}}(\omega)$ are identical to $\phi^{zz}_{\mathbf{q}}(t), \ldots,$ and therefore simplify as outlined in eqns (8.81) and (8.82). The functions $\phi^{(+)}_{\mathbf{q}}(t)$, $\phi^{(-)}_{\mathbf{q}}(t), \ldots$ do not simplify any further than as described by (8.70), (8.71), (8.72), and (8.73).

8.4. Response functions: non-Bravais lattice

In this section we generalize the results of § 8.3 to a non-Bravais lattice. This is not too difficult because in § 8.3 we were careful not to use the symmetry properties of the Bravais lattice, except in a few comments which we noted explicitly.

We begin by defining the true spin density in this model. It is

$$\hat{\mathfrak{s}}(\mathbf{r}, t) = \sum_{l,d} \tfrac{1}{2} g_d \hat{\mathfrak{s}}_{l+d}(t)\rho_d(\mathbf{r}-\mathbf{l}-\mathbf{d}) \tag{8.83}$$

where $\rho_d(\mathbf{R})$ is a normalized density centred on atom $\mathbf{l}+\mathbf{d}$ and with Fourier transform $F_d(\mathbf{q})\exp\{-W_d(\mathbf{q})\}$. Then the spin magnetization density is

$$\hat{\mathbf{M}}(\mathbf{r}, t) = -2\mu_{\mathrm{B}}\hat{\mathfrak{s}}(\mathbf{r}, t) \tag{8.84}$$

and the interaction energy with a magnetic field is

$$-\int \mathrm{d}\mathbf{r}\, \hat{\mathbf{M}}(\mathbf{r}, t)\cdot \mathbf{H}(\mathbf{r}, t).$$

Hence, if $\mathbf{H}(\mathbf{r}, t)$ is of the form $\tilde{\boldsymbol{\beta}} H^{\beta}_{\mathbf{q}}(t)\exp(\mathrm{i}\mathbf{q}\cdot\mathbf{r})$, this interaction energy becomes

$$\sum_{l,d} \mu_{\mathrm{B}} g_d F_d(\mathbf{q})\exp\{-W_d(\mathbf{q})\}\hat{S}^{\beta}_{l+d}(t)\exp\{-\mathrm{i}\mathbf{q}\cdot(\mathbf{l}+\mathbf{d})\}H^{\beta}_{\mathbf{q}}(t).$$

From these expressions we see immediately that when we ask for the response of the Fourier transform of $\hat{\mathbf{M}}(\mathbf{r}, t)$, i.e.

$$\hat{\mathbf{M}}_{\mathbf{q}}(t) = \int \mathrm{d}\mathbf{r} \exp(-\mathbf{q}\cdot\mathbf{r})\hat{\mathbf{M}}(\mathbf{r}, t)$$
$$= -\sum_{l,d} g_d \mu_{\mathrm{B}} \hat{S}_{l+d}(t) F_d(\mathbf{q}) \exp\{-W_d(\mathbf{q})\} \exp\{-\mathrm{i}\mathbf{q}\cdot(\mathbf{l}+\mathbf{d})\}, \quad (8.85)$$

to an applied field like $\tilde{\boldsymbol{\beta}} H^{\beta}_{\mathbf{q}} \exp(\mathrm{i}\mathbf{q}\cdot\mathbf{r})$, then additional factors like $F_d(\mathbf{q})\exp\{-W_d(\mathbf{q})\}$ are going to occur. This suggests that we define operators

$$\hat{T}^{\alpha}_{\mathbf{q}}(t) = \sum_{l,d} \tfrac{1}{2} g_d F_d(\mathbf{q}) \exp\{-W_d(\mathbf{q})\} \exp\{-\mathrm{i}\mathbf{q}\cdot(\mathbf{l}+\mathbf{d})\} \hat{S}^{\alpha}_{l+d} \quad (8.86)$$

and a response function

$$\bar{\phi}^{\alpha\beta}_{\mathbf{q}}(t) = \frac{(2\mu_{\mathrm{B}})^2}{N} \frac{\mathrm{i}}{\hbar} \langle [\hat{T}^{\alpha}_{\mathbf{q}}(t), \hat{T}^{\beta}_{-\mathbf{q}}] \rangle, \quad (8.87)$$

where N is the number of unit cells in the crystal and the normalization of (8.87) is chosen so as to keep a close analogy with (8.29). We notice that the response function $\bar{\phi}^{\alpha\beta}_{\mathbf{q}}(t)$ defined by (8.87) is not the same as $\phi^{\alpha\beta}_{\mathbf{q}}(t)$ defined by (8.29), but we shall find they have many symmetry properties in common.

We then define a generalized susceptibility as

$$\bar{\chi}^{\alpha\beta}_{\mathbf{q}}[\omega] = -\int_0^{\infty} \mathrm{d}t \exp(-\mathrm{i}\omega t) \bar{\phi}^{\alpha\beta}_{\mathbf{q}}(t) \quad (8.88)$$

and a relaxation function through

$$\partial_t \bar{\mathcal{R}}^{\alpha\beta}_{\mathbf{q}}(t) = \frac{-N}{(2\mu_{\mathrm{B}})^2} \bar{\phi}^{\alpha\beta}_{\mathbf{q}}(t) \quad (8.89)$$

with Fourier transform

$$\bar{\mathcal{R}}^{\alpha\beta}_{\mathbf{q}}(\omega) = \frac{1}{2\pi} \int_{-\infty}^{\infty} \mathrm{d}t \exp(-\mathrm{i}\omega t) \bar{\mathcal{R}}^{\alpha\beta}_{\mathbf{q}}(t). \quad (8.90)$$

Then in place of (8.28) we have

$$\bar{\mathcal{R}}^{\alpha\beta}_{\mathbf{q}}(t) = \int_0^{\beta} \mathrm{d}\mu \, \langle \hat{T}^{\beta}_{-\mathbf{q}}(-\mathrm{i}\hbar\mu) \hat{T}^{\alpha}_{\mathbf{q}}(t) \rangle - \beta \langle \hat{T}^{\beta}_{-\mathbf{q}} \rangle \langle \hat{T}^{\alpha}_{\mathbf{q}} \rangle. \quad (8.91)$$

In terms of these new operators, the magnetic cross-section becomes, in place of (8.24),

$$\frac{\mathrm{d}^2\sigma}{\mathrm{d}\Omega\,\mathrm{d}E'} = r_0^2 \frac{k'}{k} \sum_{\alpha\beta} (\delta_{\alpha\beta} - \tilde{\kappa}_{\alpha}\tilde{\kappa}_{\beta}) \frac{1}{2\pi\hbar} \int_{-\infty}^{\infty} \mathrm{d}t \exp(-\mathrm{i}\omega t) \langle \hat{T}^{\alpha}_{\boldsymbol{\kappa}} \hat{T}^{\beta}_{-\boldsymbol{\kappa}}(t) \rangle, \quad (8.92)$$

and in place of (8.31) we get

$$\frac{1}{2\pi\hbar}\int_{-\infty}^{\infty} \mathrm{d}t \exp(-\mathrm{i}\omega t)\langle \hat{T}^{\beta}_{-\mathbf{q}}\hat{T}^{\alpha}_{\mathbf{q}}(t)\rangle = \omega\{1+n(\omega)\}\bar{\mathcal{R}}^{\alpha\beta}_{\mathbf{q}}(\omega)+\delta(\hbar\omega)\langle \hat{T}^{\beta}_{-\mathbf{q}}\rangle\langle \hat{T}^{\alpha}_{\mathbf{q}}\rangle. \tag{8.93}$$

Now in (8.86)

$$F_d(\mathbf{q})\exp\{-W_d(\mathbf{q})\} = \int \mathrm{d}\mathbf{r} \exp\{-\mathrm{i}\mathbf{q}\cdot\mathbf{r}\}\rho(\mathbf{r})$$

$$= \langle \exp(-\mathrm{i}\mathbf{q}\cdot\hat{\mathbf{u}}_d)\rangle \int \mathrm{d}\mathbf{r} \exp(-\mathrm{i}\mathbf{q}\cdot\mathbf{r})\, |\phi(\mathbf{r})|^2. \tag{8.94a}$$

Hence

$$F_d(\mathbf{q})^*\exp\{-W_d(\mathbf{q})^*\} = F_d(-\mathbf{q})\exp\{-W_d(-\mathbf{q})\}. \tag{8.94b}$$

It follows that the Hermitian conjugate of $\hat{T}^{\alpha}_{\mathbf{q}}$ is $\hat{T}^{\alpha}_{-\mathbf{q}}$ and that all the symmetry relations for ϕ, χ, and $\mathcal{R}$ listed in (8.34) and (8.35) apply also to the functions $\bar{\phi}$, $\bar{\chi}$, $\bar{\mathcal{R}}$ defined in this section. In particular we find that (8.92) and (8.93) can be combined to give

$$\left(\frac{\mathrm{d}^2\sigma}{\mathrm{d}\Omega\,\mathrm{d}E'}\right)_{\text{inel}} = r_0^2\frac{k'}{k}\,\omega\{1+n(\omega)\}\sum_{\alpha\beta}(\delta_{\alpha\beta}-\tilde{\kappa}_\alpha\tilde{\kappa}_\beta)\bar{\mathcal{R}}^{\alpha\beta}_{\boldsymbol{\kappa}}(-\omega). \tag{8.95}$$

Now in place of (8.38) we get

$$\bar{\mathcal{R}}^{\alpha\beta}_{\boldsymbol{\kappa}}(\omega) = \frac{N}{2\pi\mathrm{i}\omega(2\mu_{\mathrm{B}})^2}\{\bar{\chi}^{\alpha\beta}_{\mathbf{q}}[\omega]-\bar{\chi}^{\beta\alpha}_{\mathbf{q}}[\omega]^*\}, \tag{8.96}$$

and hence, in place of (8.40),

$$\left(\frac{\mathrm{d}^2\sigma}{\mathrm{d}\Omega\,\mathrm{d}E'}\right)_{\text{inel}} = r_0^2\frac{k'}{k}\frac{-N}{\pi(2\mu_{\mathrm{B}})^2}\{1+n(\omega)\}\sum_{\alpha\beta}(\delta_{\alpha\beta}-\tilde{\kappa}_\alpha\tilde{\kappa}_\beta)\mathrm{Im}\{\bar{\chi}^{\alpha\beta}_{\boldsymbol{\kappa}}[-\omega]\}. \tag{8.97}$$

In place of (8.41), we now get

$$\bar{\chi}^{\alpha\beta}_{\mathbf{q}}[\omega] = -\frac{(2\mu_{\mathrm{B}})^2}{N}\bar{\mathcal{R}}^{\alpha\beta}_{\mathbf{q}}(t=0)+\frac{(2\mu_{\mathrm{B}})^2}{N}\mathrm{i}\omega\int_0^{\infty}\mathrm{d}t\exp(-\mathrm{i}\omega t)\bar{\mathcal{R}}^{\alpha\beta}_{\mathbf{q}}(t). \tag{8.98}$$

Now $\bar{\mathcal{R}}^{\alpha\beta}_{\mathbf{q}}(t=0)$ is proportional to the total isothermal susceptibility. Hence (8.98) tells us that the susceptibility $\bar{\chi}^{\alpha\beta}_{\mathbf{q}}[\omega]$ is defined as the susceptibility per unit cell. The isothermal susceptibility per unit cell is

$$\bar{\chi}^{\alpha\beta}_{\mathbf{q}} = -\bar{\chi}^{\alpha\beta}_{\mathbf{q}}[\omega=0] = \frac{(2\mu_{\mathrm{B}})^2}{N}\bar{\mathcal{R}}^{\alpha\beta}_{\mathbf{q}}(t=0). \tag{8.99}$$

In our new variables we define the retarded Green function, in place of (8.48), as

$$\bar{G}_{\boldsymbol{\kappa}}^{\alpha\beta}(t) = -\mathrm{i}\theta(t)\langle[\hat{T}_{\boldsymbol{\kappa}}^{\alpha}(t), \hat{T}_{-\boldsymbol{\kappa}}^{\beta}]\rangle \tag{8.100}$$

and then find

$$\bar{G}_{\boldsymbol{\kappa}}^{\alpha\beta}(\omega) = \frac{N\hbar}{(2\mu_{\mathrm{B}})^2}\,\chi_{\boldsymbol{\kappa}}^{\alpha\beta}[-\omega], \tag{8.101}$$

which, using (8.97), we can use to give

$$\left(\frac{\mathrm{d}^2\sigma}{\mathrm{d}\Omega\,\mathrm{d}E'}\right)_{\mathrm{inel}} = \frac{-r_0^2}{\pi\hbar}\frac{k'}{k}\{1+n(\omega)\}\sum_{\alpha\beta}(\delta_{\alpha\beta}-\tilde{\kappa}_\alpha\tilde{\kappa}_\beta)\mathrm{Im}\{\bar{G}_{\boldsymbol{\kappa}}^{\alpha\beta}(\omega)\} \tag{8.102}$$

When the total z-component of spin is a constant of motion the cross-section again splits into a longitudinal and transverse part, where (in place of (8.55))

$$\left(\frac{\mathrm{d}^2\sigma}{\mathrm{d}\Omega\,\mathrm{d}E'}\right)_{\mathrm{inel}}^{\mathrm{long}} = r_0^2\frac{k'}{k}(1-\tilde{\kappa}_z^2)\omega\{1+n(\omega)\}\bar{\mathcal{R}}_{\boldsymbol{\kappa}}^{zz}(-\omega) \tag{8.103}$$

and (in place of (8.56))

$$\left(\frac{\mathrm{d}^2\sigma}{\mathrm{d}\Omega\,\mathrm{d}E'}\right)_{\mathrm{inel}}^{\mathrm{trans}} = r_0^2\frac{k'}{k}(1+\tilde{\kappa}_z^2)\omega\{1+n(\omega)\}\bar{\mathcal{R}}_{\boldsymbol{\kappa}}^{xx}(-\omega). \tag{8.104}$$

To make analogy to the formulation in terms of raising and lowering operators we take care to define

$$\begin{aligned}\hat{T}_{\mathbf{q}}^{+}(t) &= \sum_{l,d} g_d F_d(\mathbf{q})\exp\{-W_d(\mathbf{q})\}\exp\{-\mathrm{i}\mathbf{q}\cdot(\mathbf{l}+\mathbf{d})\}\hat{S}_{l+d}^{+}\\ &= \hat{T}_{\mathbf{q}}^{x}(t)+\mathrm{i}\hat{T}_{\mathbf{q}}^{y}(t)\end{aligned} \tag{8.105}$$

and

$$\begin{aligned}\hat{T}_{\mathbf{q}}^{-}(t) &= \sum_{l,d} g_d F_d(\mathbf{q})^*\exp\{-W_d(\mathbf{q})^*\}\exp\{\mathrm{i}\mathbf{q}\cdot(\mathbf{l}+\mathbf{d})\}\hat{S}_{l+d}^{-}\\ &= \{T_{\mathbf{q}}^{+}(t)\}^{+} = \hat{T}_{-\mathbf{q}}^{x}-\mathrm{i}\hat{T}_{-\mathbf{q}}^{y}.\end{aligned} \tag{8.106}$$

Then all the discussion in § 8.3 that starts from (8.60) follows through as before: in particular, the symmetry relations in eqns (8.69)–(8.73) remain valid provided we define $\bar{\phi}_{\mathbf{q}}^{(\nu)}(t)$ and $\bar{\mathcal{R}}_{\mathbf{q}}^{(\nu)}(t)$ in terms of the $\hat{T}_{\mathbf{q}}^{\nu}(t)$ operators instead of using $\phi_{\mathbf{q}}^{(\nu)}(t)$, $\mathcal{R}_{\mathbf{q}}^{(\nu)}(t)$, and the $\hat{S}_{\mathbf{q}}^{\nu}(t)$ operators as in (8.63) and (8.66). To be quite explicit,

$$\bar{\phi}_{\mathbf{q}}^{(0)}(t)=\bar{\phi}_{\mathbf{z}}^{zz}(t), \tag{8.107a}$$

$$\bar{\phi}_{\mathbf{q}}^{(\nu)}(t)=\frac{(2\mu_{\mathrm{B}})^2}{N}\frac{\mathrm{i}}{\hbar}\langle[\hat{T}_{\mathbf{q}}^{\nu}(t),\hat{T}_{\mathbf{q}}^{-\nu}]\rangle \quad \text{for} \quad \nu=+\text{ or }-; \tag{8.107b}$$

$$\bar{\mathscr{R}}_{\mathbf{q}}^{(0)}(t)=\bar{\mathscr{R}}_{\mathbf{q}}^{zz}(t), \tag{8.107c}$$

$$\bar{\mathscr{R}}_{\mathbf{q}}^{(\nu)}(t)=\int_0^\beta \mathrm{d}\mu\,\langle\hat{T}_{\mathbf{q}}^{-\nu}(-\mathrm{i}\hbar\mu)\hat{T}_{\mathbf{q}}^{\nu}(t)\rangle-\beta\langle\hat{T}_{\mathbf{q}}^{-\nu}\rangle\langle\hat{T}_{\mathbf{q}}^{\nu}\rangle \quad \text{for} \quad \nu=+\text{ or }-. \tag{8.107d}$$

In the longitudinal cross-section (8.103) we can now work with $\bar{\mathscr{R}}_{\boldsymbol{\kappa}}^{zz}(\omega)$ directly as defined by (8.90) and (8.91) or alternatively we can express it as

$$\begin{aligned}\bar{\mathscr{R}}_{\boldsymbol{\kappa}}^{zz}(\omega)&=\mathscr{R}_{\boldsymbol{\kappa}}^{(0)}(\omega)\\ &=\frac{N}{(2\mu_{\mathrm{B}})^2}\bar{\chi}_{\boldsymbol{\kappa}}^{zz}\bar{F}_{\boldsymbol{\kappa}}^{zz}(\omega)\\ &=\frac{N}{\pi\omega(2\mu_{\mathrm{B}})^2}\mathrm{Im}\{\bar{\chi}_{\boldsymbol{\kappa}}^{(0)}[\omega]\}\\ &=\frac{1}{\pi\hbar\omega}\mathrm{Im}\{\bar{G}_{\boldsymbol{\kappa}}^{(0)}(-\omega)\}.\end{aligned} \tag{8.108}$$

In the transverse cross-section we have the following alternatives for $\bar{\mathscr{R}}_{\boldsymbol{\kappa}}^{xx}(\omega)$

$$\begin{aligned}\bar{\mathscr{R}}_{\boldsymbol{\kappa}}^{xx}(\omega)&=\frac{N}{(2\mu_{\mathrm{B}})^2}\bar{\chi}_{\boldsymbol{\kappa}}^{xx}\bar{F}_{\boldsymbol{\kappa}}^{xx}(\omega)\\ &=\frac{N}{\pi\omega(2\mu_{\mathrm{B}})^2}\mathrm{Im}\{\bar{\chi}_{\boldsymbol{\kappa}}^{xx}[\omega]\}\\ &=\frac{1}{\pi\hbar\omega}\mathrm{Im}\{\bar{G}_{\boldsymbol{\kappa}}^{xx}(-\omega)\}\\ &=\tfrac{1}{4}\{\bar{\mathscr{R}}_{\boldsymbol{\kappa}}^{(+)}(\omega)+\mathscr{R}_{-\boldsymbol{\kappa}}^{(-)}(\omega)\}\\ &=\frac{N}{16\pi\mu_{\mathrm{B}}^2\omega}\mathrm{Im}\{\bar{\chi}_{\boldsymbol{\kappa}}^{(+)}[\omega]+\bar{\chi}_{-\boldsymbol{\kappa}}^{(-)}[\omega]\}\\ &=\frac{1}{4\pi\hbar\omega}\mathrm{Im}\{\bar{G}_{\boldsymbol{\kappa}}^{(+)}(-\omega)+\bar{G}_{-\boldsymbol{\kappa}}^{(-)}(-\omega)\}.\end{aligned} \tag{8.109}$$

This is as much discussion as we need give for the general problem. It is evident that the functions used here, $\bar{\mathscr{R}}(t)$, $\bar{\mathscr{R}}(\omega)$, $\bar{\chi}[\omega]$, etc., reduce to the simpler forms $\mathscr{R}(t)$, $\mathscr{R}(\omega)$, $\chi[\omega]$, etc., in special cases when g is replaced by 2, the Debye–Waller factor is ignored, and we assume one atom per unit cell.

REFERENCES

Als-Nielsen, J. (1976). *Phase transitions and critical phenomena* (ed. C. Domb and M. S. Green) Vol. 5a. Academic Press, New York.

Lovesey, S. W. (1986). *Condensed matter physics,* Frontiers in Physics Vol. 61 Benjamin/Cummings, Reading, Mass.

Steinsvoll, O., Moon, R. M., Koehler, W. C., and Windsor, C. G. (1981). *Phys. Rev.* **B24,** 4031.

9
SPIN WAVES

9.1. Introduction

Spin-wave theory provides an adequate description of the low-energy magnetic excitations in materials that can be described by a Heisenberg exchange Hamiltonian. The excitations are constructed from a classical ground state, which is the exact ground state for simple ferromagnetic materials, and thermodynamic properties (specific heat, magnetization, etc.) are calculated in terms of a perfect fluid of (linear) spin waves, or magnons. In the low-density limit, achieved at temperatures small compared with the exchange parameter, magnon interactions are described by an effective boson Hamiltonian with a quartic coupling, which can be handled by standard many-body techniques.

For many metallic magnets it is more appropriate to use an itinerant-electron model. The apparent mismatch between the localized nature of magnetic metals, as seen, for example, in the compact distribution of the magnetization about lattice sites, and the itinerant and band-structure effects are reconciled in the Hubbard Hamiltonian which contains strong on-site repulsion between electrons of opposite spin alignment. While the low-energy collective spin excitation is akin to that derived from a Heisenberg model, band-structure effects manifest themselves at high energies. In nickel and iron, for example, the spin-wave contribution to the cross-section is over-damped at $\hbar\omega \sim 100$ meV, and this effect can be correlated with specific band-electron contributions to the cross-section. Investigations of such effects have been hampered by a strong reliance on reactor neutron sources which become rapidly less effective as the neutron energy transfer reaches ~ 100–150 meV.

Magnon–phonon hybridization and impurity-induced effects are discussed here in terms of the Heisenberg model. Both topics are approached with a view to understanding the intrinsic mechanisms responsible for observed effects.

Collective excitations in low-dimensional magnetic systems are discussed primarily in § 13.4.

9.2. Spin waves in a Heisenberg ferromagnet (Elliott and Gibson 1976; Mattis 1981; Mulder *et al.* 1982)

9.2.1. Linear spin-wave theory

The Heisenberg Hamiltonian for a ferromagnetic system with an external magnetic field H in the z-direction is

$$\hat{\mathscr{H}} = -\sum_{\mathbf{l},\mathbf{l}'} J(\mathbf{l}-\mathbf{l}')\hat{\mathbf{S}}_{\mathbf{l}} \cdot \hat{\mathbf{S}}_{\mathbf{l}'} - g\mu_{\mathrm{B}} H \sum_{\mathbf{l}} \hat{S}^z_{\mathbf{l}}. \tag{9.1}$$

In (9.1), $J(\mathbf{l}-\mathbf{l}')$ is the exchange parameter between spins situated at the Bravais lattice sites $\mathbf{l}$ and $\mathbf{l}'$ and is defined such that $J(0) = 0$.

Because the total z-component of spin commutes with $\hat{\mathscr{H}}$, it follows from our discussion in § 8.3 that the cross-section for the scattering by this system contains only the longitudinal term $\langle \hat{S}^z_{\mathbf{l}} \hat{S}^z_{\mathbf{l}'}(t)\rangle$ and the transverse terms $\langle \hat{S}^{\pm}_{\mathbf{l}} \hat{S}^{\mp}_{\mathbf{l}'}(t)\rangle$. At low temperatures the longitudinal term gives no inelastic scattering, as we shall presently confirm, and the transverse terms give rise to one-magnon creation and annihilation scattering processes. The energy of a spin wave is quantized, and the unit of energy of a spin wave is called a magnon.

On introducing the spin angular momentum raising and lowering operators $\hat{S}^{\pm} = \hat{S}^x \pm \mathrm{i}S^y$ in (9.1), the Hamiltonian can be written

$$\hat{\mathscr{H}} = -\sum_{\mathbf{l},\mathbf{l}'} J(\mathbf{l}-\mathbf{l}')\{\hat{S}^z_{\mathbf{l}} \hat{S}^z_{\mathbf{l}'} + \hat{S}^+_{\mathbf{l}} \hat{S}^-_{\mathbf{l}'}\} - g\mu_{\mathrm{B}} H \sum_{\mathbf{l}} \hat{S}^z_{\mathbf{l}}. \tag{9.2}$$

We recall that the operators $\hat{S}^{\pm}$ satisfy the commutation relations

$$[\hat{S}^+_{\mathbf{l}}, \hat{S}^-_{\mathbf{l}'}] = \delta_{\mathbf{l},\mathbf{l}'}\, 2\hat{S}^z_{\mathbf{l}} \tag{9.3a}$$

and

$$[\hat{S}^z_{\mathbf{l}}, \hat{S}^+_{\mathbf{l}'}] = \delta_{\mathbf{l},\mathbf{l}'} \hat{S}^+_{\mathbf{l}}, \qquad [\hat{S}^z_{\mathbf{l}}\, \hat{S}^-_{\mathbf{l}'}] = -\delta_{\mathbf{l},\mathbf{l}'} \hat{S}^-_{\mathbf{l}}. \tag{9.3b}$$

With the aid of these commutation relations, the equation-of-motion for the operator $\hat{S}^+$ is readily obtained. We find

$$\begin{aligned} \mathrm{i}\hbar\, \partial_t\, \hat{S}^+_{\mathbf{l}_1} &= [\hat{S}^+_{\mathbf{l}_1}, \hat{\mathscr{H}}] \\ &= g\mu_{\mathrm{B}} H \hat{S}^+_{\mathbf{l}_1} - \sum_{\mathbf{l},\mathbf{l}'} J(\mathbf{l}-\mathbf{l}')\{2\,\delta_{\mathbf{l},\mathbf{l}'} \hat{S}^+_{\mathbf{l}} \hat{S}^z_{\mathbf{l}_1} - \delta_{\mathbf{l}_1,\mathbf{l}} \hat{S}^+_{\mathbf{l}_1} \hat{S}^z_{\mathbf{l}'} - \delta_{\mathbf{l}_1,\mathbf{l}'} \hat{S}^z_{\mathbf{l}} \hat{S}^+_{\mathbf{l}_1}\}. \end{aligned} \tag{9.4}$$

If $\mathbf{l}$ and $\mathbf{l}'$ are interchanged in the third term of the second part of (9.4), then, on account of the fact that

$$J(\mathbf{l}-\mathbf{l}') = J(\mathbf{l}'-\mathbf{l}) \tag{9.5}$$

because the crystal is Bravais, we get

$$\mathrm{i}\hbar\, \partial_t \hat{S}^+_{\mathbf{l}} = g\mu_{\mathrm{B}} H \hat{S}^+_{\mathbf{l}} + 2\sum_{\mathbf{l}'} J(\mathbf{l}-\mathbf{l}')\{\hat{S}^z_{\mathbf{l}'} \hat{S}^+_{\mathbf{l}} - \hat{S}^+_{\mathbf{l}'} \hat{S}^z_{\mathbf{l}}\}. \tag{9.6}$$

This equation-of-motion is made linear in $\hat{S}^+$ by making the replacement $\hat{S}^z \rightarrow S$. Thus, in this linear approximation,

$$\mathrm{i}\hbar\, \partial_t \hat{S}^+_{\mathbf{l}} = g\mu_{\mathrm{B}} H \hat{S}^+_{\mathbf{l}} + 2S \sum_{\mathbf{l}'} J(\mathbf{l}-\mathbf{l}')(\hat{S}^+_{\mathbf{l}} - \hat{S}^+_{\mathbf{l}'}). \tag{9.7}$$

If we introduce the operators $\hat{S}_{\mathbf{q}}^{\pm}$ through the canonical transformation

$$\hat{S}_{\mathbf{l}}^{\pm} = \frac{1}{N}\sum_{\mathbf{q}} \exp(\pm i\mathbf{q}\cdot\mathbf{l})\hat{S}_{\mathbf{q}}^{\pm} \tag{9.8}$$

as in § 8.3, N being, as usual, the number of unit cells, the equation-of-motion for $\hat{S}_{\mathbf{q}}^{+}$ is, from (9.7),

$$i\hbar\,\partial_t \hat{S}_{\mathbf{q}}^{+} = g\mu_{\mathrm{B}} H\hat{S}_{\mathbf{q}}^{+} + 2S\{\mathscr{J}(0) - \mathscr{J}(\mathbf{q})\}\hat{S}_{\mathbf{q}}^{+}. \tag{9.9}$$

Here

$$\mathscr{J}(\mathbf{q}) = \sum_{\mathbf{l}} J(\mathbf{l})\exp(-i\mathbf{q}\cdot\mathbf{l}). \tag{9.10a}$$

Note that

$$\mathscr{J}(\mathbf{q}) = \mathscr{J}(-\mathbf{q}) \tag{9.10b}$$

because the lattice is Bravais, and, also,

$$\mathscr{J}(\mathbf{q}+\boldsymbol{\tau}) = \mathscr{J}(\mathbf{q}) \tag{9.10c}$$

by definition of the reciprocal lattice vectors $\boldsymbol{\tau}$ (cf. § 2.1).

To solve (9.9), set

$$\hat{S}_{\mathbf{q}}^{+}(t) = \exp(it\hat{\mathscr{H}}/\hbar)\hat{S}_{\mathbf{q}}^{+}\exp(-it\hat{\mathscr{H}}/\hbar) = \exp(-i\omega_{\mathbf{q}}t)\hat{S}_{\mathbf{q}}^{+}, \tag{9.11}$$

then, clearly,

$$\hbar\omega_{\mathbf{q}} = g\mu_{\mathrm{B}} H + 2S\{\mathscr{J}(0) - \mathscr{J}(\mathbf{q})\}, \tag{9.12}$$

and this is the dispersion relation for linear spin waves in a Heisenberg ferromagnet.

Before discussing (9.12) and its consequences, we note that the linearized equation-of-motion (9.7) could have been obtained directly by modifying the commutation relation (9.3a) to read

$$[\hat{S}_{\mathbf{l}}^{+}, \hat{S}_{\mathbf{l}'}^{-}] \simeq \delta_{\mathbf{l},\mathbf{l}'} 2S, \tag{9.13}$$

i.e. by replacing $\hat{S}^z$ by the c-number S, as is appropriate near saturation. We also need an expression for $\hat{S}^z$ itself, and an appropriate form is

$$\hat{S}_{\mathbf{l}}^{z} \simeq S - \frac{1}{2S}\hat{S}_{\mathbf{l}}^{-}\hat{S}_{\mathbf{l}}^{+}, \tag{9.14}$$

becuse (9.14) is valid both when $S_{\mathbf{l}}^{z} = S$ and when there is a single deviation on the atom at $\mathbf{l}$, i.e. when $S_{\mathbf{l}}^{z} = S - 1$. The approximations (9.13) and (9.14) preserve the commutation relations (9.3). Clearly (9.14) is exact for the special case of $S = \frac{1}{2}$. This approximation scheme is very convenient to employ in deriving linearized equations-of-motion for more complex magnetic systems than that described by (9.1) and we shall adopt it throughout for this purpose.

Let us examine the dispersion relation (9.12). First, if the exchange parameter is between nearest-neighbour spins only, and of magnitude J,

$$\mathcal{J}(\mathbf{q}) = J\sum_{\boldsymbol{\rho}} \exp(i\mathbf{q}\cdot\boldsymbol{\rho}) = rJ\gamma_{\mathbf{q}}, \tag{9.15}$$

where r is the number of nearest neighbours. Thus

$$\hbar\omega_{\mathbf{q}} = g\mu_{\mathrm{B}}H + 2rJS(1-\gamma_{\mathbf{q}}). \tag{9.16}$$

For small $\mathbf{q}$ and cubic (Bravais) lattices,

$$\gamma_{\mathbf{q}} = \frac{1}{r}\sum_{\boldsymbol{\rho}} \exp(i\mathbf{q}\cdot\boldsymbol{\rho}) = \frac{1}{r}\sum_{\boldsymbol{\rho}} \{1 + i\mathbf{q}\cdot\boldsymbol{\rho} + \tfrac{1}{2}(i\mathbf{q}\cdot\boldsymbol{\rho})^2 + \ldots\}$$

$$\simeq \frac{1}{r}\sum_{\boldsymbol{\rho}} \{1 - \tfrac{1}{2}\tfrac{1}{3}q^2\rho^2\} = 1 - \tfrac{1}{6}q^2\rho^2. \tag{9.17}$$

Now

$$\rho^2 = \begin{cases} a^2 & \text{s.c.} \quad (r=6) \\ 3a^2/4 & \text{b.c.c.} \quad (r=8) \\ a^2/2 & \text{f.c.c.} \quad (r=12) \end{cases}$$

where a is the lattice parameter, so $r\frac{1}{6}\rho^2 = a^2$ for all three cubic lattices. Hence

$$\hbar\omega_{\mathbf{q}} \simeq g\mu_{\mathrm{B}}H + Dq^2 \quad (|\mathbf{q}|\,|\boldsymbol{\rho}| \ll 1) \tag{9.18}$$

where

$$D = 2JSa^2. \tag{9.19}$$

More generally, when the exchange parameter is of longer range than just to nearest neighbours,

$$D = \tfrac{1}{3}S\sum_{\mathbf{l}} l^2 J(\mathbf{l}). \tag{9.20}$$

For terms up to $(\mathbf{q}\cdot\mathbf{l})^6$,

$$\hbar\omega_{\mathbf{q}} = g\mu_{\mathrm{B}}H + 2S\left\{\tfrac{1}{6}q^2\sum_{\mathbf{l}} l^2 J(\mathbf{l}) - \tfrac{1}{24}\sum_{\mathbf{l}} (\mathbf{q}\cdot\mathbf{l})^4 J(\mathbf{l}) + \tfrac{1}{720}\sum_{\mathbf{l}} (\mathbf{q}\cdot\mathbf{l})^6 J(\mathbf{l})\right\}. \tag{9.21}$$

We shall presently find it convenient to express the results obtained from this expansion in terms of the moments

$$\overline{l^n} = S\left\{\sum l^{n+2} J(\mathbf{l})\right\}\Big/3D. \tag{9.22}$$

To calculate the thermal average value of the z-component of spin at a temperature T we proceed as follows. From (9.8) and (9.14)

$$\begin{aligned}\langle \hat{S}_{\mathbf{l}}^{z} \rangle &= S - \frac{1}{2S} \langle \hat{S}_{\mathbf{l}}^{-} \hat{S}_{\mathbf{l}}^{+} \rangle \\ &= S - \frac{1}{2SN^2} \sum_{\mathbf{q}} \langle \hat{S}_{\mathbf{q}}^{-} \hat{S}_{\mathbf{q}}^{+} \rangle. \end{aligned} \tag{9.23}$$

By definition ($\beta = 1/k_{\mathrm{B}}T$),

$$\begin{aligned}\langle \hat{S}_{\mathbf{q}}^{-} \hat{S}_{\mathbf{q}}^{+} \rangle &= \operatorname{Tr} \exp(-\beta\hat{\mathscr{H}}) \hat{S}_{\mathbf{q}}^{-} \hat{S}_{\mathbf{q}}^{+} / \operatorname{Tr} \exp(-\beta\hat{\mathscr{H}}) \\ &= \operatorname{Tr} \exp(-\beta\hat{\mathscr{H}}) \exp(\beta\hat{\mathscr{H}}) \hat{S}_{\mathbf{q}}^{+} \exp(-\beta\hat{\mathscr{H}}) \hat{S}_{\mathbf{q}}^{-} / \operatorname{Tr} \exp(-\beta\hat{\mathscr{H}}), \end{aligned} \tag{9.24}$$

the second line following by virtue of the invariance of the trace under a cyclic permutation of operators. Since,

$$\exp(\beta\hat{\mathscr{H}}) \hat{S}_{\mathbf{q}}^{+} \exp(-\beta\hat{\mathscr{H}}) = \hat{S}_{\mathbf{q}}^{+}(-\mathrm{i}\hbar\beta), \tag{9.25}$$

it follows from (9.11) that, in the linear approximation,

$$\begin{aligned}\exp(\beta\hat{\mathscr{H}}) \hat{S}_{\mathbf{q}}^{+} \exp(-\beta\hat{\mathscr{H}}) &= \hat{S}_{\mathbf{q}}^{+} \exp\{-\mathrm{i}\omega_{\mathbf{q}}(-\mathrm{i}\hbar\beta)\} \\ &= \hat{S}_{\mathbf{q}}^{+} \exp(-\hbar\omega_{\mathbf{q}}\beta). \end{aligned} \tag{9.26}$$

Substituting (9.26) in (9.24), we get

$$\langle \hat{S}_{\mathbf{q}}^{-} \hat{S}_{\mathbf{q}}^{+} \rangle = \exp(-\hbar\omega_{\mathbf{q}}\beta) \langle \hat{S}_{\mathbf{q}}^{+} S_{\mathbf{q}}^{-} \rangle. \tag{9.27}$$

Now it is readily verified from (9.13) that $\hat{S}_{\mathbf{q}}^{\pm}$ satisfy the commutation relation

$$[\hat{S}_{\mathbf{q}}^{+}, \hat{S}_{\mathbf{q}'}^{-}] = \delta_{\mathbf{q},\mathbf{q}'} 2SN, \tag{9.28}$$

which when used in (9.27) leads to the result

$$\langle \hat{S}_{\mathbf{q}}^{-} \hat{S}_{\mathbf{q}}^{+} \rangle = \exp(-\hbar\omega_{\mathbf{q}}\beta)(2SN + \langle \hat{S}_{\mathbf{q}}^{-} \hat{S}_{\mathbf{q}}^{+} \rangle),$$

i.e.

$$\langle \hat{S}_{\mathbf{q}}^{-} \hat{S}_{\mathbf{q}}^{+} \rangle = \frac{2SN}{\exp(\hbar\omega_{\mathbf{q}}\beta) - 1} = 2SNn_{\mathbf{q}} \tag{9.29}$$

where

$$n_{\mathbf{q}} = \{\exp(\hbar\omega_{\mathbf{q}}\beta) - 1\}^{-1}.$$

Obviously $n_{\mathbf{q}}$ is the occupation number of magnons of vector $\mathbf{q}$ and, using the theory set out above, some straightforward algebra gives the results

$$\begin{aligned}\langle \hat{S}_{\mathbf{l}}^{x} \hat{S}_{\mathbf{l}+\mathbf{R}}^{x}(t) \rangle &= \langle \hat{S}_{\mathbf{l}}^{y} \hat{S}_{\mathbf{l}+\mathbf{R}}^{y}(t) \rangle \\ &= \frac{S}{2N} \sum_{\mathbf{q}} \{(n_{\mathbf{q}} + 1) \exp(-\mathrm{i}\mathbf{q} \cdot \mathbf{R} + \mathrm{i}\omega_{\mathbf{q}} t) + n_{\mathbf{q}} \exp(\mathrm{i}\mathbf{q} \cdot \mathbf{R} - \mathrm{i}\omega_{\mathbf{q}} t)\} \end{aligned}$$

and

$$\tfrac{1}{2}\langle \hat{S}_{\mathbf{l}}^x \hat{S}_{\mathbf{l}+\mathbf{R}}^y(t) + \hat{S}_{\mathbf{l}+\mathbf{R}}^y(t)\hat{S}_{\mathbf{l}}^x \rangle = \frac{S}{2N}\sum_{\mathbf{q}}(2n_{\mathbf{q}}+1)\sin(\mathbf{q}\cdot\mathbf{R}-\omega_{\mathbf{q}}t).$$

From these formulae it follows that the non-interacting spin wave is an excitation described by the correlation functions

$$\langle \hat{S}_{\mathbf{l}}^x \hat{S}_{\mathbf{l}+\mathbf{R}}^x(t)\rangle = \langle \hat{S}_{\mathbf{l}}^y \hat{S}_{\mathbf{l}+\mathbf{R}}^y(t)\rangle \sim \frac{S}{N}\cos(\mathbf{q}\cdot\mathbf{R}-\omega_{\mathbf{q}}t)$$

and

$$\tfrac{1}{2}\langle \hat{S}_{\mathbf{l}}^x \hat{S}_{\mathbf{l}+\mathbf{R}}^y(t) + \hat{S}_{\mathbf{l}+\mathbf{R}}^y(t)\hat{S}_{\mathbf{l}}^x \rangle \sim \frac{S}{N}\sin(\mathbf{q}\cdot\mathbf{R}-\omega_{\mathbf{q}}t).$$

These equations describe a precession of the spins around the z-axis, the precession propagating with wave vector $\mathbf{q}$ and frequency $\omega_{\mathbf{q}}$.

With the result (9.29) the thermal average of the z-component of spin (9.23) is

$$\begin{aligned}\langle \hat{S}_{\mathbf{l}}^z\rangle &= S - \frac{1}{N}\sum_{\mathbf{q}} n_{\mathbf{q}} \\ &= S - \frac{1}{N}\sum_{\mathbf{q}}\{\exp(\hbar\omega_{\mathbf{q}}\beta)-1\}^{-1}.\end{aligned} \tag{9.30}$$

For the quadratic dispersion law (9.18), with D in the general case given by (9.20), we may write

$$\begin{aligned}\frac{1}{N}\sum_{\mathbf{q}}\{\exp(\hbar\omega_{\mathbf{q}}\beta)-1\}^{-1} &= \frac{1}{N}\frac{V}{(2\pi)^3}\int \mathrm{d}\mathbf{q}\,\{\exp(\hbar\omega_{\mathbf{q}}\beta)-1\}^{-1} \\ &= \frac{v_0}{(2\pi)^3}\int \mathrm{d}\mathbf{q}\,[\exp\{(g\mu_{\mathrm{B}}H+Dq^2)\beta\}-1]^{-1}\end{aligned}$$

where v_0 is the volume of a unit cell. In this form it gives negligible error if, for low temperatures, the integral is extended over all space; hence

$$S - \langle \hat{S}_{\mathbf{l}}^z\rangle = \frac{v_0}{4\pi^2}\left(\frac{k_{\mathrm{B}}T}{D}\right)^{3/2}\int_0^\infty \frac{\mathrm{d}x\,x^{1/2}}{\exp(x+g\mu_{\mathrm{B}}H\beta)-1}, \tag{9.31}$$

so that it is evident that the average value of $\hat{S}^z$ decreases from its maximum value S, obtained at absolute zero, as $T^{3/2}$.

The integral

$$\zeta(n) = \frac{1}{\Gamma(n)}\int_0^\infty \frac{\mathrm{d}x\,x^{n-1}}{\mathrm{e}^x-1} \tag{9.32}$$

(where $\Gamma(n)$ is the gamma function of order n) is Riemann's zeta function.

In particular,

$$\zeta(\tfrac{3}{2}) = 2.612, \qquad \zeta(\tfrac{5}{2}) = 1.341, \quad \text{and} \quad \zeta(\tfrac{7}{2}) = 1.127. \tag{9.33}$$

Using (9.32) and $\Gamma(\frac{3}{2}) = \frac{1}{2}\sqrt{\pi}$ we obtain from (9.31) the result ($H = 0$)

$$\langle \hat{S}^z_{\mathbf{l}} \rangle = S - v_0 \left(\frac{k_B T}{4\pi D} \right)^{3/2} \zeta(\tfrac{3}{2}). \tag{9.34}$$

When H is not zero, but such that $g\mu_B H \leqslant k_B T$,

$$\langle \hat{S}^z_{\mathbf{l}} \rangle = S - v_0 \left(\frac{k_B T}{4\pi D} \right)^{3/2} F(h) \tag{9.35}$$

where $h = g\mu_B H\beta$ and

$$F(h) = \zeta(\tfrac{3}{2}) - 3.54h^{1/2} + 1.64h - 0.104h^2 + 0.00425h^3 \tag{9.36}$$

to within an accuracy of at least 1 per cent.

If in place of the simple quadratic dispersion law (9.18) we use (9.21) in (9.30), then in place of (9.34) it can be shown that

$$\langle \hat{S}^z_{\mathbf{l}} \rangle = S - v_0 \left(\frac{k_B T}{4\pi D} \right)^{3/2} \zeta(\tfrac{3}{2}) - v_0 \left(\frac{k_B T}{4\pi D} \right)^{5/2} \zeta(\tfrac{5}{2}) \frac{3\pi}{4} \overline{l^2}$$
$$- v_0 \left(\frac{k_B T}{4\pi D} \right)^{7/2} \zeta(\tfrac{7}{2}) \frac{\pi^2}{32} \{33(\overline{l^2})^2 - 8\overline{l^4} + 24(\overline{l^2})^2 \overline{\cos^4\theta}\} \tag{9.37}$$

where the moments $\overline{l^n}$ are defined by (9.22) and

$$\overline{\cos^4\theta} = \frac{\sum_{\mathbf{l},\mathbf{l}'} J(\mathbf{l}) J(\mathbf{l}') (\mathbf{l} \cdot \mathbf{l}')^4}{\left\{ \sum_{\mathbf{l}} J(\mathbf{l}) \mathbf{l}^4 \right\}^2}. \tag{9.38}$$

For the particular case when $J(\mathbf{l})$ extends only to nearest-neighbour sites,

$$\overline{l^2} = a^2 \tag{9.39}$$

and

$$\tfrac{1}{32}\{33(\overline{l^2})^2 - 8\overline{l^4} + 24(\overline{l^2})^2 \overline{\cos^4\theta}\} = a^4 \times \begin{cases} 33/32 & \text{s.c.} \\ 281/288 & \text{b.c.c.} \\ 15/16 & \text{f.c.c.} \end{cases}$$

The result (9.37) shows that the $T^{5/2}$ and $T^{7/2}$ terms in the expansion of $\langle \hat{S}^z \rangle$ do not have coefficients simply related to that of the $T^{3/2}$ term, but instead have an importance that depends upon the range of the exchange parameter $J(\mathbf{l})$, their relative importance increasing with the range of J.

9.2.2. One-magnon cross-sections

Now that we have pursued some of the physical predictions of linear spin-wave theory let us evaluate the neutron cross-section. As we have already remarked, for the system described by the Hamiltonian (9.1) we need only evaluate $\langle \hat{S}^z_{\mathbf{l}}\hat{S}^z_{\mathbf{l}'}(t)\rangle$ and $\langle \hat{S}^{\pm}_{\mathbf{l}}\hat{S}^{\mp}_{\mathbf{l}'}(t)\rangle$. However, in the linear spin-wave approximation, $\hat{S}^z(t) = \hat{S}^z(0)$, as is easily verified from (9.11) and (9.14), and, therefore, the longitudinal cross-section leads only to elastic scattering, which does not concern us here. The evaluation of the correlation functions $\langle \hat{S}^{\pm}_{\mathbf{l}}\hat{S}^{\mp}_{\mathbf{l}'}(t)\rangle$ is straightforward.

We have

$$\langle \hat{S}^-_{\mathbf{l}}\hat{S}^+_{\mathbf{l}'}(t)\rangle = \frac{1}{N^2}\sum_{\mathbf{q}} \exp\{\mathrm{i}\mathbf{q}\cdot(\mathbf{l}'-\mathbf{l})\}\langle \hat{S}^-_{\mathbf{q}}\hat{S}^+_{\mathbf{q}}(t)\rangle,$$

which on using (9.11) and (9.29) gives

$$\langle \hat{S}^-_{\mathbf{l}}\hat{S}^+_{\mathbf{l}'}(t)\rangle = \frac{2S}{N}\sum_{\mathbf{q}} \exp\{\mathrm{i}\mathbf{q}\cdot(\mathbf{l}'-\mathbf{l}) - \mathrm{i}\omega_{\mathbf{q}}t\}n_{\mathbf{q}}. \tag{9.40}$$

Also

$$\begin{aligned}\langle \hat{S}^+_{\mathbf{l}}\hat{S}^-_{\mathbf{l}'}(t)\rangle &= \frac{1}{N^2}\sum_{\mathbf{q}} \exp\{-\mathrm{i}\mathbf{q}\cdot(\mathbf{l}'-\mathbf{l})\}\langle \hat{S}^+_{\mathbf{q}}\hat{S}^-_{\mathbf{q}}(t)\rangle \\ &= \frac{2S}{N}\sum_{\mathbf{q}} \exp\{-\mathrm{i}\mathbf{q}\cdot(\mathbf{l}'-\mathbf{l}) + \mathrm{i}\omega_{\mathbf{q}}t\}(1+n_{\mathbf{q}}),\end{aligned}$$

by virtue of (9.27). Notice that the correlation functions obey the general identity,

$$\langle \hat{S}^+_{\mathbf{l}}\hat{S}^-_{\mathbf{l}'}(t)\rangle = \langle \hat{S}^-_{\mathbf{l}'}\hat{S}^+_{\mathbf{l}}(-t+\mathrm{i}\hbar\beta)\rangle. \tag{9.41}$$

From eqn (8.62) the transverse cross-section is given by

$$\begin{aligned}\frac{\mathrm{d}^2\sigma}{\mathrm{d}\Omega\,\mathrm{d}E'} &= r_0^2\frac{k'}{k}\{\tfrac{1}{2}gF(\boldsymbol{\kappa})\}^2\tfrac{1}{4}(1+\tilde{\kappa}_z^2)\exp\{-2W(\boldsymbol{\kappa})\} \\ &\quad\times\frac{1}{2\pi\hbar}\int_{-\infty}^{\infty}\mathrm{d}t\,\exp(-\mathrm{i}\omega t)\sum_{\mathbf{l},\mathbf{l}'}\exp\{\mathrm{i}\boldsymbol{\kappa}\cdot(\mathbf{l}'-\mathbf{l})\}\langle \hat{S}^+_{\mathbf{l}}\hat{S}^-_{\mathbf{l}'}(t) + \hat{S}^-_{\mathbf{l}}\hat{S}^+_{\mathbf{l}'}(t)\rangle \end{aligned} \tag{9.42}$$

$$= \left(\frac{\mathrm{d}^2\sigma}{\mathrm{d}\Omega\,\mathrm{d}E'}\right)^{(+)} + \left(\frac{\mathrm{d}^2\sigma}{\mathrm{d}\Omega\,\mathrm{d}E'}\right)^{(-)}. \tag{9.43}$$

In (9.43), $(\mathrm{d}^2\sigma/\mathrm{d}\Omega\,\mathrm{d}E')^{(+)}$ is the cross-section in which one magnon is created and $(\mathrm{d}^2\sigma/\mathrm{d}\Omega\,\mathrm{d}E')^{(-)}$ the corresponding cross-section for the annihilation process. From the results (9.40) and (9.41), (9.42) yields

$$\left(\frac{d^2\sigma}{d\Omega\, dE'}\right)^{(\pm)} = r_0^2 \frac{k'}{k}\{\tfrac{1}{2}gF(\boldsymbol{\kappa})\}^2(1+\tilde{\kappa}_z^2)\exp\{-2W(\boldsymbol{\kappa})\}\tfrac{1}{2}S$$

$$\times \frac{(2\pi)^3}{v_0}\sum_{\mathbf{q},\boldsymbol{\tau}}(n_{\mathbf{q}}+\tfrac{1}{2}\pm\tfrac{1}{2})\,\delta(\hbar\omega_{\mathbf{q}}\mp\hbar\omega)\,\delta(\boldsymbol{\kappa}\mp\mathbf{q}-\boldsymbol{\tau}). \tag{9.44}$$

Because these formulae for the cross-sections in which a single magnon is either created or annihilated contain both a momentum- and an energy-conservation condition, inelastic experiments can determine the whole spin-wave spectrum, just as the one-phonon cross-sections permit determination of phonon spectra.

There are two differences between the one-magnon cross-sections and the one-phonon cross-sections to which we draw attention. First, the nature of the interaction between the spin of the neutron and the spin of the electrons gives rise to the orientation factor $(1+\tilde{\kappa}_z^2)$ in (9.44). We remind the reader that we have defined the axes of quantization in our calculation to coincide with the z-axes. Now if a magnetic field is applied to the target sample so as to make its magnetization parallel to the scattering vector $\boldsymbol{\kappa}$, the orientation factor takes the value 2; if the field is applied so as to make the magnetization perpendicular to $\boldsymbol{\kappa}$, it has the value 1. Since the nuclear scattering is independent of an external magnetic field, the orientation factor can clearly be employed to distinguish the purely magnetic part of the total scattering.

If the target crystal is made up of several different ferromagnetic domains, then we must average the orientation factor over all possible directions of the magnetization. For a cubic crystal this averaging gives $(1+\tilde{\kappa}_z^2)_{\text{av}} = 1+\frac{1}{3} = \frac{4}{3}$.

The second point we wish to make is concerned with the range of wave vector $\mathbf{q}$ over which the spin-wave dispersion curve can be measured in the metallic ferromagnets Fe, Co, and Ni, which have high Curie temperatures. The total width of the linear spin-wave spectrum in the case of nearest-neighbour coupling is $2rJS$ and we can estimate the magnitude of this by a somewhat crude argument. The molecular field approximation gives the critical temperature T_c of a ferromagnet

$$T_c = \frac{2rJS(S+1)}{3k_B}$$

so that $2rJS = 3k_BT_c/(S+1) \simeq 2k_BT_c$ for $S=\frac{1}{2}$. If $T_c \sim 10^3$ K, as it is for Fe, the width of the spin-wave band is of order 0.2 eV, a value much in excess of the neutron-energy changes that can be accessed using reactor neutron sources. However, the neutron spectrum from advanced pulsed spallation sources is undermoderated, and therefore the intensity of energetic (epithermal) neutrons is much higher (Winsdor 1981). The width of the spin-wave bands in rare-earth (Coqblin 1977) and ionic ferromagnets (Mook 1981) is often considerably smaller than for the 3d transition metals. The latter are discussed further in § 9.3.

There are very few ionic ferromagnets, and, of these, the divalent europium chalcogenides EuO and EuS have been studied most thoroughly by neutron scattering, since they are examples of simple Heisenberg ferromagnets. In both materials the Eu^{2+} ions form a f.c.c. lattice, and nearest- and next-nearest-neighbour exchange interactions predominate; see Dietrich, Als-Nielsen, and Passell (1976) and references therein.

9.2.3. Two-magnon interactions: Hartree–Fock approximation

The spin-wave theory presented above does not take account of the interaction between two or more magnons, i.e. it is a linear theory. The inclusion of spin-wave interactions will clearly modify the dispersion law. Since neutron scattering affords a means of determining a substantial part of the dispersion curve, it is important to our understanding of the dynamical behaviour of magnetic systems that we should derive the form of the modification as it appears in the linear spectrum of the Heisenberg ferromagnet (eqn (9.12)). We only consider the effects due to two-magnon scattering processes.

The interaction betwen spin waves arises in two parts, called the kinematic and the dynamic interactions. The kinematic interaction is a consequence of spin statistics, namely that the maximum number of spin deviations that can occur at any site with a spin S is $2S$. Take as an example spins of magnitude $\frac{1}{2}$; then, clearly, two spin deviations cannot reside at the same site and the interaction that prevents this from occurring, the kinematic interaction as we have called it, is a repulsive one. The second type of interaction, the dynamic interaction, arises because it costs less energy for a spin to suffer a deviation if the spins with which it directly interacts via $J(\mathbf{l}-\mathbf{l}')$ have also undergone deviations from their fully aligned state. The dynamic interaction is, clearly, attractive.

The terminology of kinematic and dynamic interactions was introduced by Dyson (Mattis 1981) in his analysis of two-spin-wave interactions in the Heisenberg ferromagnet. He showed that at low temperatures the kinematic interaction is small. In the discussion of two-spin-wave interactions given below we shall neglect it from the outset.

To discuss the dynamical interaction it is convenient to introduce a representation of the spin operators in terms of Bose operators a and a^+.

We take

$$\hat{S}_{\mathbf{l}}^{z} = S - \hat{a}_{\mathbf{l}}^{+}\hat{a}_{\mathbf{l}},$$

$$\hat{S}_{\mathbf{l}}^{+} = (2S)^{1/2}\hat{a}_{\mathbf{l}} \quad \text{and} \quad \hat{S}_{\mathbf{l}}^{-} = (2S)^{1/2}\hat{a}_{\mathbf{l}}^{+}\left(1 - \frac{\hat{a}_{\mathbf{l}}^{+}\hat{a}_{\mathbf{l}}}{2S}\right) \tag{9.45}$$

where

$$[\hat{a}_{\mathbf{l}}, \hat{a}_{\mathbf{l}'}^{+}] = \delta_{\mathbf{l},\mathbf{l}'}. \tag{9.46}$$

The transformation (9.45) satisfies the spin angular momentum commutation rules (9.3) but does not preserve spin statistics because the Bose operators have an infinite spectrum. Thus by making the transformation (9.45) the kinematic interaction is ignored.

If (9.45) is used in the Heisenberg Hamiltonian (9.1) and the magnetic field set equal to zero, we obtain, apart from a constant term,

$$\hat{\mathscr{H}} = \sum_{\mathbf{q}} 2S\{\mathscr{J}(0) - \mathscr{J}(\mathbf{q})\}\hat{a}_{\mathbf{q}}^{+}\hat{a}_{\mathbf{q}}$$

$$-\frac{1}{N}\sum_{\mathbf{q}_1,\mathbf{q}_2,\mathbf{q}_3,\mathbf{q}_4} \hat{a}_{\mathbf{q}_1}^{+}\hat{a}_{\mathbf{q}_2}^{+}\hat{a}_{\mathbf{q}_3}\hat{a}_{\mathbf{q}_4}\,\delta_{\mathbf{q}_1+\mathbf{q}_2,\mathbf{q}_3+\mathbf{q}_4}\{\mathscr{J}(\mathbf{q}_1 - \mathbf{q}_3) - \mathscr{J}(\mathbf{q}_3)\} \tag{9.47}$$

where

$$\hat{a}_{\mathbf{l}} = \frac{1}{N^{1/2}}\sum_{\mathbf{q}} \exp(i\mathbf{q}\cdot\mathbf{l})\hat{a}_{\mathbf{q}}, \quad \text{etc.} \tag{9.48}$$

and

$$[\hat{a}_{\mathbf{q}}, \hat{a}_{\mathbf{q}'}^{+}] = \delta_{\mathbf{q},\mathbf{q}'}.$$

The first term in (9.47) represents the non-interacting spin waves, the second the interaction between them. The latter is seen to contain a product of four Bose operators, which will result in products of three operators appearing in the equation of motion for $\hat{a}_{\mathbf{q}}$, say. We can, however, obtain linear equations-of-motion if we approximate the four operators in the interaction term by products of two operators times a thermal aveage.

Explicitly, we write the product

$$\hat{a}_{\mathbf{q}_1}^{+}\hat{a}_{\mathbf{q}_2}^{+}\hat{a}_{\mathbf{q}_3}\hat{a}_{\mathbf{q}_4}$$

as the sum of all non-zero pairings of two operators, i.e.

$$\begin{aligned}\hat{a}_{\mathbf{q}_1}^{+}\hat{a}_{\mathbf{q}_2}^{+}\hat{a}_{\mathbf{q}_3}\hat{a}_{\mathbf{q}_4} \to \langle\hat{a}_{\mathbf{q}_1}^{+}\hat{a}_{\mathbf{q}_3}\rangle\hat{a}_{\mathbf{q}_2}^{+}\hat{a}_{\mathbf{q}_4} + \langle\hat{a}_{\mathbf{q}_1}^{+}\hat{a}_{\mathbf{q}_4}\rangle\hat{a}_{\mathbf{q}_2}^{+}\hat{a}_{\mathbf{q}_3} \\ + \langle\hat{a}_{\mathbf{q}_2}^{+}\hat{a}_{\mathbf{q}_3}\rangle\hat{a}_{\mathbf{q}_1}^{+}\hat{a}_{\mathbf{q}_4} + \langle\hat{a}_{\mathbf{q}_2}^{+}\hat{a}_{\mathbf{q}_4}\rangle\hat{a}_{\mathbf{q}_1}^{+}\hat{a}_{\mathbf{q}_3} \\ - \langle\hat{a}_{\mathbf{q}_1}^{+}\hat{a}_{\mathbf{q}_3}\rangle\langle\hat{a}_{\mathbf{q}_2}^{+}\hat{a}_{\mathbf{q}_4}\rangle - \langle\hat{a}_{\mathbf{q}_1}^{+}\hat{a}_{\mathbf{q}_4}\rangle\langle\hat{a}_{\mathbf{q}_2}^{+}\hat{a}_{\mathbf{q}_3}\rangle.\end{aligned} \tag{9.49}$$

The approximation of the product of four operators as represented by

(9.49) is usually referred to as the Hartree–Fock approximation. Note that the relative signs of the operator terms on the right-hand side of this expression can only be positive because the operators obey Bose statistics (Lovesey 1980).

The last two terms in (9.49) are required if we are not to count the remaining term twice, as is made evident if the thermal average of the whole expression is taken. Since, however, these terms result only in an additional constant energy in the Hamiltonian, they will not enter the equation-of-motion for $\hat{a}_{\mathbf{q}}$.

We also have the result

$$\langle \hat{a}^+_{\mathbf{q}} \hat{a}_{\mathbf{q}'} \rangle = \delta_{\mathbf{q},\mathbf{q}'} \langle \hat{a}^+_{\mathbf{q}} \hat{a}_{\mathbf{q}} \rangle. \tag{9.50}$$

With (9.50), the approximation (9.49) reduces to

$$n_{\mathbf{q}_1}(\delta_{\mathbf{q}_1,\mathbf{q}_3} \hat{a}^+_{\mathbf{q}_2} \hat{a}_{\mathbf{q}_4} + \delta_{\mathbf{q}_1,\mathbf{q}_4} \hat{a}^+_{\mathbf{q}_2} \hat{a}_{\mathbf{q}_3}) + n_{\mathbf{q}_2}(\delta_{\mathbf{q}_2,\mathbf{q}_3} \hat{a}^+_{\mathbf{q}_1} \hat{a}_{\mathbf{q}_4} + \delta_{\mathbf{q}_2,\mathbf{q}_4} \hat{a}^+_{\mathbf{q}_1} \hat{a}_{\mathbf{q}_3})$$
$$- n_{\mathbf{q}_1} n_{\mathbf{q}_2} (\delta_{\mathbf{q}_1,\mathbf{q}_3} \delta_{\mathbf{q}_2,\mathbf{q}_4} + \delta_{\mathbf{q}_1,\mathbf{q}_4} \delta_{\mathbf{q}_2,\mathbf{q}_3}) \tag{9.51}$$

where

$$n_{\mathbf{q}} = \langle \hat{a}^+_{\mathbf{q}} \hat{a}_{\mathbf{q}} \rangle = \langle \hat{n}_{\mathbf{q}} \rangle. \tag{9.52}$$

Inserting (9.51) for the product of operators in the interaction term in (9.47) the latter reduces to

$$\hat{\mathscr{H}} = \sum_{\mathbf{q}} 2S\{\mathscr{J}(0) - \mathscr{J}(\mathbf{q})\} \hat{n}_{\mathbf{q}} - \frac{2}{N} \sum_{\mathbf{q},\mathbf{q}'} \{\mathscr{J}(0) + \mathscr{J}(\mathbf{q}-\mathbf{q}') - \mathscr{J}(\mathbf{q}) - \mathscr{J}(\mathbf{q}')\} n_{\mathbf{q}} \hat{n}_{\mathbf{q}'}$$
$$+ \frac{1}{N} \sum_{\mathbf{q},\mathbf{q}'} \{\mathscr{J}(0) + \mathscr{J}(\mathbf{q}-\mathbf{q}') - \mathscr{J}(\mathbf{q}) - \mathscr{J}(\mathbf{q}')\} n_{\mathbf{q}} n_{\mathbf{q}'}. \tag{9.53}$$

If we use this approximate Hamiltonian to obtain the equation-of-motion for $\hat{a}_{\mathbf{q}}$ it will, clearly, result in an equation containing only operators of the same kind. In fact,

$$i\hbar\, \partial_t \hat{a}_{\mathbf{q}} = [\hat{a}_{\mathbf{q}}, \hat{\mathscr{H}}] = 2S\{\mathscr{J}(0) - \mathscr{J}(\mathbf{q})\} \hat{a}_{\mathbf{q}}$$
$$- \frac{2}{N} \sum_{\mathbf{q}'} \{\mathscr{J}(0) + \mathscr{J}(\mathbf{q}-\mathbf{q}') - \mathscr{J}(\mathbf{q}) - \mathscr{J}(\mathbf{q}')\} n_{\mathbf{q}'} \hat{a}_{\mathbf{q}}. \tag{9.54}$$

The spin-wave dispersion law in our approximate handling of the dynamical interaction is, therefore,

$$\hbar\omega_{\mathbf{q}} = 2S\{\mathscr{J}(0) - \mathscr{J}(\mathbf{q})\} - \frac{2}{N} \sum_{\mathbf{q}'} \{\mathscr{J}(0) + \mathscr{J}(\mathbf{q}-\mathbf{q}') - \mathscr{J}(\mathbf{q}) - \mathscr{J}(\mathbf{q}')\} n_{\mathbf{q}'}. \tag{9.55}$$

The second term on the right-hand side of (9.55) represents a temperature- and $\mathbf{q}$-dependent shift from the linear spin-wave spectrum

$2S\{\mathcal{J}(0)-\mathcal{J}(\mathbf{q})\}$. This shift in energy is purely real, so that our approximation scheme (9.51) does not predict a lifetime for the spin waves.

To evaluate the temperature dependence of the spin-wave spectrum (9.55) we must calculate the thermal average $n_{\mathbf{q}'}$. Since the operators $\hat{a}_{\mathbf{q}}$ and $\hat{a}_{\mathbf{q}}^{+}$ are bosons it follows that

$$n_{\mathbf{q}}=\{\exp(\hbar\omega_{\mathbf{q}}\beta)-1\}^{-1}, \tag{9.56}$$

and thus (9.55) and (9.56) together give a transcendental equation for the energy spectrum $\hbar\omega_{\mathbf{q}}$, and then, from (9.56), we get $n_{\mathbf{q}}$.

We also require the magnetization. From (9.45),

$$\langle\hat{S}_{\mathbf{l}}^{z}\rangle=S-\langle\hat{a}_{\mathbf{l}}^{+}\hat{a}_{\mathbf{l}}\rangle=S-\frac{1}{N}\sum_{\mathbf{q}}n_{\mathbf{q}}. \tag{9.57}$$

Before proceeding any further we introduce a simplification in the algebra for isotropic exchange. If we denote the exchange parameter between nearest neighbours by J, that between next-nearest neighbours by J' and so on, and the corresponding numbers of neighbours by r, r', etc., then

$$\mathcal{J}(\mathbf{q})=\sum_{\mathbf{l}}\exp(-i\mathbf{q}\cdot\mathbf{l})J(\mathbf{l})=rJ\gamma_{\mathbf{q}}+r'J'\gamma_{\mathbf{q}}'+\ldots. \tag{9.58}$$

In (9.58) the functions $\gamma_{\mathbf{q}}$, $\gamma_{\mathbf{q}}'$, etc. are defined by

$$\gamma_{\mathbf{q}}=\frac{1}{r}\sum_{\boldsymbol{\rho}}\exp(i\mathbf{q}\cdot\boldsymbol{\rho})=\gamma_{-\mathbf{q}},\qquad \gamma_{\mathbf{q}}'=\frac{1}{r'}\sum_{\boldsymbol{\rho}'}\exp(i\mathbf{q}\cdot\boldsymbol{\rho}')=\gamma_{-\mathbf{q}}', \tag{9.59}$$

and so on, the vectors $\boldsymbol{\rho}$, $\boldsymbol{\rho}'$, ... joining nearest neighbours, next-nearest neighbours, and so forth. For example, in the case of a body-centred cubic lattice the two functions $\gamma_{\mathbf{q}}$ and $\gamma_{\mathbf{q}}'$ are

$$\gamma_{\mathbf{q}}=\cos\tfrac{1}{2}aq_x\cos\tfrac{1}{2}aq_y\cos\tfrac{1}{2}aq_z:\qquad r=8$$

and

$$\gamma_{\mathbf{q}}'=\tfrac{1}{3}[\cos aq_x+\cos aq_y+\cos aq_z]:\quad r'=6. \tag{9.60}$$

The functions $\gamma_{\mathbf{q}}$, $\gamma_{\mathbf{q}}'$, ... defined by (9.59) all have the easily verified property that

$$\sum_{\mathbf{q}}\gamma_{\mathbf{q}-\mathbf{q}'}n_{\mathbf{q}}=\gamma_{\mathbf{q}'}\sum_{\mathbf{q}}\gamma_{\mathbf{q}}n_{\mathbf{q}}, \tag{9.61}$$

etc.

In view of this fact we can simplify the algebra involved in obtaining $\hbar\omega_{\mathbf{q}}$ and $\langle\hat{S}^{z}\rangle$ when only nearest-neighbour interactions are involved, though it is possible, of course, to carry through the analysis for an arbitrary range of $J(\mathbf{l})$.

For nearest-neighbour interactions only, (9.55) can be rewritten in the form

$$\hbar\omega_{\mathbf{q}}(T) = 2rJS(1-\gamma_{\mathbf{q}})\{1 - C(T)/S\} \tag{9.62}$$

where

$$C(T) = \frac{1}{N}\sum_{\mathbf{q}}(1-\gamma_{\mathbf{q}})n_{\mathbf{q}}. \tag{9.63}$$

It is plausible to regard (9.62) as the energy spectrum for linear spin waves with a temperature-dependent exchange integral $J(T)$, given by

$$J(T) = J\{1 - C(T)/S\}. \tag{9.64}$$

The factor $\{1 - C(T)/S\}$ is usually referred to as a renormalization factor.

For small $\mathbf{q}$ we can write for (9.62)

$$\hbar\omega_{\mathbf{q}}(T) = D(T)q^2, \tag{9.65}$$

where $D(T)$ is the D of eqn (9.19) multiplied by the renormalization factor. If we use (9.65) and also expand the factor $(1-\gamma_{\mathbf{q}})$ in (9.63) to the first non-vanishing order, the equation for $C(T)$ reads, to leading order,

$$C(T) \simeq \frac{a^2}{r}\frac{1}{N}\sum_{\mathbf{q}}\frac{q^2}{\exp\{\beta D(0)q^2\}-1} = \frac{a^2}{r}\frac{v_0}{2\pi^2}\int_0^\infty \mathrm{d}q\,\frac{q^4}{\exp\{\beta D(0)q^2\}-1}, \tag{9.66}$$

i.e.

$$C(T) \simeq \frac{v_0 a^2}{4\pi^2 r}\left\{\frac{k_{\mathrm{B}}T}{D(0)}\right\}^{5/2}\frac{3\sqrt{\pi}}{4}\zeta(\tfrac{5}{2}), \tag{9.67}$$

on using the Riemann zeta integral defined by eqn (9.32). Eqn (9.67) for $C(T)$ shows that $D(T)$ is approximately given by

$$D(T) = D\left\{1 - \frac{v_0 6a^2\pi}{rS}\left(\frac{k_{\mathrm{B}}T}{4\pi D}\right)^{5/2}\zeta(\tfrac{5}{2})\right\}. \tag{9.68}$$

We have therefore found that the dynamical interaction between the spin waves gives, to leading order, a temperature dependence to D such that it decreases with T as $T^{5/2}$. More generally, when the exchange interaction extends beyond nearest neighbours,

$$D(T) = D\left\{1 - \frac{v_0\overline{l^2}\pi}{S}\left(\frac{k_{\mathrm{B}}T}{4\pi D}\right)^{5/2}\zeta(\tfrac{5}{2})\right\} \tag{9.69}$$

where D is given by (9.20) and the moment $\overline{l^2}$ is defined by (9.22). The theory has been shown to provide a very satisfactory account of spin waves in EuO (Glinka *et al.* 1973).

Finally, we calculate the temperature dependence of $\langle\hat{S}^z\rangle$ from

(9.57). We need

$$\frac{1}{N}\sum_{\mathbf{q}} n_{\mathbf{q}} = \frac{1}{N}\sum_{\mathbf{q}} \{\exp(\hbar\omega_{\mathbf{q}}\beta)-1\}^{-1} \simeq v_0\left(\frac{k_B T}{4\pi D}\right)^{3/2} \zeta(\tfrac{3}{2})\{1-C(T)/S\}^{-3/2}$$

$$\simeq v_0\left(\frac{k_B T}{4\pi D}\right)^{3/2} \zeta(\tfrac{3}{2}) + \frac{v_0^2 9a^2\pi}{rS}\left(\frac{k_B T}{4\pi D}\right)^4 \zeta(\tfrac{3}{2})\zeta(\tfrac{5}{2}). \tag{9.70}$$

Eqn (9.70) tells us that the dynamical interaction introduces a T^4 term in the magnetization. This result is in accord with the conclusion reached by Dyson. A comparison of the coefficient of T^4 in (9.70) with that obtained by Dyson shows that we have only obtained part of the correct answer. In terms of scattering formalism, our result corresponds to the s-wave part of the correct coefficient; the remaining terms come from higher order partial waves and as such depend on the point-group symmetry of the crystal. Naturally, it is not surprising that our rather simple analysis has not given the full answer (Silberglitt and Harris 1968). The T^4 term in (9.70) when the exchange parameter extends beyond nearest neighbours is

$$v_0^2 \frac{3\overline{l^2}\pi}{2S}\left(\frac{k_B T}{4\pi D}\right)^4 \zeta(\tfrac{3}{2})\zeta(\tfrac{5}{2}). \tag{9.71}$$

We will not discuss multi-spin-wave scattering processes because their contribution to the cross-section is known to be small. It is worth remarking, however, that there is one major difference between multi-phonon and multi-magnon scattering. In the former all processes are permitted, i.e. a two-phonon process can involve the creation of two phonons, annihilation of two phonons, or creation of one and the annihilation of another. But in a multi-spin process the total z-component of spin can change at most by one unit; thus, for example, the two-spin-wave process in which two spin waves are annihilated or created is forbidden and only the process where one is created and one destroyed is allowed.

9.2.4. Magnon lifetime and bound states (Lovesey and Hood 1982)

To go beyond the approximation scheme used in § 9.2.3 in the discussion of two-spin-wave collisions and evaluate the lifetime of a spin wave it is very convenient to use thermal Green functions introduced in Chapter 8; (see Table 8.1 and eqn (8.49)).

In terms of the Boson operators introduced in the transformations (9.45) and (9.48), we define the Green functions $G_{\mathbf{q},\mathbf{q}'}(t)$ as

$$G_{\mathbf{q},\mathbf{q}'}(t) = -\mathrm{i}\theta(t)\langle[\hat{a}_{\mathbf{q}}(t), \hat{a}^{+}_{\mathbf{q}'}]\rangle. \tag{9.72}$$

Here $\theta(t)$ is the unit step function. The Fourier transform of $G_{\mathbf{q},\mathbf{q}'}(t)$

satisfies the equation-of-motion

$$\hbar\omega G_{\mathbf{q},\mathbf{q}'}(\omega) = \hbar\langle[\hat{a}_{\mathbf{q}}, \hat{a}^{+}_{\mathbf{q}'}]\rangle + \langle\langle[\hat{a}_{\mathbf{q}}, \hat{\mathscr{H}}]; \hat{a}^{+}_{\mathbf{q}'}\rangle\rangle = \hbar\,\delta_{\mathbf{q},\mathbf{q}'} + \langle\langle[\hat{a}_{\mathbf{q}}, \hat{\mathscr{H}}]; \hat{a}^{+}_{\mathbf{q}'}\rangle\rangle \quad (9.73)$$

where $\hat{\mathscr{H}}$ is given by (9.47). The value of studying $G_{\mathbf{q},\mathbf{q}'}(\omega)$ lies in the fact that its poles give the energies of the elementary excitations of the system described by $\hat{\mathscr{H}}$ (in our case magnons).

Clearly the Green function on the right-hand side of eqn (9.73) involves a Green function containing four operators. In fact for only nearest-neighbour coupling,

$$\langle\langle[\hat{a}_{\mathbf{q}}, \hat{\mathscr{H}}]; \hat{a}^{+}_{\mathbf{q}'}\rangle\rangle = 2rJS(1-\gamma_{\mathbf{q}})G_{\mathbf{q},\mathbf{q}'}(\omega)$$

$$-\frac{rJ}{N}\sum_{\mathbf{q}_1,\mathbf{q}_2,\mathbf{q}_3}\delta_{\mathbf{q}+\mathbf{q}_1,\mathbf{q}_2+\mathbf{q}_3}\{\gamma_{\mathbf{q}-\mathbf{q}_2} - 2\gamma_{\mathbf{q}_2} + \gamma_{\mathbf{q}_1-\mathbf{q}_2}\}\langle\langle\hat{a}^{+}_{\mathbf{q}_1}\hat{a}_{\mathbf{q}_2}\hat{a}_{\mathbf{q}_3}; \hat{a}^{+}_{\mathbf{q}'}\rangle\rangle. \quad (9.74)$$

If the second term on the right-hand side of (9.74) is neglected altogether, then

$$G_{\mathbf{q},\mathbf{q}'}(\omega) = \hbar\,\delta_{\mathbf{q},\mathbf{q}'}\{\hbar\omega - 2rJS(1-\gamma_{\mathbf{q}})\}^{-1} \quad (9.75)$$

and the poles of $G_{\mathbf{q},\mathbf{q}'}(\omega)$ coincide with the energy of a magnon in the linear approximation, i.e. eqn (9.16) with $H=0$.

We now focus our attention on the approximate calculation of the Green function in the second term on the right-hand side of (9.74). A change of variables reveals a separation of the interaction term. Define wave vectors $\mathbf{K}$ and $\mathbf{Q}$ through the relations

$$\mathbf{q}_2 = \tfrac{1}{2}\mathbf{K} + \mathbf{Q} \quad \text{and} \quad \mathbf{q}_3 = \tfrac{1}{2}\mathbf{K} - \mathbf{Q} \quad (9.76)$$

and, for ease of notation in subsequent working, write the Green function

$$A_{\mathbf{K}}(\mathbf{Q}) = \langle\langle\hat{a}^{+}_{\mathbf{q}_1}\hat{a}_{\mathbf{q}_2}\hat{a}_{\mathbf{q}_3}; \hat{a}^{+}_{\mathbf{q}}\rangle\rangle \quad (9.77)$$

where the dependence on $\mathbf{q}_1$ and $\mathbf{q}$ is suppressed for the moment. Using the identity (9.61) we then find

$$-\frac{rJ}{N}\sum_{\mathbf{q}_1,\mathbf{q}_2,\mathbf{q}_3}\delta_{\mathbf{q}+\mathbf{q}_1,\mathbf{q}_2+\mathbf{q}_3}\{\gamma_{\mathbf{q}-\mathbf{q}_2} - 2\gamma_{\mathbf{q}_2} + \gamma_{\mathbf{q}_1-\mathbf{q}_2}\}A_{\mathbf{K}}(\mathbf{Q})$$

$$= \frac{2rJ}{N}\sum_{\mathbf{K},\mathbf{q}_1}\delta_{\mathbf{q}+\mathbf{q}_1,\mathbf{K}}\{\gamma_{\mathbf{K}/2} - \gamma_{\mathbf{q}-\mathbf{K}/2}\}\sum_{\mathbf{Q}}\gamma_{\mathbf{Q}}\,A_{\mathbf{K}}(\mathbf{Q}), \quad (9.78)$$

which shows that the required quantity is $\sum \gamma_{\mathbf{Q}} A_{\mathbf{K}}(\mathbf{Q})$.

The equation-of-motion for the Green function (9.77) is constructed from (8.49). The quartic term in the Hamiltonian (9.47) generates the following, higher-order, Green function

$$\langle\langle[\hat{a}^{+}_{\mathbf{q}_1}\hat{a}_{\mathbf{q}_2}\hat{a}_{\mathbf{q}_3},\, \hat{a}^{+}_{\mathbf{p}_1}\hat{a}^{+}_{\mathbf{p}_2}\hat{a}_{\mathbf{p}_3}\hat{a}_{\mathbf{p}_4}];\, \hat{a}^{+}_{\mathbf{q}}\rangle\rangle$$
$$=\langle\langle\hat{a}^{+}_{\mathbf{q}_1}[\hat{a}_{\mathbf{q}_2}\hat{a}_{\mathbf{q}_3},\, \hat{a}^{+}_{\mathbf{p}_1}\hat{a}^{+}_{\mathbf{p}_2}]\hat{a}_{\mathbf{p}_3}\hat{a}_{\mathbf{p}_4};\, \hat{a}^{+}_{\mathbf{q}}\rangle\rangle+\langle\langle\hat{a}^{+}_{\mathbf{p}_2}\hat{a}^{+}_{\mathbf{p}_3}[\hat{a}^{+}_{\mathbf{q}_1},\, \hat{a}_{\mathbf{p}_3}\hat{a}_{\mathbf{p}_4}]\hat{a}_{\mathbf{q}_2}\hat{a}_{\mathbf{q}_3};\, \hat{a}^{+}_{\mathbf{q}}\rangle\rangle \quad (9.79)$$

where the development exploits the identity

$$[\hat{A},\, \hat{B}\hat{C}]\equiv[\hat{A},\, \hat{B}]\hat{C}+\hat{B}[\hat{A},\, \hat{C}]. \quad (9.80)$$

At low temperatures the most important contribution on the right-hand side is made by the first term. For if we approximate the latter by

$$\langle[\hat{a}_{\mathbf{q}_2}\hat{a}_{\mathbf{q}_3},\, \hat{a}^{+}_{\mathbf{p}_1}\hat{a}^{+}_{\mathbf{p}_2}]\rangle\langle\langle\hat{a}^{+}_{\mathbf{q}_1}\hat{a}_{\mathbf{p}_3}\hat{a}_{\mathbf{p}_4};\, \hat{a}^{+}_{\mathbf{q}}\rangle\rangle, \quad (9.81)$$

we observe that the Green function is of the same form as $A_{\mathbf{k}}(\mathbf{Q})$, required in (9.78), and the static correlation function is the weight of two magnon creation followed by annihilation. The terms omitted in (9.79), when it is approximated by (9.81), give energy shifts and damping in $A_{\mathbf{K}}(\mathbf{Q})$. The static correlation function in (9.81) is evaluated with (9.80) and, for $T=0$, we find

$$\langle[\hat{a}_{\mathbf{q}_2}\hat{a}_{\mathbf{q}_3},\, \hat{a}^{+}_{\mathbf{p}_1}\hat{a}^{+}_{\mathbf{p}_2}]\rangle=\langle[\hat{a}_{\mathbf{q}_2}\hat{a}_{\mathbf{q}_3},\, \hat{a}^{+}_{\mathbf{p}_1}]\hat{a}^{+}_{\mathbf{p}_2}\rangle=\delta_{\mathbf{p}_1,\mathbf{q}_2}\,\delta_{\mathbf{p}_2,\mathbf{q}_3}+\delta_{\mathbf{p}_1,\mathbf{q}_3}\,\delta_{\mathbf{p}_2,\mathbf{q}_2};\qquad T=0.$$

With these approximations, the equation-of-motion for $A_{\mathbf{K}}(\mathbf{Q})$ reduces to a linear integral equation, namely,

$$\{\hbar\omega-E(\mathbf{q}_1,\mathbf{q}_2,\mathbf{q}_3)\}A_{\mathbf{K}}(\mathbf{Q})=\hbar\langle[\hat{a}^{+}_{\mathbf{q}_1}\hat{a}_{\mathbf{q}_2}\hat{a}_{\mathbf{q}_3},\, \hat{a}^{+}_{\mathbf{q}}]\rangle+\frac{2rJ}{N}\{\gamma_{\mathbf{K}/2}-\gamma_{\mathbf{Q}}\}\sum_{\mathbf{Q}'}\gamma_{\mathbf{Q}'}A_{\mathbf{K}}(\mathbf{Q}') \quad (9.82)$$

where

$$E(\mathbf{q}_1,\mathbf{q}_2,\mathbf{q}_3)=\hbar(\omega_{\mathbf{q}_2}+\omega_{\mathbf{q}_3}-\omega_{\mathbf{q}_1})$$

and

$$\hbar\omega_{\mathbf{q}}=2rJS(1-\gamma_{\mathbf{q}}). \quad (9.83)$$

Eqn (9.82) is solved for $\sum\gamma_{\mathbf{Q}}A_{\mathbf{K}}(\mathbf{Q})$, which is required in (9.78), by multiplying through by $\gamma_{\mathbf{Q}}$, and performing the sum on $\mathbf{Q}$. The static correlation function in (9.82) is evaluated with (9.80). The solution involves the temperature-independent function,

$$W_{\mathbf{q}_1}(\mathbf{K},\omega)=\frac{2rJ}{N}\sum_{\mathbf{Q}}\gamma_{\mathbf{Q}}(\gamma_{\mathbf{K}/2}-\gamma_{\mathbf{Q}})\{\hbar\omega-E(\mathbf{q}_1,\tfrac{1}{2}\mathbf{K}+\mathbf{Q},\tfrac{1}{2}\mathbf{K}-\mathbf{Q})\}^{-1} \quad (9.84)$$

and the result is

$$\{1-W_{\mathbf{q}_1}(\mathbf{K},\omega)\}\sum_{\mathbf{Q}}\gamma_{\mathbf{Q}}A_{\mathbf{K}}(\mathbf{Q})=2n_{\mathbf{q}_1}\gamma_{\mathbf{q}-\mathbf{K}/2}\,\delta_{\mathbf{q}+\mathbf{q}_1,\mathbf{K}}/(\omega-\omega_{\mathbf{q}}).$$

The final step is to recognize the non-interacting Green function (9.75) on the right-hand side, which is then replaced by $G_{\mathbf{q},\mathbf{q}}(\omega)$ that includes the effects of the magnon interactions; whence

$$\{1 - W_{\mathbf{q}_1}(\mathbf{K}, \omega)\} \sum_{\mathbf{Q}} \gamma_{\mathbf{Q}} A_{\mathbf{K}}(\mathbf{Q}) = 2n_{\mathbf{q}_1} \gamma_{\mathbf{q}-\mathbf{K}/2}\, \delta_{\mathbf{q}+\mathbf{q}_1,\mathbf{K}} G_{\mathbf{q},\mathbf{q}}(\omega). \tag{9.85}$$

The result for $G_{\mathbf{q},\mathbf{q}}(\omega)$ obtained from (9.78) with (9.85) is written in the form

$$\{\hbar\omega - \hbar\omega_{\mathbf{q}} - \Sigma_{\mathbf{q}}(\omega)\} G_{\mathbf{q},\mathbf{q}}(\omega) = \hbar \tag{9.86a}$$

where the self-energy,

$$\Sigma_{\mathbf{q}}(\omega) = \frac{4rJ}{N} \sum_{\mathbf{p},\mathbf{K}} n_{\mathbf{p}}\, \delta_{\mathbf{p}+\mathbf{q},\mathbf{K}}\, \gamma_{\mathbf{q}-\mathbf{K}/2} \{\gamma_{\mathbf{K}/2} - \gamma_{\mathbf{q}-\mathbf{K}/2}\} \{1 - W_{\mathbf{p}}(\mathbf{K}, \omega)\}^{-1}. \tag{9.86b}$$

Notice that the temperature dependence of the self-energy arises from the Bose occupation function which appears explicitly in (9.86b), and that the self-energy vanishes in the limit $T \to 0$. For temperatures $T \ll (J/k_{\mathrm{B}})$ the Bose function is very sharply peaked at $\mathbf{p} = 0$ because $\omega_{\mathbf{p}}$ vanishes at the zone centre, in the absence of an external field. Consequently, for $T \to 0$, the self-energy is evaluated by taking the $\mathbf{p} \to 0$ limit of the expression multiplying the Bose function.

We will show that there are solutions of

$$1 - W_0(\mathbf{q}, \omega) = 0. \tag{9.87}$$

Therefore, there are resonant contributions to the self-energy, (9.86b). These contributions arise from the excitation of a bound state of two magnons. Indeed, (9.87) is the exact equation for the dispersion of a two-magnon bound state (Mattis 1981). The self-energy (9.86b) is not a benign function of the frequency and wave vector, and the pronounced structure as a function of ω for a given $\mathbf{q}$ is reflected in the Green function and, therefore, the one-magnon cross-section. However, this structure is confined to the zone boundary because the two-magnon bound state is damped at long wavelengths through interactions with the two-magnon continuum (Silberglitt and Harris 1968).

A general study of (9.87) is lengthy (Mattis 1981; Silberglitt and Harris 1968), and we will content ourselves with an explicit calculation for a simple cubic lattice with $\mathbf{q}$ at the zone boundary. In this instance, $\gamma_{\mathbf{q}/2} = 0$,

$$E(0, \tfrac{1}{2}\mathbf{q} + \mathbf{Q}, \tfrac{1}{2}\mathbf{q} - \mathbf{Q}) = 4rJS,$$

and (9.87) reduces to

$$1+\frac{2rJ}{N}\sum_{\mathbf{Q}}\gamma_{\mathbf{Q}}^2/(\hbar\omega-4rJS)=0,$$

or

$$\hbar\omega=4rJS\left\{1-\frac{1}{2SN}\sum_{\mathbf{Q}}\gamma_{\mathbf{Q}}^2\right\}$$
$$=4rJS\{1-1/2rS\}$$

where $r=6$ for a simple cubic lattice. Hence, the resonant contribution at the zone boundary occurs at an energy which is just below the corresponding magnon energy. Because the separation of the magnon and bound-state energies is proportional to $1/S$, the resonant contribution is most significant for low-spin systems. For arbitrary values of $\mathbf{q}$, the solutions must be obtained by numerical analysis even for a simple cubic lattice. At the zone boundary the imaginary part of $W_0(\mathbf{q},\omega)$ is zero, and hence the bound-state solution is a true resonance. However, for smaller $\mathbf{q}$ the imaginary part of $W_0(\mathbf{q},\omega)$ is non-zero, and the structure from the bound state is less significant, and quite negligible at long wavelengths.

For the latter we can legitimately study the self-energy as an expansion in $W_{\mathbf{p}}(\mathbf{K},\omega)$. The first term is independent of the frequency, namely,

$$\Sigma_{\mathbf{q}}^{(0)}(\omega)=\frac{4rJ}{N}\sum_{\mathbf{K},\mathbf{p}}n_{\mathbf{p}}\,\delta_{\mathbf{p}+\mathbf{q},\mathbf{K}}\,\gamma_{\mathbf{q}-\mathbf{K}/2}\{\gamma_{\mathbf{K}/2}-\gamma_{\mathbf{q}-\mathbf{K}/2}\}$$
$$=-\frac{2rJ}{N}(1-\gamma_{\mathbf{q}})\sum_{\mathbf{p}}n_{\mathbf{p}}(1-\gamma_{\mathbf{p}}),$$

which is the Hartree–Fock correction to the magnon frequency (cf. (9.63)).

The next term, proportional to $W_{\mathbf{p}}(\mathbf{K},\omega)$, gives rise to magnon damping. Given that the self-energy is a correction to the magnon energy, we evaluate the self-energy for $\omega=\omega_{\mathbf{q}}$. Defining a damping constant,

$$\Gamma(\mathbf{q})=-\mathrm{Im}\,\Sigma_{\mathbf{q}}^{(1)}(\omega_{\mathbf{q}}+\mathrm{i}0^+),$$

together with

$$\mathrm{Im}\{\hbar\omega+\mathrm{i}0^+-E(\mathbf{q}_1,\mathbf{q}_2,\mathbf{q}_3)\}^{-1}=-\pi\,\delta\{\hbar\omega-E(\mathbf{q}_1,\mathbf{q}_2,\mathbf{q}_3)\}$$

and

$$\delta(ax)=\frac{1}{a}\delta(x)\quad(a>0),$$

we obtain the result

$$\Gamma(\mathbf{q})=\frac{\pi rJ}{SN^2}\sum_{\mathbf{q}_1,\mathbf{q}_2}n_{\mathbf{q}_1}\delta(\gamma_{\mathbf{q}_2}+\gamma_{\mathbf{q}+\mathbf{q}_1-\mathbf{q}_2}-\gamma_{\mathbf{q}}-\gamma_{\mathbf{q}_1})\{\gamma_{\mathbf{q}-\mathbf{q}_2}+\gamma_{\mathbf{q}_1-\mathbf{q}_2}-\gamma_{\mathbf{q}}-\gamma_{\mathbf{q}_1}\}^2.\tag{9.88}$$

Note that the presence of the δ-function allows us to write

$$\{\gamma_{\mathbf{q}-\mathbf{q}_2}+\gamma_{\mathbf{q}_1-\mathbf{q}_2}-\gamma_{\mathbf{q}_2}-\gamma_{\mathbf{q}+\mathbf{q}_1-\mathbf{q}_2}\}\rightarrow\{\gamma_{\mathbf{q}-\mathbf{q}_2}+\gamma_{\mathbf{q}_1-\mathbf{q}_2}-\gamma_{\mathbf{q}}-\gamma_{\mathbf{q}_1}\}.$$

At low temperatures the Bose factor $n_{\mathbf{q}_1}$ makes the dominant contribution from the sum over $\mathbf{q}_1$ come from small $\mathbf{q}_1$. Since we have specified that $\mathbf{q}$ is small, the δ-function requires that $\mathbf{q}_2$ should also be small. Thus we require to expand $\gamma_{\mathbf{q}_1-\mathbf{q}_2}$ say, for small $\mathbf{q}_1$ and $\mathbf{q}_2$. For a simple cubic crystal with a unit cell of side a,

$$\gamma_{\mathbf{q}_1-\mathbf{q}_2}\simeq 1-\frac{a^2}{6}(q_1^2+q_2^2-2\mathbf{q}_1\cdot\mathbf{q}_2).$$

On making use of this expansion we have from (9.88) for simple cubic crystals

$$\Gamma(\mathbf{q})\simeq\frac{2\pi Ja^2}{SN^2}\sum_{\mathbf{q}_1,\mathbf{q}_2}n_{\mathbf{q}_1}(\mathbf{q}\cdot\mathbf{q}_1)^2\,\delta\{q_2^2-\mathbf{q}_2\cdot(\mathbf{q}+\mathbf{q}_1)+\mathbf{q}\cdot\mathbf{q}_1\}.$$

We now make use of the result, valid for three dimensions,

$$\frac{1}{N}\sum_{\mathbf{q}_2}\delta\{q_2^2-\mathbf{q}_2\cdot(\mathbf{q}+\mathbf{q}_1)+\mathbf{q}\cdot\mathbf{q}_1\}=\frac{v_0}{8\pi^2}|\mathbf{q}-\mathbf{q}_1|$$

to obtain

$$\Gamma(\mathbf{q})\simeq\left(\frac{Ja^2v_0}{4\pi S}\right)\frac{1}{N}\sum_{\mathbf{q}_1}n_{\mathbf{q}_1}(\mathbf{q}\cdot\mathbf{q}_1)^2\,|\mathbf{q}-\mathbf{q}_1|.$$

Although this result has been derived for the specific case of a simple cubic lattice it is in fact correct (within the initial assumptions) for all the cubic Bravais lattices (Cooke and Gersch 1967).

Proceeding as in § 9.2.1 we evaluate $\Gamma(\mathbf{q})$ to leading order in temperature and find

$$\Gamma(\mathbf{q})\simeq\left(\frac{6}{\nu r}\right)^3\left(\frac{J}{S}\right)(aq)^3\left\{\frac{k_\mathrm{B}T}{8\pi JS}\right\}^{5/2}\zeta(\tfrac{5}{2});\qquad \hbar\omega_\mathbf{q}\gg k_\mathrm{B}T \tag{9.89}$$

where

$$\nu=\begin{cases}1 & \text{s.c.}\\ 2^{1/3} & \text{f.c.c.}\\ 3\times 2^{-4/3} & \text{b.c.c.}\end{cases}$$

Thus at low temperatures and long wavelengths the damping of a spin wave is proportional to $q^3T^{5/2}$.

In neutron-scattering experiments the condition $\hbar\omega_\mathbf{q}\gg k_\mathrm{B}T$ is nearly always satisfied. However, it is worthwhile to note the form of the damping Γ in the opposite limit $\hbar\omega_\mathbf{q}\ll k_\mathrm{B}T$. In this instance Harris

(1968, 1969) has shown that

$$\Gamma(\mathbf{q}) \propto q^4 T^2 \left\{ \tfrac{1}{6} \ln^2\left(\frac{k_B T}{2JSa^2q^2}\right) + \tfrac{5}{9} \ln\left(\frac{k_B T}{2JSa^2q^2}\right) - 0.05 \right\}. \tag{9.90}$$

A knowledge of the Green function $G_{\mathbf{q},\mathbf{q}}(\omega)$ does not provide a complete expression for the transverse one-magnon cross-section, since the latter contains the spin correlation function

$$\langle \hat{S}_{\mathbf{l}}^- \hat{S}_{\mathbf{l}'}^+(t)\rangle = 2S\langle \hat{a}_{\mathbf{l}}^+ \hat{a}_{\mathbf{l}'}(t)\rangle - \langle \hat{a}_{\mathbf{l}}^+ \hat{a}_{\mathbf{l}}^+ \hat{a}_{\mathbf{l}} \hat{a}_{\mathbf{l}'}(t)\rangle.$$

The correlation function involving four Bose operators is calculated by defining an appropriate Green function and following the argument which leads to (9.86). The result is proportional to $G_{\mathbf{q},\mathbf{q}}(\omega)$, which means that the same resonant structure is present in both correlation functions in $\langle \hat{S}^- \hat{S}^+ \rangle$.

The calculations described are based on the Boson Hamiltonian (9.47), which does not take account of the kinematical interaction. The modifications introduced by the latter are insignificant for $T \ll (J/k_B)$ (Cooke and Hahn 1970), as might be expected. However, the two-spin correlation functions in the one-magnon cross-sections are proportional to the magnetization, $\langle \hat{S}^z \rangle$, which vanishes at the critical point. This feature is not contained in the approximate results derived here, since they are low-temperature approximations, and we might anticipate kinematic interactions to be important in the dynamic response when $\langle \hat{S}^z \rangle$ deviates substantially from its saturation value.

Our discussion and calculations thus far are valid for three-dimensional ferromagnets in the limit of low temperatures. However, several highly anisotropic compounds are known which display features expected of one-dimensional magnets. Perhaps the most telling feature is that the neutron diffraction pattern is constant in planes perpendicular to the axis of the magnetic chains, rather than exhibiting a sequence of Bragg peaks characteristic of a three-dimensional magnetic structure. Three-dimensional ordering will occur at a sufficiently low temperature, and the onset of the ordering can be monitored through the diffraction pattern. Quasi-one-dimensional magnets are discussed in Chapter 13. Here we draw attention to some interesting aspects of the magnon interaction in one dimension.

The result (9.88) for the magnon damping constant is not defined in one dimension, as may be readily verified. To understand this behaviour we return to (9.87) for the dispersion of the two-magnon bound state, and observe that, for one dimension, the bound-state energy is less than the magnon energy for all wave vectors (Mattis 1981). Hence, the magnon interaction, described by the quartic term in (9.47), is not a small perturbation for any value of the wave vector, and a simple perturbative approach is inadequate.

The result (9.86b), from which (9.87) is derived, corresponds to an infinite-order perturbation theory, with terms characterized by the power of $W_{\mathbf{p}}(\mathbf{K}, \omega)$. The self-energy is finite and well behaved in one dimension because the two-magnon bound state is properly accounted for in the theory. The function $W_{\mathbf{p}}(\mathbf{K}, \omega)$ can be evaluated analytically, and the remaining integration in the self-energy is straightforward numerically.

In the limit $T \ll (J/k_B)$ the integral is accomplished analytically by expanding the integrand about $\mathbf{p} = 0$. For wave vectors not close to the zone boundary,

$$\Gamma(q) = \frac{2.16J}{S}\sin(aq)\left(\frac{k_B T}{4JS}\right)^{3/2}; \qquad aq < \pi, \qquad k_B T \ll J \qquad (9.91)$$

so that $\Gamma(q)$ is linear in q near the zone centre, cf. (9.89). The zone-boundary value of the damping is proportional to $\{T/(S-1)\}^2$, and therefore highly spin-dependent (Glaus, Lovesey, and Stoll 1983). Numerical results confirm that, for the extreme values $S = \frac{1}{2}$ and $S \to \infty$ (classical spins), the damping is a minimum at the zone boundary, whereas for $S = 1$ there is a large maximum.

9.3. Itinerant-electron model of ferromagnetism†

The Heisenberg model of ferromagnetism discussed in the preceding section represents the localized theory of magnetism in solids. It is now our task to discuss the itinerant or band theory of ferromagnetism. Such a theory would certainly seem more appropriate than the Heisenberg model for the transition-metal elements and their alloys. Our discussion is designed to bring out the fundamental aspects of the itinerant model and we shall for this purpose adopt many approximations that are open to argument. In particular, we consider only a single band of electrons. This might well seem odd in view of the fact that we always have in the back of our minds the transition metals with their five d bands. However, the arguments we present below for a single band differ little from those that would have to be employed in the case of degenerate bands, and the latter are algebraically more complicated (Cooke *et al.* 1980).

We must first derive a Hamiltonian that we believe incorporates the salient features of electrons in a narrow energy band. Having done this we can investigate its properties, paying particular regard to the possibility of a collective mode of electron spin propagating in the band, i.e. to

† The author is indebted to Dr J. Hubbard for access to unpublished lectures on ferromagnetism in narrow energy bands given at Advanced Nato Summer School on Magnetism, McGill University, Canada (1967). Director, A. J. Freeman.

the existence of a spin wave or magnon. Before embarking on this scheme we ask what physically we must expect an itinerant-electron model of ferromagnetism to contain. Experiments show that the d electrons of transition metals have a behaviour that in some instances makes them appear well localized and in other cases itinerant. For example, long-wavelength spin waves in ferromagnetic metals obey the fundamental predictions of the Heisenberg ferromagnet, while, on the other hand, the large d-electron contribution to the low-temperature specific heat and the occurrence of magnetic moments per atom that are far from integral numbers of Bohr magnetons μ_B are properties that are explainable by band theory. The key to understanding these seemingly opposing characteristics of localized and itinerant electron states is correlation between the electrons (Olés and Stollhoff 1984). For if, in spite of their band motion, the d electrons on an atom are strongly correlated with each other but only weakly with electrons on other atoms, then such intra-atomic correlation leads inevitably to a behaviour that is to some degree characteristic of an atomic electron model. We shall in fact find that our approximate Hamiltonian for electrons in a narrow band contains a repulsive interaction that operates between electrons of *opposite* spin on the same atom. The correlation caused by this 'on-site' repulsion is that which leads to the possibility of a ferromagnetic state and the existence of spin-wave modes.

A calculation of the cross-section for a realistic model of itinerant magnetism is therefore a formidable task, since it requires both a complete band-structure calculation, and a prescription for the strong electron correlations. Here we direct our attention to the consequences of the electron correlations. Although this can be achieved with a simplified band structure, to the extent of demonstrating the existence of a spin wave and its damping at short wavelengths, some features are absent which are important for a detailed confrontation of theory and experiment. Extensive, calculations from first principles for realistic models of nickel and iron, at zero temperature, demonstrate the importance of the band structure in the spin dynamics through matrix elements that depend on the band index and wave vector. The importance of these band-structure effects is borne out by comparison of numerical results with experimental findings (Cooke *et al.* 1980; Mook and Paul 1985; Callaway *et al.* 1983). A striking effect, not present in a single-band model, is the existence of an optic spin wave (Blackman *et al.* 1985). According to the calculations, the acoustic spin wave can show strong dispersion and damping at short wavelengths caused by a repulsion with the optic spin wave. Hence, spin-wave damping in metals can be ascribed to an interaction with the continuum of spin-flip, or Stoner, modes, or an interaction with an optic spin wave created from the Stoner modes by strong electron correlations. The intensity of the optic spin wave in the neutron cross-section vanishes in the limit of long wavelengths. This behaviour is required by a sum rule which shows that, in the limit of long wavelengths, all the intensity is contained in the acoustic spin-wave branch.

The temperature dependence of spin waves in iron and nickel has been studied extensively by Lynn (1975) and Lynn and Mook (1981). An unusual polarization dependence of the transverse neutron cross-section observed by Lowde *et al.* (1983) has been attributed to the itinerancy of the unpaired electrons. Above the critical temperature, the spectrum of scattered neutrons for a fixed wave vector is a Lorentizian function of ω, centred at $\omega = 0$, for long wavelengths (Wicksted *et al.* 1984; Martinez *et al.* 1985). The width of the Lorentzian vanishes in the long wavelength limit. This result, together with an Ornstein–Zernike form for the wave vector dependent susceptibility, means that the spectra of scattered neutrons for constant wave vector and constant energy transfer are distinctly different (Lynn and Mook 1981), because the constant-ω spectrum will vanish at long wavelengths. Multi-phonon scattering is significant in these studies of high T_c magnets and it is troublesome in data analysis, so much so that polarization analysis, with the attendant intensity penalty, has been extensively utilized (Wicksted *et al.* 1984; Martinez *et al.* 1985; Mook and Lynn 1985).

9.3.1. Hubbard Hamiltonian (Hubbard 1963)

The Hamiltonian of electrons in a solid is of the form

$$\hat{\mathscr{H}} = \sum_i \frac{1}{2m_e}\hat{\mathbf{p}}_i^2 + \hat{V}(\mathbf{r}_i) + \sum_{i \neq j} \frac{e^2}{|\mathbf{r}_i - \mathbf{r}_j|}. \tag{9.92}$$

Here $\hat{\mathbf{p}}_i$ and $\mathbf{r}_i$ are, respectively, the momentum operator and position vector of the ith electron. The first part of (9.92) consists of the kinetic energy of the ith electron and the periodic crystal potential $\hat{V}(\mathbf{r})$ due to the ion cores. As we noted in Chapter 7, the eigenfunctions of this part of the Hamiltonian, the band-theory Hamiltonian, are Bloch functions $\varphi_{\mathbf{k}\sigma}(\mathbf{r})$. The second part of (9.92) is the Coulomb interaction of the electrons. It is this part that we seek to approximate and to do this we exploit the fact that the band is very narrow.

As in Chapter 7, we employ the second quantization formalism to write $\hat{\mathscr{H}}$ in the form

$$\hat{\mathscr{H}} = \sum_\sigma \sum_{\mathbf{l},\mathbf{l}'} T(\mathbf{l}-\mathbf{l}')\hat{c}^+_{\mathbf{l}\sigma}\hat{c}_{\mathbf{l}'\sigma} + \tfrac{1}{2}\sum_{\mathbf{l},\mathbf{m}}\sum_{\mathbf{l}',\mathbf{m}'}\sum_{\sigma,\sigma'} (\mathbf{lm}\,|1/r|\,\mathbf{l}'\mathbf{m})\hat{c}^+_{\mathbf{l}\sigma}\hat{c}^+_{\mathbf{m}\sigma'}\hat{c}_{\mathbf{m}'\sigma'}\hat{c}_{\mathbf{l}'\sigma}. \tag{9.93}$$

We recall that the matrix $T(\mathbf{l}-\mathbf{l}')$ is

$$T(\mathbf{l}-\mathbf{l}') = \int \mathrm{d}\mathbf{r}\, \phi^*(\mathbf{r}-\mathbf{l})\left\{\frac{1}{2m_e}\hat{\mathbf{p}}^2 + \hat{V}(\mathbf{r})\right\}\phi(\mathbf{r}-\mathbf{l}')$$

and

$$(\mathbf{lm}\,|1/r|\,\mathbf{l'm'}) = e^2\int \mathrm{d}\mathbf{r}\int \mathrm{d}\mathbf{r}'\,\frac{\phi^*(\mathbf{r}-\mathbf{l})\phi(\mathbf{r}-\mathbf{l}')\phi^*(\mathbf{r}'-\mathbf{m})\phi(\mathbf{r}'-\mathbf{m}')}{|\mathbf{r}-\mathbf{r}'|}. \tag{9.94}$$

In (9.94) the $\phi(\mathbf{r}-\mathbf{l})$ are a set of orthonormal wave functions centred about the lattice sites $\mathbf{l}$. The operators $\hat{c}^+_{\mathbf{l}\sigma}$ and $\hat{c}_{\mathbf{l}\sigma}$, respectively create and destroy an electron in the state $\phi(\mathbf{r}-\mathbf{l})$ with spin σ.

In a narrow energy band, the overlap of the functions ϕ centred about different sites is small. Therefore, the largest of the matrix elements $(\mathbf{lm}\,|1/r|\,\mathbf{l'm'})$ is that with $\mathbf{l}=\mathbf{m}=\mathbf{l}'=\mathbf{m}'$; we denote this matrix element by I and neglect all others in (9.93). Hence our approximate Hamiltonian is

$$\mathscr{H} = \sum_{\mathbf{l},\mathbf{l}',\sigma} T(\mathbf{l}-\mathbf{l}')\hat{c}^+_{\mathbf{l}\sigma}\hat{c}_{\mathbf{l}'\sigma} + \tfrac{1}{2}I\sum_{\mathbf{l},\sigma,\sigma'} \hat{c}^+_{\mathbf{l}\sigma}\hat{c}^+_{\mathbf{l}\sigma'}\hat{c}_{\mathbf{l}\sigma'}\hat{c}_{\mathbf{l}\sigma}. \tag{9.95}$$

The sum over the four operators in the second term of (9.95) can be simplified by the following manipulations. Remembering that the $\hat{c}$s are Fermion operators,

$$\hat{c}^+_{\sigma}\hat{c}^+_{\sigma'}\hat{c}_{\sigma'}\hat{c}_{\sigma} = -\hat{c}^+_{\sigma}\hat{c}^+_{\sigma'}\hat{c}_{\sigma}\hat{c}_{\sigma'} = -\hat{c}^+_{\sigma}(\delta_{\sigma,\sigma'} - \hat{c}_{\sigma}\hat{c}^+_{\sigma'})\hat{c}_{\sigma'}.$$

Thus,

$$\begin{aligned}\sum_{\sigma,\sigma'} \hat{c}^+_{\sigma}\hat{c}^+_{\sigma'}\hat{c}_{\sigma'}\hat{c}_{\sigma} &= -\hat{n}_{\uparrow} - \hat{n}_{\downarrow} + \hat{n}_{\uparrow}\hat{n}_{\downarrow} + \hat{n}_{\downarrow}\hat{n}_{\downarrow} + \hat{n}_{\uparrow}\hat{n}_{\uparrow} + \hat{n}_{\downarrow}\hat{n}_{\uparrow}\\ &= \hat{n}_{\uparrow}\hat{n}_{\downarrow} + \hat{n}_{\downarrow}\hat{n}_{\uparrow}\\ &\equiv \sum_{\sigma} \hat{n}_{\sigma}\hat{n}_{-\sigma}.\end{aligned} \tag{9.96}$$

In reaching the second line of (9.96) we have made use of the property

$$\hat{n}_{\sigma}\hat{n}_{\sigma} = \hat{c}^+_{\sigma}\hat{c}_{\sigma}\hat{c}^+_{\sigma}\hat{c}_{\sigma} = \hat{c}^+_{\sigma}\hat{c}_{\sigma} = \hat{n}_{\sigma}.$$

Using (9.96) in (9.95) we finally obtain the desired result

$$\hat{\mathscr{H}} = \sum_{\mathbf{l},\mathbf{l}'\sigma} T(\mathbf{l}-\mathbf{l}')\hat{c}^+_{\mathbf{l}\sigma}\hat{c}_{\mathbf{l}'\sigma} + \tfrac{1}{2}I\sum_{\mathbf{l},\sigma} \hat{n}_{\mathbf{l}\sigma}\hat{n}_{\mathbf{l}-\sigma}. \tag{9.97}$$

The first term in (9.97) is

$$\sum_{\mathbf{k},\sigma} \mathscr{E}_{\mathbf{k}}\hat{c}^+_{\mathbf{k}\sigma}\hat{c}_{\mathbf{k}\sigma} \tag{9.98}$$

and represents the band energy. The second term represents the Coulomb interaction of electrons of opposite spin at the same site only.

In deriving (9.97) we have neglected all the matrix elements

$$(\mathbf{lm}\,|1/r|\,\mathbf{l'm'})$$

except the one for which $\mathbf{l}=\mathbf{m}=\mathbf{l}'=\mathbf{m}'$ and it is only reasonable to question how good such an approximation is. First, I has the order of magnitude 20 eV for 3d electrons in transition metals, and these have a bandwidth of order 4–5 eV. The largest of the neglected terms is $(\mathbf{lm}\,|1/r|\,\mathbf{lm})$, where $\mathbf{l}$ and $\mathbf{m}$ are nearest-neighbour lattice sites. These integrals are of order 6 eV, but this figure is reduced by the screening of the interactions of electrons on different atoms by the conduction-band electrons. It is probable that a more realistic figure for these matrix elements is 2–3 eV, and therefore neglecting them in comparison with I would appear quite reasonable. There are two effects, however, that will certainly go some way to reducing I in an actual metal from the figure of 20 eV. The first of these effects arises from the production of a correlation hole in the conduction-electron gas about an atom with an extra d electron. The presence of the correlation hole reduces the electrostatic potential at the atom by about 5 eV, which is the same as reducing I by the same figure. Also, correlation between the d electrons at any site will go towards reducing I. The total reduction of I by these effects is a matter of some debate, but it is probable that if we choose an 'effective' I of order 5 eV, this would not be too far from the truth. We therefore regard the Hamiltonian (9.97) as an effective Hamiltonian for the description of electrons in a narrow energy band and investigate, in a simple manner, some of its consequences, treating I as an adjustable parameter of the order of several eV.

The interaction term in (9.97) is the product of four Fermion operators and thus to derive a linear equation of motion for $\hat{c}_{\mathbf{k}\sigma}$, say, we invoke the same approximation scheme (the Hartree–Fock approximation) as we employed in the previous section when discussing the dynamical interaction of ferromagnetic spin waves. That is to say, we take

$$\hat{n}_{\sigma}\hat{n}_{-\sigma} \Rightarrow \langle\hat{n}_{\sigma}\rangle\hat{n}_{-\sigma} + \hat{n}_{\sigma}\langle\hat{n}_{-\sigma}\rangle - \langle\hat{n}_{\sigma}\rangle\langle\hat{n}_{-\sigma}\rangle. \tag{9.99}$$

The terms on the right-hand side of (9.99) represent all the non-zero pairings of two operators that can be found from the product $\hat{n}_{\sigma}\hat{n}_{-\sigma}$. It is a relatively simple matter to see that we have chosen the correct relative sign of the two operators on the right-hand side. Furthermore, we shall in the main part of this section be concerned only with the zero-temperature properties of the system described by (9.97).

With the approximation (9.99) (we neglect the constant term) the Hamiltonian (9.97) becomes

$$\hat{\mathcal{H}} = \sum_{\mathbf{l},\mathbf{l}',\sigma} T(\mathbf{l}-\mathbf{l}')\hat{c}^{+}_{\mathbf{l}\sigma}\hat{c}_{\mathbf{l}'\sigma} + I\sum_{\mathbf{l},\sigma}\langle\hat{n}_{\mathbf{l}-\sigma}\rangle\hat{n}_{\mathbf{l}\sigma}. \tag{9.100}$$

We now assume $\langle\hat{n}_{\mathbf{l}-\sigma}\rangle$ to be independent of the site label $\mathbf{l}$, in which case (9.100), in terms of Bloch state operators $\hat{c}^{+}_{\mathbf{k}\sigma}$ and $\hat{c}_{\mathbf{k}\sigma}$, is

$$\hat{\mathcal{H}} = \sum_{\mathbf{k},\sigma}\{\mathscr{E}_{\mathbf{k}} + I\langle\hat{n}_{-\sigma}\rangle\}\hat{c}^{+}_{\mathbf{k}\sigma}\hat{c}_{\mathbf{k}\sigma}. \tag{9.101}$$

This Hamiltonian simply represents a band of non-interacting electrons with spin-dependent energies

$$\mathscr{E}_{\mathbf{k}} + I\langle \hat{n}_{-\sigma} \rangle. \tag{9.102}$$

Let $g(\mathscr{E}.)$ be the density of states per atom for the Bloch states. (9.102) tells us that the effective density of states for σ-spin electrons in the presence of the interaction, $\rho_\sigma(\mathscr{E})$, is given by

$$\rho_\sigma(\mathscr{E}) = g(\mathscr{E} - I\langle \hat{n}_{-\sigma} \rangle). \tag{9.103}$$

If μ is the chemical potential, then at absolute zero

$$\langle \hat{n}_\sigma \rangle = \int_{-\infty}^{\mu} \rho_\sigma(\mathscr{E})\, \mathrm{d}\mathscr{E} = \int_{-\infty}^{\mu - I\langle n_{-\sigma} \rangle} g(\mathscr{E})\, \mathrm{d}\mathscr{E} = \int_{-\infty}^{\mu - In + I\langle n_\sigma \rangle} g(\mathscr{E})\, \mathrm{d}\mathscr{E}, \tag{9.104}$$

since $\langle \hat{n}_\sigma \rangle + \langle \hat{n}_{-\sigma} \rangle = n$, where n is the number of itinerant electrons per atom.

In the paramagnetic state there are as many spins up as there are down at any site, i.e. $\langle \hat{n}_\sigma \rangle = \frac{1}{2}n$. However, this state may be unstable to a ferromagnetic ordering. For if $\langle \hat{n}_\sigma \rangle = \frac{1}{2}n + \delta n$, say, the right-hand side of (9.104) changes by

$$I\, \delta n g(\mathscr{E}_{\mathrm{f}})$$

where $\mathscr{E}_{\mathrm{f}} = \mu - \frac{1}{2}In$, and thus the paramagnetic state will be *magnetically* stable if

$$Ig(\mathscr{E}_{\mathrm{f}}) = 1.$$

On the other hand it is unstable to ferromagnetic order if

$$Ig(\mathscr{E}_{\mathrm{f}}) > 1. \tag{9.105}$$

(9.105) is the Stoner condition for ferromagnetism. It shows that conditions favourable to ferromagnetism are achieved with a large density of states at the Fermi energy $\mathscr{E}_{\mathrm{f}}$ and a large I. When (9.105) holds, we seek $\langle \hat{n}_\uparrow \rangle$ and $\langle \hat{n}_\downarrow \rangle$ that satisfy $\langle \hat{n}_\uparrow \rangle + \langle \hat{n}_\downarrow \rangle = n$, the band now splitting into two bands, one of spin-up electrons and one of spin-down electrons separated by an energy $\Delta = I(\langle \hat{n}_\downarrow \rangle - \langle \hat{n}_\uparrow \rangle)$.

9.3.2. *Transverse cross-section*

In calculating the cross-section for the scattering of neutrons by the band electrons described by the Hamiltonian (9.97), we need only consider the scattering that arises from the spin part of the neutron–electron interaction because in Chapter 7 we found, in the limit of a vanishing scattering vector, the amplitude for the scattering via that part of the interaction arising from the motion of the electrons to be smaller than the spin part for electrons in narrow energy bands. Furthermore, we saw in Chapter 8 that the spin-only scattering cross-section for band electrons can be expressed in terms of transverse $\chi_{\mathbf{k}}^{(\pm)}[\omega]$ and longitudinal $\chi_{\mathbf{k}}^{(0)}[\omega]$ generalized susceptibilities. Thus the calculation of the cross-section for the scattering by the electrons described by (9.97) reduces to calculating the linear response of the latter to an external magnetic field. We shall calculate first the transverse susceptibility $\chi_{-\boldsymbol{\kappa}}^{(+)}[\omega]$.

To this end we add to the Hamiltonian describing the electrons the interaction term

$$g\mu_{\mathrm{B}} \sum_{\mathbf{l},\nu} (\hat{S}_{\mathbf{l}}^{\nu})^{+} H_{\mathbf{l}}^{\nu}(t) = g\mu_{\mathrm{B}} \sum_{\mathbf{l},\nu} \hat{S}_{\mathbf{l}}^{-\nu} H_{\mathbf{l}}^{\nu}(t) \tag{9.106}$$

where $H_{\mathbf{l}}^{\nu}(t)$ is the νth component ($\nu = +, -$) of a magnetic field that varies in both space and time. Remembering that $\hat{S}_{\mathbf{l}}^{+} = \hat{c}_{\mathbf{l}\uparrow}^{+}\hat{c}_{\mathbf{l}\downarrow}$ and $\hat{S}_{\mathbf{l}}^{-} = \hat{c}_{\mathbf{l}\downarrow}^{+}\hat{c}_{\mathbf{l}\uparrow}$ and with the approximate form (9.100) of the Hamiltonian (9.97), we now have the total Hamiltonian

$$\hat{\mathscr{H}} = \sum_{\mathbf{l},\mathbf{l}',\sigma} T(\mathbf{l}-\mathbf{l}')\hat{c}_{\mathbf{l}\sigma}^{+}\hat{c}_{\mathbf{l}'\sigma} + I\sum_{\mathbf{l},\sigma} \langle \hat{n}_{-\sigma}\rangle \hat{n}_{\mathbf{l}\sigma} + g\mu_{\mathrm{B}} \sum_{\mathbf{l}} \hat{c}_{\mathbf{l}\uparrow}^{+}\hat{c}_{\mathbf{l}\downarrow} H_{\mathbf{l}}^{-}(t) + \hat{c}_{\mathbf{l}\downarrow}^{+}\hat{c}_{\mathbf{l}\uparrow} H_{\mathbf{l}}^{+}(t). \tag{9.107}$$

If

$$H_{\mathbf{l}}^{+}(t) = \frac{1}{N}\sum_{\mathbf{q}} \exp(\mathrm{i}\mathbf{q}\cdot\mathbf{l}) H_{\mathbf{q}}^{+}(t), \tag{9.108}$$

then (9.107) in terms of Bloch state operators

$$\hat{c}_{\mathbf{k}\sigma} = \frac{1}{N^{1/2}} \sum_{\mathbf{l}} \exp(-\mathrm{i}\mathbf{k}\cdot\mathbf{l})\hat{c}_{\mathbf{l}\sigma}, \quad \text{etc.}$$

is

$$\hat{\mathscr{H}} = \sum_{\mathbf{k},\sigma} \{\mathscr{E}_{\mathbf{k}} + I\langle \hat{n}_{-\sigma}\rangle\}\hat{c}_{\mathbf{k}\sigma}^{+}\hat{c}_{\mathbf{k}\sigma} + g\mu_{\mathrm{B}}\frac{1}{N}\sum_{\mathbf{k},\mathbf{q}} \hat{c}_{\mathbf{k}+\mathbf{q}\downarrow}^{+}\hat{c}_{\mathbf{k}\uparrow} H_{\mathbf{q}}^{+} + \text{c.c.} \tag{9.109}$$

The equation-of-motion for the operator

$$\hat{S}_{-\mathbf{k}}^{+} = \sum_{\mathbf{q}} \hat{c}_{\mathbf{k}+\mathbf{q}\uparrow}^{+}\hat{c}_{\mathbf{q}\downarrow}$$

is obtained using the identity

$$[\hat{A}\hat{B}, \hat{C}\hat{D}] = -[\hat{A}, \hat{C}]_{+}\hat{B}\hat{D} - \hat{C}[\hat{A}, \hat{D}]_{+}\hat{B} + \hat{A}[\hat{B}, \hat{C}]_{+}\hat{D} + \hat{C}\hat{A}[\hat{B}, \hat{D}]_{+}. \tag{9.110}$$

We find

$$\mathrm{i}\hbar\,\partial_t \hat{c}_{\mathbf{k}+\mathbf{q}\uparrow}^{+}\hat{c}_{\mathbf{q}\downarrow} = [\hat{c}_{\mathbf{k}+\mathbf{q}\uparrow}^{+}\hat{c}_{\mathbf{q}\downarrow}, \hat{\mathscr{H}}]$$

$$= \{\mathscr{E}_{\mathbf{q}} - \mathscr{E}_{\mathbf{k}+\mathbf{q}} - \Delta\}\hat{c}_{\mathbf{k}+\mathbf{q}\uparrow}^{+}\hat{c}_{\mathbf{q}\downarrow} + \frac{g\mu_{\mathrm{B}}}{N}\sum_{\mathbf{q}'} H_{\mathbf{q}'}^{+}(t)(\hat{c}_{\mathbf{k}+\mathbf{q}\uparrow}^{+}\hat{c}_{\mathbf{q}-\mathbf{q}'\uparrow} - \hat{c}_{\mathbf{k}+\mathbf{q}+\mathbf{q}'\downarrow}^{+}\hat{c}_{\mathbf{q}\downarrow}).$$

Here

$$\Delta = I(\langle \hat{n}_\downarrow \rangle - \langle \hat{n}_\uparrow \rangle)$$

is the splitting of the two bands.

For the particular choice of $H_{\mathbf{l}}^{+}(t)$

$$H_{\mathbf{l}}^{+}(t) = \exp(-\mathrm{i}\mathbf{k}\cdot\mathbf{l})H^{+}(t), \tag{9.111}$$

we have

$$H^{+}_{\mathbf{q}}(t) = N\,\delta_{\mathbf{k},\mathbf{q}}H^{+}(t).$$

Hence

$$\mathrm{i}\hbar\,\partial_t\,\hat{c}^{+}_{\mathbf{k}+\mathbf{q}\uparrow}\hat{c}_{\mathbf{q}\downarrow} = \{\mathscr{E}_{\mathbf{q}} - \mathscr{E}_{\mathbf{k}+\mathbf{q}} - \Delta\}\hat{c}^{+}_{\mathbf{k}+\mathbf{q}\uparrow}\hat{c}_{\mathbf{q}\downarrow} + g\mu_{\mathrm{B}}H^{+}(t)(\hat{n}_{\mathbf{k}+\mathbf{q}\uparrow} - \hat{n}_{\mathbf{q}\downarrow}), \tag{9.112}$$

where

$$\hat{n}_{\mathbf{k}\sigma} = \hat{c}^{+}_{\mathbf{k}\sigma}\hat{c}_{\mathbf{k}\sigma}.$$

We now choose the time dependence of the external field $H^{+}(t)$ to be

$$H^{+}(t) = H\exp(\mathrm{i}\omega t + \epsilon t). \tag{9.113}$$

In (9.113) ϵ is a small positive constant which is included to make $H^{+}(-\infty) = 0$.

To first order in the applied field H'', $\hat{n}_{\mathbf{k}\sigma}$ is a constant independent of time. Since we only require the solution of (9.112) to terms linear in H^{+} this means that in the second term the time-dependent operators can be replaced by their values in the absence of the field. The solution of the resultant equation that is zero for $t \to -\infty$ is then readily shown to be

$$\hat{c}^{+}_{\mathbf{k}+\mathbf{q}\uparrow}\hat{c}_{\mathbf{q}\downarrow} = \frac{-g\mu_{\mathrm{B}}\exp(\mathrm{i}\omega t + \epsilon t)H(\hat{n}_{\mathbf{k}+\mathbf{q}\uparrow} - \hat{n}_{\mathbf{q}\downarrow})}{\mathscr{E}_{\mathbf{q}} - \mathscr{E}_{\mathbf{k}+\mathbf{q}} - \Delta + \hbar\omega - \mathrm{i}\epsilon} \tag{9.114}$$

and from this one can get

$$\begin{aligned} M^{+}_{-\mathbf{k}}(t) &= -g\mu_{\mathrm{B}}\sum_{\mathbf{l}}\exp(\mathrm{i}\mathbf{k}\cdot\mathbf{l})\mathrm{Tr}\,\hat{\rho}(t)\hat{S}^{+}_{\mathbf{l}} \\ &= -g\mu_{\mathrm{B}}\sum_{\mathbf{l}}\exp(\mathrm{i}\mathbf{k}\cdot\mathbf{l})\mathrm{Tr}\,\hat{\rho}_0\hat{S}^{+}_{\mathbf{l}}(t) \\ &= -g\mu_{\mathrm{B}}\,\mathrm{Tr}\,\hat{\rho}_0\sum_{\mathbf{q}}\hat{c}^{+}_{\mathbf{k}+\mathbf{q}\uparrow}(t)\hat{c}_{\mathbf{q}\downarrow}(t) \end{aligned}$$

where $\hat{\rho}_0$ is the density matrix for the unperturbed system.

Now by definition (eqn (8.26)) the transverse component of the magnetization is given in terms of $\chi^{(+)}_{\mathbf{k}}[\omega]$ by

$$M^{+}_{\mathbf{k}}(t) = -\chi^{(+)}_{\mathbf{k}}[\omega]H^{+}_{\mathbf{k}}(t), \tag{9.115}$$

so that

$$\chi^{(+)}_{-\mathbf{k},0}[\omega]=(g\mu_B)^2\frac{1}{N}\sum_{\mathbf{q}}\frac{\langle\hat{n}_{\mathbf{k}+\mathbf{q}\uparrow}\rangle-\langle\hat{n}_{\mathbf{q}\downarrow}\rangle}{\mathscr{E}_{\mathbf{k}+\mathbf{q}}-\mathscr{E}_{\mathbf{q}}+\Delta-\hbar\omega+\mathrm{i}\epsilon}. \tag{9.116}$$

In (9.116) we may use

$$\langle\hat{n}_{\mathbf{k}\uparrow}\rangle=f_{\mathbf{k}\uparrow} \tag{9.117}$$

where $f_{\mathbf{k}\uparrow}$ is the Fermi–Dirac distribution function for the Bloch state $\mathbf{k}\uparrow$.

We have attached a suffix 0 to $\chi^{(+)}$ given in eqn (9.116) because it represents what we shall call the *non-interacting* susceptibility. The reason for this is as follows. In the Hamiltonian (9.107) we have employed the approximation (9.99) for $\hat{n}_\sigma\hat{n}_{-\sigma}$. However, with the addition of the applied field $H_{\mathbf{l}}^{\nu}(t)$ the total z-component of spin is no longer a constant of motion and thus terms such as $\langle\hat{c}^+_{\mathbf{l}\uparrow}\hat{c}_{\mathbf{l}\downarrow}\rangle$ are no longer zero; this is clearly so because from (9.114) we can get the first-order, or non-interacting, value of $\langle\hat{c}^+_{\mathbf{l}\uparrow}\hat{c}_{\mathbf{l}\downarrow}\rangle$. We have therefore to return to (9.99) and take into account the terms $\langle\hat{c}^+_{\mathbf{l}\uparrow}\hat{c}_{\mathbf{l}\downarrow}\rangle$ and $\langle\hat{c}^+_{\mathbf{l}\downarrow}\hat{c}_{\mathbf{l}\uparrow}\rangle$ that are non-zero in the presence of $H_{\mathbf{l}}^{\nu}(t)$ and include them in the approximation scheme for $\hat{n}_{\mathbf{l}\sigma}\hat{n}_{\mathbf{l}-\sigma}$ to be used in the Hamiltonian consisting of the sum of (9.97) and (9.106).

In place of (9.99) we now write for $\hat{n}_\uparrow\hat{n}_\downarrow$, say,

$$\hat{c}^+_\uparrow\hat{c}_\uparrow\hat{c}^+_\downarrow\hat{c}_\downarrow=\hat{n}_\uparrow\langle\hat{n}_\downarrow\rangle+\langle\hat{n}_\uparrow\rangle\hat{n}_\downarrow-\langle\hat{n}_\uparrow\rangle\langle\hat{n}_\downarrow\rangle+\hat{c}^+_\uparrow\hat{c}_\downarrow\langle\hat{c}_\uparrow\hat{c}^+_\downarrow\rangle+\langle\hat{c}^+_\uparrow\hat{c}_\downarrow\rangle\hat{c}_\uparrow\hat{c}^+_\downarrow-\langle\hat{c}^+_\uparrow\hat{c}_\downarrow\rangle\langle\hat{c}_\uparrow\hat{c}^+_\downarrow\rangle, \tag{9.118}$$

so that

$$\sum_\sigma\hat{n}_{\mathbf{l}\sigma}\hat{n}_{\mathbf{l}-\sigma}\Rightarrow-2\langle\hat{n}_{\mathbf{l}\uparrow}\rangle\langle\hat{n}_{\mathbf{l}\downarrow}\rangle+2\sum_\sigma\hat{n}_{\mathbf{l}\sigma}\langle\hat{n}_{\mathbf{l}-\sigma}\rangle$$
$$+2\{\hat{c}^+_{\mathbf{l}\uparrow}\hat{c}_{\mathbf{l}\downarrow}\langle\hat{c}_{\mathbf{l}\downarrow}\hat{c}^+_{\mathbf{l}\uparrow}\rangle+\langle\hat{c}^+_{\mathbf{l}\uparrow}\hat{c}_{\mathbf{l}\downarrow}\rangle\hat{c}_{\mathbf{l}\downarrow}\hat{c}^+_{\mathbf{l}\uparrow}-\langle\hat{c}^+_{\mathbf{l}\uparrow}\hat{c}_{\mathbf{l}\downarrow}\rangle\langle\hat{c}_{\mathbf{l}\uparrow}\hat{c}^+_{\mathbf{l}\downarrow}\rangle\}. \tag{9.19}$$

We can again ignore the dependence of $\langle\hat{n}_{\mathbf{l}-\sigma}\rangle$ on $\mathbf{l}$ because the first corrections to $\langle\hat{n}_{\mathbf{l}-\sigma}\rangle$ due to the external field H^{ν} are of second order in H^{ν}.

Taking the approximation (9.119) in (9.97) we have in the interacting case the Hamiltonian (apart from constant terms)

$$\mathscr{H}=\sum_{\mathbf{l},\mathbf{l}',\sigma}T(\mathbf{l}-\mathbf{l}')\hat{c}^+_{\mathbf{l}\sigma}\hat{c}_{\mathbf{l}'\sigma}+I\sum_{\mathbf{l},\sigma}\langle\hat{n}_{-\sigma}\rangle\hat{n}_{\mathbf{l}\sigma}-I\sum_{\mathbf{l}}\{\hat{c}^+_{\mathbf{l}\uparrow}\hat{c}_{\mathbf{l}\downarrow}\langle\hat{c}^+_{\mathbf{l}\downarrow}\hat{c}_{\mathbf{l}\uparrow}\rangle+\hat{c}^+_{\mathbf{l}\downarrow}\hat{c}_{\mathbf{l}\uparrow}\langle\hat{c}^+_{\mathbf{l}\uparrow}\hat{c}_{\mathbf{l}\downarrow}\rangle\}. \tag{9.120}$$

It follows that, compared to (9.112), we get an additional contribution on the right-hand side of the equation-of-motion for $\hat{c}^+_{\mathbf{k}+\mathbf{q}\uparrow}\hat{c}_{\mathbf{q}\downarrow}$ which is

$$-I[\hat{c}^+_{\mathbf{k}+\mathbf{q}\uparrow}\hat{c}_{\mathbf{q}\downarrow},\sum_{\mathbf{l}}\hat{c}^+_{\mathbf{l}\downarrow}\hat{c}_{\mathbf{l}\uparrow}\langle\hat{c}^+_{\mathbf{l}\uparrow}\hat{c}_{\mathbf{l}\downarrow}\rangle].$$

Now

$$M_{\mathbf{l}}^{+} = -g\mu_{\mathrm{B}}\langle \hat{S}_{\mathbf{l}}^{+}\rangle = -g\mu_{\mathrm{B}}\langle \hat{c}_{\mathbf{l}\uparrow}^{+}\hat{c}_{\mathbf{l}\downarrow}\rangle$$

and also, by definition of the interacting susceptibility $\chi_{\mathbf{q}}^{(+)}[\omega]$,

$$\begin{aligned} M_{\mathbf{l}}^{+} &= \frac{1}{N}\sum_{\mathbf{q}} \exp(\mathrm{i}\mathbf{q}\cdot\mathbf{l}) M_{\mathbf{q}}^{+}(t) \\ &= \frac{-1}{N}\sum_{\mathbf{q}} \exp(\mathrm{i}\mathbf{q}\cdot\mathbf{l})\chi_{\mathbf{q}}^{(+)}[\omega] H_{\mathbf{q}}^{+}\exp(\mathrm{i}\omega t + \epsilon t). \end{aligned} \tag{9.121}$$

Therefore,

$$\begin{aligned} \langle \hat{c}_{\mathbf{l}\uparrow}\hat{c}_{\mathbf{l}\downarrow}\rangle &= \frac{1}{Ng\mu_{\mathrm{B}}}\sum_{\mathbf{q}} \exp(\mathrm{i}\mathbf{q}\cdot\mathbf{l})\chi_{\mathbf{q}}^{(+)}[\omega] H_{\mathbf{q}}^{+}\exp(\mathrm{i}\omega t + \epsilon t) \\ &= \frac{1}{g\mu_{\mathrm{B}}}\exp(-\mathrm{i}\mathbf{k}\cdot\mathbf{l})\chi_{-\mathbf{k}}^{(+)}[\omega] H^{+}\exp(\mathrm{i}\omega t + \epsilon t). \end{aligned} \tag{9.122}$$

Hence the additional term in the equation-of-motion is

$$\begin{aligned} &\frac{-I}{Ng\mu_{\mathrm{B}}}\chi_{-\mathbf{k}}^{(+)}[\omega] H^{+}\exp(\mathrm{i}\omega t + \epsilon t)\sum_{\mathbf{l},\mathbf{k}',\mathbf{q}'} \exp\{-\mathrm{i}\mathbf{l}\cdot(\mathbf{k}+\mathbf{k}'-\mathbf{q}')\}\,[\hat{c}_{\mathbf{k}+\mathbf{q}\uparrow}^{+}\hat{c}_{\mathbf{q}\downarrow},\, \hat{c}_{\mathbf{k}'\downarrow}^{+}\hat{c}_{\mathbf{q}'\uparrow}] \\ &\quad = \frac{-I}{g\mu_{\mathrm{B}}}\chi_{-\mathbf{k}}^{(+)}[\omega] H^{+}\exp(\mathrm{i}\omega t + \epsilon t)\sum_{\mathbf{k}'} [\hat{c}_{\mathbf{k}+\mathbf{q}\uparrow}^{+}\hat{c}_{\mathbf{q}\downarrow},\, \hat{c}_{\mathbf{k}'\downarrow}^{+}\hat{c}_{\mathbf{k}+\mathbf{k}'\uparrow}] \\ &\quad = \frac{-I}{g\mu_{\mathrm{B}}}\chi_{-\mathbf{k}}^{(+)}[\omega] H^{+}\exp(\mathrm{i}\omega t + \epsilon t)(\hat{n}_{\mathbf{k}+\mathbf{q}\uparrow} - \hat{n}_{\mathbf{q}\downarrow}). \end{aligned} \tag{9.123}$$

Adding this term to (9.112), we notice that the analysis follows through as before provided $H^{+}(t)$ is replaced by

$$\left\{1 - \frac{I}{(g\mu_{\mathrm{B}})^{2}}\chi_{-\mathbf{k}}^{(+)}[\omega]\right\} H^{+}(t).$$

Hence we get

$$\chi_{\mathbf{k}}^{(+)}[\omega] = \left\{1 - \frac{I}{(g\mu_{\mathrm{B}})^{2}}\chi_{\mathbf{k}}^{(+)}[\omega]\right\}\chi_{\mathbf{k},0}^{(+)}[\omega]. \tag{9.124}$$

Solving (9.124) for $\chi_{\mathbf{k}}^{(+)}[\omega]$, we obtain

$$\chi_{\mathbf{k}}^{(+)}[\omega] = \frac{\chi_{\mathbf{k},0}^{(+)}[\omega]}{1 + \{I/(g\mu_{\mathrm{B}})^{2}\}\chi_{\mathbf{k},0}^{(+)}[\omega]}. \tag{9.125}$$

In terms of $\chi_{\boldsymbol{\kappa}}^{(+)}[\omega]$, the transverse part of the neutron cross-section for the scattering from the band electrons is given in eqn (8.80) as

$$\frac{\mathrm{d}^{2}\sigma}{\mathrm{d}\Omega\,\mathrm{d}E'} = r_{0}^{2}\frac{k'}{k}\{F(\boldsymbol{\kappa})\}^{2}\tfrac{1}{4}(1+\tilde{\kappa}_{z}^{2})\{N/(g\mu_{\mathrm{B}})^{2}\}\{1+n(\omega)\}\,\frac{1}{\pi}\,\mathrm{Im}\{\chi_{-\boldsymbol{\kappa}}^{(+)}[\omega] - \chi_{\boldsymbol{\kappa}}^{(+)}[-\omega]\}. \tag{9.126}$$

Since $\chi_{\mathbf{k}}^{(+)}[\omega]$ is expressed in terms of the non-interacting susceptibility $\chi_{\mathbf{k},0}^{(+)}[\omega]$, we begin our discussion of the transverse part of the neutron cross-section by first discussing the properties of $\chi_{\mathbf{k},0}^{(+)}[\omega]$.

9.3.3. *Non-interacting susceptibility*

First we recall that the poles of the generalized susceptibility functions $\chi^{(\pm)}$ determine the energy spectra of the excitations in the system. Let us therefore answer the question, what types of excitations can we obtain in our model of a single band split into two sub-bands of spin ↓ and spin ↑, as shown in Fig. 9.1? The simplest type of excitation is that in which a ↓ spin electron is merely promoted in energy. In Fig. 9.1 this is denoted as process (1). In constructing the energy spectrum for these excitations (Fig. 9.2) it must be borne in mind that because of the Pauli exclusion principle the electrons can only be promoted to empty states, i.e. the final energy of the spin must be greater than $\mathscr{E}_{f\sigma}$. This type of excitation does not involve a change of spin state and does not therefore interest us.

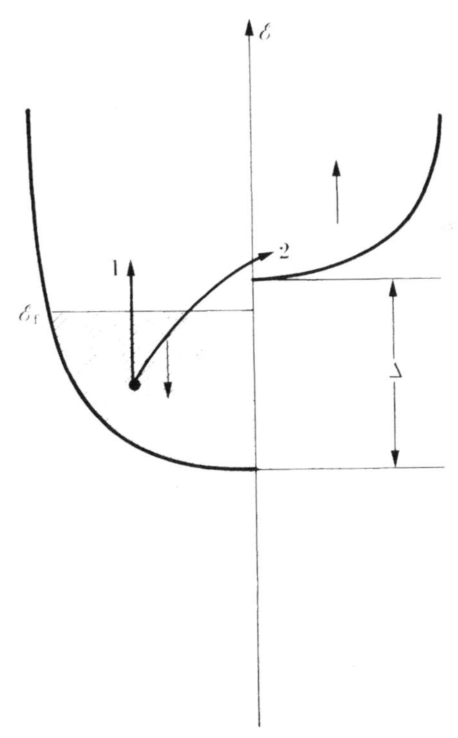

FIG. 9.1. Schematic diagram of two bands of ↓ and ↑ spin in the ferromagnetic state, for the particular case when the two bands do not overlap ($\Delta > \mathscr{E}_f$). The excitation of type 1 is non-spin-flip, i.e. ↓ → ↓, whereas type 2 is spin-flip, namely ↓ → ↑.

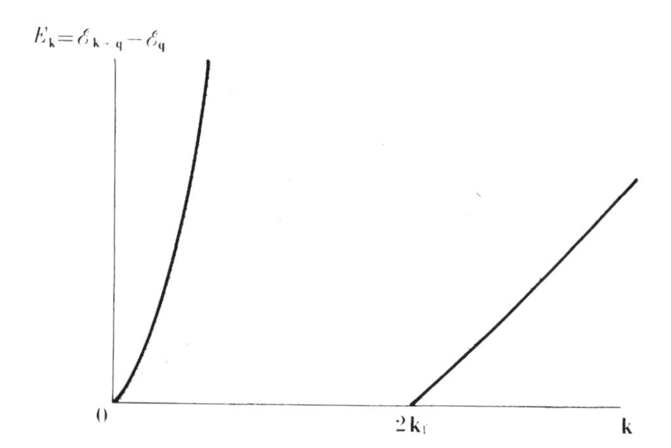

FIG. 9.2. Excitation of type 1 shown in Fig. 9.1 (non-spin-flip). The energy spectrum is $E_{\mathbf{k}}=\mathscr{E}_{\mathbf{k}+\mathbf{q}}-\mathscr{E}_{\mathbf{q}}$.

However, another type of process is that in which a ↓ spin electron is excited into the ↑ spin band. The minimum energy of this excitation is equal to the band-splitting Δ and the energy spectrum of the excitations is $E_{\mathbf{k}}=\mathscr{E}_{\mathbf{k}+\mathbf{q}}-\mathscr{E}_{\mathbf{q}}+\Delta$. In Fig. 9.3(a), (b) we show the excitation spectra, the former figure being for the case when the ↑ and ↓ spin bands do not overlap, as in Fig. 9.1, and Fig. 9.3(b) for the case when the two bands overlap ($\Delta<\mathscr{E}_{\mathrm{f}}$). The spin-flip excitations depicted in Fig. 9.3 are often called Stoner modes. If we now refer back to the expression we derived for $\chi^{(+)}_{\mathbf{k},0}[\omega]$ (eqn (9.116)), we can immediately understand its structure in the light of this discussion.

The denominator of the integrand is seen to have zeros when $\hbar\omega$ is equal to the energy of the type 2 excitation of Fig. 9.1, and the numerator gives the appropriate weight. Also note that

$$\operatorname{Im}\chi^{(+)}_{-\mathbf{k},0}[\omega]=\pi(g\mu_{\mathrm{B}})^2\frac{1}{N}\sum_{\mathbf{q}}(f_{\mathbf{q}\downarrow}-f_{\mathbf{k}+\mathbf{q}\uparrow})\,\delta(\mathscr{E}_{\mathbf{k}+\mathbf{q}}-\mathscr{E}_{\mathbf{q}}+\Delta-\hbar\omega),\tag{9.127}$$

as follows from (9.116) and the identity

$$\operatorname{Im}\lim_{\epsilon\to0^+}\frac{1}{x+\mathrm{i}\epsilon}=-\pi\,\delta(x).\tag{9.128}$$

The above discussion and its relationship to the non-interacting transverse susceptibility is perhaps best illustrated for the particular case of

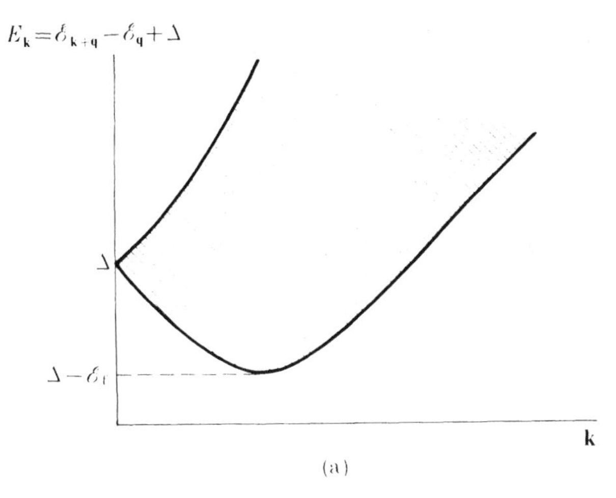

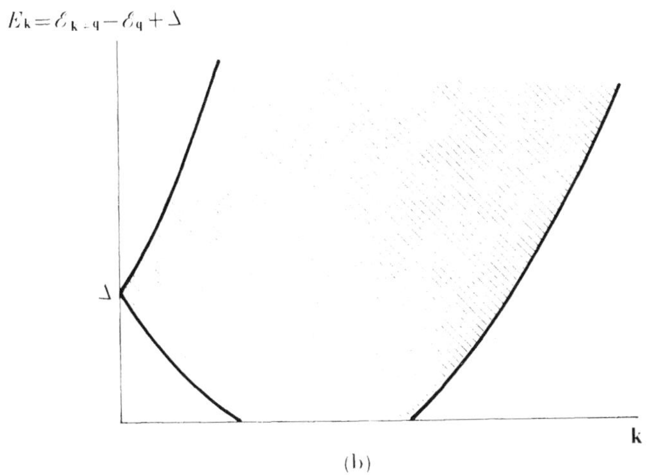

FIG. 9.3. These two figures show the excitation spectra for processes labelled type 2 in Fig. 9.1 (spin-flip excitations). (a) Corresponds to the configuration shown in Fig. 9.1 in which the two bands do not overlap ($\Delta > \mathscr{E}_{\mathrm{f}}$). (b) Shows the excitation spectrum for overlapping bands ($\Delta < \mathscr{E}_{\mathrm{f}}$).

$\mathbf{k} = 0$. From (9.127) we obtain for $\chi^{(+)}_{\mathbf{k}=0,0}$ the result

$$\operatorname{Im} \chi^{(+)}_{\mathbf{k}=0,0}[\omega] = \pi (g\mu_{\mathrm{B}})^2\, \delta(\hbar\omega - \Delta) \frac{1}{N} \sum_{\mathbf{q}} (f_{\mathbf{q}\downarrow} - f_{\mathbf{q}\uparrow})$$
$$= \pi (g\mu_{\mathrm{B}})^2\, \delta(\hbar\omega - \Delta)(\langle \hat{n}_{\downarrow} \rangle - \langle \hat{n}_{\uparrow} \rangle),$$

or

$$\operatorname{Im} \chi^{(+)}_{\mathbf{k}=0,0}[\omega] = \pi (g\mu_{\mathrm{B}})^2\, \delta(\hbar\omega - \Delta) \frac{\Delta}{I}, \tag{9.129}$$

on using the definition $\Delta = I(\langle \hat{n}_\downarrow \rangle - \langle \hat{n}_\uparrow \rangle)$. Thus the particular case $\mathbf{k}=0$ merely corresponds to promoting an electron of spin $\downarrow$ to the bottom of the spin $\uparrow$ band, the energy cost to the external field being $\hbar\omega = \Delta$ (cf. Fig. 9.1).

9.3.4. *Spins waves*

Now let us discuss the properties of $\chi^{(+)}_{-\mathbf{k}}[\omega]$ in eqn (9.124). Its poles are determined by the equation

$$1 = I\frac{-1}{(g\mu_B)^2}\,\mathrm{Re}\,\chi^{(+)}_{-\mathbf{k},0}[\omega]$$

and this reads in detail

$$1 = \frac{I}{N}\,\mathrm{Re}\sum_{\mathbf{q}} \frac{f_{\mathbf{q}\downarrow} - f_{\mathbf{k}+\mathbf{q}\uparrow}}{\mathscr{E}_{\mathbf{k}+\mathbf{q}} - \mathscr{E}_{\mathbf{q}} + \Delta - \hbar\omega + \mathrm{i}\epsilon}. \tag{9.130}$$

If we set $\omega = \Omega_{\mathbf{k}}$, say, then for $\mathbf{k}=0$ the secular equation (9.130) for the excitation energy Ω_0 is

$$1 = I\frac{1}{N}\sum_{\mathbf{q}} \frac{f_{\mathbf{q}\downarrow} - f_{\mathbf{q}\uparrow}}{\Delta - \hbar\Omega_0} = \frac{I}{\Delta - \hbar\Omega_0}\frac{\Delta}{I},$$

which is satisfied if $\Omega_0 = 0$. Thus a solution of (9.130) exists which has the property $\lim_{\mathbf{k}\to 0}\Omega_{\mathbf{k}} = 0$ and this solution corresponds to a spin-wave excitation in the narrow energy band of electrons. If we plot the solutions of eqn (9.130), we find they appear as in Fig. 9.4, i.e., in addition to solutions of the type shown in Fig. 9.3 (so-called Stoner modes), we have now the spin-wave mode occurring in the gap around $\mathbf{k}=0$.

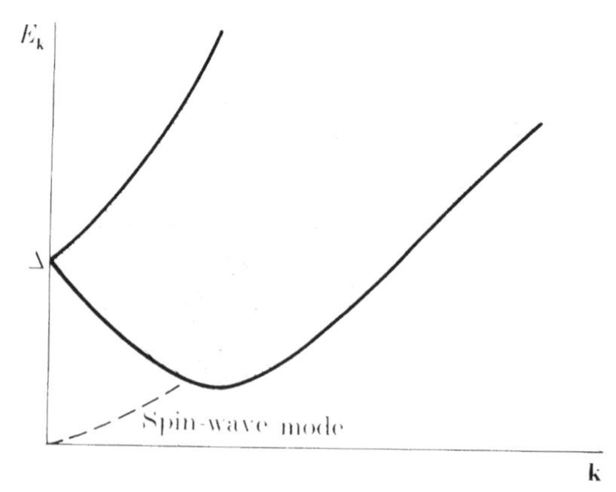

FIG. 9.4. Schematic diagram of spin-wave dispersion curve around $\mathbf{k}=0$ for a narrow s band of electrons described by a Hubbard Hamiltonian (eqn (9.97)).

For small values of $\mathbf{k}$ we expect the dispersion relation for the spin-wave mode in a cubic crystal to be of the form

$$\hbar\Omega_{\mathbf{k}} = Dk^2 \tag{9.131}$$

and we can determine the coefficient D as follows. We write

$$\frac{1}{N}\sum_{\mathbf{q}} \frac{f_{\mathbf{q}\downarrow} - f_{\mathbf{k}+\mathbf{q}\uparrow}}{\mathscr{E}_{\mathbf{k}+\mathbf{q}} - \mathscr{E}_{\mathbf{q}} + \Delta - \hbar\Omega_{\mathbf{k}}} = \frac{1}{N}\sum_{\mathbf{q}} \frac{f_{\mathbf{q}\downarrow}}{\mathscr{E}_{\mathbf{k}+\mathbf{q}} - \mathscr{E}_{\mathbf{q}} + \Delta - \hbar\Omega_{\mathbf{k}}} - \frac{1}{N}\sum_{\mathbf{q}} \frac{f_{\mathbf{k}+\mathbf{q}\uparrow}}{\mathscr{E}_{\mathbf{k}+\mathbf{q}} - \mathscr{E}_{\mathbf{q}} + \Delta - \hbar\Omega_{\mathbf{k}}}$$

and, since

$$\frac{1}{N}\sum_{\mathbf{q}} \frac{f_{\mathbf{k}+\mathbf{q}\uparrow}}{\mathscr{E}_{\mathbf{k}+\mathbf{q}} - \mathscr{E}_{\mathbf{q}} + \Delta - \hbar\Omega_{\mathbf{k}}} = \frac{1}{N}\sum_{\mathbf{q}} \frac{f_{\mathbf{q}\uparrow}}{\mathscr{E}_{\mathbf{q}} - \mathscr{E}_{\mathbf{q}-\mathbf{k}} + \Delta - \hbar\Omega_{\mathbf{k}}},$$

we have, on expanding $\mathscr{E}_{\mathbf{q}+\mathbf{k}}$ and $\mathscr{E}_{\mathbf{q}-\mathbf{k}}$ in $\mathbf{k}$,

$$1 = I\frac{1}{N}\sum_{\mathbf{q}} \frac{f_{\mathbf{q}\downarrow}}{\mathbf{k}\cdot\nabla_{\mathbf{q}}\mathscr{E}_{\mathbf{q}} + \frac{1}{2}(\mathbf{k}\cdot\nabla_{\mathbf{q}})^2\mathscr{E}_{\mathbf{q}} - \hbar\Omega_{\mathbf{k}} + \Delta} + \frac{f_{\mathbf{q}\uparrow}}{-\mathbf{k}\cdot\nabla_{\mathbf{q}}\mathscr{E}_{\mathbf{q}} + \frac{1}{2}(\mathbf{k}\cdot\nabla_{\mathbf{q}})^2\mathscr{E}_{\mathbf{q}} + \hbar\Omega_{\mathbf{k}} - \Delta}. \tag{9.132}$$

Expanding in powers of $\mathbf{k}$ gives

$$1 = \frac{I}{\Delta N}\sum_{\mathbf{q}} (f_{\mathbf{q}\downarrow} - f_{\mathbf{q}\uparrow}) - \frac{I}{\Delta N}\sum_{\mathbf{q}} (f_{\mathbf{q}\downarrow} + f_{\mathbf{q}\uparrow})\frac{1}{2\Delta}(\mathbf{k}\cdot\nabla_{\mathbf{q}})^2\mathscr{E}_{\mathbf{q}} + \frac{I}{\Delta N}\sum_{\mathbf{q}} (f_{\mathbf{q}\downarrow} - f_{\mathbf{q}\uparrow})\frac{\hbar\Omega_{\mathbf{k}}}{\Delta} + \frac{I}{\Delta N}\sum_{\mathbf{q}} (f_{\mathbf{q}\downarrow} - f_{\mathbf{q}\uparrow})\frac{1}{\Delta^2}\{\mathbf{k}\cdot\nabla_{\mathbf{q}}\mathscr{E}_{\mathbf{q}}\}^2.$$

Now

$$\sum_{\mathbf{q}} f_{\mathbf{q}\downarrow} = N\langle\hat{n}_{\downarrow}\rangle$$

and, remembering that $\Delta = I(\langle\hat{n}_{\downarrow}\rangle - \langle\hat{n}_{\uparrow}\rangle)$, we get

$$\hbar\Omega_{\mathbf{k}} = \frac{I}{2\Delta}\frac{1}{N}\sum_{\mathbf{q}} \{(\mathbf{k}\cdot\nabla_{\mathbf{q}})^2\mathscr{E}_{\mathbf{q}}\}(f_{\mathbf{q}\uparrow} + f_{\mathbf{q}\downarrow}) + \frac{I}{\Delta^2}\frac{1}{N}\sum_{\mathbf{q}} (\mathbf{k}\cdot\nabla_{\mathbf{q}}\mathscr{E}_{\mathbf{q}})^2(f_{\mathbf{q}\uparrow} - f_{\mathbf{q}\downarrow}). \tag{9.133}$$

Thus for cubic lattices we find that D in eqn (9.131) is given by the expression

$$D = \frac{I}{6\Delta}\frac{1}{N}\sum_{\mathbf{q}} \{\nabla_{\mathbf{q}}^2\mathscr{E}_{\mathbf{q}}\}(f_{\mathbf{q}\uparrow} + f_{\mathbf{q}\downarrow}) + \frac{I}{3\Delta^2}\frac{1}{N}\sum_{\mathbf{q}} (\nabla_{\mathbf{q}}\mathscr{E}_{\mathbf{q}})^2(f_{\mathbf{q}\uparrow} - f_{\mathbf{q}\downarrow}). \tag{9.134}$$

If we assume that $\mathscr{E}_{\mathbf{q}}$ can be represented by

$$\mathscr{E}_{\mathbf{q}} = \frac{\hbar^2}{2m^*} q^2 \tag{9.135}$$

where m^* is the effective mass for the band, then it is straightforward to show that in this instance eqn (9.134) for D becomes

$$D = \frac{\hbar^2 In}{2m^*\Delta} \left\{1 - \frac{4(\mathscr{E}_{\mathrm{f}\downarrow}\langle \hat{n}_\downarrow \rangle - \mathscr{E}_{\mathrm{f}\uparrow}\langle \hat{n}_\uparrow \rangle)}{5\Delta n}\right\}. \tag{9.136}$$

Turning now to the calculation of the cross-section (9.126), we obtain from eqn (9.125) the imaginary part of $\chi_{\mathbf{k}}^{(+)}[\omega]$, namely,

$$\operatorname{Im} \chi_{\mathbf{k}}^{(+)}[\omega] = \frac{\operatorname{Im} \chi_{\mathbf{k},0}^{(+)}[\omega]}{[1 + \{I/(g\mu_{\mathrm{B}})^2\}\operatorname{Re} \chi_{\mathbf{k},0}^{(+)}[\omega]]^2 + [\{I/(g\mu_{\mathrm{B}})^2\}\operatorname{Im} \chi_{\mathbf{k},0}^{(+)}[\omega]]^2}. \tag{9.137}$$

Now it follows from (9.127) that $\operatorname{Im} \chi_{-\boldsymbol{\kappa},0}^{(+)}[\omega]$ is non-zero only for values of ω equal to the energies of the Stoner modes of Fig. 9.3. Hence it follows that, when $\omega = \Omega_{\mathbf{k}}$, $\operatorname{Im} \chi_{-\boldsymbol{\kappa},0}^{(+)}$ is zero, except when $\Omega_{\boldsymbol{\kappa}}$ merges into the continuum of the Stoner modes. Leaving aside this particular case, we can in (9.137) take the limit $\operatorname{Im} \chi_{\mathbf{k},0}^{(+)} \to 0^-$ and, using the result (9.128), obtain

$$\frac{I}{(g\mu_{\mathrm{B}})^2} \operatorname{Im} \chi_{-\boldsymbol{\kappa}}^{(+)}[\omega] = \pi\, \delta\left\{1 + \frac{I}{(g\mu_{\mathrm{B}})^2} \operatorname{Re} \chi_{-\boldsymbol{\kappa},0}^{(+)}[\omega]\right\}. \tag{9.138}$$

But, for $\boldsymbol{\kappa} \to 0$,

$$1 + \frac{I}{(g\mu_{\mathrm{B}})^2} \operatorname{Re} \chi_{-\boldsymbol{\kappa},0}^{(+)}[\omega] \simeq \hbar(\Omega_{\boldsymbol{\kappa}} - \omega)/\Delta \tag{9.139}$$

and thus

$$\operatorname{Im} \chi_{-\boldsymbol{\kappa}}^{(+)}[\omega] = \pi (g\mu_{\mathrm{B}})^2 \frac{\Delta}{I}\, \delta(\hbar\omega - \hbar\Omega_{\mathbf{k}}). \tag{9.140}$$

Using (9.140) and (9.126) leads to the expression, valid for $\boldsymbol{\kappa} \to 0$,

$$\frac{\mathrm{d}^2\sigma}{\mathrm{d}\Omega\, \mathrm{d}E'} = r_0^2 \frac{k'}{k} \{F(\boldsymbol{\kappa})\}^2 \tfrac{1}{4}(1 + \tilde{\kappa}_z^2) N\, \Delta/I$$
$$\times [n(\Omega_{\boldsymbol{\kappa}})\, \delta(\hbar\omega + \hbar\Omega_{\boldsymbol{\kappa}}) + \{1 + n(\Omega_{\boldsymbol{\kappa}})\}\, \delta(\hbar\omega - \hbar\Omega_{\boldsymbol{\kappa}})] \tag{9.141}$$

for the transverse neutron cross-section. We immediately note the similarity of this expression for the scattering by the spin-wave mode in a narrow band of electrons with that for the scattering by spin waves in a Heisenberg ferromagnet, eqn (9.44). In fact, we can identify each factor in

(9.141) with a corresponding factor in (9.44) when we recall that $\Delta/I = \langle \hat{n}_{\downarrow}\rangle - \langle \hat{n}_{\uparrow}\rangle$ is proportional to the spin moment per atom.

The spin-wave scattering described by (9.141) occurs near each Bragg point, although the intensity decreases quite rapidly with increasing $\boldsymbol{\kappa}$ because of the atomic form factor.

The cross-section (9.141) does not include the contribution arising from the scattering by Stoner modes, i.e. that determined by $\chi^{(+)}_{\boldsymbol{\kappa},0}[\omega]$. According to our model, the spin wave is damped only when the dispersion energy merges with the continuum of Stoner modes. In a realistic, multi-band model the damping is not confined to a small energy–wave vector domain, as implied in Fig. 9.4, and the Stoner continuum extends over a very wide domain reaching down to small energies. It is, therefore, tempting to link the strong spin-wave damping at short wavelengths observed in nickel and iron with structure in the Stoner continuum. For both metals the damping at $\hbar\omega \sim 100$ meV is comparable with the energy, and this renders the spin wave unobservable, i.e. the spin-wave signal in the cross-section disappears into the background. However, the band splitting Δ for nickel and iron is of order 0.5 and 2.0 eV, respectively. The severe mismatch between the band splitting and the maximum observed spin-wave energy indicates that linking the disappearance phenomenon to the Stoner continuum is, at best, over-simplified reasoning. This suspicion is confirmed by numerical calculations for realistic band-structure models of nickel and iron (Cooke *et al.* 1980), which show no significant structure in the Stoner continua in the vicinity of the maximum observed spin-wave energy.

The over-simplification arises from neglect of band-structure coefficients (cf. § 7.5) and matrix elements that occur in a multi-band formulation of the transverse susceptibility for a Hubbard Hamiltonian. Numerical calculations of the appropriately weighted Stoner continua for nickel and iron display significant features at the spin-wave energy that grow in intensity with increasing wave vector to a value which renders the spin-wave intensity negligible for an energy $\hbar\omega \sim 100$ meV. Calculated values of D in the spin-wave dispersion (9.131) are in good agreement with measured values: $D = 0.40$ and $0.28\,\mathrm{eV\,Å^2}$ for nickel and iron, respectively. The correlation between the observations and the numerical results appear to validate the multi-band model and the handling of the strong-electron interaction in the Hubbard Hamiltonian.

To evaluate the temperature dependence of the coefficient D in (9.131) requires, for a detailed answer, the use of thermodynamic perturbation theory (Kishore 1979). Such an approach would lead us too far afield and we therefore content ourselves with an argument based on quite general considerations that shows what the form of the dependence on T must be.

In the absence of spin-wave interactions, the total energy of the spin waves in the itinerant-electron model is

$$\sum_{\mathbf{k}} \hat{n}_{\mathbf{k}} \hbar \Omega_{\mathbf{k}}.$$

If now we imagine the interaction terms to have been evaluated, the exact energy will involve terms linear in $\hat{n}_{\mathbf{k}}$ plus terms quadratic in $\hat{n}_{\mathbf{k}}$ plus higher-order terms. Hence

$$\hat{\mathscr{H}} = \sum_{\mathbf{k}} \hat{n}_{\mathbf{k}} \hbar \Omega_{\mathbf{k}} - \sum_{\mathbf{k},\mathbf{q}} \hat{n}_{\mathbf{k}} \hat{n}_{\mathbf{q}} c_{\mathbf{k},\mathbf{q}} \tag{9.142a}$$

where $c_{\mathbf{k},\mathbf{q}}$ is a coefficient to be determined. The only assumption included in (9.142a) is that the state of the crystal can be specified by just the occupation number operators $\hat{n}_{\mathbf{k}}$, i.e. the interactions of spin waves with other types of excitations in the crystal have been neglected.

If we now write for $\hat{n}_{\mathbf{k}}\hat{n}_{\mathbf{q}}$ in (9.142a) $\hat{n}_{\mathbf{k}}\langle \hat{n}_{\mathbf{q}} \rangle + \langle \hat{n}_{\mathbf{k}} \rangle \hat{n}_{\mathbf{q}}$, then

$$\hat{\mathscr{H}} = \sum_{\mathbf{k}} \left\{ \hbar \Omega_{\mathbf{k}} - 2 \sum_{\mathbf{q}} c_{\mathbf{k},\mathbf{q}} \langle \hat{n}_{\mathbf{q}} \rangle \right\} \hat{n}_{\mathbf{k}} = \sum_{\mathbf{k}} E_{\mathbf{k}} \hat{n}_{\mathbf{k}}. \tag{9.142b}$$

In the interaction term in $E_{\mathbf{k}}$ the only dependence on $\mathbf{k}$ arises from the coefficient $c_{\mathbf{k},\mathbf{q}}$. This coefficient must go to zero as $\mathbf{k}$ tends to zero; other $E_{\mathbf{k}}$ would possess an energy gap at $\mathbf{k} = 0$ and it is known that the uniform excitation mode ($\mathbf{k} = 0$) has zero energy. Because we are also confident that the energy–momentum relation is quadratic at small $\mathbf{k}$, it follows that the leading term in $c_{\mathbf{k},\mathbf{q}}$ must be of order k^2 at most at small $\mathbf{k}$. By definition $c_{\mathbf{k},\mathbf{q}}$ is symmetric in $\mathbf{k}$ and $\mathbf{q}$ so if it has a factor k^2 it must also have a factor q^2. Thus it follows that

$$N c_{\mathbf{k},\mathbf{q}} = k^2 q^2 c. \tag{9.143}$$

On taking $\hbar \Omega_{\mathbf{k}} = D k^2$ as in (9.131), the relation (9.143) shows, by arguments identical to those used in § 9.2, that

$$\begin{aligned} E_{\mathbf{k}} &= k^2 \left\{ D - \frac{2c}{N} \sum_{\mathbf{q}} q^2 \langle \hat{n}_{\mathbf{q}} \rangle \right\} \\ &= k^2 D \left\{ 1 - \frac{v_0 4\pi c}{3\sqrt{3D}} \left(\frac{k_B T}{4\pi D} \right)^{5/2} \zeta(\tfrac{5}{2}) \right\} \end{aligned} \tag{9.144}$$

to first order in c.

Thus, we find the temperature dependence of D in the itinerant model to be the same as that in the Heisenberg model for ferromagnetism. On examining the argument leading to (9.144), we see that this conclusion is to be expected, since the assumptions made are equally valid for either model.

One of these assumptions was that the spin waves do not interact with any other excitation in the solid. Let us now consider the effect of the interaction between the spin waves and the electrons excited out of their zero-temperature configuration. If we denote the number operator for the thermally excited itinerant electrons by $\delta\hat{n}_{\mathbf{k}}$, then to (9.142a) we must add a term $\sum \mathscr{E}_{\mathbf{k}}\, \delta\hat{n}_{\mathbf{k}}$ corresponding to their kinetic energy, an interaction term of the form $-\sum \bar{c}_{\mathbf{k},\mathbf{q}}\, \delta\hat{n}_{\mathbf{k}}\, \delta\hat{n}_{\mathbf{q}}$, and a further term $-\sum c'_{\mathbf{k},\mathbf{q}}\hat{n}_{\mathbf{q}}\, \delta\hat{n}_{\mathbf{k}}$ representing the interaction of the electrons with the spin waves. Thus in place of the dispersion relation given in (9.142b) for the spin waves we now have the result

$$E_{\mathbf{k}} = \hbar\Omega_{\mathbf{k}} - 2\sum_{\mathbf{q}} c_{\mathbf{k},\mathbf{q}}\langle \hat{n}_{\mathbf{q}}\rangle - \sum_{\mathbf{q}} c'_{\mathbf{k},\mathbf{q}}\langle \delta\hat{n}_{\mathbf{q}}\rangle$$

and it may be easily verified that the last term gives rise to T^2 dependence. Hence we expect for cubic metals that the coefficients D should have a temperature dependence of the form

$$D(T) = D_0 - D_1 T^2 - D_2 T^{5/2}. \tag{9.145}$$

The evaluation of the coefficent D_1 in simple cases is discussed by Herring (1966).

9.3.5. *Longitudinal susceptibility* (*Edwards 1980*)

The longitudinal cross-section for the itinerant-electron model behaves very differently to (9.141). In fact, just as we found for the Heisenberg ferromagnet, it does not yield spin-wave scattering. However, it would be important for a consideration of paramagnetic and critical scattering and, since its derivation is akin to that for the transverse cross-section, it is logical for us to present the calculation at this juncture.

We follow the same scheme in calculating the longitudinal susceptibility $\chi^{(0)}_{\mathbf{k}}[\omega]$ as we adopted for the transverse susceptibility, namely, we calculate first the non-interacting value $\chi^{(0)}_{\mathbf{k},0}[\omega]$ and then its value in the interacting case. To the Hamiltonian (9.100) we therefore add the term

$$g\mu_{\mathrm{B}} \sum_{\mathbf{l}} H_{\mathbf{l}}(t)\hat{S}^z_{\mathbf{l}}(t) = \tfrac{1}{2} g\mu_{\mathrm{B}} \sum_{\mathbf{l},\sigma} H_{\mathbf{l}}(t)\sigma\hat{n}_{\mathbf{l}\sigma} \tag{9.146}$$

and hence the analogue of the Hamiltonian (9.107) is

$$\mathscr{H} = \sum_{\mathbf{l},\mathbf{l}',\sigma} T(\mathbf{l}-\mathbf{l}')\hat{c}^+_{\mathbf{l}\sigma}\hat{c}_{\mathbf{l}'\sigma} + I\sum_{\mathbf{l},\sigma} \langle \hat{n}_{-\sigma}\rangle \hat{n}_{\mathbf{l}\sigma} + \tfrac{1}{2} g\mu_{\mathrm{B}} \sum_{\mathbf{l},\sigma} H_{\mathbf{l}}(t)\sigma\hat{n}_{\mathbf{l}\sigma}. \tag{9.147}$$

If

$$H_{\mathbf{l}}(t) = \tfrac{1}{2} H\{\exp(-\mathrm{i}\mathbf{k}\cdot\mathbf{l} + \mathrm{i}\omega t) + \mathrm{c.c.}\}\exp \epsilon t, \tag{9.148}$$

then the Hamiltonian (9.147) in terms of Bloch state operators is

$$\mathscr{H} = \sum_{\mathbf{k}',\sigma} \{\mathscr{E}_{\mathbf{k}'} + I\langle \hat{n}_{-\sigma}\rangle\}\hat{c}^+_{\mathbf{k}'\sigma}\hat{c}_{\mathbf{k}'\sigma}$$

$$+\tfrac{1}{4}g\mu_B H \sum_{\mathbf{k}',\sigma} \sigma\{\exp(\mathrm{i}\omega t)\hat{c}^+_{\mathbf{k}'\sigma}\hat{c}_{\mathbf{k}+\mathbf{k}'\sigma} + \exp(-\mathrm{i}\omega t)\hat{c}^+_{\mathbf{k}+\mathbf{k}'\sigma}\hat{c}_{\mathbf{k}'\sigma}\}. \qquad (9.149)$$

The equation-of-motion for $\hat{c}^+_{\mathbf{q}\sigma}$ is, therefore,

$$\mathrm{i}\hbar\,\partial_t\hat{c}^+_{\mathbf{q}\sigma} = -E_{\mathbf{q}\sigma}\hat{c}^+_{\mathbf{q}\sigma} - \tfrac{1}{4}g\mu_B H\sigma\{\exp(\mathrm{i}\omega t)\hat{c}^+_{\mathbf{q}-\mathbf{k}\sigma} + \exp(-\mathrm{i}\omega t)\hat{c}^+_{\mathbf{q}+\mathbf{k}\sigma}\}\exp \epsilon t \qquad (9.150)$$

where we find it convenient to use the notation

$$E_{\mathbf{q}\sigma} = \mathscr{E}_{\mathbf{q}} + I\langle \hat{n}_{-\sigma}\rangle. \qquad (9.151)$$

If $H=0$, then the solution of (9.150) is

$$\hat{c}^+_{\mathbf{q}\sigma}(t) = \hat{c}^+_{\mathbf{q}\sigma}\exp(\mathrm{i}tE_{\mathbf{q}\sigma}/\hbar).$$

Because we require the solution of (9.150) only to first order in H we can use this result for the operators appearing in the coefficient of H in (9.150). Thus,

$$\mathrm{i}\hbar\,\partial_t \exp(-\mathrm{i}tE_{\mathbf{q}\sigma}/\hbar)\hat{c}^+_{\mathbf{q}\sigma} = -\tfrac{1}{4}g\mu_B H\sigma\{\exp(\mathrm{i}tE_{\mathbf{q}-\mathbf{k}\sigma}/\hbar + \mathrm{i}\omega t)\hat{c}^+_{\mathbf{q}-\mathbf{k}\sigma}$$
$$+\exp(\mathrm{i}tE_{\mathbf{q}+\mathbf{k}\sigma}/\hbar - \mathrm{i}\omega t)\hat{c}^+_{\mathbf{q}+\mathbf{k}\sigma}\}\exp(-\mathrm{i}tE_{\mathbf{q}\sigma}/\hbar + \epsilon t),$$

which when integrated leads to the results

$$\hat{c}^+_{\mathbf{q}\sigma}(t) = \exp(\mathrm{i}tE_{\mathbf{q}\sigma}/\hbar)\hat{c}^+_{\mathbf{q}\sigma}$$
$$-\tfrac{1}{4}g\mu_B H\sigma\left\{\frac{\exp(\mathrm{i}tE_{\mathbf{q}-\mathbf{k}\sigma}/\hbar + \mathrm{i}\omega t)\hat{c}^+_{\mathbf{q}-\mathbf{k}\sigma}}{\mathscr{E}_{\mathbf{q}} - \mathscr{E}_{\mathbf{q}-\mathbf{k}} - \hbar\omega + \mathrm{i}\epsilon} + \frac{\exp(\mathrm{i}tE_{\mathbf{q}+\mathbf{k}\sigma}/\hbar - \mathrm{i}\omega t)\hat{c}^+_{\mathbf{q}+\mathbf{k}\sigma}}{\mathscr{E}_{\mathbf{q}} - \mathscr{E}_{\mathbf{q}+\mathbf{k}} + \hbar\omega + \mathrm{i}\epsilon}\right\} \qquad (9.152a)$$

and

$$\hat{c}_{\mathbf{q}'\sigma}(t) = \exp(-\mathrm{i}tE_{\mathbf{q}'\sigma}/\hbar)\hat{c}_{\mathbf{q}'\sigma}$$
$$-\tfrac{1}{4}g\mu_B H\sigma\left\{\frac{\exp(-\mathrm{i}tE_{\mathbf{q}-\mathbf{k}\sigma}/\hbar - \mathrm{i}\omega t)\hat{c}^+_{\mathbf{q}-\mathbf{k}\sigma}}{\mathscr{E}_{\mathbf{q}'} - \mathscr{E}_{\mathbf{q}'-\mathbf{k}} - \hbar\omega - \mathrm{i}\epsilon} + \frac{\exp(-\mathrm{i}tE_{\mathbf{q}+\mathbf{k}\sigma}/\hbar + \mathrm{i}\omega t)\hat{c}^+_{\mathbf{q}+\mathbf{k}\sigma}}{\mathscr{E}_{\mathbf{q}'} - \mathscr{E}_{\mathbf{q}'+\mathbf{k}} + \hbar\omega - \mathrm{i}\epsilon}\right\} \qquad (9.152b)$$

Now

$$\langle \hat{S}^z_{\mathbf{l}}(t)\rangle = \frac{1}{N}\sum_{\mathbf{q},\mathbf{q}'\sigma} \tfrac{1}{2}\sigma\langle \hat{c}^+_{\mathbf{q}\sigma}(t)\hat{c}_{\mathbf{q}'\sigma}(t)\rangle\exp\{\mathrm{i}\mathbf{l}\cdot(\mathbf{q}'-\mathbf{q})\}, \qquad (9.153)$$

and we find from (9.152) by a straightforward calculation

$$\langle \hat{S}^z_{\mathbf{l}}(t)\rangle = \frac{1}{N}\sum_{\mathbf{q},\sigma} \tfrac{1}{2}\sigma\langle \hat{n}_{\mathbf{q}\sigma}\rangle - \tfrac{1}{4}g\mu_B H\frac{1}{N}\sum_{\mathbf{q},\sigma}\{\langle \hat{n}_{\mathbf{q}\sigma}\rangle - \langle \hat{n}_{\mathbf{k}+\mathbf{q}\sigma}\rangle\}$$
$$\times \mathrm{Re}\left\{\frac{\exp(-\mathrm{i}\mathbf{k}\cdot\mathbf{l} + \mathrm{i}\omega t)}{\mathscr{E}_{\mathbf{k}+\mathbf{q}} - \mathscr{E}_{\mathbf{q}} - \hbar\omega + \mathrm{i}\epsilon}\right\}, \qquad (9.154)$$

where

$$\langle \hat{n}_{\mathbf{k}\sigma} \rangle \equiv \langle \hat{c}^{+}_{\mathbf{k}\sigma} \hat{c}_{\mathbf{k}\sigma} \rangle = f_{\mathbf{k}\sigma}.$$

If the system were in a paramagnetic state, the first term in (9.154) would be zero and the second independent of spin, the sum over σ giving just an additional factor of 2.

By definition the z-component of magnetization $M^{z}_{\mathbf{l}}(t)$ is given in terms of $\chi^{(0)}_{\mathbf{k}}[\omega]$ through the relation

$$M^{z}_{\mathbf{l}}(t) = -\frac{1}{2}\frac{g\mu_{\mathrm{B}}}{N}\sum_{\mathbf{q},\sigma} \sigma f_{\mathbf{q}\sigma} - \frac{H}{N}\operatorname{Re}\{\exp(-\mathrm{i}\mathbf{k}\cdot\mathbf{l} + \mathrm{i}\omega t)\chi^{(0)}_{\mathbf{k}}[\omega]\}. \tag{9.155}$$

Hence, by comparing (9.155) with $(-g\mu_{\mathrm{B}})$ times (9.154) we have the non-interacting longitudinal susceptibility

$$\chi^{(0)}_{\mathbf{k},0}[\omega] = -(\tfrac{1}{2}g\mu_{\mathrm{B}})^{2}\frac{1}{N}\sum_{\mathbf{q},\sigma}\frac{f_{\mathbf{k}+\mathbf{q}\sigma} - f_{\mathbf{q}\sigma}}{\mathscr{E}_{\mathbf{k}+\mathbf{q}} - \mathscr{E}_{\mathbf{q}} - \hbar\omega + \mathrm{i}\epsilon}. \tag{9.156}$$

In order to calculate $\chi^{(0)}_{\mathbf{k}}[\omega]$, the longitudinal susceptibility in the interacting case, $\langle \hat{n}_{-\sigma}\rangle$ in (9.147) must be allowed to be a function of position, as it must be in the presence of the external field $H_{\mathbf{l}}(t)$. The result (9.152) tells us that to first order in H, $\langle \hat{n}_{\mathbf{l}-\sigma}\rangle$ is given by

$$\langle \hat{n}_{\mathbf{l}-\sigma}\rangle = \frac{1}{N}\sum_{\mathbf{q}} f_{\mathbf{q}-\sigma} - \tfrac{1}{2}g\mu_{\mathrm{B}}H\sigma\frac{1}{N}\sum_{\mathbf{q}}(f_{\mathbf{k}+\mathbf{q}-\sigma} - f_{\mathbf{q}-\sigma})\operatorname{Re}\left\{\frac{\exp(-\mathrm{i}\mathbf{k}\cdot\mathbf{l} + \mathrm{i}\omega t)}{\mathscr{E}_{\mathbf{k}+\mathbf{q}} - \mathscr{E}_{\mathbf{q}} - \hbar\omega + \mathrm{i}\epsilon}\right\}.$$

Hence its first-order correction *in the paramagnetic region* is

$$\frac{I}{g\mu_{\mathrm{B}}}H\sigma\operatorname{Re}\{\exp(-\mathrm{i}\mathbf{k}\cdot\mathbf{l} + \mathrm{i}\omega t)\chi^{(0)}_{\mathbf{k},0}[\omega]\}. \tag{9.157}$$

In view of this result, the perturbation term to be added to the unperturbed Hamiltonian is now

$$H\sum_{\mathbf{l},\sigma}\sigma\hat{n}_{\mathbf{l}\sigma}\operatorname{Re}\left\{\left(\tfrac{1}{2}g\mu_{\mathrm{B}} + \frac{I}{g\mu_{\mathrm{B}}}\chi^{(0)}_{\mathbf{k}}[\omega]\right)\exp(-\mathrm{i}\mathbf{k}\cdot\mathbf{l} + \mathrm{i}\omega t)\right\}\exp \epsilon t \tag{9.158}$$

in place of (9.146). The calculation of $\hat{c}^{+}_{\mathbf{q}\sigma}(t)$ with (9.158) added to the unperturbed Hamiltonian proceeds just as in the non-interacting case. It is readily shown that in the interacting case the susceptibility is

$$\chi^{(0)}_{\mathbf{k}}[\omega] = \frac{\chi^{(0)}_{\mathbf{k},0}[\omega]}{1 - 2/(g\mu_{\mathrm{B}})^{2} I\chi^{(0)}_{\mathbf{k},0}[\omega]}. \tag{9.159}$$

It is clear from the structure of $\chi^{(0)}_{\mathbf{k},0}[\omega]$ (eqn (9.156)) that the paramagnetic longitudinal susceptibility (9.159) leads to scattering that is nothing like that arising from the transverse susceptibility.

9.4. Heisenberg ferromagnet with dipole interaction

In many cases of practical importance the Heisenberg Hamiltonian (9.1) is inadequate to describe the properties of systems with localized magnetic moments because it does not include the dipole term ($\mathbf{r} \neq 0$)

$$\hat{\mathscr{H}}_{\rm d} = \tfrac{1}{2}(g\mu_{\rm B})^2 \sum_{\mathbf{l},\mathbf{r}} \left\{ \hat{\mathbf{S}}_{\mathbf{l}} \cdot \hat{\mathbf{S}}_{\mathbf{l}+\mathbf{r}} - 3\frac{(\hat{\mathbf{S}}_{\mathbf{l}} \cdot \mathbf{r})(\hat{\mathbf{S}}_{\mathbf{l}+\mathbf{r}} \cdot \mathbf{r})}{r^2} \right\} \frac{1}{r^3}. \tag{9.160}$$

To take this into account in the linear spin-wave approximation we first write

$$\mathbf{r} = \left\{ \tfrac{1}{2}(r^+ + r^-), \frac{1}{2\mathrm{i}}(r^+ - r^-), z \right\}, \tag{9.161}$$

so that

$$\mathbf{S}_{\mathbf{l}} \cdot \mathbf{r} = z\hat{S}^z_{\mathbf{l}} + \tfrac{1}{2}(r^+\hat{S}^-_{\mathbf{l}} + r^-\hat{S}^+_{\mathbf{l}}) \tag{9.162}$$

and similarly for $\hat{\mathbf{S}}_{\mathbf{l}+\mathbf{r}} \cdot \mathbf{r}$. We then find, for Bravais crystal structures,

$$\begin{aligned}\hat{\mathscr{H}}_{\rm d} = \tfrac{1}{2}(g\mu_{\rm B})^2 \sum_{\mathbf{l},\mathbf{r}} \frac{1}{r^3} \{ &\hat{S}^z_{\mathbf{l}}\hat{S}^z_{\mathbf{l}+\mathbf{r}}(1 - 3z^2/r^2) + \hat{S}^+_{\mathbf{l}}\hat{S}^-_{\mathbf{l}+\mathbf{r}}(1 - 3r^+r^-/2r^2) \\ &- (3z/r^2)(r^+\hat{S}^z_{\mathbf{l}}\hat{S}^-_{\mathbf{l}+\mathbf{r}} + r^-\hat{S}^z_{\mathbf{l}}\hat{S}^+_{\mathbf{l}+\mathbf{r}}) - (3/4r^2)(r^{+2}\hat{S}^-_{\mathbf{l}}\hat{S}^-_{\mathbf{l}+\mathbf{r}} + r^{-2}\hat{S}^+_{\mathbf{l}}\hat{S}^+_{\mathbf{l}+\mathbf{r}})\}. \end{aligned} \tag{9.163}$$

If we use the representation (9.14) for $\hat{S}^z$, as is appropriate for linear spin-wave theory, it is clear that the third set of terms in (9.163) give rise only to terms independent of $\hat{S}^{\pm}$ and terms cubic in these operators. We may therefore neglect this term in a linear theory.

Using the commutation rule (9.13) we then obtain for the commutator $[\hat{S}^+_{\mathbf{l}}, \hat{\mathscr{H}}_{\rm d}]$, which occurs in the equation-of-motion for $\hat{S}^+_{\mathbf{l}}$ when the Hamiltonian $\hat{\mathscr{H}}_{\rm d}$ is added to be the Heisenberg exchange Hamiltonian (9.1), the result

$$\begin{aligned}[\hat{S}^+_{\mathbf{l}}, \hat{\mathscr{H}}_{\rm d}] = S(g\mu_{\rm B})^2 \sum_{\mathbf{r}} \frac{1}{r^3} \{ &-\hat{S}^+_{\mathbf{l}}(1 - 3z^2/r^2) + \hat{S}^+_{\mathbf{l}+\mathbf{r}}(1 - 3r^+r^-/2r^2) \\ &-(3/2r^2)r^{+2}\hat{S}^-_{\mathbf{l}+\mathbf{r}}\}. \end{aligned} \tag{9.164}$$

If we now write this commutator in terms of the operators $\hat{S}^{\pm}_{\mathbf{q}}$, we have, on adding the result to the right-hand side of (9.9), the equation-of-motion for $\hat{S}^+_{\mathbf{q}}$. The resulting equation-of-motion can be written

$$\mathrm{i}\hbar\, \partial_t \hat{S}^+_{\mathbf{q}} = A_{\mathbf{q}}\hat{S}^+_{\mathbf{q}} + B^*_{\mathbf{q}}\hat{S}^-_{-\mathbf{q}} \tag{9.165}$$

where

$$\begin{aligned} A_{\mathbf{q}} = g\mu_{\rm B}H &+ 2S\{\mathscr{J}(0) - \mathscr{J}(\mathbf{q})\} \\ &+ S(g\mu_{\rm B})^2\left[\sum_{\mathbf{r}} \frac{\exp(\mathrm{i}\mathbf{q}\cdot\mathbf{r})}{r^3}\{1 - 3(x^2 + y^2)/2r^2\} - \sum_{\mathbf{r}} \frac{1}{r^3}(1 - 3z^2/r^2)\right] \end{aligned} \tag{9.166a}$$

and

$$B_{\mathbf{q}} = -\frac{3S}{2}(g\mu_{\mathrm{B}})^2 \sum_{\mathbf{r}} \frac{\exp(-i\mathbf{q}\cdot\mathbf{r})}{r^5}(x-iy)^2. \tag{9.166b}$$

The Hamiltonian (9.160) is that derived from classical electromagnetics. However, as an indirect result of spin–orbit coupling there are usually other terms of the same structure as the classical dipole term. These are usually referred to a 'pseudodipolar' terms and their inclusion in (9.160) would still result in an equation-of-motion of the form (9.165) but with modified values of the coefficients $A_{\mathbf{q}}$ and $B_{\mathbf{q}}$. We shall not explicitly consider these pseudodipolar terms. The evaluation of the coefficients $A_{\mathbf{q}}$ and $B_{\mathbf{q}}$, including the pseudodipolar terms, is given in full by Keffer (1967).

We notice that

$$A_{-\mathbf{q}} = A_{\mathbf{q}} = A^*_{\mathbf{q}}$$

and

$$B_{-\mathbf{q}} = B_{\mathbf{q}} \neq B^*_{\mathbf{q}}. \tag{9.167}$$

We look for a solution of the form

$$\hat{S}^+_{\mathbf{q}} = u_{\mathbf{q}}\hat{\alpha}_{\mathbf{q}} + v_{\mathbf{q}}\hat{\alpha}^+_{-\mathbf{q}}, \tag{9.168a}$$

and hence

$$\hat{S}^-_{\mathbf{q}} = u^*_{\mathbf{q}}\hat{\alpha}^+_{\mathbf{q}} + v^*_{\mathbf{q}}\hat{\alpha}_{-\mathbf{q}}, \tag{9.168b}$$

and demand that the transformation diagonalizes the linear Hamiltonian and that $\hat{\alpha}_{\mathbf{q}}$, $\hat{\alpha}^+_{\mathbf{q}}$ satisfy the commutation rule

$$[\hat{\alpha}_{\mathbf{q}}, \hat{\alpha}^+_{\mathbf{q}'}] = \delta_{\mathbf{q},\mathbf{q}'}. \tag{9.169}$$

Inverting (9.168) we get

$$\hat{\alpha}_{\mathbf{q}} = (u^*_{-\mathbf{q}}\hat{S}^+_{\mathbf{q}} - v_{\mathbf{q}}\hat{S}^-_{-\mathbf{q}})/(u_{\mathbf{q}}u^*_{-\mathbf{q}} - v_{\mathbf{q}}v^*_{-\mathbf{q}}). \tag{9.170}$$

Then

$$i\hbar\,\partial_t\hat{\alpha}_{\mathbf{q}} = \hbar\omega_{\mathbf{q}}\hat{\alpha}_{\mathbf{q}}$$

gives

$$u^*_{-\mathbf{q}}(A_{\mathbf{q}} - \hbar\omega_{\mathbf{q}}) + v_{\mathbf{q}}B_{\mathbf{q}} = 0$$

and

$$u^*_{-\mathbf{q}}B^*_{\mathbf{q}} + v_{\mathbf{q}}(A_{\mathbf{q}} + \hbar\omega_{\mathbf{q}}) = 0. \tag{9.171}$$

Hence

$$\hbar\omega_{\mathbf{q}} = \{A^2_{\mathbf{q}} - |B_{\mathbf{q}}|^2\}^{1/2}. \tag{9.172}$$

We can choose to make

$$u_{\mathbf{q}} = u_{-\mathbf{q}} = u^*_{\mathbf{q}} = u^*_{-\mathbf{q}}. \tag{9.173}$$

Hence

$$v_{\mathbf{q}} = v_{-\mathbf{q}} = \frac{-u_{\mathbf{q}} B_{\mathbf{q}}^{*}}{(A_{\mathbf{q}} + \hbar\omega_{\mathbf{q}})}. \tag{9.174}$$

(9.169), (9.170), and (9.28) together give

$$u_{\mathbf{q}}^{2} - |v_{\mathbf{q}}|^{2} = 2SN \tag{9.175}$$

and (9.174) and (9.175) together give

$$u_{\mathbf{q}}^{2} = 2SN \frac{(A_{\mathbf{q}} + \hbar\omega_{\mathbf{q}})}{2\hbar\omega_{\mathbf{q}}}$$

and

$$2v_{\mathbf{q}} u_{\mathbf{q}} = \frac{-2SNB_{\mathbf{q}}^{*}}{\hbar\omega_{\mathbf{q}}}. \tag{9.176}$$

This completes the information we need on the transformation. We now note that

$$\begin{aligned}
\hat{S}_{\mathbf{l}}^{x}(t) &= \tfrac{1}{2}\{\hat{S}_{\mathbf{l}}^{+}(t) + \hat{S}_{\mathbf{l}}^{-}(t)\} \\
&= \frac{1}{2N} \sum_{\mathbf{q}} \exp(\mathrm{i}\mathbf{q}\cdot\mathbf{l})\{\hat{S}_{\mathbf{q}}^{+}(t) + \hat{S}_{-\mathbf{q}}^{-}(t)\} \\
&= \frac{1}{2N} \sum_{\mathbf{q}} \exp(\mathrm{i}\mathbf{q}\cdot\mathbf{l})\{(u_{\mathbf{q}} + v_{\mathbf{q}}^{*})\hat{\alpha}_{\mathbf{q}}(t) + (u_{\mathbf{q}} + v_{\mathbf{q}})\hat{\alpha}_{-\mathbf{q}}^{+}(t)\} \\
&= \frac{1}{2N} \sum_{\mathbf{q}} \exp(\mathrm{i}\mathbf{q}\cdot\mathbf{l})\{(u_{\mathbf{q}} + v_{\mathbf{q}}^{*})\exp(-\mathrm{i}\omega_{\mathbf{q}}t)\hat{\alpha}_{\mathbf{q}} + (u_{\mathbf{q}} + v_{\mathbf{q}})\exp(\mathrm{i}\omega_{\mathbf{q}}t)\hat{\alpha}_{-\mathbf{q}}^{+}\}
\end{aligned} \tag{9.177}$$

and

$$\begin{aligned}
\hat{S}_{\mathbf{l}}^{y}(t) &= \frac{1}{2\mathrm{i}}\{\hat{S}_{\mathbf{l}}^{+}(t) - \hat{S}_{\mathbf{l}}^{-}(t)\} \\
&= \frac{1}{2\mathrm{i}N} \sum_{\mathbf{q}} \exp(\mathrm{i}\mathbf{q}\cdot\mathbf{l})\{\hat{S}_{\mathbf{q}}^{+}(t) - \hat{S}_{-\mathbf{q}}^{-}(t)\} \\
&= \frac{1}{2\mathrm{i}N} \sum_{\mathbf{q}} \exp(\mathrm{i}\mathbf{q}\cdot\mathbf{l})\{(u_{\mathbf{q}} - v_{\mathbf{q}}^{*})\exp(-\mathrm{i}\omega_{\mathbf{q}}t)\hat{\alpha}_{\mathbf{q}} - (u_{\mathbf{q}} - v_{\mathbf{q}})\exp(\mathrm{i}\omega_{\mathbf{q}}t)\hat{\alpha}_{-\mathbf{q}}^{+}\}.
\end{aligned} \tag{9.178}$$

Hence

$$\begin{aligned}
&\langle \hat{S}_{\mathbf{l}}^{x}\hat{S}_{\mathbf{l}'}^{x}(t)\rangle \\
&= \frac{1}{4N^{2}} \sum_{\mathbf{q}} \exp\{\mathrm{i}\mathbf{q}\cdot(\mathbf{l}-\mathbf{l}')\}\,|u_{\mathbf{q}} + v_{\mathbf{q}}|^{2}\,\{\exp(-\mathrm{i}\omega_{\mathbf{q}}t)n_{\mathbf{q}} + \exp(\mathrm{i}\omega_{\mathbf{q}}t)(n_{\mathbf{q}} + 1)\},
\end{aligned} \tag{9.179a}$$

$$\langle \hat{S}_{\mathbf{l}}^{y}\hat{S}_{\mathbf{l}'}^{y}(t)\rangle$$
$$=\frac{1}{4N^2}\sum_{\mathbf{q}}\exp\{\mathrm{i}\mathbf{q}\cdot(\mathbf{l}-\mathbf{l}')\}\,|u_{\mathbf{q}}-v_{\mathbf{q}}|^2\,\{\exp(-\mathrm{i}\omega_{\mathbf{q}}t)n_{\mathbf{q}}+\exp(\mathrm{i}\omega_{\mathbf{q}}t)(n_{\mathbf{q}}+1)\},\tag{9.179b}$$

and

$$\langle \hat{S}_{\mathbf{l}}^{x}\hat{S}_{\mathbf{l}'}^{y}(t)+\hat{S}_{\mathbf{l}}^{y}\hat{S}_{\mathbf{l}'}^{x}(t)\rangle$$
$$=\frac{1}{4\mathrm{i}N^2}\sum_{\mathbf{q}}\exp\{\mathrm{i}\mathbf{q}\cdot(\mathbf{l}-\mathbf{l}')\}2u_{\mathbf{q}}(v_{\mathbf{q}}-v_{\mathbf{q}}^{*})\{\exp(\mathrm{i}\omega_{\mathbf{q}}t)(n_{\mathbf{q}}+1)+\exp(-\mathrm{i}\omega_{\mathbf{q}}t)n_{\mathbf{q}}\}.\tag{9.180}$$

Putting all these results together gives, finally,

$$\frac{\mathrm{d}^2\sigma}{\mathrm{d}\Omega\,\mathrm{d}E'}=r_0^2\frac{k'}{k}\{\tfrac{1}{2}gF(\boldsymbol{\kappa})\}^2\exp\{-2W(\boldsymbol{\kappa})\}\frac{S}{2}\frac{(2\pi)^3}{v_0}$$
$$\times\sum_{\mathbf{q},\boldsymbol{\tau}}\left\{(1+\tilde{\kappa}_z^2)\frac{A_{\mathbf{q}}}{\hbar\omega_{\mathbf{q}}}+(1-\tilde{\kappa}_z^2)\frac{|B_{\mathbf{q}}|}{\hbar\omega_{\mathbf{q}}}\cos 2(\phi+\phi_{\mathbf{q}})\right\}$$
$$\times\{n_{\mathbf{q}}\,\delta(\hbar\omega+\hbar\omega_{\mathbf{q}})\,\delta(\boldsymbol{\kappa}+\mathbf{q}-\boldsymbol{\tau})+(n_{\mathbf{q}}+1)\,\delta(\hbar\omega-\hbar\omega_{\mathbf{q}})\,\delta(\boldsymbol{\kappa}-\mathbf{q}-\boldsymbol{\tau})\}\tag{9.181}$$

where we have defined

$$B_{\mathbf{q}}=|B_{\mathbf{q}}|\exp(2\mathrm{i}\phi_{\mathbf{q}}),$$

and

$$\tilde{\boldsymbol{\kappa}}=(\sin\theta\cos\phi,\,\sin\theta\sin\phi,\,\cos\theta)\tag{9.182}$$

is the unit scattering vector.

If $|B_{\mathbf{q}}|$ is small, then the cross-section (9.181) is seen to retain the same, simple dependence $1+\tilde{\kappa}_z^2$ as for the cross-section in the absence of dipolar forces. However, in the general case the dipole forces greatly complicate the angular dependence of the cross-section.

9.5. Rare earths (Taylor and Darby 1972; Coqblin 1977; Lindgård 1978)

The discussion in § 9.2 of the scattering by spin waves in a Heisenberg ferromagnet was restricted to structures possessing a Bravais-crystal lattice. We have previously remarked that the Heisenberg model ought to furnish a reasonable description of the rare-earth metals, and these commonly possess a hexagonal close-packed structure. In this section we therefore consider the scattering by spin waves in a h.c.p. lattice, taking into account the single-ion anisotropy terms appropriate to terbium. The analysis is, of course, also applicable to other h.c.p. ferromagnets (e.g. Co) in as far as they are adequately described by the Heisenberg model of ferromagnetism.

The h.c.p. lattice is constructed from two hexagonal Bravais lattices as described in § 2.1. We denote the spin operators on one sublattice by $\hat{S}$ and those associated with the ions on the other by $\hat{T}$. If the exchange interaction between the spins on different sublattices is denoted by J and that within a sublattice by J', then the exchange part of the Heisenberg Hamiltonian is

$$\mathscr{H}_{\text{ex}} = -\sum_{\mathbf{m},\mathbf{n}} 2J(\mathbf{m}-\mathbf{n})\hat{\mathbf{S}}_{\mathbf{m}} \cdot \hat{\mathbf{T}}_{\mathbf{n}} - \sum_{\mathbf{m},\mathbf{m}'} J'(\mathbf{m}-\mathbf{m}')\hat{\mathbf{S}}_{\mathbf{m}} \cdot \hat{\mathbf{S}}_{\mathbf{m}'} - \sum_{\mathbf{n},\mathbf{n}'} J'(\mathbf{n}-\mathbf{n}')\hat{\mathbf{T}}_{\mathbf{n}} \cdot \hat{\mathbf{T}}_{\mathbf{n}'}. \tag{9.183}$$

In (9.183) we have adopted the notation of $\mathbf{m}$ for position vectors defining the sublattice with spin operators $\hat{\mathbf{S}}$ and $\mathbf{n}$ for the position vectors defining the sublattice with spin operators $\hat{\mathbf{T}}$.

In the rare earths the exchange interactions arise from an indirect interaction between the highly localized 4f electrons via the conduction electrons. This interaction has the characteristic of being long-ranged and oscillatory and for this reason alone the evaluation of the cross-section for comparison with a measured cross-section is complicated.

The exchange interaction is not sufficient to describe the magnetic state of rare earths because they have, in general, unquenched orbital angular momentum. This means that the 4f electrons of terbium, for instance, with $S = L = 3$ and $J = 6$, do not form a spherical distribution but are expanded in a plane normal to the total angular momentum J.† This results in large single-ion anisotropy terms which have to be taken into account in order to describe adequately the spin-wave modes. We assume the anisotropy to give rise to the following additional terms in the total Hamiltonian,

$$\mathscr{H}_{\text{an}} = B\sum_{\mathbf{m}} (\hat{S}^z_{\mathbf{m}})^2 + B\sum_{\mathbf{n}} (\hat{T}^z_{\mathbf{n}})^2$$

$$-\tfrac{1}{2}G\sum_{\mathbf{m}} \{(\hat{S}^x_{\mathbf{m}} + i\hat{S}^y_{\mathbf{m}})^6 + (\hat{S}^x_{\mathbf{m}} - i\hat{S}^y_{\mathbf{m}})^6\} - \tfrac{1}{2}G\sum_{\mathbf{n}} \{(\hat{T}^x_{\mathbf{n}} + i\hat{T}^y_{\mathbf{n}})^6 + (\hat{T}^x_{\mathbf{n}} - i\hat{T}^y_{\mathbf{n}})^6\}. \tag{9.184}$$

† The lattice parameters of Tb are $a = 3.599$ Å and $c = 5.696$ Å at room temperature. It is a ferromagnetic below 216 K, in the temperature range 216–226 K it forms an anti-ferromagnetic spiral structure, and is paramagnetic above 226 K. We also note that the mean radius of the 4f shell wave function is believed to be 0.35 Å, which is an order of magnitude less than the interatomic spacing.

In (9.184) $\hat{S}^z$ and $\hat{T}^z$ are the components of $\hat{\mathbf{S}}$ and $\hat{\mathbf{T}}$ in the z-direction. The first terms on the right-hand side account for the tendency of the magnetic moments to lie in the hexagonal basal plane and the second set of terms is responsible for the preferred direction within the basal plane. With both B and G positive the crystal-field anisotropy energy has a minimum value when the moments are aligned in the x-direction. In view of this latter fact,

$$\hat{S}^{\pm} = \hat{S}^{y} \pm \mathrm{i}\hat{S}^{z}. \tag{9.185}$$

In deriving the spin-wave modes in the system described by the Hamiltonian formed from the sum of (9.183) and (9.184), we shall use the modified commutation relation (9.13) to obtain the necessary linear equations-of-motion. It follows from (9.185) that

$$(\hat{S}^z)^2 = -\tfrac{1}{4}\{(\hat{S}^+)^2 + (\hat{S}^-)^2 - 2S - 2\hat{S}^-\hat{S}^+\}. \tag{9.186}$$

To terms quadratic in $\hat{S}^+$ and $\hat{S}^-$ it is easy to show that

$$(\hat{S}^x + \mathrm{i}\hat{S}^y)^6 + (\hat{S}^x - \mathrm{i}\hat{S}^y)^6 \simeq S^6\left\{2 - \frac{6}{S^2}\,\hat{S}^-\hat{S}^+ - \frac{15}{2S^2}(\hat{S}^+ + \hat{S}^-)^2\right\} \tag{9.187}$$

on using the representation

$$\hat{S}^x = \hat{S} - \frac{1}{2S}\,\hat{S}^-\hat{S}^+$$

as appropriate for linear spin-wave theory (cf. eqn (9.14)).

With (9.186) and (9.187) the linear equation-of-motion for the operator $\hat{S}^+_{\mathbf{m}}$ is

$$\begin{aligned} \mathrm{i}\hbar\,\partial_t\hat{S}^+_{\mathbf{m}} = {} & 2S\sum_{\mathbf{m}'} J'(\mathbf{m}-\mathbf{m}')(\hat{S}^+_{\mathbf{m}} - \hat{S}^+_{\mathbf{m}'}) + 2S\sum_{\mathbf{n}} J(\mathbf{m}-\mathbf{n})(\hat{S}^+_{\mathbf{m}} - \hat{T}^+_{\mathbf{n}}) \\ & + \hat{S}^+_{\mathbf{m}}(SB + 21S^5G) + \hat{S}^-_{\mathbf{m}}(15S^5G - SB) \end{aligned} \tag{9.188}$$

and we shall henceforth denote the coefficient of $\hat{S}^+_{\mathbf{m}}$ in the third term of (9.188) by A and that of $\hat{S}^-_{\mathbf{m}}$ in the fourth by A'.

We now transform to operators $\hat{S}^{\pm}_{\mathbf{q}}$ and $\hat{T}^{\pm}_{\mathbf{q}}$ with the convention defined by eqn (9.8) to write (9.188) as

$$\mathrm{i}\hbar\,\partial_t\hat{S}^+_{\mathbf{q}} = \mathscr{A}_{\mathbf{q}}\hat{S}^+_{\mathbf{q}} + A'\hat{S}^-_{-\mathbf{q}} - 2S\mathscr{J}(\mathbf{q})\hat{T}^+_{\mathbf{q}}, \tag{9.189}$$

where the coefficient $\mathscr{A}_{\mathbf{q}}$ is defined by

$$\mathscr{A}_{\mathbf{q}} = 2S\{\mathscr{J}'(0) - \mathscr{J}'(\mathbf{q})\} + A + 2S\mathscr{J}(0). \tag{9.190}$$

In (9.189) and (9.190)

$$\mathscr{J}(\mathbf{q}) = \sum_{\mathbf{m}} J(\mathbf{m})\exp(-\mathrm{i}\mathbf{q}\cdot\mathbf{m}) = \mathscr{J}^*(-\mathbf{q}) \tag{9.191a}$$

and

$$\mathscr{J}'(\mathbf{q}) = \sum_{\mathbf{m}} J'(\mathbf{m})\exp(-\mathrm{i}\mathbf{q}\cdot\mathbf{m}) = \mathscr{J}'(-\mathbf{q}). \tag{9.191b}$$

$\mathscr{J}(\mathbf{q})$ is complex because, as we noted in § 2.1, the vector joining sites on different sublattices in h.c.p. is not a lattice vector, i.e. ions are not centres of symmetry with respect to ions on the opposite sublattice. Of course $\mathscr{J}'(\mathbf{q})$ is real because each sublattice is Bravais.

Since the equation-of-motion for $\hat{S}^+_{\mathbf{q}}$ contains in addition to $\hat{S}^+_{\mathbf{q}}$ the operators $\hat{S}^-_{-\mathbf{q}}$ and $\hat{T}^+_{\mathbf{q}}$, we have also to form their equations-of-motion. In fact we find that these equations-of-motion contain $\hat{T}^-_{-\mathbf{q}}$ so that all told we require four equations-of-motion including (9.189). The remaining three are readily shown to be

$$\mathrm{i}\hbar\,\partial_t\hat{S}^-_{-\mathbf{q}} = -\mathscr{A}_{\mathbf{q}}\hat{S}^-_{-\mathbf{q}} - A'\hat{S}^+_{\mathbf{q}} + 2S\mathscr{J}(\mathbf{q})\hat{T}^-_{-\mathbf{q}}, \tag{9.192a}$$

$$\mathrm{i}\hbar\,\partial_t\hat{T}^+_{\mathbf{q}} = \mathscr{A}_{\mathbf{q}}\hat{T}^+_{\mathbf{q}} + A'\hat{T}^-_{-\mathbf{q}} - 2S\mathscr{J}^*(\mathbf{q})\hat{S}^+_{\mathbf{q}}, \tag{9.192b}$$

and

$$\mathrm{i}\hbar\,\partial_t\hat{T}^-_{-\mathbf{q}} = -\mathscr{A}_{\mathbf{q}}\hat{T}^-_{-\mathbf{q}} - A'\hat{T}^+_{\mathbf{q}} + 2S\mathscr{J}^*(\mathbf{q})\hat{S}^-_{-\mathbf{q}}. \tag{9.192c}$$

The equations-of-motion (9.189) and (9.192) are derivable from the quadratic Hamiltonian

$$\begin{aligned}\hat{\mathscr{H}} = \frac{1}{2SN}\sum_{\mathbf{q}} & \tfrac{1}{2}\mathscr{A}_{\mathbf{q}}(\hat{S}^-_{\mathbf{q}}\hat{S}^+_{\mathbf{q}} + \hat{S}^+_{\mathbf{q}}\hat{S}^-_{\mathbf{q}} + \hat{T}^-_{\mathbf{q}}\hat{T}^+_{\mathbf{q}} + \hat{T}^+_{\mathbf{q}}\hat{T}^-_{\mathbf{q}}) \\ & + \tfrac{1}{2}A'(\hat{S}^-_{\mathbf{q}}\hat{S}^-_{-\mathbf{q}} + \hat{S}^+_{\mathbf{q}}\hat{S}^+_{-\mathbf{q}} + \hat{T}^-_{\mathbf{q}}\hat{T}^-_{-\mathbf{q}} + \hat{T}^+_{\mathbf{q}}\hat{T}^+_{-\mathbf{q}}) \\ & - S[\mathscr{J}^*(\mathbf{q})\{\hat{S}^+_{\mathbf{q}}\hat{T}^-_{\mathbf{q}} + \hat{S}^-_{-\mathbf{q}}\hat{T}^+_{-\mathbf{q}}\} + \mathscr{J}(\mathbf{q})\{\hat{S}^-_{\mathbf{q}}\hat{T}^+_{\mathbf{q}} + \hat{S}^+_{-\mathbf{q}}\hat{T}^-_{-\mathbf{q}}\}].\end{aligned} \tag{9.193}$$

This may easily be verified with the aid of the commutation rule (9.28).

We can proceed in two completely equivalent ways; we may either solve the four coupled equations-of-motion formed by (9.189) and (9.192) or seek the transformation that diagonalizes the Hamiltonian (9.193). Since in § 9.1 we have already given an example of the equation-of-motion method we here opt for the second alternative and find it convenient to use a matrix notation.

We write the Hamiltonian (9.193) as

$$\hat{\mathscr{H}} = \frac{1}{2SN}\sum_{\mathbf{q}}\hat{\mathscr{H}}_{\mathbf{q}} \tag{9.194}$$

where $\hat{\mathscr{H}}_{\mathbf{q}}$ is given in terms of the matrices $\hat{X}$ and $\mathscr{H}$. The column matrix operator $\hat{X}$ is defined to be

$$\hat{X} = \begin{pmatrix} \hat{S}^{+}_{\mathbf{q}} \\ \hat{S}^{-}_{-\mathbf{q}} \\ \hat{T}^{+}_{\mathbf{q}} \\ \hat{T}^{-}_{-\mathbf{q}} \end{pmatrix}, \tag{9.195}$$

and

$$\mathscr{H} = \begin{pmatrix} h & h_1^* \\ h_1 & h \end{pmatrix} \tag{9.196a}$$

where

$$h = \begin{pmatrix} \mathscr{A}_{\mathbf{q}} & A' \\ A' & \mathscr{A}_{\mathbf{q}} \end{pmatrix} \quad \text{and} \quad h_1 = \begin{pmatrix} C_{\mathbf{q}} & 0 \\ 0 & C_{\mathbf{q}} \end{pmatrix} \tag{9.196b}$$

with

$$C_{\mathbf{q}} = -2S\mathscr{J}^*(\mathbf{q}). \tag{9.197}$$

In terms of $\hat{X}$ and $\mathscr{H}$,

$$\hat{\mathscr{H}}_{\mathbf{q}} = \tfrac{1}{2}\hat{X}^{+}\mathscr{H}\hat{X}. \tag{9.198}$$

If

$$\hat{X} = T\hat{Y} \tag{9.199}$$

where

$$\hat{Y} = \begin{pmatrix} \hat{\alpha}_{\mathbf{q}} \\ \hat{\alpha}^{+}_{-\mathbf{q}} \\ \hat{\beta}_{\mathbf{q}} \\ \hat{\beta}^{+}_{-\mathbf{q}} \end{pmatrix}, \tag{9.200}$$

then we can find the matrix T such that $\hat{\mathscr{H}}_{\mathbf{q}}$ is diagonal in the operators $\hat{\alpha}$ and $\hat{\beta}$, i.e. such that

$$\hat{\mathscr{H}}_{\mathbf{q}} = \tfrac{1}{2}\hat{Y}^{+}E\hat{Y} \tag{9.201}$$

where E is a diagonal matrix.

As may be verified by straightforward, though perhaps tedious, matrix algebra the matrix T is given by

$$T = \begin{pmatrix} t_0 & t_1 \\ ct_0 & -ct_1 \end{pmatrix} \tag{9.202a}$$

where

$$t_a = \begin{pmatrix} p_a & -m_a \\ -m_a & p_a \end{pmatrix}, \quad \text{with} \quad a = 0, 1 \tag{9.202b}$$

and

$$c = C_{\mathbf{q}}/|C_{\mathbf{q}}| = -\mathscr{J}^*(\mathbf{q})/|\mathscr{J}(\mathbf{q})|. \tag{9.203}$$

The elements p_a and m_a of the matrix t_a are

$$m_a^2 = (2SN)\frac{1}{4E_a}\{(E_a^2 + A'^2)^{1/2} - E_a\} \tag{9.204}$$

and

$$m_a p_a = (2SN)A'/4E_a \tag{9.205}$$

where the energies E_0 and E_1 are given by

$$E_{a=0}^2(\mathbf{q}) = -A'^2 + (\mathscr{A}_\mathbf{q} + |C_\mathbf{q}|)^2\text{: optical,} \tag{9.206}$$

$$E_{a=1}^2(\mathbf{q}) = -A'^2 + (\mathscr{A}_\mathbf{q} - |C_\mathbf{q}|)^2\text{: acoustic.} \tag{9.207}$$

We choose to call E_0 the optical magnon branch and E_1 the acoustic magnon branch by analogy with the terminology used in phonon theory.

Thus

$$\hat{\mathscr{H}}_\mathbf{q} = \tfrac{1}{2}\hat{X}^+\mathscr{H}\hat{X} = \tfrac{1}{2}(T\hat{Y})^+\mathscr{H}T\hat{Y} = \tfrac{1}{2}\hat{Y}^+(T^+\mathscr{H}T)\hat{Y}$$

where

$$T^+\mathscr{H}T = E = (2SN)\begin{pmatrix} E_0(\mathbf{q}) & 0 & 0 & 0 \\ 0 & E_0(\mathbf{q}) & 0 & 0 \\ 0 & 0 & E_1(\mathbf{q}) & 0 \\ 0 & 0 & 0 & E_1(\mathbf{q}) \end{pmatrix} \tag{9.208}$$

and

$$\left(\frac{1}{2SN}\right)\tfrac{1}{2}\hat{Y}^+E\hat{Y} = E_0(\mathbf{q})(\hat{\alpha}_\mathbf{q}^+\hat{\alpha}_\mathbf{q} + \tfrac{1}{2}) + E_1(\mathbf{q})(\hat{\beta}_\mathbf{q}^+\hat{\beta}_\mathbf{q} + \tfrac{1}{2}),$$

i.e.

$$\hat{\mathscr{H}} = \sum_\mathbf{q}\{E_0(\mathbf{q})(\hat{\alpha}_\mathbf{q}^+\hat{\alpha}_\mathbf{q} + \tfrac{1}{2}) + E_1(\mathbf{q})(\hat{\beta}_\mathbf{q}^+\hat{\beta}_\mathbf{q} + \tfrac{1}{2})\}. \tag{9.209}$$

The total x-component of spin $\sum \hat{S}^x + \sum \hat{T}^x$ does not commute with the Hamiltonian (9.193). Thus the cross-section for the scattering by the optical and acoustic magnons will contain terms other than those occurring in the transverse inelastic cross-section (9.42), i.e. the correlation functions

$$\langle \hat{S}_\mathbf{m}^+\hat{S}_{\mathbf{m}'}^+(t)\rangle \quad\text{and}\quad \langle \hat{S}_\mathbf{m}^-\hat{S}_{\mathbf{m}'}^-(t)\rangle \tag{9.210}$$

and the equivalent ones in $\hat{T}_\mathbf{n}^\pm$, are non-zero.

If we denote the position of any site in the h.c.p. structure by $\mathbf{R}$, so that $\mathbf{R}$ takes $2N$ values, where N is the number of ions on one hexagonal sublattice, the correlation function required to calculate the cross-section for the creation or annihilation of a magnon is

$$\sum_{\alpha,\beta}(\delta_{\alpha\beta}-\tilde{\kappa}_\alpha\tilde{\kappa}_\beta)\sum_{\mathbf{R},\mathbf{R}'}\exp\{\mathrm{i}\boldsymbol{\kappa}\cdot(\mathbf{R}'-\mathbf{R})\}\langle\hat{S}^\alpha_{\mathbf{R}}\hat{S}^\beta_{\mathbf{R}'}(t)\rangle \tag{9.211}$$

where $\alpha, \beta = y$ and z. Because

$$\langle\hat{S}^+_{\mathbf{R}}\hat{S}^+_{\mathbf{R}'}(t)-\hat{S}^-_{\mathbf{R}}\hat{S}^-_{\mathbf{R}'}(t)\rangle=0,$$

(9.211) contains no terms involving $\tilde{\kappa}_y\tilde{\kappa}_z$. Thus in (9.211)

$$\sum_{\alpha,\beta}(\delta_{\alpha\beta}-\tilde{\kappa}_\alpha\tilde{\kappa}_\beta)\langle\hat{S}^\alpha_{\mathbf{R}}\hat{S}^\beta_{\mathbf{R}'}(t)\rangle$$
$$=\tfrac{1}{4}(1+\tilde{\kappa}_x^2)\langle\hat{S}^+_{\mathbf{R}}\hat{S}^-_{\mathbf{R}'}(t)+\hat{S}^-_{\mathbf{R}}\hat{S}^+_{\mathbf{R}'}(t)\rangle+\tfrac{1}{4}(\tilde{\kappa}_z^2-\tilde{\kappa}_y^2)\langle\hat{S}^+_{\mathbf{R}}\hat{S}^+_{\mathbf{R}'}(t)+\hat{S}^-_{\mathbf{R}}\hat{S}^-_{\mathbf{R}'}(t)\rangle. \tag{9.212}$$

The sums over $\mathbf{R}$ and $\mathbf{R}'$ in (9.211) must each be split into a sum over $\mathbf{m}$ and a sum over $\mathbf{n}$ and so mixed correlation functions of the type $\langle\hat{S}^+_{\mathbf{m}}\hat{T}^+_{\mathbf{n}}(t)\rangle$ also occur in the evaluation of the cross-section, in addition to those of the type (9.210) and $\langle\hat{S}^\pm_{\mathbf{m}}\hat{S}^\mp_{\mathbf{m}'}(t)\rangle$ and $\langle\hat{T}^\pm_{\mathbf{n}}\hat{T}^\mp_{\mathbf{n}'}(t)\rangle$. Each of these correlation functions can be calculated in terms of the operators $\hat{\alpha}$, $\hat{\beta}$ via the transformation (9.199) and, since these diagonalize $\hat{\mathcal{H}}$, it follows that

$$\langle\hat{\alpha}\hat{\beta}\rangle=\langle\hat{\alpha}^+\hat{\beta}^+\rangle=\langle\hat{\alpha}^+\hat{\alpha}^+\rangle=\langle\hat{\alpha}\hat{\alpha}\rangle=\langle\hat{\beta}\hat{\beta}\rangle=\langle\hat{\beta}^+\hat{\beta}^+\rangle=0.$$

The result of the calculation is

$$\tfrac{1}{4}(1+\tilde{\kappa}_x^2)\sum_{\mathbf{R},\mathbf{R}'}\exp\{\mathrm{i}\boldsymbol{\kappa}\cdot(\mathbf{R}'-\mathbf{R})\}\langle\hat{S}^+_{\mathbf{R}}\hat{S}^-_{\mathbf{R}'}(t)+\hat{S}^-_{\mathbf{R}}\hat{S}^+_{\mathbf{R}'}(t)\rangle$$
$$+\tfrac{1}{4}(\tilde{\kappa}_z^2-\tilde{\kappa}_y^2)\sum_{\mathbf{R},\mathbf{R}'}\exp\{\mathrm{i}\boldsymbol{\kappa}\cdot(\mathbf{R}'-\mathbf{R})\}\langle\hat{S}^+_{\mathbf{R}}\hat{S}^+_{\mathbf{R}'}(t)+\hat{S}^-_{\mathbf{R}}\hat{S}^-_{\mathbf{R}'}(t)\rangle \tag{9.213}$$
$$=\frac{1}{N^2}\sum_{\mathbf{q},a}\{(n_{\mathbf{q},a}+1)\exp(\mathrm{i}\omega_{\mathbf{q},a}t)+n_{\mathbf{q},a}\exp(-\mathrm{i}\omega_{\mathbf{q},a}t)\}$$
$$\times\{\tfrac{1}{4}(1+\tilde{\kappa}_x^2)(m_a^2+p_a^2)+\tfrac{1}{2}(\tilde{\kappa}_y^2-\tilde{\kappa}_z^2)m_ap_a\}$$
$$\times[2\sum_{\mathbf{m},\mathbf{m}'}\exp\{\mathbf{i}(\boldsymbol{\kappa}-\mathbf{q})\cdot(\mathbf{m}'-\mathbf{m})\}$$
$$+(-1)^a\sum_{\mathbf{m},\mathbf{n}}c\exp\{\mathrm{i}(\boldsymbol{\kappa}-\mathbf{q})\cdot(\mathbf{m}-\mathbf{n})\}+c^*\exp\{-\mathrm{i}(\boldsymbol{\kappa}-\mathbf{q})\cdot(\mathbf{m}-\mathbf{n})\}].$$

In (9.213), the integer a takes the two values 0 ($\equiv$optical) and 1 ($\equiv$acoustic) and $n_{\mathbf{q},a}$ is the Bose factor

$$n_{\mathbf{q},a}=\{\exp(\hbar\omega_{\mathbf{q},a}\beta)-1\}^{-1} \tag{9.214}$$

where

$$\hbar\omega_{\mathbf{q},a}=E_a(\mathbf{q}).$$

Consider the sum

$$\sum_{\mathbf{m},\mathbf{n}}\exp\{\mathrm{i}(\boldsymbol{\kappa}-\mathbf{q})\cdot(\mathbf{m}-\mathbf{n})\} \tag{9.215}$$

that occurs in (9.213). If we denote the vector joining an ion on one sublattice to its nearest neighbour on the other sublattice by $\boldsymbol{\rho}$, (9.215) can be written

$$\exp\{i\boldsymbol{\rho}\cdot(\boldsymbol{\kappa}-\mathbf{q})\}\sum_{\mathbf{m},\mathbf{m}'}\exp\{i(\boldsymbol{\kappa}-\mathbf{q})\cdot(\mathbf{m}-\mathbf{m}')\}$$

$$=\exp\{i\boldsymbol{\rho}\cdot(\boldsymbol{\kappa}-\mathbf{q})\}N\frac{(2\pi)^3}{v_0}\sum_{\boldsymbol{\tau}}\delta(\boldsymbol{\kappa}-\mathbf{q}-\boldsymbol{\tau}) \quad (9.216)$$

where $\boldsymbol{\tau}$ are the reciprocal lattice vectors of a *sublattice*. For the particular case of a h.c.p. structure,

$$\boldsymbol{\rho}=\tfrac{1}{3}\mathbf{a}_1+\tfrac{1}{3}\mathbf{a}_2+\tfrac{1}{2}\mathbf{a}_3 \quad (9.217)$$

and the basic vectors $\mathbf{a}_j$, and also the reciprocal lattice vectors $\boldsymbol{\tau}$, are given in § 2.1.

Because $|c|=1$ we may write in (9.213)

$$c=\exp i\phi. \quad (9.218)$$

Thus, using (9.216) and (9.218) in (9.213), the latter becomes

$$\tfrac{1}{2}S\frac{(2\pi)^3}{v_0}\sum_{\boldsymbol{\tau},\mathbf{q}}\delta(\boldsymbol{\kappa}-\mathbf{q}-\boldsymbol{\tau})\sum_{a}\{(n_{\mathbf{q},a}+1)\exp(i\omega_{\mathbf{q},a}t)+n_{\mathbf{q},a}\exp(-i\omega_{\mathbf{q},a}t)\}$$

$$\times\{1+(-1)^a\cos(\boldsymbol{\tau}\cdot\boldsymbol{\rho}+\phi)\}\left\{(1+\tilde{\kappa}_x^2)\frac{(E_a^2+A'^2)^{1/2}}{E_a}+(\tilde{\kappa}_y^2-\tilde{\kappa}_z^2)\frac{A'}{E_a}\right\}. \quad (9.219)$$

From (9.219) we can immediately calculate the neutron cross-section. This is

$$\frac{d^2\sigma}{d\Omega\,dE'}=r_0^2\frac{k'}{k}\{\tfrac{1}{2}gF(\boldsymbol{\kappa})\}^2\exp\{-2W(\boldsymbol{\kappa})\}\tfrac{1}{2}S\frac{(2\pi)^3}{v_0}\sum_{\boldsymbol{\tau},\mathbf{q}}\delta(\boldsymbol{\kappa}-\mathbf{q}-\boldsymbol{\tau})$$

$$\times\sum_{a}[(n_{\mathbf{q},a}+1)\,\delta\{\hbar\omega-E_a(\mathbf{q})\}+n_{\mathbf{q},a}\,\delta\{\hbar\omega+E_a(\mathbf{q})\}]$$

$$\times\{1+(-1)^a\cos(\boldsymbol{\tau}\cdot\boldsymbol{\rho}+\phi)\}\left\{(1+\tilde{\kappa}_x^2)\frac{(E_a^2+A'^2)^{1/2}}{E_a}+(\tilde{\kappa}_y^2-\tilde{\kappa}_z^2)\frac{A'}{E_a}\right\}. \quad (9.220)$$

The cross-section (9.220) is that for the inelastic scattering by the spin-wave modes of the Hamiltonian (9.193) and consists of the sum of two terms, one corresponding to magnon creation and the other to magnon annihilation of the optical and acoustic magnons with energies $E_0(\mathbf{q})$ and $E_1(\mathbf{q})$, respectively. If we set $A=A'=C_{\mathbf{q}}=0$, the Hamiltonian (9.193) is just twice that for linear spin waves in an isotropic Heisenberg ferromagnet and (9.220) becomes identical with the cross-section for this system (eqn (9.44)).

The factor $\{1+(-1)^{\alpha}\cos(\boldsymbol{\tau}\cdot\boldsymbol{\rho}+\phi)\}$ in (9.220) is usually referred to as the magnetic-structure factor since it originates from the two atoms per unit cell of the h.c.p. structure. In general ϕ is not zero nor a multiple of π, except in special symmetry directions.

9.6. Heisenberg antiferromagnets

While there are very few known non-metallic ferromagnets, there exist several simple ionic antiferromagnets that are well described by a Heisenberg exchange Hamiltonian with the addition of some single-ion anisotropy terms. In this section we consider the magnetic properties of Heisenberg antiferromagnets on much the same lines as the discussion of the Heisenberg ferromagnets in § 9.2. We shall, however, find it convenient to work with a more general Hamiltonian than that required to describe antiferromagnets, the Hamiltonian being that also appropriate for the description of ferrimagnets. The antiferromagnet is obtained as a limiting case; namely, when all the ions are identical. We shall discuss ferrimagnets in § 9.7.

We limit our discussion to those antiferromagnets and ferrimagnets that can be described in terms of two identical interpenetrating sublattices, with lattices sites given by the lattice vectors **m** and **n**. Often we distinguish between the two sublattices by referring to one as the 'up' sublattice (lattice vectors **m**) and the other as the 'down' sublattice (**n**).

9.6.1. *Linear spin-wave theory*

For an antiferromagnet, the Heisenberg Hamiltonian is of the form

$$\hat{\mathscr{H}} = 2J\sum_{\mathbf{m},\boldsymbol{\rho}} \hat{\mathbf{S}}_{\mathbf{m}}\cdot\hat{\mathbf{S}}_{\mathbf{m}+\boldsymbol{\rho}}. \tag{9.221}$$

In (9.221) only a nearest-neighbour exchange is included for the present; we shall subsequently allow the exchange to have an arbitrary range.

There is a fundamental difference between calculating the spin-wave spectrum of (9.221) as compared to that for the Heisenberg ferromagnet, namely, that the exact ground state of the Hamiltonian is not known. This means that spin-wave theory is constructed from the approximate ground state, this state being that of lowest classical energy, or Néel state, in which the spins on one sublattice are oppositely directed to those on the other. As may be easily verified, this state is not, in fact, even an eigenstate of the Hamiltonian (9.221).

The spin-wave theory constructed from the Néel state predicts a correction to the latter which is manifest in a zero-point spin deviation. This situation is the same as that for the Heisenberg ferromagnet in the presence of dipolar terms, because in this case the fully aligned state, from which the spin-wave theory is constructed, is no longer the ground state and there are zero-point spin fluctuations.

We are not concerned here with the ground-state properties of the antiferromagnet, or those of the ferrimagnet, but are concerned with the excitations that occur as the spins deviate from their assumed ground-state values. Experiments show that spin-wave theory gives an extremely good description of these states. Also, with essentially the same, simple treatment of two-spin-wave interactions as we gave for the ferromagnet, the theory is found to apply remarkably well over a large temperature range.

Consider the Hamiltonian

$$\mathscr{H}=\sum_{\mathbf{m},\mathbf{r}} J(\mathbf{r})\hat{\mathbf{S}}_{\mathbf{m}}\cdot\hat{\mathbf{S}}_{\mathbf{m}+\mathbf{r}}+\sum_{\mathbf{n},\mathbf{r}} J(\mathbf{r})\hat{\mathbf{S}}_{\mathbf{n}}\cdot\hat{\mathbf{S}}_{\mathbf{n}+\mathbf{r}}+\sum_{\mathbf{m},\mathbf{r}} J_1(\mathbf{R})\hat{\mathbf{S}}_{\mathbf{m}}\cdot\hat{\mathbf{S}}_{\mathbf{m}+\mathbf{R}}+\sum_{\mathbf{n},\mathbf{R}} J_2(\mathbf{R})\hat{\mathbf{S}}_{\mathbf{n}}\cdot\hat{\mathbf{S}}_{\mathbf{n}+\mathbf{r}}$$
$$-g_1\mu_{\mathrm{B}}(H+H_{\mathrm{A},1})\sum_{\mathbf{m}}\hat{S}^z_{\mathbf{m}}-g_2\mu_{\mathrm{B}}(H-H_{\mathrm{A},2})\sum_{\mathbf{n}}\hat{S}^z_{\mathbf{n}} \quad (9.222)$$

where $\mathbf{r}$ connects sites on opposite sublattices and $\mathbf{R}$ those on the same sublattice. In (9.222) the spins on the $\mathbf{m}$ sublattice have magnitude S_1 and those on $\mathbf{n}$ magnitude S_2, and the exchange parameters within each sublattice are denoted by J_1 and J_2 respectively. A uniaxial anisotropy has been incorporated in (9.222) via effective magnetic fields $H_{\mathrm{A},1}$ and $H_{\mathrm{A},2}$, the gyromagnetic ratios for the two types of ion being denoted by g_1 and g_2.

In calculating the spin-wave modes of the Hamiltonian (9.222) it is convenient to apply a canonical transformation to the spin operators associated with ions on the $\mathbf{n}$ sublattice; we rotate the axes of the $\mathbf{n}$ sublattice through π about the x-axes. Thus

$$\begin{aligned}
&\hat{S}^x_{\mathbf{n}}=\hat{T}^x_{\mathbf{n}}, \qquad \hat{S}^+_{\mathbf{n}}=\hat{T}^-_{\mathbf{n}},\\
&\hat{S}^y_{\mathbf{n}}=-\hat{T}^y_{\mathbf{n}}, \qquad \hat{S}^-_{\mathbf{n}}=\hat{T}^+_{\mathbf{n}},\\
&\hat{S}^z_{\mathbf{n}}=-\hat{T}^z_{\mathbf{n}}.
\end{aligned} \quad (9.223)$$

Also the Fourier transforms are defined

$$\begin{aligned}
\hat{S}^+_{\mathbf{m}}&=\frac{1}{N}\sum_{\mathbf{q}}\exp(\mathrm{i}\mathbf{q}\cdot\mathbf{m})\hat{S}^+_{\mathbf{q}},\\
\hat{S}^-_{\mathbf{m}}&=\frac{1}{N}\sum_{\mathbf{q}}\exp(-\mathrm{i}\mathbf{q}\cdot\mathbf{m})\hat{S}^-_{\mathbf{q}},\\
\hat{T}^+_{\mathbf{n}}&=\frac{1}{N}\sum_{\mathbf{q}}\exp(-\mathrm{i}\mathbf{q}\cdot\mathbf{n})\hat{T}^+_{\mathbf{q}},\\
\hat{T}^-_{\mathbf{n}}&=\frac{1}{N}\sum_{\mathbf{q}}\exp(\mathrm{i}\mathbf{q}\cdot\mathbf{n})\hat{T}^-_{\mathbf{q}}
\end{aligned} \quad (9.224)$$

where $\mathbf{q}$ are restricted to the first Brillouin zone of one sublattice, each sublattice having N lattice sites.

Using the commutation rule (9.13) and neglecting terms involving three raising and lowering operators, e.g. $\hat{S}^-_{\mathbf{m}}\hat{S}^+_{\mathbf{m}}\hat{T}^-_{\mathbf{m}+\mathbf{r}}$, the linear equation-of-motion for $\hat{S}^+_{\mathbf{m}}$ is found to be

$$\begin{aligned} i\hbar\,\partial_t\hat{S}^+_{\mathbf{m}} &= 2S_1\sum_{\mathbf{r}} J(\mathbf{r})\{\sigma\hat{S}^+_{\mathbf{m}} + \hat{T}^-_{\mathbf{m}+\mathbf{r}}\} \\ &\quad + 2S_1\sum_{\mathbf{R}} J_1(\mathbf{R})\{-\hat{S}^+_{\mathbf{m}} + \hat{S}^+_{\mathbf{m}+\mathbf{R}}\} + g_1\mu_B(H+H_{A,1})\hat{S}^+_{\mathbf{m}}. \end{aligned} \tag{9.225}$$

In this equation the parameter σ is

$$\sigma = S_2/S_1 \tag{9.226}$$

and it arises in (9.225) because the term

$$\hat{S}^+_{\mathbf{m}}\hat{T}^z_{\mathbf{m}+\mathbf{r}} = \hat{S}^+_{\mathbf{m}}\left(S_2 - \frac{1}{2S_2}\hat{T}^-_{\mathbf{m}+\mathbf{r}}\hat{T}^+_{\mathbf{m}+\mathbf{r}}\right)$$

in the equation-of-motion for $\hat{S}^+_{\mathbf{m}}$ is replaced by

$$S_2\hat{S}^+_{\mathbf{m}}$$

in the linear approximation.

In terms of $S^+_{\mathbf{q}}$ and $\hat{T}^-_{\mathbf{q}}$, (9.225) reads

$$i\hbar\,\partial_t\hat{S}^+_{\mathbf{q}} = [2S_1\sigma\mathcal{J}(0) - 2S_1\{\mathcal{J}_1(0) - \mathcal{J}_1(\mathbf{q})\} + g_1\mu_B(H+H_{A,1})]\hat{S}^+_{\mathbf{q}} + 2S_1\mathcal{J}(\mathbf{q})\hat{T}^-_{\mathbf{q}}. \tag{9.227}$$

The equation-of-motion for $\hat{T}^-_{\mathbf{q}}$ is

$$\begin{aligned} i\hbar\,\partial_t\hat{T}^-_{\mathbf{q}} = -[2S_1\mathcal{J}(0) - 2S_1\sigma\{\mathcal{J}_2(0) - \mathcal{J}_2(\mathbf{q})\} - g_2\mu_B(H-H_{A,2})]\hat{T}^-_{\mathbf{q}} \\ - 2S_1\sigma\mathcal{J}(\mathbf{q})\hat{S}^+_{\mathbf{q}}. \end{aligned} \tag{9.228}$$

The solution of (9.227) and (9.228) for $\hat{S}^+_{\mathbf{q}}$ and $\hat{T}^-_{\mathbf{q}}$ parallels that of eqns (9.165).

Let

$$\hat{S}^+_{\mathbf{q}} = u_{\mathbf{q}}\hat{\alpha}_{\mathbf{q}} + v_{\mathbf{q}}\hat{\beta}^+_{\mathbf{q}}$$

and

$$\hat{T}^-_{\mathbf{q}} = \sigma^{1/2}\{u_{\mathbf{q}}\hat{\beta}^+_{\mathbf{q}} + v_{\mathbf{q}}\hat{\alpha}_{\mathbf{q}}\}. \tag{9.229}$$

For this case we may assume $u_{\mathbf{q}}$ and $v_{\mathbf{q}}$ are real. Then we have the relation

$$u^2_{\mathbf{q}} - v^2_{\mathbf{q}} = 2S_1N \tag{9.230}$$

in order to preserve the commutation relations

$$[\hat{S}^{+}_{\mathbf{m}}, \hat{S}^{-}_{\mathbf{m}'}] = 2S_1\, \delta_{\mathbf{m},\mathbf{m}'}$$

and

$$[\hat{T}^{+}_{\mathbf{n}}, \hat{T}^{-}_{\mathbf{n}'}] = 2S_2\, \delta_{\mathbf{n},\mathbf{n}'}. \tag{9.231}$$

We demand that $\hat{\alpha}_{\mathbf{q}}$ satisfies

$$\hbar\omega_{\mathbf{q},0}\hat{\alpha}_{\mathbf{q}} = [\hat{\alpha}_{\mathbf{q}}, \hat{\mathscr{H}}]$$

and that

$$\hbar\omega_{\mathbf{q},1}\hat{\beta}_{\mathbf{q}} = [\hat{\beta}_{\mathbf{q}}, \hat{\mathscr{H}}]. \tag{9.22}$$

From the former we immediately obtain, using (9.227), (9.228), and (9.229), the following pair of homogeneous equations for the coefficients $u_{\mathbf{q}}$ and $v_{\mathbf{q}}$

$$\begin{aligned}(\hbar\omega_{\mathbf{q},0} - a_1 - b_1)u_{\mathbf{q}} - 2S_1\sigma^{1/2}\mathscr{J}(\mathbf{q})v_{\mathbf{q}} = 0;\\ 2S_1\sigma^{1/2}\mathscr{J}(\mathbf{q})u_{\mathbf{q}} + (\hbar\omega_{\mathbf{q},0} - a_2 + b_2)v_{\mathbf{q}} = 0.\end{aligned} \tag{9.233}$$

The functions a_1, a_2, b_1, and b_2 are given by

$$a_1 = g_1\mu_{\mathrm{B}}H, \qquad a_2 = g_2\mu_{\mathrm{B}}H$$

and

$$\begin{aligned}* \; b_1 = 2S_1\sigma\mathscr{J}(0) - 2S_1\{\mathscr{J}_1(0) - \mathscr{J}_1(\mathbf{q})\} + g_1\mu_{\mathrm{B}}H_{\mathrm{A},1},\\ * \; b_2 = 2S_1\mathscr{J}(0) - 2S_1\sigma\{\mathscr{J}_2(0) - \mathscr{J}_2(\mathbf{q})\} + g_2\mu_{\mathrm{B}}H_{\mathrm{A},2}.\end{aligned} \tag{9.234}$$

Eqn (9.233) possess a solution for

$$2\hbar\omega_{\mathbf{q},0} = (a_1 + a_2 + b_1 - b_2) + 2\Omega(\mathbf{q})$$

where

$$2\Omega(\mathbf{q}) = [(a_1 - a_2 + b_1 + b_2)^2 - S_1S_2\{4\mathscr{J}(\mathbf{q})\}^2]^{1/2}. \tag{9.235a}$$

Further, the second equation in (9.232) gives, in the same way,

$$2\hbar\omega_{\mathbf{q},1} = -(a_1 + a_2 + b_1 - b_2) + 2\Omega(\mathbf{q}). \tag{9.235b}$$

These equations mean that the Hamiltonian (9.222) possesses two linear spin-wave modes, one associated with the operators $\hat{\alpha}^{+}_{\mathbf{q}}$, $\hat{\alpha}_{\mathbf{q}}$ of energy $\hbar\omega_{\mathbf{q},0}$ and another associated with the operators $\hat{\beta}^{+}_{\mathbf{q}}$, $\hat{\beta}_{\mathbf{q}}$ of energy $\hbar\omega_{\mathbf{q},1}$. In general these two modes are not degenerate.

The transformation coefficients $u_{\mathbf{q}}$ and $v_{\mathbf{q}}$, both of which are purely real, are easily obtained from eqns (9.233). It is found that $u_{\mathbf{q}}$ and $v_{\mathbf{q}}$ satisfy

$$* \qquad u^2_{\mathbf{q}} = (2S_1N)(\hbar\omega_{\mathbf{q},0} - a_2 + b_2)/2\Omega(\mathbf{q})$$

and

$$* \quad u_{\mathbf{q}}v_{\mathbf{q}} = -(2S_1N)S_1\sigma^{1/2}\mathscr{J}(\mathbf{q})/\Omega(\mathbf{q}). \tag{9.236}$$

It is instructive to show that the transformation (9.229), with the coefficients $u_{\mathbf{q}}$ and $v_{\mathbf{q}}$ satisfying (9.236), does indeed diagonalize the linearized Hamiltonian that corresponds to the equations-of-motion (9.227) and (9.228), i.e. to show that the Hamiltonian

$$\hat{\mathscr{H}} = \frac{1}{N}\sum_{\mathbf{q}}\left(\frac{a_1+b_1}{2S_1}\right)\hat{S}^-_{\mathbf{q}}\hat{S}^+_{\mathbf{q}} - \left(\frac{a_2-b_2}{2S_2}\right)\hat{T}^-_{\mathbf{q}}\hat{T}^+_{\mathbf{q}} + \mathscr{J}(\mathbf{q})(\hat{S}^-_{\mathbf{q}}\hat{T}^-_{\mathbf{q}} + \hat{S}^+_{\mathbf{q}}\hat{T}^+_{\mathbf{q}}) \quad (9.237a)$$

from which (9.227) and (9.228) can be derived, is equivalent to

$$\hat{\mathscr{H}} = \sum_{\mathbf{q}} \hbar\omega_{\mathbf{q},0}\hat{\alpha}^+_{\mathbf{q}}\hat{\alpha}_{\mathbf{q}} + \hbar\omega_{\mathbf{q},1}\hat{\beta}^+_{\mathbf{q}}\hat{\beta}_{\mathbf{q}}. \quad (9.237b)$$

We now specialize to the case of antiferromagnet, returning to the ferrimagnet in § 9.7.

9.6.2. *Antiferromagnets*

For an antiferromagnet, $g_1 = g_2 = g$, $H_{A,1} = H_{A,2} = H_A$, $S_1 = S_2 = S$, and $\mathscr{J}_1 = \mathscr{J}_2 = \mathscr{J}'$. This means that the functions a_1, a_2, b_1, and b_2 satisfy $a_1 = a_2 = g\mu_B H$ and

$$b_1 = b_2 = 2S\mathscr{J}(0) - 2S\{\mathscr{J}'(0) - \mathscr{J}'(\mathbf{q})\} + g\mu_B H_A = b, \quad \text{say},$$

so

$$\Omega(\mathbf{q}) \equiv [b^2 - \{2S\mathscr{J}(\mathbf{q})\}^2]^{1/2} \quad (9.238)$$

with

$$\hbar\omega_{\mathbf{q},a} = (-1)^a g\mu_B H + \Omega(\mathbf{q}) \quad (a = 0, 1). \quad (9.239)$$

For small values of $\mathbf{q}$ and cubic crystals,

$$\mathscr{J}(\mathbf{q}) = \sum_{\mathbf{r}} \exp(-i\mathbf{q}\cdot\mathbf{r})J(\mathbf{r}) \simeq \mathscr{J}(0) - \tfrac{1}{6}q^2\sum_{\mathbf{r}} r^2 J(\mathbf{r})$$

and we define

$$J^{(2)} = \sum_{\mathbf{r}} r^2 J(\mathbf{r}), \quad (9.240)$$

i.e.

$$\mathscr{J}(0) - \mathscr{J}(\mathbf{q}) \simeq \tfrac{1}{6}q^2 J^{(2)}. \quad (9.241)$$

Similarly,

$$\mathscr{J}'(0) - \mathscr{J}'(\mathbf{q}) \simeq \tfrac{1}{6}q^2 J'^{(2)}. \quad (9.242)$$

(In (9.22) we defined the moments $\overline{l^n}$ through $\overline{l^n} = \sum_{\mathbf{l}} l^{n+2}J(\mathbf{l})/\sum_{\mathbf{l}} l^2 J(\mathbf{l})$. In the notation of (9.240) this would read $J^{(n+2)}/J^{(2)}$.)

Though (9.241) and (9.242) hold rigorously at small $\mathbf{q}$ only for cubic crystals, it is often a good approximation to use them in other cases. The use of (9.241) and (9.242) in (9.239) yields, for $\mathbf{q} \to 0$,

$$\hbar\omega_{\mathbf{q},a} = [g\mu_B H_A\{4S\mathscr{J}(0) + g\mu_B H_A\} + \tfrac{1}{3}q^2 S\{4S\mathscr{J}(0)(J^{(2)} - J'^{(2)}) - 2g\mu_B H_A J'^{(2)}\}]^{1/2} + (-1)^a g\mu_B H \quad (9.243)$$

and this simplifies to

$$\hbar\omega_{\mathbf{q},a} = q\left\{\frac{4S^2}{3}\mathscr{J}(0)(J^{(2)} - J'^{(2)})\right\}^{1/2} + (-1)^a g\mu_B H \tag{9.244}$$

when $H_A = 0$.

In many antiferromagnets it is found that the nearest-neighbour exchange coupling is the dominant one. If this is denoted by J, then from (9.239) with $H = 0$ the spin-wave spectrum is degenerate with

$$\hbar\omega_{\mathbf{q}} = 2rJS\{(1 + h_A)^2 - \gamma_{\mathbf{q}}^2\}^{1/2} = 2rJS\mathscr{E}_{\mathbf{q}} \tag{9.245}$$

where we have defined a reduced anisotropy field $h_A = g\mu_B H_A/2rJS$.

It is seen that the anisotropy field H_A creates a gap in the spin-wave spectrum at $\mathbf{q} = 0$, i.e. it suppresses the low-energy excitations. One important consequence of this is that, particularly for large h_A, when the temperature is high enough to excite spin waves, there is a significant fraction with large values of $\mathbf{q}$.

Note that for the particular case when only nearest-neighbour coupling is taken into account the transformation coefficients have the particularly simple form

$$\left(\frac{1}{2SN}\right)u_{\mathbf{q}}^2 = (1 + h_A + \mathscr{E}_{\mathbf{q}})/2\mathscr{E}_{\mathbf{q}},$$

$$\left(\frac{1}{2SN}\right)v_{\mathbf{q}}^2 = (1 + h_A - \mathscr{E}_{\mathbf{q}})/2\mathscr{E}_{\mathbf{q}},$$

and

$$\left(\frac{1}{2SN}\right)u_{\mathbf{q}}v_{\mathbf{q}} = -\gamma_{\mathbf{q}}/2\mathscr{E}_{\mathbf{q}}. \tag{9.246}$$

Here, $\mathscr{E}_{\mathbf{q}}$ is defined by (9.245) and the expressions for $u_{\mathbf{q}}^2$, $v_{\mathbf{q}}^2$, and $u_{\mathbf{q}}v_{\mathbf{q}}$ follows directly from (9.236).

Let us now calculate some of the thermodynamic properties of the simple antiferromagnet with the spin-wave spectrum (9.245).

First, there are two sets of spin waves, one set corresponding to the operators $\hat{\alpha}_{\mathbf{q}}^+$, $\hat{\alpha}_{\mathbf{q}}$ the other to the operators $\hat{\beta}_{\mathbf{q}}^+$, $\hat{\beta}_{\mathbf{q}}$, and these are degenerate in the absence of a magnetic field. These spin waves can be considered as states in which the spins on the two sublattices precess in the same sense but with unequal amplitude. The doubly degenerate spectrum then corresponds to clockwise and anticlockwise precession of the spins about the anisotropy field direction; in our case this direction coincides with the z-axis.

The sublattice magnetization can be easily calculated in terms of the operators $\hat{\alpha}_{\mathbf{q}}$ and $\hat{\beta}_{\mathbf{q}}$. For it follows from (9.229) that

$$\begin{aligned}\langle \hat{S}_{\mathbf{q}}^{-}\hat{S}_{\mathbf{q}}^{+}\rangle &= \langle (u_{\mathbf{q}}\hat{\alpha}_{\mathbf{q}}^{+}+v_{\mathbf{q}}\hat{\beta}_{\mathbf{q}})(u_{\mathbf{q}}\hat{\alpha}_{\mathbf{q}}+v_{\mathbf{q}}\hat{\beta}_{\mathbf{q}}^{+})\rangle \\ &= u_{\mathbf{q}}^{2}n_{\mathbf{q}}+v_{\mathbf{q}}^{2}(n_{\mathbf{q}}+1)\end{aligned} \tag{9.247}$$

and

$$\begin{aligned}\langle \hat{S}_{\mathbf{m}}^{z}\rangle &= S-\frac{1}{2S}\frac{1}{N^{2}}\sum_{\mathbf{q}}\langle \hat{S}_{\mathbf{q}}^{-}\hat{S}_{\mathbf{q}}^{+}\rangle \\ &= S-\frac{1}{2N}\sum_{\mathbf{q}}\left\{\left(\frac{1+h_{\mathrm{A}}}{\mathscr{E}_{\mathbf{q}}}\right)\coth(\tfrac{1}{2}\hbar\omega_{\mathbf{q}}\beta)-1\right\}.\end{aligned} \tag{9.248}$$

The last line of (9.248) follows immediately from (9.247) on using the expressions (9.246) for $u_{\mathbf{q}}^{2}$ and $v_{\mathbf{q}}^{2}$. It is worthwhile to note that the $\hat{\alpha}_{\mathbf{q}}^{+}$ spin-wave contribution to (9.247) contains $n_{\mathbf{q}}$, the $\hat{\beta}_{\mathbf{q}}^{+}$ spin-wave contribution the factor $1+n_{\mathbf{q}}$; at zero temperature only the latter is finite, indicating that the $\hat{\beta}_{\mathbf{q}}^{+}$ mode decreases the sublattice magnetization. The reverse is true for the down sublattice.

Taking the limit $\beta\rightarrow\infty$ $(T=0)$ in (9.248) we see that the average value of $\hat{S}^{z}$ is less than its value in the Néel ground state S by an amount

$$\delta\langle \hat{S}^{z}\rangle_{0}=\frac{1}{2N}\sum_{\mathbf{q}}\left\{\left(\frac{1+h_{\mathrm{A}}}{\mathscr{E}_{\mathbf{q}}}\right)-1\right\}. \tag{9.249}$$

$\delta\langle \hat{S}^{z}\rangle_{0}$, the zero-point spin deviation, has a value $\simeq 0.06$ for a body-centred antiferromagnet with $h_{\mathrm{A}}=0$, and increases in magnitude as the number of neighbours decreases, e.g. for a quadratic layer $(r=4)$ it is $\simeq 0.2$.

At a finite temperature we write for $\langle \hat{S}^{z}\rangle$

$$\langle \hat{S}^{z}\rangle = S-\delta\langle \hat{S}^{z}\rangle_{0}-\frac{1}{N}\sum_{\mathbf{q}}\left(\frac{1+h_{\mathrm{A}}}{\mathscr{E}_{\mathbf{q}}}\right)\frac{1}{\exp(\hbar\omega_{\mathbf{q}}\beta)-1}. \tag{9.250}$$

For small q and $h_{\mathrm{A}}=0$ (see eqn (9.17))

$$\mathscr{E}_{\mathbf{q}}=q(\rho^{2}/3)^{1/2}=Dq/2rJS, \tag{9.251}$$

which defines the parameter D for an antiferromagnet. Hence

$$\begin{aligned}&\frac{1}{N}\sum_{\mathbf{q}}\frac{1}{(1-\gamma_{\mathbf{q}}^{2})^{1/2}}\frac{1}{\exp(\hbar\omega_{\mathbf{q}}\beta)-1} \\ &\quad\simeq\frac{1}{N}\frac{V}{(2\pi)^{3}}\int_{0}^{\infty}\mathrm{d}q\,4\pi q^{2}\frac{\sqrt{3}}{\rho q}\frac{1}{\exp(\beta Dq)-1}=v_{0}\frac{\sqrt{3}}{2\rho}\left(\frac{k_{\mathrm{B}}T}{\pi D}\right)^{2}\zeta(2)\end{aligned} \tag{9.252}$$

where v_{0} is the volume per ion. We therefore see that the sublattice magnetization of an antiferromagnet decreases as T^{2}, in contrast to the $T^{3/2}$ dependence of the magnetization in a ferromagnet.

It is also of interest to calculate the exchange energy. A straightforward calculation shows that

$$\begin{aligned} E_{\text{exch}} &= \langle \hat{\mathscr{H}} \rangle - \langle \text{anisotropy energy} \rangle \\ &= -2rJS(S+1)N + 2rJS \sum_{\mathbf{q}} \left\{ \frac{1 + h_{\text{A}} - \gamma_{\mathbf{q}}^2}{\mathscr{E}_{\mathbf{q}}} \right\} \coth(\tfrac{1}{2}\hbar\omega_{\mathbf{q}}\beta) \end{aligned} \quad (9.253)$$

for a nearest-neighbour model.

In the limit $\beta \to \infty$, E_{exch} can be written

$$E_{\text{exch}} = -2rJS^2N\{1 + e_0/S\}: \qquad T = 0, \quad (9.254)$$

where

$$e_0 = \frac{1}{N} \sum_{\mathbf{q}} \{1 - (1 + h_{\text{A}} - \gamma_{\mathbf{q}}^2)/\mathscr{E}_{\mathbf{q}}\}. \quad (9.255)$$

Note that increasing the anisotropy field *increases* ground-state energy, tending in the limit $h_{\text{A}} \to \infty$ to the ground-state energy for the corresponding Ising model, $-2rJS^2N$. For $h_{\text{A}} \gg 0$,

$$E_{\text{exch}} \simeq -2rJS^2N\{1 + 1/rS(1 + h_{\text{A}})\}: \qquad T = 0 \quad (9.256)$$

and, for $h_{\text{A}} = 0$ and a body-centred orthorhombic crystal,

$$E_{\text{exch}} = -2rJS^2N\{1 + 0.582/rS\}: \qquad T = 0. \quad (9.257)$$

As one would anticipate, e_0 for a given value of h_{A} increases as the number of nearest neighbours decreases. Several quasi-one-dimensional antiferromagnets have been studied with neutron spectroscopy, e.g. $CsMnBr_3$ reported by Collins and Gaulin (1984).

9.6.3. *One-magnon cross-sections*

Now that we have familiarized ourselves with the fundamental properties of the antiferromagnet, we turn to the calculation of the neutron cross-section for inelastic scattering involving the creation or annihilation of single magnons. Because the *total* z-component of spin commutes with the Hamiltonian (9.222), the cross-section contains only the terms $\langle \hat{S}^-\hat{S}^+ \rangle$ and $\langle \hat{S}^+\hat{S}^- \rangle$.

If by $\mathbf{R}$ we mean the position vector of *any* ion in the antiferromagnet, a straightforward calcuation gives

$$\begin{aligned} &\sum_{\mathbf{R},\mathbf{R}'} \exp\{\mathrm{i}\boldsymbol{\kappa} \cdot (\mathbf{R}' - \mathbf{R})\} \langle \hat{S}^+_{\mathbf{R}} \hat{S}^-_{\mathbf{R}'}(t) \rangle \\ &= \frac{1}{N^2} \sum_{\mathbf{q}} \{(1 + n_{\mathbf{q},0})\exp(\mathrm{i}\omega_{\mathbf{q},0}t) + n_{\mathbf{q},1} \exp(-\mathrm{i}\omega_{\mathbf{q},1}t)\} \\ &\quad \times \left\{ (u_{\mathbf{q}}^2 + v_{\mathbf{q}}^2) \left| \sum_{\mathbf{m}} \exp\{\mathrm{i}(\boldsymbol{\kappa} - \mathbf{q}) \cdot \mathbf{m}\} \right|^2 + u_{\mathbf{q}}v_{\mathbf{q}} \sum_{\mathbf{m},\mathbf{n}} [\exp\{\mathrm{i}(\boldsymbol{\kappa} - \mathbf{q}) \cdot (\mathbf{m} - \mathbf{n})\} + \text{c.c.}] \right\} \end{aligned}$$

(9.258)

where

$$n_{\mathbf{q},a} = \{\exp(\hbar\omega_{\mathbf{q},a}\beta) - 1\}^{-1}; \quad (a = 0, 1).$$

We now utilize the result (cf. § 9.5),

$$\sum_{\mathbf{m},\mathbf{n}} \exp\{\mathrm{i}(\boldsymbol{\kappa} - \mathbf{q}) \cdot (\mathbf{m} - \mathbf{n})\} = \exp\{\mathrm{i}\boldsymbol{\rho} \cdot (\boldsymbol{\kappa} - \mathbf{q})\} \left| \sum_{\mathbf{m}} \exp\{\mathrm{i}(\boldsymbol{\kappa} - \mathbf{q}) \cdot \mathbf{m}\} \right|^2, \tag{9.259}$$

where $\boldsymbol{\rho}$ is the vector joining an ion on one sublattice with its nearest neighbour on the opposite sublattice, to write (9.258) as

$$\frac{1}{N^2} \sum_{\mathbf{q}} \left| \sum_{\mathbf{m}} \exp\{\mathrm{i}(\boldsymbol{\kappa} - \mathbf{q}) \cdot \mathbf{m}\} \right|^2 \{(1 + n_{\mathbf{q},0})\exp(\mathrm{i}\omega_{\mathbf{q},0}t) + n_{\mathbf{q},1} \exp(-\mathrm{i}\omega_{\mathbf{q},1}t)\}$$
$$\times \{u_{\mathbf{q}}^2 + v_{\mathbf{q}}^2 + 2u_{\mathbf{q}}v_{\mathbf{q}} \cos \boldsymbol{\rho} \cdot (\boldsymbol{\kappa} - \mathbf{q})\}. \tag{9.260}$$

Since

$$\langle \hat{S}_{\mathbf{R}}^{-} \hat{S}_{\mathbf{R}'}^{+}(t) \rangle = \langle \hat{S}_{\mathbf{R}'}^{+} \hat{S}_{\mathbf{R}}^{-}(-t + \mathrm{i}\hbar\beta) \rangle, \tag{9.261}$$

the function

$$\sum_{\mathbf{R},\mathbf{R}'} \exp\{\mathrm{i}(\mathbf{R}' - \mathbf{R}) \cdot \boldsymbol{\kappa}\} \langle \hat{S}_{\mathbf{R}}^{-} \hat{S}_{\mathbf{R}'}^{+}(t) \rangle$$

is easily calculated from (9.260). Hence, the cross-sections for creation and annihilation of a single magnon are

$$\left(\frac{\mathrm{d}^2\sigma}{\mathrm{d}\Omega\,\mathrm{d}E'}\right)^{(\pm)} = r_0^2 \frac{k'}{k} \{\tfrac{1}{2}gF(\boldsymbol{\kappa})\}^2 \tfrac{1}{4}(1 + \tilde{\kappa}_z^2)\exp\{-2W(\boldsymbol{\kappa})\} \frac{(2\pi)^3}{Nv_0}$$
$$\times \sum_{a=0,1} \sum_{\mathbf{q},\boldsymbol{\tau}} (n_{\mathbf{q},a} + \tfrac{1}{2} \pm \tfrac{1}{2})\, \delta(\hbar\omega_{\mathbf{q},a} \mp \hbar\omega)\, \delta(\boldsymbol{\kappa} \mp \mathbf{q} - \boldsymbol{\tau})\{u_{\mathbf{q}}^2 + v_{\mathbf{q}}^2 + 2u_{\mathbf{q}}v_{\mathbf{q}} \cos \boldsymbol{\rho} \cdot \boldsymbol{\tau}\}. \tag{9.262}$$

The reciprocal lattice vectors $\boldsymbol{\tau}$ are those of a sublattice. The cross-section (9.262) applies for any two-sublattice antiferromagnet; the coefficients $u_{\mathbf{q}}$ and $v_{\mathbf{q}}$ are given by (9.230) and (9.236) with $g_1 = g_2$, $H_{A,1} = H_{A,2}$, $\mathcal{J}_1 = \mathcal{J}_2$, and $S_1 = S_2$. When the external field exceeds a critical value, determined by the anisotropy, a transition to a spin-flop configuration occurs. The corresponding cross-section is quite different to (9.262) (Lovesey 1981*a*).

If $2\boldsymbol{\rho} = \mathbf{m}$, where $\mathbf{m}$ is some lattice vector of a sublattice, it follows by definition of the reciprocal lattice vectors $\boldsymbol{\tau}$ that

$$\boldsymbol{\rho} \cdot \boldsymbol{\tau} = \tfrac{1}{2}\mathbf{m} \cdot \boldsymbol{\tau} = \pi \times \text{integer} \tag{9.263}$$

Thus

$$\cos \boldsymbol{\tau} \cdot \boldsymbol{\rho} = \pm 1 \tag{9.264}$$

according to whether the integer is even (+) or odd (−).

For instance, consider a b.c.c. lattice. In this case the sublattices are simple cubic lattices, $\boldsymbol{\rho}=\frac{1}{2}a(1, 1, 1)$ and the reciprocal lattice vectors $\boldsymbol{\tau}$ are given by

$$\boldsymbol{\tau}=\frac{2\pi}{a}(\tau_1, \tau_2, \tau_3)$$

where τ_1, τ_2, τ_3 are arbitrary integers. The reciprocal lattice vectors for a b.c.c. structure, however, are given by

$$\boldsymbol{\tau}_n=\frac{2\pi}{a}(\tau_1, \tau_2, \tau_3)$$

where $\tau_1+\tau_2+\tau_3$ is an even integer. Hence

$$\boldsymbol{\rho}\cdot\boldsymbol{\tau}_n=\pi\times\text{even integer} \tag{9.265}$$

and

$$\boldsymbol{\rho}\cdot(\boldsymbol{\tau}_n+\mathbf{w})=\pi\times\text{odd integer}, \tag{9.266}$$

if

$$\mathbf{w}=\frac{2\pi}{a}(w_1, w_2, w_3)$$

where $w_1+w_2+w_3$ is an odd integer. Thus we may write

$$\sum_{\boldsymbol{\tau}}\delta(\boldsymbol{\kappa}-\mathbf{q}-\boldsymbol{\tau})(u_{\mathbf{q}}^2+v_{\mathbf{q}}^2+2u_{\mathbf{q}}v_{\mathbf{q}}\cos\boldsymbol{\rho}\cdot\boldsymbol{\tau})\equiv\sum_{\boldsymbol{\tau}_n}(u_{\mathbf{q}}\pm v_{\mathbf{q}})^2\begin{cases}\delta(\boldsymbol{\kappa}-\mathbf{q}-\boldsymbol{\tau}_n)\\ \delta(\boldsymbol{\kappa}-\mathbf{q}-\boldsymbol{\tau}_n-\mathbf{w}),\end{cases} \tag{9.267}$$

that is to say, the scattering occurs at all the reciprocal lattice vectors, but near the peaks that coincide with the nuclear reflections the intensity of scattering is small $[\sim(u_{\mathbf{q}}+v_{\mathbf{q}})^2]$ whereas near the *superlattice* reflections the intensity of scattering is large $[\sim(u_{\mathbf{q}}-v_{\mathbf{q}})^2]$.

Also note that for a b.c.c. lattice the vectors $\mathbf{R}$ are given by

$$\mathbf{R}=\tfrac{1}{2}a(R_1, R_2, R_3)$$

where the integers R_1, R_2, R_3 are either all even or all odd. Now $\exp(i\mathbf{w}\cdot\mathbf{R})=1$ when all R_i are even and equals -1 when all R_i are odd, i.e. $\exp(i\mathbf{w}\cdot\mathbf{R})$ alternates in sign as $\mathbf{R}$ goes from one sublattice to another.

Similarly, a s.c. lattice can be constructed from two sublattices corresponding to the two cases $\mathbf{m}=a(m_1, m_2, m_3)$ with $m_1+m_2+m_3$ an even integer and $\mathbf{n}=a(n_1, n_2, n_3)$ with $n_1+n_2+n_3$ an odd integer. For this lattice,

$$\boldsymbol{\tau}_n=\frac{2\pi}{a}(\tau_1, \tau_2, \tau_3)$$

and we can choose $\mathbf{w}=(\pi/a)(1, 1, 1)$.

For the case when only a nearest-neighbour exchange is important (9.246) shows that

$$
\begin{aligned}
(u_{\mathbf{q}} \pm v_{\mathbf{q}})^2 &= 2SN(1 + h_{\mathrm{A}} \mp \gamma_{\mathbf{q}})/\mathscr{E}_{\mathbf{q}} \\
&\equiv 2SN(1 + h_{\mathrm{A}} \mp \gamma_{\mathbf{q}})/\{(1 + h_{\mathrm{A}})^2 - \gamma_{\mathbf{q}}^2\}^{1/2}.
\end{aligned}
$$

9.6.4. Two-magnon interactions

To study the effects of the two-spin-wave dynamical interaction on the spin-wave dispersion relation of an antiferromagnet we use a slight extension of the method of § 9.2.3 and, for brevity, we consider from the outset only a nearest-neighbour exchange coupling. The discussion can easily be extended to cover the case of a longer-range coupling.

The calculation of the spin-wave damping in antiferromagnets is quite complicated, and it is not pursued here. A comparative study of theory and measurements for $RbMnF_3$ is given in Rezende and White (1978).

For the spins on the 'up' sublattice,

$$
\hat{S}^z_{\mathbf{m}} = S - \hat{a}^+_{\mathbf{m}}\hat{a}_{\mathbf{m}}, \qquad \hat{S}^+_{\mathbf{m}} = (2S)^{1/2}\hat{a}_{\mathbf{m}}, \qquad \hat{S}^-_{\mathbf{m}} = (2S)^{1/2}\hat{a}^+_{\mathbf{m}}(1 - \hat{a}^+_{\mathbf{m}}\hat{a}_{\mathbf{m}}/2S), \tag{9.268}
$$

and, in view of (9.223), for the spins on the 'down' sublattice,

$$
\hat{S}^z_{\mathbf{n}} = -S + \hat{b}^+_{\mathbf{n}}\hat{b}_{\mathbf{n}}, \qquad \hat{S}^+_{\mathbf{n}} = (2S)^{1/2}\hat{b}^+_{\mathbf{n}}, \qquad \hat{S}^-_{\mathbf{n}} = (2S)^{1/2}(1 - \hat{b}^+_{\mathbf{n}}\hat{b}_{\mathbf{n}}/2S)\hat{b}_{\mathbf{n}}. \tag{9.269}
$$

The Hamiltonian we use is obtained from (9.221) with the addition of a uniaxial anisotropy

$$
\hat{\mathscr{H}} = 2J \sum_{\mathbf{m},\boldsymbol{\rho}} \hat{\mathbf{S}}_{\mathbf{m}} \cdot \hat{\mathbf{S}}_{\mathbf{m}+\boldsymbol{\rho}} - g\mu_{\mathrm{B}}H_{\mathrm{A}} \sum_{\mathbf{m}} \hat{S}^z_{\mathbf{m}} + g\mu_{\mathrm{B}}H_{\mathrm{A}} \sum_{\mathbf{n}} \hat{S}^z_{\mathbf{n}}. \tag{9.270}
$$

Inserting the transformations (9.268) and (9.269), and defining

$$
\hat{a}_{\mathbf{m}} = \frac{1}{\surd N} \sum_{\mathbf{q}} \exp(\mathrm{i}\mathbf{q} \cdot \mathbf{m})\hat{a}_{\mathbf{q}}
$$

and

$$
\hat{b}^+_{\mathbf{n}} = \frac{1}{\surd N} \sum_{\mathbf{q}} \exp(\mathrm{i}\mathbf{q} \cdot \mathbf{n})\hat{b}^+_{\mathbf{q}}, \tag{9.271}
$$

the Hamiltonian (9.270) becomes, apart from constant terms,

$$
\begin{aligned}
\frac{1}{2rJS}\hat{\mathscr{H}} &= \sum_{\mathbf{q}} \{(1 + h_{\mathrm{A}})(\hat{a}^+_{\mathbf{q}}\hat{a}_{\mathbf{q}} + \hat{b}^+_{\mathbf{q}}\hat{b}_{\mathbf{q}}) + \gamma_{\mathbf{q}}(\hat{a}_{\mathbf{q}}\hat{b}_{\mathbf{q}} + \hat{a}^+_{\mathbf{q}}\hat{b}^+_{\mathbf{q}})\} \\
&\quad - \frac{1}{NS} \sum_{\mathbf{1},\mathbf{2},\mathbf{3},\mathbf{4}} \delta_{\mathbf{1}+\mathbf{4},\mathbf{2}+\mathbf{3}}(\gamma_{\mathbf{3}-\mathbf{4}}\hat{a}^+_{\mathbf{1}}\hat{a}_{\mathbf{2}}\hat{b}^+_{\mathbf{3}}\hat{b}_{\mathbf{4}} + \tfrac{1}{2}\gamma_{\mathbf{1}}\hat{a}_{-\mathbf{1}}\hat{b}^+_{\mathbf{2}}\hat{b}_{-\mathbf{3}}\hat{b}_{\mathbf{4}} + \tfrac{1}{2}\gamma_{\mathbf{2}}\hat{a}^+_{\mathbf{1}}\mathrm{b}^+_{\mathbf{2}}\hat{a}^+_{-\mathbf{3}}\hat{a}_{-\mathbf{4}}).
\end{aligned} \tag{9.272}
$$

In the second term in (9.272) we have used the shorthand notation $\mathbf{q}_1 = \mathbf{1}$, $\mathbf{q}_2 = \mathbf{2}$, etc.

The first part of (9.272) represents the non-interacting spin waves and is brought into diagonal form by using the transformation (cf. (9.229))

$$\hat{a}_{\mathbf{q}} = \bar{u}_{\mathbf{q}}\hat{\alpha}_{\mathbf{q}} + \bar{v}_{\mathbf{q}}\hat{\beta}^+_{\mathbf{q}}, \qquad \hat{b}^+_{\mathbf{q}} = \bar{u}_{\mathbf{q}}\hat{\beta}^+_{\mathbf{q}} + \bar{v}_{\mathbf{q}}\hat{\alpha}_{\mathbf{q}}. \tag{9.273}$$

The coefficients $\bar{u}_{\mathbf{q}}$ and $\bar{v}_{\mathbf{q}}$ differ from those given by eqn (9.246) only in the factor $(2SN)^{-1}$. This difference arises because of the different normalization adopted for the Bose operators $\hat{a}_{\mathbf{q}}$, $\hat{a}^+_{\mathbf{q}}$ as compared to that used for the operators $\hat{S}^{\pm}_{\mathbf{q}}$. We recall that the non-interacting spin waves have the energy spectrum

$$\hbar\omega_{\mathbf{q}} = 2rJS\{(1+h_{\mathrm{A}})^2 - \gamma^2_{\mathbf{q}}\}^{1/2} \equiv 2rJS\mathscr{E}_{\mathbf{q}}. \tag{9.274}$$

The procedure we adopt to approximate the products of four Bose operators in the second part of (9.272) by products of two operators is just the same as that used in § 9.2.3. Take first the term $\hat{a}^+_{\mathbf{1}}\hat{a}_{\mathbf{2}}\hat{b}^+_{\mathbf{3}}\hat{b}_{\mathbf{4}}$. Because the terms $\langle\hat{a}^+\hat{b}\rangle$, $\langle\hat{a}\hat{b}^+\rangle$, etc. are zero,

$$\begin{aligned}\hat{a}^+_{\mathbf{1}}\hat{a}_{\mathbf{2}}\hat{b}^+_{\mathbf{3}}\hat{b}_{\mathbf{4}} \rightarrow\ & \delta_{\mathbf{1},\mathbf{2}}\langle\hat{a}^+_{\mathbf{1}}\hat{a}_{\mathbf{1}}\rangle\hat{b}^+_{\mathbf{3}}\hat{b}_{\mathbf{4}} + \delta_{\mathbf{3},\mathbf{4}}\hat{a}^+_{\mathbf{1}}\hat{a}_{\mathbf{2}}\langle\hat{b}^+_{\mathbf{3}}\hat{b}_{\mathbf{3}}\rangle \\ &+ \delta_{\mathbf{1},\mathbf{3}}\langle\hat{a}^+_{\mathbf{1}}\hat{b}^+_{\mathbf{1}}\rangle\hat{a}_{\mathbf{2}}\hat{b}_{\mathbf{4}} + \delta_{\mathbf{2},\mathbf{4}}\hat{a}^+_{\mathbf{1}}\hat{b}^+_{\mathbf{3}}\langle\hat{a}_{\mathbf{2}}\hat{b}_{\mathbf{2}}\rangle.\end{aligned} \tag{9.275}$$

In (9.275) we have dropped the constant terms. We denote the thermal averages

$$\langle\hat{a}^+_{\mathbf{1}}\hat{a}_{\mathbf{1}}\rangle = \langle\hat{b}^+_{\mathbf{1}}\hat{b}_{\mathbf{1}}\rangle = \eta_{\mathbf{1}}$$

and

$$\langle\hat{a}^+_{\mathbf{1}}\hat{b}^+_{\mathbf{1}}\rangle = \langle\hat{a}_{\mathbf{1}}\hat{b}_{\mathbf{1}}\rangle = \xi_{\mathbf{1}}. \tag{9.276}$$

Thus, from (9.275) and (9.276),

$$\begin{aligned}\hat{a}^+_{\mathbf{1}}\hat{a}_{\mathbf{2}}\hat{b}^+_{\mathbf{3}}\hat{b}_{\mathbf{4}}\,\delta_{\mathbf{1}+\mathbf{4},\mathbf{2}+\mathbf{3}}\,\gamma_{\mathbf{3}-\mathbf{4}} \rightarrow\ & \delta_{\mathbf{1},\mathbf{2}}\,\delta_{\mathbf{3},\mathbf{4}}(\eta_{\mathbf{1}}\hat{b}^+_{\mathbf{3}}\hat{b}_{\mathbf{3}} + \eta_{\mathbf{3}}\hat{a}^+_{\mathbf{1}}\hat{a}_{\mathbf{1}}) \\ &+ \delta_{\mathbf{1},\mathbf{3}}\,\delta_{\mathbf{2},\mathbf{4}}(\xi_{\mathbf{1}}\hat{a}_{\mathbf{2}}\hat{b}_{\mathbf{2}} + \xi_{\mathbf{2}}\hat{a}^+_{\mathbf{1}}\hat{b}^+_{\mathbf{1}})\gamma_{\mathbf{1}-\mathbf{2}}.\end{aligned} \tag{9.277}$$

Similarly,

$$\begin{aligned}&\hat{a}_{-\mathbf{1}}\hat{b}^+_{\mathbf{2}}\hat{b}_{-\mathbf{3}}\hat{b}_{\mathbf{4}}\,\delta_{\mathbf{1}+\mathbf{4},\mathbf{2}+\mathbf{3}}\,\gamma_{\mathbf{1}} \\ &\quad\rightarrow (\xi_{\mathbf{1}}\hat{b}^+_{\mathbf{2}}\hat{b}_{\mathbf{2}} + \eta_{\mathbf{2}}\hat{a}_{-\mathbf{1}}\hat{b}_{-\mathbf{1}})\gamma_{\mathbf{1}}(\delta_{\mathbf{1},\mathbf{3}}\,\delta_{\mathbf{2},\mathbf{4}} + \delta_{\mathbf{2},-\mathbf{3}}\,\delta_{\mathbf{1},-\mathbf{4}})\end{aligned} \tag{9.278}$$

and

$$\begin{aligned}&\hat{a}^+_{\mathbf{1}}\hat{b}^+_{\mathbf{2}}\hat{a}^+_{-\mathbf{3}}\hat{a}_{-\mathbf{4}}\,\delta_{\mathbf{1}+\mathbf{4},\mathbf{2}+\mathbf{3}}\gamma_{\mathbf{2}} \\ &\quad\rightarrow \delta_{\mathbf{1},\mathbf{2}}\,\delta_{\mathbf{3},\mathbf{4}}(\xi_{\mathbf{1}}\hat{a}^+_{-\mathbf{3}}\hat{a}_{-\mathbf{3}} + \eta_{\mathbf{3}}\hat{a}^+_{\mathbf{1}}\hat{b}^+_{\mathbf{1}})\gamma_{\mathbf{1}} + \delta_{\mathbf{1},-\mathbf{4}}\,\delta_{\mathbf{2},-\mathbf{3}}\,(\eta_{\mathbf{1}}\hat{b}^+_{\mathbf{2}}\hat{a}^+_{\mathbf{2}} + \xi_{\mathbf{2}}\hat{a}^+_{\mathbf{1}}\hat{a}_{\mathbf{1}}).\end{aligned} \tag{9.279}$$

If (9.277), (9.278), and (9.279) are inserted into the second part of the Hamiltonian (9.272), it reduces to

$$-\frac{1}{SN}\sum_{\mathbf{1},\mathbf{2}}(\hat{a}_\mathbf{1}^+\hat{a}_\mathbf{1}+\hat{b}_\mathbf{1}^+\hat{b}_\mathbf{1})(\eta_\mathbf{2}+\gamma_\mathbf{2}\xi_\mathbf{2})+(\hat{a}_\mathbf{1}\hat{b}_\mathbf{1}+\hat{a}_\mathbf{1}^+\hat{b}_\mathbf{1}^+)(\gamma_{\mathbf{1}-\mathbf{2}}\xi_\mathbf{2}+\gamma_\mathbf{1}\eta_\mathbf{2}),$$

which, when added to the first part of (9.272), the Hamiltonian of the non-interacting spin waves, yields

$$\frac{1}{2rJS}\mathscr{H}=\sum_\mathbf{q}(\hat{a}_\mathbf{q}^+\hat{a}_\mathbf{q}+\hat{b}_\mathbf{q}^+\hat{b}_\mathbf{q})\{1+h_\mathrm{A}-C(T)\}+(\hat{a}_\mathbf{q}\hat{b}_\mathbf{q}+\hat{a}_\mathbf{q}^+\hat{b}_\mathbf{q}^+)\gamma_\mathbf{q}\{1-C(T)\}. \tag{9.280}$$

Here

$$C(T)=\frac{1}{NS}\sum_\mathbf{q}(\eta_\mathbf{q}+\gamma_\mathbf{q}\xi_\mathbf{q}). \tag{9.281}$$

In arriving at (9.280) we have made use of (9.61) to write

$$\sum_\mathbf{2}\gamma_{\mathbf{1}-\mathbf{2}}\xi_\mathbf{2}=\gamma_\mathbf{1}\sum_\mathbf{2}\gamma_\mathbf{2}\xi_\mathbf{2}. \tag{9.282}$$

The Hamiltonian (9.280) has exactly the same form as for non-interacting spin waves except that $1+h_\mathrm{A}$ has been replaced by $1+h_\mathrm{A}-C(T)$ and $\gamma_\mathbf{q}$ by $\gamma_\mathbf{q}\{1-C(T)\}$, and hence we can immediately write down the quantities of interest. The transformation coefficients $\bar{u}_\mathbf{q}$ and $\bar{v}_\mathbf{q}$ in (9.273) for the diagonalization of (9.280) satisfy

$$\bar{u}_\mathbf{q}^2-\bar{v}_\mathbf{q}^2=1, \qquad \bar{u}_\mathbf{q}^2=\frac{1+h_\mathrm{A}-C(T)+\mathscr{E}_\mathbf{q}(T)}{2\mathscr{E}_\mathbf{q}(T)} \tag{9.283}$$

and

$$2\bar{u}_\mathbf{q}\bar{v}_\mathbf{q}=-\gamma_\mathbf{q}\{1-C(T)\}/\mathscr{E}_\mathbf{q}(T).$$

In (9.283)

$$\begin{aligned}\mathscr{E}_\mathbf{q}(T)&=[\{1+h_\mathrm{A}-C(T)\}^2-\gamma_\mathbf{q}^2\{1-C(T)\}^2]^{1/2}\\&=\{1-C(T)\}[\{1+h_\mathrm{A}(T)\}^2-\gamma_\mathbf{q}^2]^{1/2}\end{aligned} \tag{9.284}$$

where

$$h_\mathrm{A}(T)=\frac{h_\mathrm{A}}{1-C(T)}\simeq h_\mathrm{A}\{1+C(T)\}, \tag{9.285}$$

i.e. the dispersion relation is

$$\hbar\omega_\mathbf{q}(T)=2rJS\{1-C(T)\}[\{1+h_\mathrm{A}(T)\}^2-\gamma_\mathbf{q}^2]^{1/2}. \tag{9.286}$$

Thus, as we found in the ferromagnet, the dynamical interaction has led to a renormalization of the dispersion relation, but for the antiferromagnet there is the additional feature of renormalizing the anisotropy field.

For the function

$$C(T)=\frac{1}{NS}\sum_{\mathbf{q}}\{\langle\hat{a}_{\mathbf{q}}^{+}\hat{a}_{\mathbf{q}}\rangle+\gamma_{\mathbf{q}}\langle\hat{a}_{\mathbf{q}}\hat{b}_{\mathbf{q}}\rangle\},$$

we eventually obtain the integral equation

$$C(T)=\frac{1}{2NS}\sum_{\mathbf{q}}\left[\frac{\{1-C(T)\}(1-\gamma_{\mathbf{q}}^{2})+h_{\mathrm{A}}}{\mathscr{E}_{\mathbf{q}}(T)}\right]\coth\{\tfrac{1}{2}\hbar\omega_{\mathbf{q}}(T)\beta\}-1. \tag{9.287}$$

And, for $\langle\hat{S}^{z}\rangle$,

$$\begin{aligned}\langle\hat{S}^{z}\rangle&=S-\frac{1}{N}\sum_{\mathbf{q}}\langle\hat{a}_{\mathbf{q}}^{+}\hat{a}_{\mathbf{q}}\rangle\\&=S-\frac{1}{2N}\sum_{\mathbf{q}}\left\{\frac{1+h_{\mathrm{A}}-C(T)}{\mathscr{E}_{\mathbf{q}}(T)}\right\}\coth\{\tfrac{1}{2}\hbar\omega_{\mathbf{q}}(T)\beta\}-1.\end{aligned} \tag{9.288}$$

To calculate the dispersion relation (9.286) as a function of temperature or to compute the sublattice magnetization (9.288), we have to solve a complicated integral equation and this in general is a numerical problem. However, the equations undergo a great simplification for $h_{\mathrm{A}}=0$ and from these it is relatively straightforward to calculate the leading-order corrections introduced by the dynamical interaction.

Setting $h_{\mathrm{A}}=0$ in (9.287), $C(T)$ is given by

$$C(T)=\frac{1}{2NS}\sum_{\mathbf{q}}\{(1-\gamma_{\mathbf{q}}^{2})^{1/2}-1\}+\frac{1}{NS}\sum_{\mathbf{q}}(1-\gamma_{\mathbf{q}}^{2})^{1/2}\frac{1}{\exp\{\hbar\omega_{\mathbf{q}}(T)\beta\}-1} \tag{9.289}$$

where now

$$\hbar\omega_{\mathbf{q}}(T)=2rJS\{1-C(T)\}(1-\gamma_{\mathbf{q}}^{2})^{1/2}. \tag{9.290}$$

For small values of $|\mathbf{q}|$,

$$\hbar\omega_{\mathbf{q}}(T)=D(T)q \tag{9.291}$$

where

$$D(T)=D\{1-C(T)\} \tag{9.292}$$

and D is given by eqn (9.251). Using the relation (9.291) and approximating $(1-\gamma_{\mathbf{q}}^{2})^{1/2}$ by $\rho q/\sqrt{3}$ we obtain for

$$\frac{1}{NS}\sum_{\mathbf{q}}(1-\gamma_{\mathbf{q}}^{2})^{1/2}\frac{1}{\exp\{\hbar\omega_{\mathbf{q}}(T)\beta\}-1}$$

the result

$$\frac{v_0}{2\pi^2 S}\frac{\rho}{\sqrt{3}}\left\{\frac{k_B T}{D(T)}\right\}^4 \Gamma(4)\zeta(4) \qquad (9.293)$$

and $\Gamma(4)=3!$

If we denote the constant (Keffer 1967)

$$C_0 = \frac{1}{N}\sum_{\mathbf{q}}\{(1-\gamma_{\mathbf{q}}^2)^{1/2}-1\} \qquad (9.294)$$

and treat $C(T)-C_0/2S$ as a small quantity, (9.293) leads to the result

$$C(T)-C_0/2S = \frac{3v_0}{\pi^2 S(1-C_0/2S)^4}\left(\frac{\sqrt{3}}{\rho}\right)^3\left(\frac{k_B T}{2rJS}\right)^4 \zeta(4) + \frac{4}{(1-C_0/2S)^9}\left\{\frac{3v_0}{\pi^2 S}\left(\frac{\sqrt{3}}{\rho}\right)^3 \zeta(4)\right\}^2\left(\frac{k_B T}{2rJS}\right)^8. \qquad (9.295)$$

From this we conclude that $D(T)$ for an antiferromagnet decreases with temperature, because of the dynamical interaction of the spin waves, as T^4. The renormalized anisotropy field, $h_A(T)\simeq h_A\{1+C(T)\}$, *increases* as T^4. The effect on the dispersion curve is to decrease the overall bandwidth, the anisotropy-field gap

$$(h_A^2+2h_A)^{1/2} - \frac{C(T)}{2(h_A^2+2h_A)^{1/2}} \qquad (9.296)$$

decreasing with temperature like T^4.

Finally, we examine the sublattice magnetization to find the leading-order correction arising from the two-spin-wave dynamical interaction. For $h_A=0$,

$$\frac{1}{N}\sum_{\mathbf{q}}\langle \hat{a}_{\mathbf{q}}^+ \hat{a}_{\mathbf{q}}\rangle = \delta\langle \hat{S}^z\rangle_0 + \frac{1}{N}\sum_{\mathbf{q}}\frac{1}{(1-\gamma_{\mathbf{q}}^2)^{1/2}}\frac{1}{\exp\{\hbar\omega_{\mathbf{q}}(T)\beta\}-1}, \qquad (9.297)$$

which for small $\mathbf{q}$ becomes

$$\delta\langle \hat{S}^z\rangle_0 + \frac{v_0}{2\pi^2}\left(\frac{\sqrt{3}}{\rho}\right)^3\left(\frac{k_B T}{2rJS}\right)^2 \frac{\zeta(2)}{\{1-C(T)\}^2}.$$

This evaluated to the first-order correction yields (Keffer 1967)

$$\langle \hat{S}^z\rangle = S - \delta\langle \hat{S}^z\rangle_0 - \frac{v_0}{2\pi^2(1-C_0/2S)^2}\left(\frac{\sqrt{3}}{\rho}\right)^3\left(\frac{k_B T}{2rJS}\right)^2 \zeta(2) - \frac{3}{S(1-C_0/2S)^7}\left\{\frac{v_0}{\pi^2}\left(\frac{\sqrt{3}}{\rho}\right)^3\right\}^2\left(\frac{k_B T}{2rJS}\right)^6 \zeta(2)\zeta(4). \qquad (9.298)$$

From (9.298) we see that the dynamical interaction enters the sublattice magnetization as T^6, and modifies the non-interacting term by a factor $(1-C_0/2S)^{-2}$.

9.7. Heisenberg ferrimagnets

Our discussion of the properties of ferrimagnets will be less detailed than that given for ferromagnets and antiferromagnets. In part this is due to the complexity of the attendant algebra; a detailed discussion would involve numerical computations and the number of parameters makes this unreasonable. We shall in fact, content ourselves with writing down the cross-section in general form and then briefly survey the fundamental properties in simple cases. The analysis set up in § 9.6 is sufficient to treat a particular case in detail except, perhaps, in respect of the anisotropy terms considered, since these were taken to be uniaxial.

As we stated in § 9.6, the Hamiltonian (9.222) commutes with the *total* z-component of spin; thus we have only to calculate $\langle \hat{S}^+\hat{S}^+\rangle$ to evaluate the neutron cross-sections for the processes involving the creation or annihilation of a single magnon. These correlation functions are simply calculated with the transformation (9.229).

If we denote a general lattice site by $\mathbf{R}$, as in § 9.6, a straightforward calculation gives

$$\sum_{\mathbf{R},\mathbf{R}'} \exp\{\mathrm{i}\boldsymbol{\kappa}\cdot(\mathbf{R}'-\mathbf{R})\}\langle \hat{S}^+_{\mathbf{R}}\hat{S}_{\mathbf{R}'}(t)\rangle$$
$$=\frac{1}{N^2}\sum_{\mathbf{q},\boldsymbol{\tau}} N\frac{(2\pi)^3}{v_0}\,\delta(\boldsymbol{\kappa}-\mathbf{q}-\boldsymbol{\tau})$$
$$\times[(1+n_{\mathbf{q},0})\exp(\mathrm{i}\omega_{\mathbf{q},0}t)(u_{\mathbf{q}}^2+\sigma v_{\mathbf{q}}^2)+n_{\mathbf{q},1}\exp(-\mathrm{i}\omega_{\mathbf{q},1}t)(\sigma u_{\mathbf{q}}^2+v_{\mathbf{q}}^2)$$
$$+2u_{\mathbf{q}}v_{\mathbf{q}}\sigma^{1/2}\cos\boldsymbol{\rho}\cdot\boldsymbol{\tau}\{(1+n_{\mathbf{q},0})\exp(\mathrm{i}\omega_{\mathbf{q},0}t)+n_{\mathbf{q},1}\exp(-\mathrm{i}\omega_{\mathbf{q},1}t)\}]. \quad (9.299)$$

We remind the reader that the reciprocal lattice vectors $\boldsymbol{\tau}$ in (9.299) are those belonging to a sublattice and N is the number of ions situated on one sublattice. Also, the suffixes 0 and 1 refer to the two spin-wave modes in the ferrimagnet; these have energy spectra given by (9.235a) (corresponding to $a=0$) and (9.235b) (corresponding to $a=1$).

We can again use (9.261) to obtain

$$\sum_{\mathbf{R},\mathbf{R}'} \exp\{\mathrm{i}\boldsymbol{\kappa}\cdot(\mathbf{R}'-\mathbf{R})\}\langle \hat{S}^+_{\mathbf{R}}\hat{S}_{\mathbf{R}'}(t)+\hat{S}_{\mathbf{R}}\hat{S}^+_{\mathbf{R}'}(t)\rangle$$
$$=\frac{(2\pi)^3}{Nv_0}\sum_{\mathbf{q},\boldsymbol{\tau}}\delta(\boldsymbol{\kappa}-\mathbf{q}-\boldsymbol{\tau})[\{(1+n_{\mathbf{q},0})\exp(\mathrm{i}\omega_{\mathbf{q},0}t)+n_{\mathbf{q},0}\exp(-\mathrm{i}\omega_{\mathbf{q},0}t)\}$$
$$\times\{u_{\mathbf{q}}^2+\sigma v_{\mathbf{q}}^2+2u_{\mathbf{q}}v_{\mathbf{q}}\sigma^{1/2}\cos\boldsymbol{\rho}\cdot\boldsymbol{\tau}\}+$$
$$\{(1+n_{\mathbf{q},1})\exp(\mathrm{i}\omega_{\mathbf{q},1}t)+n_{\mathbf{q},1}\exp(-\mathrm{i}\omega_{\mathbf{q},1}t)\}\{\sigma u_{\mathbf{q}}^2+v_{\mathbf{q}}^2+2u_{\mathbf{q}}v_{\mathbf{q}}\sigma^{1/2}\cos\boldsymbol{\rho}\cdot\boldsymbol{\tau}\}]. \quad (9.300)$$

Note that the result (9.300) shows that the two spin-wave modes have different structure factors associated with them.

We can immediately calculate from (9.300) the cross-sections in which a magnon of type $a=0$ is created or annihilated and similarly for a magnon of type $a=1$. If we neglect the difference in both the Debye–Waller factors for the two types of ion in the ferrimagnet and their atomic form factors, these cross-sections are

$$\left(\frac{d^2\sigma}{d\Omega\, dE'}\right)^{(\pm)}_{a=0} = r_0^2\frac{k'}{k}\{\tfrac{1}{2}gF(\boldsymbol{\kappa})\}^2\tfrac{1}{4}(1+\tilde{\kappa}_z^2)\exp\{-2W(\boldsymbol{\kappa})\}\frac{(2\pi)^3}{Nv_0}$$

$$\times\sum_{\mathbf{q},\boldsymbol{\tau}}(n_{\mathbf{q},0}+\tfrac{1}{2}\pm\tfrac{1}{2})\delta(\hbar\omega_{\mathbf{q},0}\mp\hbar\omega)\delta(\boldsymbol{\kappa}\mp\mathbf{q}-\boldsymbol{\tau})\,(u_{\mathbf{q}}^2+\sigma v_{\mathbf{q}}^2+2u_{\mathbf{q}}v_{\mathbf{q}}\sigma^{1/2}\cos\boldsymbol{\rho}\cdot\boldsymbol{\tau}) \tag{9.301}$$

and

$$\left(\frac{d^2\sigma}{d\Omega\, dE'}\right)^{(\pm)}_{a=1} = r_0^2\frac{k'}{k}\{\tfrac{1}{2}gF(\boldsymbol{\kappa})\}^2\tfrac{1}{4}(1+\tilde{\kappa}_z^2)\exp\{-2W(\boldsymbol{\kappa})\}\frac{(2\pi)^3}{Nv_0}$$

$$\times\sum_{\mathbf{q},\boldsymbol{\tau}}(n_{\mathbf{q},1}+\tfrac{1}{2}\pm\tfrac{1}{2})\,\delta(\hbar\omega_{\mathbf{q},1}\mp\hbar\omega)\,\delta(\boldsymbol{\kappa}\mp\mathbf{q}-\boldsymbol{\tau})\,(\sigma u_{\mathbf{q}}^2+v_{\mathbf{q}}^2+2u_{\mathbf{q}}v_{\mathbf{q}}\sigma^{1/2}\cos\boldsymbol{\rho}\cdot\boldsymbol{\tau}). \tag{9.302}$$

In general we might expect $\hbar\omega_{\mathbf{q},1}$ to be greater than $\hbar\omega_{\mathbf{q},0}$ in the limit $\mathbf{q}\to 0$, so by analogy with the phonon problem, it is usual to refer to (9.301) as giving the cross-sections for the creation and annihilation of the acoustic magnons and (9.302) the corresponding cross-sections for the optical magnons.

Under certain conditions, a ferrimagnet possesses thermodynamic properties akin to the ferromagnet. To see this set $H_{A,1}=H_{A,2}=0$, $H=0$, and in addition $\mathcal{J}_1=\mathcal{J}_2=0$. The dispersion relations for the two spin-wave modes are in this case

$$\hbar\omega_{\mathbf{q},a}=(-1)^a\mathcal{J}(0)(S_2-S_1)+[\{\mathcal{J}(0)(S_1+S_2)\}^2-4S_1S_2\mathcal{J}^2(\mathbf{q})]^{1/2}. \tag{9.303}$$

This result follows immediately from (9.235).

If we now write (cf. eqn (9.240))

$$\mathcal{J}(0)-\mathcal{J}(\mathbf{q})\simeq\tfrac{1}{6}q^2J^{(2)}, \tag{9.304}$$

then

$$\mathcal{J}^2(\mathbf{q})\simeq\mathcal{J}^2(0)-\tfrac{1}{3}q^2\mathcal{J}(0)J^{(2)}.$$

From this it follows that, for $\mathbf{q}\to 0$,

$$\begin{aligned}\hbar\omega_{\mathbf{q},a}&\simeq-\mathcal{J}(0)(S_1-S_2)(-1)^a+[\{\mathcal{J}(0)(S_1-S_2)\}^2+\tfrac{16}{3}S_1S_2q^2\mathcal{J}(0)J^{(2)}]^{1/2}\\&\simeq\mathcal{J}(0)(S_1-S_2)\{1+(-1)^{a+1}\}+q^2\frac{\tfrac{2}{3}S_1S_2J^{(2)}}{(S_1-S_2)}\end{aligned}$$

and since, at low temperatures, only the mode for which $a = 0$ (acoustic mode) is of importance we have

$$\hbar\omega_{\mathbf{q},0} \simeq q^2 \left\{ \frac{2}{3} \frac{S_1 S_2 J^{(2)}}{(S_1 - S_2)} \right\}. \tag{9.305}$$

Thus, because of the q^2-dependence at small $\mathbf{q}$, the thermodynamic properties of a ferrimagnet (under the conditions specified above) are analogous to those of a ferromagnet, e.g. the magnetization decreases with temperature $T^{3/2}$. Note, however, that if $S_1 \sim S_2$ the effective D in the ferrimagnet can be quite large. For nearest-neighbour coupling only we have from (9.305) that D for a ferrimagnet is given by

$$D = \frac{4S_1 S_2 a^2 J}{S_1 - S_2}. \tag{9.306}$$

9.8. **Magnon–phonon hybridization** (Coqblin 1977; Lovesey 1976; Stinchcombe 1977; Thalmeier and Fulde 1982)

Magnon–phonon interactions are most apparent in neutron spectroscopy when they generate mixing, or hybridization, of magnon and phonon modes. The gaps introduced at the nominal interections of the dispersion relations can be as large as a few meV. Experimental results indicate that the occurrence of hybridization is linked to significant magneto-elastic interactions, for hybridization has been observed, with neutron spectroscopy, in terbium and dysprosium, and ferrous compounds, but not in gadolinium and manganese compounds which contain s-state (half-filled shell) magnetic ions. Acoustic phonon and magnon dispersions are likely to cross in magnetic compounds that possess anisotropy which creates a gap in the magnon dispersion at the zone centre but, clearly, the crossing is not sufficient to guarantee hybridization. For example, the rutile compounds MnF_2 and FeF_2 possess gaps at the zone centre (approximately 1 and 7 meV, respectively) and hybridization is readily observed in the latter only. The essential difference between the two compounds, in this respect, is that Mn^{2+} ($3d^5$) is an s-state ion, whereas the ferrous ion, Fe^{2+} ($3d^6$), possesses a single electron outside the half-filled shell which is sensitive to vibrations of the ligand-crystal field.

Another magnon–phonon interaction, which is present in all exchanged coupled magnetic compounds to some extent, arises from the position dependence of the exchange parameters. Consider the simple Hamiltonian (9.1) in which the spins are coupled by an exchange parameter $J(\mathbf{l} - \mathbf{l}')$. In a crystal the ions do not form a rigid lattice but execute vibrations about the rigid-lattice equilibrium configuration. Provided the vibrations have a small amplitude their effect on the exchange energy can be derived by replacing the site vector $\mathbf{l}$ by $\mathbf{l}+\mathbf{u}(\mathbf{l})$, where $\mathbf{u}(\mathbf{l})$ is the displacement and

$$J(\mathbf{l}+\mathbf{u}-\mathbf{l}'-\mathbf{u}') \doteq J(\mathbf{l}-\mathbf{l}')+\mathbf{K}\cdot(\mathbf{u}-\mathbf{u}') \tag{9.307}$$

Here, the constant $\mathbf{K}$ is related, formally, to the spatial derivative of the exchange parameter. Inserting (9.307) in the exchange Hamiltonian (9.1) generates a two-ion coupling of the lattice vibrations and spins which contributes to the magnon damping. However, it is usually relatively unimportant for compounds that possess strong magneto-elastic interactions which generate magnon–phonon hybridization.

We consider the interpretation of magnon–phonon hybridization observed in ferrous compounds. To be concrete, we focus the discussion on the rutile antiferromagnet FeF_2, for which the spin-wave theory of § 9.6.2 with a nearest-neighbour exchange coupling suffices (Lovesey 1972). The magnetic anisotropy comes largely from the spin–orbit coupling

$$\hat{\mathscr{H}}_{\mathrm{so}} = \lambda\hat{\mathbf{L}}\cdot\hat{\mathbf{S}}, \tag{9.308}$$

where $\hat{\mathbf{L}}$ is the orbital angular momentum operator and λ is the spin-orbit coupling parameter.

The ligand-crystal field experienced by a ferrous ion is taken to arise from the six nearest fluorine ions, each carrying a charge Ze, at positions defined by the vectors $\mathbf{R}_\delta$ (see Fig. 12.9). The electrostatic energy experienced by the 3d electron at $\mathbf{r}$ is

$$Ze^2\sum_{\boldsymbol{\delta}}|\mathbf{r}-\mathbf{R}_\delta|^{-1}. \tag{9.309}$$

With $\mathbf{R}_\delta=\boldsymbol{\delta}+\mathbf{u}(\boldsymbol{\delta})$, the linear modulation of the ligand-crystal field is

$$\hat{\mathscr{H}}_{\mathrm{lo}} = Ze^2\sum_{\boldsymbol{\delta}}\{\hat{\mathbf{u}}-\hat{\mathbf{u}}(\boldsymbol{\delta})\}\cdot\boldsymbol{\nabla}_r\,|\mathbf{r}-\boldsymbol{\delta}|^{-1}. \tag{9.310}$$

The static component of the crystal field (9.309) is diagonalized in terms of states $|L, M_L\rangle$ where $L=2$ and $M_L=-L, -(L-1),\ldots, L$. The interaction

$$\hat{\mathscr{H}}_1 = \hat{\mathscr{H}}_{\mathrm{so}}+\hat{\mathscr{H}}_{\mathrm{lo}} \tag{9.311}$$

is treated by perturbation theory within the crystal-field states to produce an effective spin Hamiltonian. Those terms in the spin–orbit coupling alone contribute to the single-ion anisotropy and are not of immediate interest. Terms of order λZ vanish by symmetry. Of the various terms in the third-order contribution $\lambda^2 Z$, some contain the operators $\hat{S}^z\hat{S}^y$ and $\hat{S}^z\hat{S}^x$ which, at low temperatures, are replaced by $S\hat{S}^y$ and $S\hat{S}^x$, to a good approximation. Because these terms are linear in the spins and lattice vibrations, they lead to hybridization of the magnon and phonon modes.

Adding the terms linear in the spins and displacements to the antiferromagnet magnon Hamiltonian (9.237b) gives the required spin Hamiltonian including hybridization. The calculation of the spin correlation functions is straightforward since the Hamiltonian is quadratic in magnon and phonon operators.

The corresponding magnetic neutron cross-section is more complicated than might be expected, since the correlation functions $\langle \hat{S}^{+}\hat{S}^{+}(t)\rangle$ are non-zero as a result of the magnon-phonon coupling, i.e. the total z-component of the spin is not a constant of motion. The correlation functions $\langle \hat{S}^{\pm}\hat{S}^{\mp}(t)\rangle$ are associated with the orientation factor $(1+\tilde{\kappa}_z^2)$, whereas $\langle \hat{S}^{\pm}\hat{S}^{\pm}(t)\rangle$ have factors $(\tilde{\kappa}_y \pm \mathrm{i}\tilde{\kappa}_x)^2$.

In the absence of magnon–phonon coupling, the one-magnon cross-sections are derived from the spin correlation function (9.260) and lead to the result (9.262). For the magnon creation term, for example, we now obtain the result

$$\tfrac{1}{2}(1+\tilde{\kappa}_z^2-2\tilde{\kappa}_x\tilde{\kappa}_y)\,\delta\{\hbar\omega-\hbar\omega_{\mathbf{q}}\} \\ +(1+\tilde{\kappa}_z^2+2\tilde{\kappa}_x\tilde{\kappa}_y)[W_{+}(\mathbf{q})\,\delta\{\hbar\omega-E_{+}(\mathbf{q})\}+W_{-}(\mathbf{q})\,\delta\{\hbar\omega-E_{-}(\mathbf{q})\}]. \quad (9.312)$$

Here, $\omega_{\mathbf{q}}$ is given by (9.245), and $W_{\pm}(\mathbf{q})$ and $E_{\pm}(\mathbf{q})$ are the weights and energies of the hybridized modes. With no coupling, $W_{-}=0$, $W_{+}=\frac{1}{2}$, $E_{+}=\hbar\omega_{\mathbf{q}}$, and (9.312) reverts to the result used in (9.262). This result is appropriate near the zone centre since the magnon–phonon coupling is proportional to q^2 at long wavelengths. The situation is reversed at the zone boundary, where the phonon energy exceeds the magnon energy, and $W_{+} \ll W_{-}$ with $E_{-} \sim \hbar\omega_{\mathbf{q}}$. The difference $\{E_{-}(\mathbf{q})-E_{+}(\mathbf{q})\}$ at the nominal intersection of the magnon and phonon dispersions is a direct measure of the coupling strength, as might be expected. In the vicinity of the nominal intersection, and for a sufficiently large coupling strength, there are three distinct contributions to the cross-section as given by (9.312). The latter applies for specific directions of the wave vector, and for other directions the contribution from the unperturbed magnon is absent. For a relatively simple compound like FeF_2, the few parameters in the cross-section (λ and the crystal-field energies) can be correlated with independent measurements, and the interpretation of the data in terms of the spin–orbit mechanism can be made with confidence.

The spectrum of modes is changed significantly by the application of a modest magnetic field parallel to the axis of quantization. To understand the origin of this effect, recall that the degeneracy of the unperturbed magnon modes is lifted by a magnetic field (cf. (9.239)). Moreover, the nondegenerate modes affect the neutron spin, and an initially unpolarized neutron beam is partially polarized through scattering. Since neutron-spin polarization is not created by scattering from lattice vibrations, the degree of polarization is a direct measure of the magnetic character of the hybridized modes.

9.9. **Mixed magnetic systems** (Ziman 1979; Elliott, Krumhansl, and Leath 1974)

We will discuss the effects of magnetic impurities on the inelastic cross-section in terms of the Heisenberg model for ferromagnetic and antiferromagnetic materials. The problem of a single impurity can be solved exactly, in the linear spin-wave approximation and, for nearest-neighbour interactions, the cross-section is relatively simple. An impurity can produce resonance modes in the mixed magnet which are significant features of the cross-section when the host density-of-states is small. In particular, a resonance mode whose energy exceeds the maximum spin-wave energy is not damped.

Results for the single-impurity problem are useful for the interpretation of data for materials containing small concentrations of impurities. However, the host spin waves are not damped in this approximation. We therefore calculate the spin-wave response of a system that contains a small finite concentration of impurities.

All our calculations are based on the linear spin-wave approximation. It is convenient to use a spin Green function defined in (8.75), namely,

$$G(m, n; \omega) = \int_{-\infty}^{\infty} \mathrm{d}t \exp(\mathrm{i}\omega t)\{-\mathrm{i}\theta(t)\langle[\hat{S}^+_m(t), \hat{S}^-_n]\rangle\}$$
$$\equiv \langle\langle \hat{S}^+_m; \hat{S}^-_n \rangle\rangle$$

where m and n are site labels.

The transverse cross-section (9.42) contains the correlation function

$$\int_{-\infty}^{\infty} \mathrm{d}t \exp(-\mathrm{i}\omega t)\langle \hat{S}^+_m \hat{S}^-_n(t)\rangle = \int_{-\infty}^{\infty} \mathrm{d}t \exp(\mathrm{i}\omega t)\langle \hat{S}^+_m(t)\hat{S}^-_n\rangle$$
$$= -2\{1+n(\omega)\}\mathrm{Im}\, G(m, n; \omega) \quad (9.314)$$

and the imaginary part of the Green function on the right-hand side is calculated with $\omega \to \omega + \mathrm{i}\delta$ and $\delta \to 0^+$. The result (9.314) is valid for models in which each site is a centre of inversion symmetry, since it requires the correlation functions to be invariant under an interchange of the site labels. A correlation function depends on the relative position of the sites, $\mathbf{m}-\mathbf{n}$, but for Bravais lattices, and other lattices in which each site is a centre of inversion symmetry, the positional dependence reduces to $|\mathbf{m}-\mathbf{n}|$. To prove (9.314) we form the imaginary part of $G(m, n; \omega)$

and use the identity (9.41)

$$2\,\mathrm{Im}\,G(m,n;\omega)=-\int_0^{\infty}\mathrm{d}t\exp(\mathrm{i}\omega t)\langle[\hat{S}^+_m(t),\hat{S}^-_n]\rangle+\int_{-\infty}^{0}\mathrm{d}t\exp(\mathrm{i}\omega t)\langle[\hat{S}^-_m,\hat{S}^+_n(t)]\rangle$$

$$=-\{1-\exp(-\hbar\omega\beta)\}\int_{-\infty}^{\infty}\mathrm{d}t\exp(\mathrm{i}\omega t)\langle\hat{S}^+_m(t)\hat{S}^-_n\rangle.$$

Taking $g=2$ in (9.42), and incorporating the Debye–Waller factor in a site-dependent form factor, we arrive at the result

$$\left(\frac{\mathrm{d}^2\sigma}{\mathrm{d}\Omega\,\mathrm{d}E'}\right)^{(\pm)}=\mp\frac{1}{4\pi\hbar}r_0^2\frac{k'}{k}(1+\tilde{\kappa}_z^2)\{n(\omega)+1\}$$

$$\times\sum_{m,n}F_m(\boldsymbol{\kappa})F_n(\boldsymbol{\kappa})\exp\{\mathrm{i}\boldsymbol{\kappa}\cdot(\mathbf{m}-\mathbf{n})\}\,\mathrm{Im}\,G(m,n;\pm\omega).\tag{9.315}$$

Thus, the calculation of the neutron cross-section has reduced essentially to the calculation of a Green function for the mixed system; $G(m,n;\omega)$ is determined by its equation-of-motion

$$\hbar\omega G(m,n;\omega)=\hbar\langle[\hat{S}^+_m,\hat{S}^-_m]\rangle+\langle\langle[\hat{S}^+_m,\hat{\mathscr{H}}];\hat{S}^-_n\rangle\rangle.\tag{9.316}$$

The total exchange Hamiltonian for a single impurity at the site $m=s$, coupled to the host with an exchange coupling $\pm J'$, is

$$\hat{\mathscr{H}}=\hat{\mathscr{H}}_0+\hat{\mathscr{H}}_1\tag{9.317}$$

where

$$\hat{\mathscr{H}}_1=2J\sum_{l(s)}\hat{\mathbf{S}}_s\cdot\hat{\mathbf{S}}_l\mp 2J'\sum_{l(s)}\hat{\mathbf{S}}'_s\cdot\hat{\mathbf{S}}_l\tag{9.318}$$

and $\hat{\mathscr{H}}_0$ is the exchange interaction in (9.1) with a nearest-neighbour exchange parameter J. The notation $l(s)$ means those r ions at sites l that are nearest neighbours to the site s. With the upper sign in (9.318) the ground state is, clearly, that in which the impurity spin is fully aligned with the host spins. This we refer to as the ferromagnetic impurity case. If the other possible sign of the impurity–host exchange is taken, then the fully aligned state is, of course, no longer the ground state. Instead we use the approximate ground state in which the impurity is aligned antiparallel to the host spins. We refer to this as the antiferromagnetic impurity.

The Green function on the right-hand side of eqn (9.316) is of a higher order than G, because the commutator $[\hat{S}^+,\hat{\mathscr{H}}]$ contain $\hat{S}^z\hat{S}^+$, so that an approximation scheme must be introduced to reduce this Green function to an expression involving only G. If we make the replacement

$$\langle\langle \hat{S}_l^z \hat{S}_m^+; \hat{S}_n^- \rangle\rangle \rightarrow S_l G(m, n; \omega) \tag{9.319}$$

where S_l is the magnitude of the spin at the lth site (a c-number), then our approximation scheme is equivalent to the linear spin-wave approximation. An alternative way of regarding (9.319) is to replace the commutator $[\hat{S}^+, \hat{S}^-] = 2\hat{S}^z$ by $[\hat{S}^+, \hat{S}^-] = 2S$ (cf. § 9.2).

To illustrate the approximation scheme we calculate the Green function for the unperturbed host $P(m, n; \omega)$, i.e. the function that satisfies (9.316) with $\mathscr{H} = \mathscr{H}_0$. With the use of (9.319),

$$\hbar\omega P(m, n; \omega) = \hbar(2S)\,\delta_{m,n} - 2JS \sum_{l(m)} \{P(l, n; \omega) - P(m, n; \omega)\}$$

or

$$(\mathscr{E} - 1)P(m, n; \mathscr{E}) + \frac{1}{r}\sum_{l(m)} P(l, n; \mathscr{E}) = \hbar\frac{2S}{2rJS}\,\delta_{m,n} = \alpha S\,\delta_{m,n} \tag{9.320}$$

where the reduced energy $\mathscr{E}$ is defined by

$$\mathscr{E} = \frac{\hbar\omega}{2rJS}. \tag{9.321}$$

If

$$P(m, n; \mathscr{E}) = \frac{1}{N}\sum_{\mathbf{q}} P_{\mathbf{q}}(\mathscr{E})\exp\{-i\mathbf{q}\cdot(\mathbf{m}-\mathbf{n})\},$$

then, from (9.320)

$$P_{\mathbf{q}}(\mathscr{E}) = \frac{\alpha S}{\mathscr{E} - 1 + \gamma_{\mathbf{q}}} = \frac{\alpha S}{\mathscr{E} - \mathscr{E}_{\mathbf{q}}} \tag{9.322}$$

where

$$\gamma_{\mathbf{q}} = \frac{1}{r}\sum_{l(0)} \exp(i\mathbf{q}\cdot\mathbf{l}).$$

We recognize

$$\hbar\omega_{\mathbf{q}} = 2rJS\mathscr{E}_{\mathbf{q}} = 2rJS(1 - \gamma_{\mathbf{q}}) \tag{9.323}$$

as the energy spectrum of linear spin waves. Also

$$\operatorname{Im} P_{\mathbf{q}}(\mathscr{E}) = \alpha S \lim_{\delta\to 0^+} \operatorname{Im}(\mathscr{E} + i\delta - \mathscr{E}_{\mathbf{q}})^{-1} = -\pi\alpha S\,\delta(\mathscr{E} - \mathscr{E}_{\mathbf{q}}), \tag{9.324}$$

so that for the pure host the formula (9.315) for the cross-section reduces to

$$\left(\frac{d^2\sigma}{d\Omega\, dE'}\right)^{(\pm)} = \mp\left(\frac{1}{4\pi\hbar}\right)\left(-\pi\hbar\frac{2S}{2rJS}\right)r_0^2\frac{k'}{k}(1+\tilde{\kappa}_z^2)F^2(\boldsymbol{\kappa})$$

$$\times\{n(\omega)+1\}\frac{1}{N}\sum_{\mathbf{q}}\left|\sum_m \exp\{\mathrm{i}\mathbf{m}\cdot(\boldsymbol{\kappa}\pm\mathbf{q})\}\right|^2 \delta(\mathscr{E}\mp\mathscr{E}_{\mathbf{q}})$$

$$= r_0^2\frac{k'}{k}F^2(\boldsymbol{\kappa})(1+\tilde{\kappa}_z^2)\tfrac{1}{2}S\frac{(2\pi)^3}{v_0}\sum_{\mathbf{q},\boldsymbol{\tau}}\{n(\omega_{\mathbf{q}})\pm\tfrac{1}{2}+\tfrac{1}{2}\}\,\delta(\hbar\omega_{\mathbf{q}}\mp\hbar\omega)\,\delta(\boldsymbol{\kappa}\mp\mathbf{q}-\boldsymbol{\tau}) \tag{9.325}$$

in agreement with the expression (9.44).

9.9.1. *Ferromagnetic impurity*

Taking the upper sign in eqn (9.318) for $\hat{\mathscr{H}}_1$ and using the decoupling scheme (9.319), the equation-of-motion for $G(m,n;\omega)$ (eqn (9.316)) becomes

$$(\mathscr{E}-1)G(m,n;\mathscr{E})+\frac{1}{r}\sum_{l(m)}G(l,n;\mathscr{E})=\alpha\,\delta_{m,n}S_n+\sum_l V_{m,l}G(l,n;\mathscr{E}). \tag{9.326}$$

Here

$$S_n = S+\delta_{n,s}(S'-S) \tag{9.327}$$

and the perturbation matrix $V_{m,n}$, of dimension $(r+1)\times(r+1)$, is

$$V_{m,n} = \epsilon\,\delta_{m,s}\,\delta_{s,n}+\frac{1}{r}\sum_{l(s)}\{\rho\,\delta_{m,l}\,\delta_{l,n}-\rho\,\delta_{m,s}\,\delta_{l,n}-\epsilon\,\delta_{m,l}\,\delta_{s,n}\} \tag{9.328}$$

$$\equiv r^{-1}\begin{pmatrix} r\epsilon & -\rho & -\rho & \cdots & -\rho \\ -\epsilon & \rho & 0 & \cdots & 0 \\ -\epsilon & 0 & \rho & \cdots & 0 \\ -\epsilon & 0 & 0 & \cdots & 0 \\ \cdot & \cdot & \cdot & \cdots & \cdot \\ -\epsilon & 0 & 0 & \cdots & \rho \end{pmatrix}$$

where the perturbation parameters ϵ and ρ are defined to be

$$\epsilon = \frac{J'}{J}-1 \quad\text{and}\quad \rho=\frac{J'S'}{JS}-1. \tag{9.329}$$

(9.326) can be written in terms of P, which satisfies the same equation with $V=0$, namely,

$$G(m,n;\mathscr{E})=\alpha S_n P(m-n;\mathscr{E})+(PVG)_{m,n}. \tag{9.330}$$

Because the perturbation $V_{m,n}$ is not diagonal in m and n the algebra for this problem is more complicated than for the mass-defect problem (cf. § 4.7): also the perturbation appears in (9.330) not only in the second term on the right-hand side but also in the first term, because S_n has a different value on the impurity site from the value it has on the host. We shall, therefore, concern ourselves with a discussion of the many-impurity problem for a particular set of perturbation parameters that allow us to use analogues between the present problem and the mass-defect problem. For the present we discuss the solution of (9.330) and, in particular, the impurity-energy modes.

The poles of $G(m, n; \mathscr{E})$ are determined by

$$\operatorname{Re} \det|\mathscr{I} - PV| = 0. \tag{9.331a}$$

The matrix PV has the symmetry of the crystal lattice and the solution of eqn (9.331a) is greatly simplified by making use of this fact. For if the unitary matrix U is constructed from the vectors that transform as the irreducible representations of the point group of the crystal structure, then U^+PVU is diagonal and the criterion (9.331a) reads (Elliott and Dawber 1979)

$$\operatorname{Re} \det|\mathscr{I} - PV| = \operatorname{Re} \det|U^+(\mathscr{I} - PV)U| = \prod_{\xi=1}^{r+1} \operatorname{Re} D_\xi(\mathscr{E}_\xi) = 0 \tag{9.331b}$$

where D_ξ are the elements of the diagonal matrix $U^+(\mathscr{I} - PV)U$. These elements are usually labelled s, p, d, or f because the associated functions of the various irreducible representations of the point group O_h, to which they belong, are of the same form as the wave functions of the hydrogen atom. In Table 9.1 the irreducible representations of the O_h group are given for each of the three cubic lattices together with the degeneracies of the various factors in (9.331b).

For all three cubic Bravais lattices the s-like factor has exactly the same form

$$D_s^{(+)}(\mathscr{E}) = 1 + \epsilon\{1 - \mathscr{E}P(\mathscr{E})\} + \rho\mathscr{E}R(\mathscr{E}) \tag{9.332}$$

where

$$P(\mathscr{E}) \equiv P(0, 0, 0; \mathscr{E})$$

Table 9.1

Lattice	Γ_1(s-like)	Γ_4'(p-like)	Γ_3(d-like)	Γ_5(d-like)	Γ_2'(f-like)	Γ_5'(f-like)
s.c.	2	3	2			
b.c.c.	2	3		3	1	
f.c.c.	2	3	2	3		3

and

$$R(\mathscr{E}) = 1-(\mathscr{E}-1)P(\mathscr{E}) = \frac{1}{r}\sum_{l(0)} P(l;\mathscr{E}). \tag{9.333}$$

The remaining factors are

$$\text{s.c.}\begin{cases} D_{\text{p}}(\mathscr{E}) = \dfrac{1}{r}[P(0,0,0;\mathscr{E}) - P(2,0,0;\mathscr{E})] - \rho^{-1} \\ D_{\text{d}}(\mathscr{E}) = \dfrac{1}{r}[P(0,0,0;\mathscr{E}) - 2P(1,1,0;\mathscr{E}) + P(2,0,0;\mathscr{E})] - \rho^{-1} \end{cases} \tag{9.334}$$

$$\text{b.c.c.}\begin{cases} D_{\text{p}}(\mathscr{E}) = \dfrac{1}{r}[P(0,0,0;\mathscr{E}) + P(1,0,0;\mathscr{E}) - P(1,1,0;\mathscr{E}) \\ \qquad\qquad - P(1,1,1;\mathscr{E})] - \rho^{-1} \\ D_{\text{d}}(\mathscr{E}) = \dfrac{1}{r}[P(0,0,0;\mathscr{E}) - P(1,0,0;\mathscr{E}) - P(1,1,0;\mathscr{E}) \\ \qquad\qquad + P(1,1,1;\mathscr{E})] - \rho^{-1} \\ D_{\text{f}}(\mathscr{E}) = \dfrac{1}{r}[P(0,0,0;\mathscr{E}) - 3P(1,0,0;\mathscr{E}) + 3P(1,1,0;\mathscr{E}) \\ \qquad\qquad - P(1,1,1;\mathscr{E})] - \rho^{-1} \end{cases} \tag{9.335}$$

$$\text{f.c.c.}\begin{cases} D_{\text{p}}(\mathscr{E}) = \dfrac{1}{r}[P(0,0,0;\mathscr{E}) + 2P(\tfrac{1}{2},\tfrac{1}{2},0;\mathscr{E}) - P(1,1,0;\mathscr{E}) \\ \qquad\qquad - 2P(1,\tfrac{1}{2},\tfrac{1}{2};\mathscr{E})] - \rho^{-1} \\ D_{\text{d}_1}(\mathscr{E}) = \dfrac{1}{r}[P(0,0,0;\mathscr{E}) - 2P(\tfrac{1}{2},\tfrac{1}{2},0;\mathscr{E}) + 2P(1,0,0;\mathscr{E}) \\ \qquad\qquad + P(1,1,0;\mathscr{E}) - 2P(1,\tfrac{1}{2},\tfrac{1}{2};\mathscr{E})] - \rho^{-1} \\ D_{\text{d}_2}(\mathscr{E}) = \dfrac{1}{r}[P(0,0,0;\mathscr{E}) - 2P(1,0,0;\mathscr{E}) + P(1,1,0;\mathscr{E})] - \rho^{-1} \\ D_{\text{f}}(\mathscr{E}) = \dfrac{1}{r}[P(0,0,0;\mathscr{E}) - 2P(\tfrac{1}{2},\tfrac{1}{2},0;\mathscr{E}) - P(1,1,0;\mathscr{E}) \\ \qquad\qquad + 2P(1,\tfrac{1}{2},\tfrac{1}{2};\mathscr{E})] - \rho^{-1}. \end{cases} \tag{9.336}$$

In eqns (9.334), (9.335), and (9.336), $P(l;\mathscr{E})$ is the crystal Green function appropriate to the particular lattice and r is the number of nearest neighbours.

It can be shown that the spin-wave density of states $N(\mathscr{E})$ is given by

(Callaway 1974)

$$N(\mathscr{E})=-\frac{1}{\pi}\left[P''(\mathscr{E})+\frac{1}{\pi N}\operatorname{Im}\frac{\mathrm{d}}{\mathrm{d}\mathscr{E}}\left\{\ln\prod_{\xi}D_{\xi}(\mathscr{E})\right\}\right].$$

For the simple cubic lattice this is

$$N(\mathscr{E})=N_0(\mathscr{E})-\frac{1}{\pi N}\operatorname{Im}\frac{\mathrm{d}}{\mathrm{d}\mathscr{E}}\{D_{\mathrm{s}}^{(+)}(\mathscr{E})D_{\mathrm{p}}^{3}(\mathscr{E})D_{\mathrm{d}}^{2}(\mathscr{E})\}.$$

For $\mathscr{E}\sim\mathscr{E}_{\xi}$, where $\mathscr{E}_{\xi}$ is the solution of

$$\operatorname{Re}D_{\xi}(\mathscr{E}_{\xi})=0\quad(\xi=\mathrm{s,\,p,\,or\,d\,(s.c.)}),$$

there is a contribution to the density-of-states

$$\delta N_{\xi}\simeq\frac{1}{\pi N}\frac{\Gamma_{\xi}}{(\mathscr{E}-\mathscr{E}_{\xi})^2+(\Gamma_{\xi})^2}$$

where

$$\Gamma_{\xi}=\operatorname{Im}D_{\xi}(\mathscr{E}_{\xi})\Big/\left[\operatorname{Re}\left\{\frac{\mathrm{d}}{\mathrm{d}\mathscr{E}}D_{\xi}(\mathscr{E})\right\}\Big|_{\mathscr{E}=\mathscr{E}_{\xi}}\right]$$

characterizes the width of the contribution to the density-of-states.

Let us now examine the locations of the various impurity modes as a function of the parameters J'/J and $\sigma=S'/S$. Fig. 9.5 shows the two functions

$$\operatorname{Re}\frac{2}{r}[P(0,0,0;E)-P(2,0,0;E)]\quad(\mathrm{p})$$

and

$$\operatorname{Re}\frac{2}{r}[P(0,0,0;E)-2P(1,1,0;E)+P(2,0,0;E)]\quad(\mathrm{d})$$

for the simple cubic lattice. From eqn (9.334) it follows that if we equate $2/\rho$ to the first we obtain the location of the p-like modes, whereas the second gives the positions of the d-like modes. The parameter $2/\rho$ has values in the range $-\infty$ to -2 and 0 to $+\infty$; from Fig. 9.5 it follows that there are no p or d modes near the bottom of the spin-wave band, a characteristic feature of those modes in all the cubic lattices.

Eqn (9.330) for $G(m,n;\mathscr{E})$ is, explicitly,

$$\begin{aligned}G(m,n;\mathscr{E})&=\alpha S_nP(m-n;\mathscr{E})\\&\quad+P(m-s;\mathscr{E})\left\{\epsilon\mathscr{E}G(s,n;\mathscr{E})-\frac{\rho}{r}\sum_{l(s)}G(l,n;\mathscr{E})\right\}\\&\quad-\epsilon\,\delta_{m,s}G(s,n;\mathscr{E})+\frac{\rho}{r}\sum_{l(s)}P(m-l;\mathscr{E})G(l,n;\mathscr{E}).\end{aligned}\qquad(9.337)$$

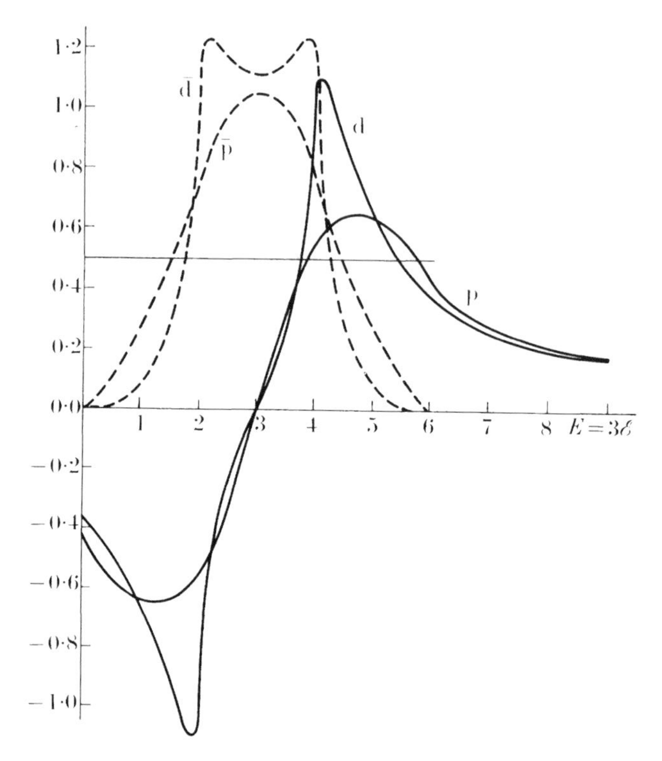

FIG. 9.5. The two solid lines are

$$\mathrm{Re}\,\frac{2}{r}[P(0,0,0;E)-P(2,0,0;E)] \quad \text{(p)}$$

and

$$\mathrm{Re}\,\frac{2}{r}[P(0,0,0;E)-2P(1,1,0;E)+P(2,0,0;E)] \quad \text{(d)}$$

as functions of $E = 3\mathscr{E}$. The intersection of these curves with $2/\rho$ gives the locations of the p and d type modes (the host spin-wave band runs from 0 to 6). The dashed lines, $\bar{\mathrm{p}}$ and $\bar{\mathrm{d}}$, are the imaginary parts of these two functions, from which the sign of $\Gamma_{\mathscr{E}}$ can be determined. For a true resonance state $\Gamma_{\mathscr{E}} > 0$, so that these occur only near the top of the band.

If we set $m = s$ and then m equal to a nearest neighbour to s, the two functions occurring on the right-hand side, $G(s, n; \mathscr{E})$ and $\sum G(l, n; \mathscr{E})$, can be solved for with the aid of the relation

$$\frac{1}{r}\sum_{l(m)} P(l-n;\mathscr{E}) = \delta_{m,n} - (\mathscr{E}-1)P(m-n;\mathscr{E}). \tag{9.338}$$

For instance

$$G(s,n;\mathscr{E}) = \alpha S_n \frac{[(1+\rho)P(s-n;\mathscr{E}) + \rho\{1-\mathscr{E}P(\mathscr{E})\}\,\delta_{s,n}]}{D_s^{(+)}(\mathscr{E})}. \tag{9.339}$$

The r nearest-neighbour Green functions occurring in the last term of (9.337) are conveniently obtained by forming a matrix equation for them. If (g, f) denote the r nearest-neighbour positions to s, then from (9.337)

$$-\rho G(g, n; \mathscr{E}) = \sum_{f=1}^{r} (M^{-1})_{g,f} \left[\alpha S_n P(f-n; \mathscr{E}) + P(f-n; \mathscr{E}) \times \left\{ \epsilon \mathscr{E} G(s, n; \mathscr{E}) - \frac{\rho}{r} \sum_{l(s)} G(l, n; \mathscr{E}) \right\}\right]. \tag{9.340}$$

M is a symmetric matrix whose elements are functions of the crystal Green function P, and has the symmetry of the crystal lattice. The latter property means that $U^{+}MU$ is a diagonal matrix; hence M^{-1} is easily calculated. For a simple cubic host lattice

$$\begin{aligned}\frac{1}{r} \sum_{g,f=1}^{r} (M^{-1})_{g,f} \exp(\mathrm{i}\mathbf{q}\cdot\mathbf{g} - \mathrm{i}\mathbf{q}'\cdot\mathbf{f}) &= \gamma_{\mathbf{q}}\gamma_{\mathbf{q}'}\{-(\mathscr{E}-1)R(\mathscr{E}) - \rho^{-1}\}^{-1} + \frac{3A^{(\mathrm{p})}(\mathbf{q}, \mathbf{q}')}{D_{\mathrm{p}}(\mathscr{E})} + \frac{2A^{(\mathrm{d})}(\mathbf{q}, \mathbf{q}')}{D_{\mathrm{d}}(\mathscr{E})} \\ &= \gamma_{\mathbf{q}}\gamma_{\mathbf{q}'}\{-(\mathscr{E}-1)R(\mathscr{E}) - \rho^{-1}\}^{-1} + H(\mathbf{q}, \mathbf{q}'; \mathscr{E})\end{aligned} \tag{9.341}$$

where

$$\begin{aligned} A^{(\mathrm{p})}(\mathbf{q}, \mathbf{q}') &= \tfrac{2}{3}(\sin u \sin u' + \sin v \sin v' + \sin w \sin w'), \\ A^{(\mathrm{d})}(\mathbf{q}, \mathbf{q}') &= \tfrac{1}{3}\{2(\cos u \cos u' + \cos v \cos v' + \cos w \cos w') \\ &\quad - \cos u(\cos v' + \cos w') - \cos v(\cos u' + \cos w') \\ &\quad - \cos w(\cos u' + \cos v')\}. \end{aligned} \tag{9.342}$$

In (9.342), $u = aq_x$, $v = aq_y$, and $w = aq_z$, where a is the lattice spacing.

If (9.341) is used in (9.337), then the latter can be brought into the form

$$\frac{1}{\alpha S_n} G(m, n; \mathscr{E}) = P(m-n; \mathscr{E}) + \mathscr{L}^{(+)}(m, n; \mathscr{E}) - \frac{1}{N^2} \sum_{\mathbf{q},\mathbf{q}'} P_{\mathbf{q}}(\mathscr{E}) P_{\mathbf{q}'}(\mathscr{E}) \exp(-\mathrm{i}\mathbf{q}\cdot\mathbf{m} + \mathrm{i}\mathbf{q}'\cdot\mathbf{n}) H(\mathbf{q}, \mathbf{q}'; \mathscr{E}) \tag{9.343}$$

where

$$D_s^{(+)}(\mathscr{E}) \mathscr{L}^{(+)}(m, n; \mathscr{E}) = \{\mathscr{E} P(m-s; \mathscr{E}) - \delta_{m,s}\} \times \{(\rho\mathscr{E} - \rho + \epsilon) P(s-n; \mathscr{E}) - \rho\, \delta_{s,n}\}.$$

The inelastic one-magnon cross-section eqn (9.315) contains

$$\sum F_m F_n G(m, n; \mathscr{E}) \exp\{\mathrm{i}\boldsymbol{\kappa}\cdot(\mathbf{m}-\mathbf{n})\}.$$

For a single impurity at $\mathbf{m}=\mathbf{s}$,

$$F_m(\boldsymbol{\kappa})=F(\boldsymbol{\kappa})+\delta_{m,s}\{F'(\boldsymbol{\kappa})-F(\boldsymbol{\kappa})\}$$

where $F(\boldsymbol{\kappa})$ is the form factor associated with the host ions and $F'(\boldsymbol{\kappa})$ that of the impurity. The functions $G(s,s;\mathscr{E})$ and $\sum G(s,n;\mathscr{E})$ are easily calculated from (9.339). With some tedious algebra we obtain

$$\begin{aligned}\sum_{m,n} & F_m(\boldsymbol{\kappa})F_n(\boldsymbol{\kappa})\exp\{\mathrm{i}\boldsymbol{\kappa}\cdot(\mathbf{m}-\mathbf{n})\}G(m,n;\mathscr{E})\\ &=\alpha[SF^2\{NP_{\boldsymbol{\kappa}}(\mathscr{E})+P^2_{\boldsymbol{\kappa}}(\mathscr{E})T(\boldsymbol{\kappa},\mathscr{E})\}\\ &\quad+2\sigma SF(F'-F)\{(1+\epsilon)P_{\boldsymbol{\kappa}}(\mathscr{E})+\rho R(\mathscr{E})-\epsilon P(\mathscr{E})\}/D_{\mathrm{s}}^{(+)}(\mathscr{E})\\ &\quad+\sigma S(F'-F)^2\{P(\mathscr{E})+\rho R(\mathscr{E})\}/D_{\mathrm{s}}^{(+)}(\mathscr{E})]\end{aligned}\tag{9.344}$$

where

$$\begin{aligned}T(\boldsymbol{\kappa},\mathscr{E})=\frac{1}{D_{\mathrm{s}}^{(+)}(\mathscr{E})}&[(\rho\mathscr{E}-\rho+\epsilon)\{\mathscr{E}-2P_{\boldsymbol{\kappa}}^{-1}(\mathscr{E})\}\\ &+P_{\boldsymbol{\kappa}}^{-2}(\mathscr{E})\{\sigma\rho-(\sigma-1)P(\mathscr{E})(\rho\mathscr{E}-\rho+\epsilon)\}]-H(\boldsymbol{\kappa},\boldsymbol{\kappa};\mathscr{E}).\end{aligned}\tag{9.345}$$

From (9.344) it is clear that the neutron cross-section will display peaks whenever the neutron energy change is such that for a given set of perturbations, the real parts of $D_{\mathrm{s}}^{(+)}$, D_{p}, or D_{d} are zero, i.e. whenever the neutrons excite an impurity-energy mode. If the resonance mode lies outside the spin-wave bands, a localized mode, then the cross-section behaves like a delta function in energy at the position of the impurity mode. If, on the other hand, a virtual mode is excited, then this will have a certain width associated with it, $\Gamma_{\mathscr{E}}$. Because the width is proportional to the density of spin-wave states in the pure host only those virtual states near the bottom of the band will give a sharp peak in the cross-section. Only the s states exist near the bottom of the band so that these dominate the behaviour of the cross-section when $\hbar\omega$ is less than the width of the spin-wave band $4rJS$.

The expression (9.344) contains a diffuse contribution whose energy dependence is determined by $D_{\mathrm{s}}^{(+)}(\mathscr{E})$, i.e. it depends only on the s mode. From symmetry conditions this is to be expected. The diffuse part of (9.344) is

$$\frac{S}{D_{\mathrm{s}}^{(+)}(\mathscr{E})}[\sigma F'^2\{\rho R(\mathscr{E})+P(\mathscr{E})\}-2\sigma FF'(1+\epsilon)P(\mathscr{E})+F^2(\rho\mathscr{E}+1+\epsilon)P(\mathscr{E})].\tag{9.346}$$

When $F=F'$, $S=S'$, and $J=J'$, this term is zero. Also, when $F=F'$, $S=S'$, it reduces to the very simple result

$$\frac{\epsilon SF^2}{D_{\mathrm{s}}^{(+)}(\mathscr{E})}.\tag{9.347}$$

Now let us examine the contributions from the p and d modes, which are contained in $H(\boldsymbol{\kappa}, \boldsymbol{\kappa}; \mathscr{E})$. Because the impurity ion does not participate in these modes, the form factor and spin associated with this term must be those of the host ions, as is seen to be the case. Also, these terms contribute only to the coherent part of the cross-section, reflecting as they do the symmetry of the crystal lattice, in this instance simple cubic. In contrast to this, the diffuse part of (9.344) has exactly the same form for all three cubic lattices. $H(\boldsymbol{\kappa}, \boldsymbol{\kappa}; \mathscr{E})$ is zero at the magnetic Bragg positions, i.e. when $\boldsymbol{\kappa}$ coincides with a reciprocal lattice vector. For the Bragg peaks are a consequence of the translational symmetry of the crystal lattice and this is destroyed by the introduction of a magnetic impurity; $A^{(\mathrm{p})}$ and $A^{(\mathrm{d})}$ have zero amplitude at the Bragg positions.

The neutron cross-section for a finite concentration of magnetic impurities is well represented by the cross-section for a single impurity multiplied by Nc, provided the neutron energy change is such that only localized modes contribute to the cross-section. The cross-section for the single impurity is, in essence, the imaginary part of (9.344).

The approximate expression for the one-magnon cross-section outside the spin-wave band, i.e. for $\hbar\omega > 4rJS$, and Nc impurities, is, from eqns (9.315) and (9.344),

$$\left(\frac{\mathrm{d}^2\sigma}{\mathrm{d}\Omega\,\mathrm{d}E'}\right)^{(+)} = \frac{Nc}{4rJS} r_0^2 \frac{k'}{k} (1+\tilde{\kappa}_z^2)\{n(\omega)+1\}$$
$$\times\Bigg[\delta(\mathscr{E}-\mathscr{E}_\mathrm{s})\frac{1}{\{\mathrm{d}D_\mathrm{s}^{(+)}(\mathscr{E})/\mathrm{d}\mathscr{E}\}}[\![[(\rho\mathscr{E}-\rho+\epsilon)\{\mathscr{E}P_\mathrm{k}(\mathscr{E})-2\}P_\kappa(\mathscr{E})$$
$$+\{\sigma\rho-(\sigma-1)(\rho\mathscr{E}-\rho+\epsilon)P(\mathscr{E})\}]SF^2(\boldsymbol{\kappa})$$
$$+2\sigma SF(\boldsymbol{\kappa})\{F'(\boldsymbol{\kappa})-F(\boldsymbol{\kappa})\}\{(1+\epsilon)P_\kappa(\mathscr{E})+\rho R(\mathscr{E})-\epsilon P(\mathscr{E})\}$$
$$+\sigma S\{F'(\boldsymbol{\kappa})-F(\boldsymbol{\kappa})\}^2[P(\mathscr{E})+\rho R(\mathscr{E})]]\!]$$
$$-SF^2(\boldsymbol{\kappa})P_\kappa^2(\mathscr{E})\Bigg\{\delta(\mathscr{E}-\mathscr{E}_\mathrm{p})\frac{1}{\{\mathrm{d}D_\mathrm{p}(\mathscr{E})/\mathrm{d}\mathscr{E}\}}3A^{(\mathrm{p})}(\boldsymbol{\kappa},\boldsymbol{\kappa})$$
$$+\delta(\mathscr{E}-\mathscr{E}_\mathrm{d})\frac{1}{\{\mathrm{d}D_\mathrm{d}(\mathscr{E})/\mathrm{d}\mathscr{E}\}}2A^{(\mathrm{d})}(\boldsymbol{\kappa},\boldsymbol{\kappa})\Bigg\}\Bigg]. \qquad (9.348)$$

In (9.348) we have retained the use of the reduced energy

$$\mathscr{E} = \hbar\omega/2rJS$$

for simplicity. The expression is for a simple cubic host lattice. The analogous result for the body-centred and face-centred host lattices has the same structure, the diffuse parts being identical in form for all three cubic lattices. For $F(\boldsymbol{\kappa}) = F'(\boldsymbol{\kappa})$ and $\sigma = S'/S = 1$, the coefficient of

$\delta(\mathscr{E}-\mathscr{E}_s)$ in (9.348) reduces to

$$\left\{\frac{1}{\epsilon N^{-1}\sum_{\mathbf{q}}\mathscr{E}_{\mathbf{q}}^2/(\mathscr{E}-\mathscr{E}_{\mathbf{q}})^2}\right\}\frac{SF^2(\boldsymbol{\kappa})\epsilon\mathscr{E}_{\boldsymbol{\kappa}}^2}{(\mathscr{E}-\mathscr{E}_{\boldsymbol{\kappa}})^2}$$

$$=SF^2(\boldsymbol{\kappa})P_{\boldsymbol{\kappa}}^2(\mathscr{E})\mathscr{E}_{\boldsymbol{\kappa}}^2\left\{\frac{1}{N}\sum_{\mathbf{q}}P_{\mathbf{q}}^2(\mathscr{E})\mathscr{E}_{\mathbf{q}}^2\right\}^{-1}. \quad (9.349)$$

We further note that

$$D_{\rm p}(\mathscr{E})=-\rho^{-1}+\frac{1}{rN}\sum_{\mathbf{q}}P_{\mathbf{q}}(\mathscr{E})A^{(\rm p)}(\mathbf{q},\mathbf{q})$$

and

$$D_{\rm d}(\mathscr{E})=-\rho^{-1}+\frac{1}{rN}\sum_{\mathbf{q}}P_{\mathbf{q}}(\mathscr{E})A^{(\rm d)}(\mathbf{q},\mathbf{q}). \quad (9.350a)$$

So

$$\frac{\mathrm{d}D_{\rm p}(\mathscr{E})}{\mathrm{d}\mathscr{E}}=-\frac{1}{rN}\sum_{\mathbf{q}}P_{\mathbf{q}}^2(\mathscr{E})A^{(\rm p)}(\mathbf{q},\mathbf{q})$$

and

$$\frac{\mathrm{d}D_{\rm d}(\mathscr{E})}{\mathrm{d}\mathscr{E}}=-\frac{1}{rN}\sum_{\mathbf{q}}P_{\mathbf{q}}^2(\mathscr{E})A^{(\rm d)}(\mathbf{q},\mathbf{q}). \quad (9.350b)$$

9.9.2. Ferromagnetic impurities—low-concentration theory

For the particular choice $\sigma=S'/S=1$, eqn (9.330) reduces to a Dyson equation. Also, if $F(\boldsymbol{\kappa})=F'(\boldsymbol{\kappa})$, which is consistent with $\sigma=1$, the inelastic neutron cross-section for a small, finite concentration c of impurities, involves only $\sum_{m,n}\exp\{i\boldsymbol{\kappa}\cdot(\mathbf{m}-\mathbf{n})\}\bar{G}(m,n;\mathscr{E})$ where $\bar{G}(m,n;\omega)$ is the configurationally averaged spin Green function. The fact that for this choice of parameters G satisfies a Dyson equation means that we can use directly the results of § 4.7.3 on mass defects in harmonic crystals to calculate $\bar{G}$.

From eqn (9.344)

$$\frac{1}{\alpha N}\sum_{m,n}\exp\{i\mathbf{q}\cdot(\mathbf{m}-\mathbf{n})\}G(m,n;\mathscr{E})=P_{\mathbf{q}}(\mathscr{E})+\frac{1}{N}P_{\mathbf{q}}^2(\mathscr{E})T(\mathbf{q},\mathscr{E}) \quad (9.351)$$

where, for $S'/S=1$, the t-matrix is

$$T(\mathbf{q},\mathscr{E})=\frac{1}{D_{\rm s}^{(+)}(\mathscr{E})}\epsilon\mathscr{E}_{\mathbf{q}}^2-H(\mathbf{q},\mathbf{q};\mathscr{E}). \quad (9.352)$$

If we now regard $1/N$ as c, our discussion in § 4.7.3 leads us to the result, to first order in c,

$$\bar{G}(\mathbf{q})=\frac{\alpha P_{\mathbf{q}}(\mathscr{E})}{1-cP_{\mathbf{q}}(\mathscr{E})T(\mathbf{q},\mathscr{E})}=\frac{\alpha}{\mathscr{E}-\mathscr{E}_{\mathbf{q}}-cT(\mathbf{q},\mathscr{E})} \quad (9.353)$$

for the Fourier transform of the Green function of the system with a small number of randomly distributed magnetic impurities in a simple cubic host lattice. If

$$cT(\mathbf{q}, \mathscr{E}) = \delta(\mathbf{q}, \mathscr{E}) - \mathrm{i}\gamma(\mathbf{q}, \mathscr{E}), \tag{9.354}$$

then

$$\operatorname{Im} \bar{G}(\mathbf{q}) = \frac{-\gamma(\mathbf{q}, \mathscr{E})\alpha}{\{\mathscr{E} - \mathscr{E}_{\mathbf{q}} - \delta(\mathbf{q}, \mathscr{E})\}^2 + \{\gamma(\mathbf{q}, \mathscr{E})\}^2}. \tag{9.355}$$

For small $\mathbf{q}$ the width γ has the form (Callaway 1974)

$$\gamma(\mathbf{q}, \omega) \simeq 2c\pi\left(\frac{2\epsilon}{1+\Lambda\epsilon}\right)^2 N(\omega_{\mathbf{q}})(2JSq^2a^2)^2 \tag{9.356}$$

where Λ is a constant that depends on the host lattice; for the simple cubic lattice, $\Lambda = 0.21$. The energy spectrum of the spin waves can be approximately calculated for small $\mathbf{q}$ from the expression

$$\mathscr{E} \simeq \mathscr{E}_{\mathbf{q}} + \delta(\mathbf{q}, \mathscr{E}_{\mathbf{q}}).$$

We find

$$\hbar\omega \simeq a^2q^2 2JS\left(1 + c\frac{2\epsilon}{1+\Lambda\epsilon}\right). \tag{9.357}$$

The one-magnon cross-section, for $\hbar\omega < 2rJS$ follows immediately from (9.355):

$$\left(\frac{\mathrm{d}^2\sigma}{\mathrm{d}\Omega\,\mathrm{d}E'}\right)^{(+)} = \frac{N}{4\pi rJS} r_0^2 \frac{k'}{k} SF^2(\boldsymbol{\kappa})(1+\tilde{\kappa}_z^2)\{n(\omega)+1\}$$

$$\times \frac{\gamma(\boldsymbol{\kappa}, \mathscr{E})}{\{\mathscr{E} - \mathscr{E}_{\kappa} - \delta(\boldsymbol{\kappa}, \mathscr{E})\}^2 + \{\gamma(\boldsymbol{\kappa}, \mathscr{E})\}^2}, \tag{9.358}$$

where, as in previous formulae, we have retained the use of the reduced energy $\mathscr{E} = \hbar\omega/2rJS$.

The results of this subsection have been applied to the interpretation of scattering from mixed quasi-one-dimensional magnets (Lovesey 1981*b*). For one dimension the host Green function is quite simple, and the t-matrix is obtained in analytic form. A confrontation between experiment, computer simulations, and more sophisticated theories for quasi-two-dimensional mixed magnets has provoked several theoretical developments (Coombs and Cowley 1975; Cowley *et al.* 1980).

9.9.3. Antiferromagnetic impurity (Rezende 1983)

The classical ground state of the system consisting of a single antiferromagnetic impurity in a Heisenberg ferromagnet is that in which the impurity spin is aligned antiparallel to those of the host. We shall seek spin-wave excitations from this ground state. As in the spin-wave theory of a two-sublattice Heisenberg antiferromagnet, the approximate nature of this ground state manifests itself in the appearance of a non-zero spin deviation at absolute zero.

We rotate the axes of the impurity through π about the x-axis, which constitutes a canonical transformation, and use the decoupling scheme

$$\langle\langle \hat{S}_l^z \hat{S}_m^+ ; \hat{S}_n^- \rangle\rangle = S_l G(m, n; \omega)$$

where

$$S_l = S - \delta_{l,s}(S' + S). \tag{9.359}$$

From

$$[\hat{S}_m^+, \mathscr{H}_1] = -2J \sum_{l(s)} (\delta_{m,s} - \delta_{m,l})(\hat{S}_s^+ \hat{S}_l^z - \hat{S}_s^z \hat{S}_l^+)$$
$$+ 2J' \sum_{l(s)} \delta_{m,s}(\hat{S}_s^+ \hat{S}_l^z + \hat{S}_s^z \hat{S}_l^-) + \delta_{m,l}(\hat{S}_s^z \hat{S}_l^+ + \hat{S}_s^- \hat{S}_l^z), \tag{9.360}$$

it is evident that the equation-of-motion for $G(m, n; \omega)$ contains a new Green function $\langle\langle \hat{S}_m^- ; \hat{S}_n^- \rangle\rangle$. An equation-of-motion for this can be formed with the adjoint of eqn (9.360), and hence an equation for $G(m, n; \omega)$ obtained. The mathematical analysis then parallels that given in § 9.9.2. The result is

$$G(m, n; \mathscr{E})$$
$$= \alpha S_n \Big[P(m-n; \mathscr{E}) + \frac{1}{D_s^{(-)}(-\mathscr{E})} [\mathscr{E}(2+\rho+\epsilon-\rho\mathscr{E}) P(m-s; \mathscr{E}) P(s-n; \mathscr{E})$$
$$+ (\rho\mathscr{E} - 1 - \epsilon)\{P(m-s; \mathscr{E})\,\delta_{n,s} + \delta_{m,s} P(s-n; \mathscr{E})\}$$
$$- \delta_{m,s}\,\delta_{s,n}(\rho\mathscr{E} - 1 - \epsilon) P(\mathscr{E})] + \delta_{m,s}\,\delta_{s,n} \left\{ \frac{P(-\mathscr{E}) + \rho R(-\mathscr{E})}{D_s^{(-)}(\mathscr{E})} \right\}$$
$$- \frac{1}{N^2} \sum_{\mathbf{q},\mathbf{q}'} P_{\mathbf{q}}(\mathscr{E}) P_{\mathbf{q}'}(\mathscr{E}) H(\mathbf{q}, \mathbf{q}'; \mathscr{E}) \exp(-i\mathbf{q}\cdot\mathbf{m} + i\mathbf{q}'\cdot\mathbf{n}) \Big] \tag{9.361}$$

where

$$D_s^{(-)}(\mathscr{E}) = \mathscr{E}\{P(-\mathscr{E}) + \rho R(-\mathscr{E})\} + (1+\epsilon)\{1 + \mathscr{E} P(-\mathscr{E})\}. \tag{9.362}$$

The contribution from the p and d modes is exactly the same as that for the ferromagnetic impurity, because the impurity spin does not participate in these modes. A major difference between the present case and the ferromagnetic impurity is the appearance of two s-like modes, one corresponding to the solutions of

$$\operatorname{Re} D_s^{(-)}(\mathscr{E}_0) = 0 \quad (\mathrm{s}_0\text{-mode}), \tag{9.362}$$

and another corresponding to the solutions of

$$\text{Re}\, D_s^{(-)}(-\mathscr{E}_1) = 0 \quad (s_1\text{-mode}). \tag{9.364}$$

There is a solution of first condition for all ϵ and ρ, see Lovesey (1967). This mode, denoted by s_0, cannot decay into the spin wave energy band of the host, i.e. it is a localized mode. (Clearly, $\text{Im}\, P(-\mathscr{E}) = 0$ for all $\mathscr{E} > 0$). The s_1 mode, whose energy is determined by the condition (9.364) can be either virtual or localized, but the virtual modes occur only at the top of the spin-wave band. Another point to be noted is that in the limit $J'/J = S'/S = 1$, eqn (9.361) does not go over to the unperturbed Green function, because, even when the magnitudes of the exchange couplings and spins are equal, the spin at the site $m = s$ is still an impurity since it is oppositely aligned to the host spins.

From (9.361),

$$\begin{aligned}\alpha^{-1} \sum_{m,n} F_m F_n G(m, n; \mathscr{E}) \exp\{i\boldsymbol{\kappa} \cdot (\mathbf{m} - \mathbf{n})\} \\ = NSF^2(\boldsymbol{\kappa}) P_{\boldsymbol{\kappa}}(\mathscr{E}) - \frac{S'\{F'(\boldsymbol{\kappa})\}^2}{D_s^{(-)}(\mathscr{E})} \{P(-\mathscr{E}) + \rho R(-\mathscr{E})\} \\ + \frac{SF^2(\boldsymbol{\kappa})}{D_s^{(-)}(-\mathscr{E})} \{\mathscr{E}(2 + \rho + \epsilon - \rho\mathscr{E}) P_{\boldsymbol{\kappa}}^2(\mathscr{E}) + 2(\rho\mathscr{E} - 1 - \epsilon) P_{\boldsymbol{\kappa}}(\mathscr{E}) \\ - (\rho\mathscr{E} - 1 - \epsilon) P(\mathscr{E})\} - SF^2(\boldsymbol{\kappa}) H(\boldsymbol{\kappa}, \boldsymbol{\kappa}; \mathscr{E}).\end{aligned} \tag{9.365}$$

The first and last terms in this expression are identical to the first two terms in (9.344). The significant fact about the remaining two terms in (9.365) is that one can be attributed to the impurity and the other to the perturbed host spins. This is even more evident when it is noted that

$$G(s, s; \mathscr{E}) = -\frac{\alpha S'\{P(-\mathscr{E}) + \rho R(-\mathscr{E})\}}{D_s^{(-)}(\mathscr{E})}. \tag{9.366}$$

To understand this result fully it is necessary to examine the physical nature of the s_0 and s_1 modes in more detail than we have hitherto. The s_0 mode is strictly localized (the fact that $G(m, s; \mathscr{E}) \propto \delta_{m,s}$ is indicative of this) and corresponds to a state in which the impurity is precessing in a natural sense while the host spins, driven through the exchange coupling, precess in an unnatural sense. The precessional senses are reversed in the s_1 mode. It is therefore to be expected that the contributions to the neutron cross-section from the s_0 and s_1 modes should appear as coming from the impurity and perturbed host spins respectively, and that the former should give only a diffuse contribution because it cannot propagate in the host. Since neither the p and d modes nor the s_1 modes occur near the bottom of the spin-wave band the additional structure in the neutron cross-section due to the impurity for $\hbar\omega < 2rJS$ will be determined by the second term in eqn (9.365). For a small concentration c of impurities this term will contribute, to a very good approximation,

$$\left(\frac{\mathrm{d}^2\sigma}{\mathrm{d}\Omega\,\mathrm{d}E'}\right)^{(+)} = -\frac{Nc}{4rJS} r_0^2 \frac{k'}{k} (1+\tilde{\kappa}_z^2) S' F'^2(\boldsymbol{\kappa})\{n(\omega_0)+1\}\delta(\mathscr{E}-\mathscr{E}_0)$$
$$\times\{P(-\mathscr{E}_0)+\rho R(-\mathscr{E}_0)\}\Big/\left\{\frac{\mathrm{d}}{\mathrm{d}\mathscr{E}} D_s^{(-)}(\mathscr{E})\right\}\Bigg|_{\mathscr{E}=\mathscr{E}_0} \qquad (9.367)$$

where $\mathscr{E} = \hbar\omega/2rJS$.

9.10. Incommensurably modulated magnets

Spin wave spectra of systems in which the average moment varies from site to site are distinctly different from those for the simple magnetically ordered phases considered in previous sections. Examples include the sinusoidally modulated phases of neodymium, praseodymium and $CeAl_2$; for a review of experimental work see, for example, Stirling and McEwen (1987). The basic effect present with incommensurate modulation is that the magnetic structure lacks translational invariance and therefore the excitation wave vector q is not a good quantum number.

In a material like neodymium competing exchange interactions and single-site anisotropy terms contrive to stabilize at an elevated temperature a modulated ground state. The modulation wave vector Q is determined by exchange and anistropy parameters. Calculations of ground state properties are performed using a molecular field approximation, and in perhaps the simplest example the average moment is

$$\langle \hat{S}_l^z \rangle = S\cos(\boldsymbol{Q}\cdot\boldsymbol{l}), \qquad (9.368)$$

and Q parallel with the z-axis. Linear spin dynamics is derived from the equation of motion for the transverse spin operator by replacing the operator $\hat{S}^z$ by its average (9.368), as described in § 9.2, for a pure ferromagnet. The feature peculiar to a modulated magnet is that the equation connects the Fourier components $\hat{S}^+_{q+nQ}$ with $n=0, \pm 1$, and it is a member of an infinite set of difference equations labelled by $n=0$, $\pm 1, \pm 2 \ldots$

Because the dynamics is based on a linear spin equation many quantities of interest can be calculated analytically, including the static susceptibility and frequency moments (Lovesey 1987). Moreover, good numerical methods exist to calculate the spin wave spectrum from the difference equations. Ziman and Lindgård (1986) show in this way that the spectrum consists of a series of bands. Because the difference equation is one-dimensional the Van Hove singularities are an inverse square root, and the spectrum is highly structured. The latter can be understood by studying periodic structures $Q = (2\pi/N)$ with $N = 2$, 3, . . . and for which the spin response can be calculated analytically. For example, it turns out that Ziman and Lindgård (1986) perform calculations for $Q = 0.26\pi$ and their results and those for $Q = (\pi/4)$ are essentially the same, as might be expected. The analytic work provides relations between band edge positions (seven for $Q = (\pi/4)$) and parameters of the model (Lovesey and Megann 1988).

REFERENCES

Blackman, J. A., Morgan, T., and Cooke, J. F. (1985). *Phys. Rev. Lett.* **55,** 2814.

Callaway, J. (1974). *Quantum theory of the solid state,* Part B. Academic Press, New York.

——, Chatterjee, A. K., Singhal, S. P., and Ziegler, A. (1983). *Phys. Rev.* **B28,** 3818.

Collins, M. F. and Gaulin, B. D. (1984) *J. Appl. Phys.* **55**(6), 1869.

Cooke, J. F. and Gersch, H. A. (1967). *Phys. Rev.* **153,** 641.

—— and Hahn, H. H. (1970). *Phys. Rev.* **B1,** 1243.

——, Lynn, J. W., and Davis, H. L. (1980). *Phys. Rev.* **B21,** 4118.

Coombs, G. J. and Cowley, R. A. (1975). *J. Phys.* **C8,** 1889.

Coqblin, B. (1977). *The electronic structure of rare-earth metals and alloys: the magnetic heavy rare-earths.* Academic Press, New York.

Cowley, R. A., Birgeneau, R. J., and Shirane, G. (1980). *Ordering in strongly fluctuating condensed matter sytems* (ed. T. Riste). Plenum Press, New York.

Dietrich, O. W., Als-Nielsen, J., and Passell, L. (1976). *Phys. Rev.* **B14,** 4932.

Edwards, D. M. (1980). *J. Mag. Mag. Mat.* **15–18,** 262.

Elliott, J. P. and Dawber, P. G. (1979). *Symmetry in physics,* Vol. 1, The Macmillan Press, London.

Elliott, R. J. and Gibson, A. F. (1976). *Solid state physics.* The Macmillan Press, London.

——, Krumhansl, J. A., and Leath, P. L. (1974). *Rev. mod. Phys.* **46,** 465.

Glaus, U., Lovesey, S. W., and Stoll, E. (1983). *Phys. Rev.* **B27,** 4369.

Glinka, C. J., Minkiewicz, V. J. and Passell, L. (1973). *AIP Conference Proc.* **18,** 1060.

Harris, A. B. (1968). *Phys. Rev.* **175,** 674.

——, (1969). *Phys. Rev.* **184,** 606.

Herring, C. (1966). *Magnetism,* Vol. IV. Academic Press, New York.

Hubbard, J. (1963). *Proc. R. Soc.* **A276,** 238.

—— (1971). *J. Phys.* **C4,** 53.

Keffer, F. (1967). *Encyclopedia of physics,* Vol. XVIII/2. Springer-Verlag, Heidelberg.

Kishore, R. (1979). *Phys. Rev.* **B19,** 3822.

Lindgård, P-A. (1978). *Inst. Phys. Conf. Ser.* **37**.

Lovesey, S. W. (1967). *Proc. Phys. Soc.* **91,** 658.

Lovesey, S. W. (1972). *J. Phys.* **C5,** 2769.

—— (1976). *Comments Solid State Phys.* **7,** 117.

—— (1986). *Condensed matter physics,* Frontiers in Physics Vol. 61. Benjamin/Cummings, Reading, Mass.
—— (1987). *Z. Physik.* **B67,** 525.
—— (1981*a*). *Z. Physik.* **B42,** 307.
—— (1981*b*). *Solid State Comm.* **38,** 953.
—— and Hood, M. (1982). *Z. Phys.* **B47,** 327.
—— and Megann, A. P. (1988). *Proceeding of the Fifth International Conference on Recent Progress in Many-Body Theories* (ed. E. Pajanne). Plenum Press, New York.
Lowde, R. D., Moon, R. M., Pagonis, B., Perry, C. H., Sokoloff, J. B., Vaughan-Watkins, R. S., Wiltshire, M. C. K., and Crangle, J. (1983) *J. Phys.* **F13,** 249.
Lowde, R. D., Moon, R. M., Pagonis, B., Perry, C. H., Sokoloff, J. B., Vaughan-Watkins, R. S., Wiltshire, M. C. K., and Crangle, J. (1983). *J. Phys.* **F13,** 249.
Lynn, J. W. (1975). *Phys. Rev.* **B11,** 2624.
—— (1983). *Phys. Rev.* **B28,** 6550.
—— Mook, H. A. (1981). *Phys. Rev.* **B23,** 198.
Martinez, J. L., Böni, P., and Shirane. G. (1985) *Phys. Rev.* **B32,** 7037.
Mattis, D. C. (1981). *The theory of magnetism.* Springer-Verlag, Heidelberg.
Mook, H. A. (1981). *Phys. Rev. Lett.* **46,** 508.
—— and Lynn, J. W. (1985). *J. appl. Phys.* **57,** 3006.
—— and Paul, D. McK. (1985). *Phys. Rev. Lett.* **54,** 227.
Mulder, C. A. M., Chapel, H. W., and Perk, J. H. H. (1982). *Physica* **112B,** 147.
Olés, A. M. and Stollhoff, G. (1984). *Phys. Rev.* **B29,** 314.
Rezende, S. M. (1983). *Phys. Rev.* **B27,** 3032.
—— and White, R. M. (1978). *Phys. Rev.* **B18,** 2346.
Silberglitt, R. and Harris, A. B. (1968). *Phys. Rev.* **174,** 640.
Stinchcombe, R. B. (1977). *Electron–phonon interactions and phase transitions* (ed. T. Riste). Plenum, Press, New York.
Stirling, W. G. and McEwan, K. A. (1987). Chapter 20 in *Neutron Scattering* (eds. K. Sköld and D. L. Price). Academic Press, New York.
Taylor, K. N. R. and Darby, M. I. (1972). *Physics of rare earth solids.* Chapman and Hall, London.
Thalmeier, P. and Fulde, P. (1982). *Phys. Rev. Lett.* **49,** 1588.
Wicksted, J. P., Böni, P. and Shirane, G. (1984). *Phys. Rev.* **B30,** 3655.
Windsor, C. G. (1981). *Pulsed neutron scattering.* Taylor and Francis.
Ziman, J. M. (1979). *Models of disorder.* Cambridge University Press, Cambridge.
Ziman, T. and Lindgård, P.-A. (1986). *Phys. Rev.* **B33,** 1976.

10
POLARIZATION ANALYSIS

To specify completely the physical state of a beam of neutrons we must give both its momentum and spin state. Up to now we have always taken the spins of the incident neutrons to be randomly orientated, i.e. unpolarized; neither have we asked for the spin state of the scattered beam. Clearly, however, if we use incident neutrons whose spins have some preferred axis, i.e. the beam is, to some degree, polarized, and in addition analyse not only the spatial and energy distribution of the scattered neutrons but also their spin states, we must gain more information about the nature of the scattering process than if we did neither. In such a scheme we are introducing two new variables into the scattering process, the polarization of the incident beam and the polarization of the scattered beam. We must, therefore, first ask how the cross-section for the scattering process depends on the polarization of the incident neutrons and secondly, relate the polarization of the scattered neutrons to the properties of the target system. In some instances the scattered neutrons may be polarized even though the incident neutrons are unpolarized. Thus, in this instance, polarization is created in the scattering process and this polarization must be intimately connected with the physical properties of the target system sensed by the neutrons.

Quite generally we can see that, within the Born approximation (cf. Appendix A), the cross-section for the scattering of polarized neutrons will be independent of the polarization if there is no preferred axis in the target system itself. Thus, pure nuclear scattering is independent of polarization because nuclear spins are always randomly orientated at temperatures of interest to us. Even so, it is of interest to ask how the scattering affects the polarization if it is initially present, and we discuss this in detail. Because a ferromagnet does have a preferred axis (when the domains are aligned by an external magnetic field), the cross-section for scattering from a ferromagnet depends on the polarization. We find the purely nuclear and purely magnetic cross-sections are as before, but a new term due to interference between nuclear and magnetic scattering appears. Also, in the polarization of neutrons inelastically scattered by magnetic target systems we find created polarization, the orientation of which depends on whether it is associated with a scattering process in which an elementary excitation is created or annihilated. Throughout this chapter we assume that the nuclei are randomly oriented. Studies of ordered systems of nuclei are discussed in Abragam and Bleaney (1983), Abragam and Goldman (1982), and Fano (1983).

10.1. Description of a polarized beam (Joachain 1983)

First, what do we mean when we say a beam of neutrons possesses a certain polarization? We define the polarization of a beam of neutrons as twice the average value of the spin of the neutrons in the beam, viz.

$$\mathbf{P} = 2\langle \hat{\mathbf{s}} \rangle = \langle \hat{\boldsymbol{\sigma}} \rangle. \tag{10.1}$$

In (10.1), $\hat{\sigma}^\alpha$ are the Pauli matrices

$$\hat{\sigma}^x = \begin{pmatrix} 0 & 1 \\ 1 & 0 \end{pmatrix}, \quad \hat{\sigma}^y = \begin{pmatrix} 0 & -\mathrm{i} \\ \mathrm{i} & 0 \end{pmatrix}, \quad \hat{\sigma}^z = \begin{pmatrix} 1 & 0 \\ 0 & -1 \end{pmatrix}. \tag{10.2}$$

For an *unpolarized beam*, $\mathbf{P} = 0$; for a *completely polarized beam* $|\mathbf{P}| = 1$. By a *partially polarized beam* we mean that $0 < |\mathbf{P}| < 1$.

If a beam of neutrons is partially polarized, we cannot have sufficient information about it to give a complete quantum mechanical description, i.e. we cannot assign a wave function to the spin state of the beam. For if the spin state of a neutron is described by a wave function χ, which for a spin $\frac{1}{2}$ particle is a two-component spinor, it follows that in some direction in space the neutron has a definite spin value, i.e. it is either $\pm\frac{1}{2}$, and is therefore completely polarized. Hence, a partially polarized beam of neutrons can only be defined by some probability distribution. The physical properties of such a system, usually called a 'mixed' system, can be described in terms of a density matrix operator $\hat{\rho}$. We choose to introduce the density matrix operator by first considering a single neutron with a spin wave χ and then construct the density matrix for the beam of incident neutrons by averaging over the density matrices of the individual neutrons.

The most general form of the spin-wave function χ for a neutron is

$$\chi = a\chi_\uparrow + b\chi_\downarrow \tag{10.3}$$

where $\chi_\uparrow$ and $\chi_\downarrow$ are the eigenfunctions of $\hat{\sigma}^z$. The coefficients a and b must satisfy

$$|a|^2 + |b|^2 = 1 \tag{10.4}$$

for χ to be normalized to unity.

Clearly the quantities $|a|^2$ and $|b|^2$ are, respectively, the probabilities that a measurement of the z-component of spin will show that the spin is parallel ($\uparrow$) or antiparallel ($\downarrow$) to the z-direction. There will always exist a direction defined by the unit vector $\tilde{\boldsymbol{\xi}}$ such that

$$(\hat{\boldsymbol{\sigma}} \cdot \tilde{\boldsymbol{\xi}})\chi = \chi,$$

i.e. the spin is *fully* aligned in the direction $\tilde{\boldsymbol{\xi}}$.

We define the density matrix operator $\hat{\rho}$ as

$$\hat{\rho} = \chi\chi^{+} = \begin{pmatrix} |a|^2 & ab^* \\ ba^* & |b|^2 \end{pmatrix}. \quad (10.5)$$

$\hat{\rho}$ is Hermitian and has unit trace, i.e.

$$\mathrm{Tr}\,\hat{\rho} = 1. \quad (10.6)$$

The average value of an arbitrary operator $\hat{O}$ in the state described by χ is $\chi^{+}\hat{O}\chi$. From (10.3) and (10.5) we see that this may be written in terms of $\hat{\rho}$, since

$$\langle \hat{O} \rangle = \chi^{+}\hat{O}\chi = \mathrm{Tr}\,\hat{O}\hat{\rho} = \mathrm{Tr}\,\hat{\rho}\hat{O}. \quad (10.7)$$

The last equality follows from the invariance of the trace to a cyclic permutation of the operators.

Because $\hat{\rho}$ is a 2×2 Hermitian matrix it can be expanded in terms of the unit matrix $\mathscr{I}$ and the Pauli matrices $\hat{\sigma}^{\alpha}$ with real expansion coefficients. In fact

$$\hat{\rho} = \tfrac{1}{2}(\mathscr{I} + \mathbf{P}\cdot\hat{\boldsymbol{\sigma}}) \quad (10.8)$$

where we have defined

$$P_x = 2\,\mathrm{Re}(a^*b),$$
$$P_y = 2\,\mathrm{Im}(a^*b),$$
$$P_z = |a|^2 - |b|^2. \quad (10.9)$$

That the vector $\mathbf{P}$ in (10.8) coincides with the definition of the polarization vector given in (10.1) follows immediately from (10.7) if we make use of

$$\mathrm{Tr}\,\hat{\sigma}^{\alpha} = 0 \quad \text{for all } \alpha, \quad (10.10)$$

$$\mathrm{Tr}\,\mathscr{I} = 2, \quad (10.11)$$

and, also, the identity

$$\mathrm{Tr}\,\hat{\sigma}^{\alpha}\hat{\sigma}^{\beta} = 2\delta_{\alpha,\beta}. \quad (10.12)$$

Of course, in the present case $\mathbf{P}$ has unit magnitude because the neutron is fully polarized. This follows directly from (10.9) or by noting that from (10.5) we must have

$$\hat{\rho}^2 = \chi\chi^{+}\chi\chi^{+} = \chi\chi^{+} = \hat{\rho}. \quad (10.13)$$

Consider now a beam of neutrons. Each individual neutron will have its own spin state described by a density matrix. However, in general, the beam is not in a definite spin state. We therefore define the polarization of the beam as the average over the polarizations of each neutron. Thus if the jth neutron has a polarization $\mathbf{P}_j$ and there are $\mathcal{N}$ neutrons,

$$\mathbf{P}=\frac{1}{\mathcal{N}}\sum_j \mathbf{P}_j. \tag{10.14}$$

The magnitude of $\mathbf{P}$ must lie between zero and unity and the density matrix of the beam is, clearly,

$$\hat{\rho}=\tfrac{1}{2}(\mathcal{I}+\mathbf{P}\cdot\hat{\boldsymbol{\sigma}}) \quad (0\leqslant|\mathbf{P}|\leqslant 1). \tag{10.15}$$

We note that

$$\langle\hat{\rho}\rangle=\mathrm{Tr}\,\hat{\rho}\hat{\rho}=\tfrac{1}{2}\{1+|\mathbf{P}|^2\}, \tag{10.16}$$

from which it follows that

$$\langle\hat{\rho}\rangle\geqslant\tfrac{1}{2} \tag{10.17}$$

and, since $|\mathbf{P}|^2\leqslant 1$,

$$\langle\hat{\rho}\rangle\leqslant 1. \tag{10.18}$$

The equality in (10.17) is satisfied when the neutron beam is unpolarized ($\mathbf{P}=0$), whereas the equality in (10.18) is satisfied when the beam is completely polarized.

It is also worthwhile to note that an unpolarized beam can be regarded as the sum of two completely polarized beams that possess opposite polarizations. Let the two beams have polarizations in the β-direction, then the above statement merely says that the density matrix

$$\{\tfrac{1}{2}(\mathcal{I}+P_\beta\hat{\sigma}^\beta)+\tfrac{1}{2}(\mathcal{I}-P_\beta\hat{\sigma}^\beta)\}/\mathrm{Tr}\{\tfrac{1}{2}(\mathcal{I}+P_\beta\hat{\sigma}^\beta)+\tfrac{1}{2}(\mathcal{I}-P_\beta\hat{\sigma}^\beta)\}=\tfrac{1}{2}\mathcal{I}, \tag{10.19}$$

as required for an unpolarized beam.

The vector nature of the polarization of a neutron beam is preserved in an experiment provided the magnetic guide field on the target sample is not too strong. For a strong magnetic field makes the precessing component of the polarization, that is perpendicular to the field, average very rapidly to zero. The latter situation is usually unavoidable with ferromagnetic target samples which depolarize the neutron beam unless they are magnetized to saturation. In this instance, the polarization of the incident beam is parallel with the strong, saturating magnetic field, and only this component of the polarization of the scattered beam can be measured.

10.2. Cross-section and polarization of scattered beam

The partial differential cross-section is given by

$$\frac{\mathrm{d}^2\sigma}{\mathrm{d}\Omega\,\mathrm{d}E'}=\frac{k'}{k}\sum_{\lambda,\sigma}p_\lambda p_\sigma\sum_{\lambda',\sigma'}\langle\lambda,\sigma|\,\hat{V}^+(\boldsymbol{\kappa})\,|\lambda',\sigma'\rangle \times\langle\lambda',\sigma'|\,\hat{V}(\boldsymbol{\kappa})\,|\lambda,\sigma\rangle\,\delta(\hbar\omega+E_\lambda-E_{\lambda'}). \tag{10.20}$$

In (10.20), $\hat{V}(\boldsymbol{\kappa})$ is the Fourier transform of the interaction potential between the incident neutron and the target system multiplied by $(m/2\pi\hbar^2)$; it has the dimension of length.

For purely nuclear scattering from an array of rigid nuclei,

$$\hat{V}_{\mathrm{N}}(\boldsymbol{\kappa}) = \sum_{\mathbf{l},\mathbf{d}} \exp(\mathrm{i}\boldsymbol{\kappa} \cdot \mathbf{R}_{ld}) \hat{b}_{ld}. \tag{10.21}$$

Here $\hat{b}_{ld}$ is the scattering amplitude operator (cf. (1.57))

$$\hat{b}_{ld} = A_{ld} + \tfrac{1}{2} B_{ld} \hat{\boldsymbol{\sigma}} \cdot \hat{\mathbf{i}}_{ld}, \tag{10.22}$$

where

$$A = \{(i+1)b^{(+)} + ib^{(-)}\}/(2i+1)$$

and

$$B = 2(b^{(+)} - b^{(-)})/(2i+1)$$

and $\hat{\mathbf{i}}_{ld}$ is the angular momentum operator for the nucleus at lattice site $\mathbf{R}_{ld} = \mathbf{l} + \mathbf{d}$.

For purely magnetic scattering

$$\hat{V}_{\mathrm{M}}(\boldsymbol{\kappa}) = r_0 \hat{\boldsymbol{\sigma}} \cdot \hat{\mathbf{Q}}_{\perp} = r_0 \hat{\boldsymbol{\sigma}} \cdot \sum_i \exp(\mathrm{i}\boldsymbol{\kappa} \cdot \mathbf{r}_i) \left\{ \tilde{\boldsymbol{\kappa}} \times (\hat{\mathbf{s}}_i \times \tilde{\boldsymbol{\kappa}}) - \frac{\mathrm{i}}{\hbar\,|\boldsymbol{\kappa}|} \tilde{\boldsymbol{\kappa}} \times \hat{\mathbf{p}}_i \right\}. \tag{10.23}$$

where $r_0 = -0.54 \cdot 10^{-12}$ cm.

In a later section of this chapter we calculate the elastic cross-section for the scattering of neutrons by the electric field produced by the nuclei and atomic electrons in a solid. The interaction potential for this process is there shown to be

$$\hat{V}_{\mathrm{E}}(\boldsymbol{\kappa}) = \left(\frac{m_{\mathrm{e}}}{2m}\right) r_0 \{\mathrm{i} \cot(\tfrac{1}{2}\theta) \tilde{\mathbf{n}} \cdot \hat{\boldsymbol{\sigma}} - 1\} \times \left\{ \sum_{\mathbf{l},\mathbf{d}} \exp(\mathrm{i}\boldsymbol{\kappa} \cdot \mathbf{R}_{ld}) Z_d - \sum_i \exp(\mathrm{i}\boldsymbol{\kappa} \cdot \mathbf{r}_i) \right\} \tag{10.24}$$

where

$$\mathbf{k}' \times \mathbf{k} = k^2 \sin\theta\, \tilde{\mathbf{n}} \tag{10.25}$$

defines the unit vector $\tilde{\mathbf{n}}$ and eZ_d is the charge associated with the nucleus of site $\mathbf{d}$ within a unit cell.

An examination of the three types of interaction potential given above shows that they all have the form

$$\hat{v} = \hat{\beta} + \hat{\boldsymbol{\alpha}} \cdot \hat{\boldsymbol{\sigma}} \tag{10.26}$$

where the operators $\hat{\beta}$ and $\hat{\boldsymbol{\alpha}}$ refer to the target system and have the dimension of length. It is therefore sufficient in studying the modification to the cross-section due to polarization, to consider just the general form of interaction potential (10.26). Let us also, for the moment, consider that part of the cross-section (10.20) that depends on the neutron spin, namely,

$$\sum_{\sigma,\sigma'} p_\sigma \langle\sigma|\, \hat{v}^+ \,|\sigma'\rangle\langle\sigma'|\, \hat{v} \,|\sigma\rangle. \tag{10.27}$$

The sum over σ' in (10.27) can be done by closure, to give

$$\sum_{\sigma} p_\sigma \langle\sigma|\, \hat{v}^+\hat{v} \,|\sigma\rangle. \tag{10.28}$$

This is valid only if there is no phase correlation between the states labelled with the quantum number σ, i.e. only if with respect to these states σ the density matrix is diagonal. But in this case the probability p_σ is just the diagonal element $\langle\sigma|\, \hat{\rho} \,|\sigma\rangle$, so (10.28) can be rewritten

$$\sum_{\sigma} \langle\sigma|\, \hat{\rho} \,|\sigma\rangle\langle\sigma|\, \hat{v}^+\hat{v} \,|\sigma\rangle. \tag{10.29}$$

Furthermore, if $\hat{\rho}$ is diagonal,

$$\langle\sigma'|\, \hat{\rho} \,|\sigma\rangle = \delta_{\sigma,\sigma'}\langle\sigma|\, \hat{\rho} \,|\sigma\rangle$$

so that (10.29) becomes

$$\sum_{\sigma,\sigma'} \langle\sigma|\, \hat{v}^+\hat{v} \,|\sigma'\rangle\langle\sigma'|\, \hat{\rho} \,|\sigma\rangle$$

and the sum over σ' can be done by closure to give for (10.27)

$$\sum_{\sigma,\sigma'} p_\sigma \langle\sigma|\, \hat{v}^+ \,|\sigma'\rangle\langle\sigma'|\, \hat{v} \,|\sigma\rangle = \sum_{\sigma} \langle\sigma|\, \hat{v}^+\hat{v}\hat{\rho} \,|\sigma\rangle \equiv \mathrm{Tr}\, \hat{v}^+\hat{v}\hat{\rho} \equiv \mathrm{Tr}\, \hat{\rho}\hat{v}^+\hat{v}. \tag{10.30}$$

This final form is independent of the representation that is chosen to label the states and hence for this last form it does not matter whether or not $\hat{\rho}$ is diagonal. We conclude therefore that a formula which is much more general than (10.19) is

$$\frac{\mathrm{d}^2\sigma}{\mathrm{d}\Omega\,\mathrm{d}E'} = \frac{k'}{k} \sum_{\lambda,\lambda'} p_\lambda \mathrm{Tr}\, \hat{\rho} \langle\lambda|\, \hat{V}^+(\boldsymbol{\kappa}) \,|\lambda'\rangle\langle\lambda'|\, \hat{V}(\boldsymbol{\kappa}) \,|\lambda\rangle\, \delta(\hbar\omega + E_\lambda - E_{\lambda'}), \tag{10.31}$$

where it is understood that the trace is to be taken with respect only to the neutron spin coordinates. Before we examine the structure of the spin-dependent part of this cross-section we derive an expression for the polarization of the scattered beam, $\mathbf{P}'$. The formula for $\mathbf{P}'$ must represent the transformation of the spin state of the incident neutron beam, defined by $\mathbf{P}$, due to the interaction with the target system; we must average the initial spin state of the beam over all possible scattering processes and sum over all possible final states. Thus

$$\mathbf{P}' \propto \mathrm{Tr}\,\hat{\rho}\hat{v}^{+}\hat{\boldsymbol{\sigma}}\hat{v}.$$

The constant of proportionality is determined by normalization, viz.

$$\mathbf{P}' = \mathrm{Tr}\,\hat{\rho}\hat{v}^{+}\hat{\boldsymbol{\sigma}}\hat{v}/\mathrm{Tr}\,\hat{\rho}\hat{v}^{+}\hat{v}. \tag{10.32}$$

In full we have

$$\mathbf{P}'\left(\frac{\mathrm{d}^2\sigma}{\mathrm{d}\Omega\,\mathrm{d}E'}\right) = \frac{k'}{k}\sum_{\lambda,\lambda'} p_\lambda\,\mathrm{Tr}\,\hat{\rho}\langle\lambda|\,\hat{V}^{+}(\boldsymbol{\kappa})\,|\lambda'\rangle\hat{\boldsymbol{\sigma}}\langle\lambda'|\,\hat{V}(\boldsymbol{\kappa})\,|\lambda\rangle\delta(\hbar\omega + E_\lambda - E_{\lambda'}). \tag{10.33}$$

Let us examine the structure of eqns (10.30) and (10.32) with the general form of $\hat{v}$ given by eqn (10.26). For the cross-section we need to evaluate (10.30),

$$\begin{aligned}\mathrm{Tr}\,\hat{\rho}\hat{v}^{+}\hat{v} &= \tfrac{1}{2}\mathrm{Tr}(\mathcal{I} + \mathbf{P}\cdot\hat{\boldsymbol{\sigma}})(\hat{\beta}^{+} + \hat{\boldsymbol{\alpha}}^{+}\cdot\hat{\boldsymbol{\sigma}})(\hat{\beta} + \hat{\boldsymbol{\alpha}}\cdot\hat{\boldsymbol{\sigma}})\\ &= \tfrac{1}{2}\mathrm{Tr}(\mathcal{I} + \mathbf{P}\cdot\hat{\boldsymbol{\sigma}})(\hat{\beta}^{+}\hat{\beta} + \hat{\beta}^{+}\hat{\boldsymbol{\alpha}}\cdot\hat{\boldsymbol{\sigma}} + \hat{\boldsymbol{\alpha}}^{+}\cdot\hat{\boldsymbol{\sigma}}\hat{\beta} + \hat{\boldsymbol{\alpha}}^{+}\cdot\hat{\boldsymbol{\sigma}}\hat{\boldsymbol{\alpha}}\cdot\hat{\boldsymbol{\sigma}}).\end{aligned} \tag{10.34}$$

The evaluation of the trace of products of Pauli operators is facilitated by making use of the identity

$$\hat{\sigma}^{\alpha}\hat{\sigma}^{\beta} = \delta_{\alpha,\beta}\mathcal{I} + \mathrm{i}\sum_{\gamma}\epsilon^{\alpha\beta\gamma}\hat{\sigma}^{\gamma} \tag{10.35}$$

where $\epsilon^{\alpha\beta\gamma}$ is the completely antisymmetrical tensor with three indices, i.e.

$$\begin{aligned}\epsilon^{\alpha\beta\gamma} &= +1 \quad \text{if } \alpha,\ \beta, \text{ and } \gamma \text{ are in cyclic order,}\\ &= -1 \quad \text{if } \alpha,\ \beta, \text{ and } \gamma \text{ are not in cyclic order,}\\ &= 0 \quad\ \ \text{otherwise.}\end{aligned}$$

This identity is derived by using the commutation relation for angular momentum operators,

$$[\hat{L}^{\alpha}, \hat{L}^{\beta}] = \mathrm{i}\sum_{\gamma}\epsilon^{\alpha\beta\gamma}\hat{L}^{\gamma}$$

and the fact that Pauli operators anticommute,

$$\hat{\sigma}^{\alpha}\hat{\sigma}^{\beta} + \hat{\sigma}^{\beta}\hat{\sigma}^{\alpha} = 2\delta_{\alpha\beta}\mathcal{I}.$$

Because $\mathrm{Tr}\,\hat{\sigma}^{\gamma}$ is zero, we obtain from (10.35)

$$\mathrm{Tr}\,\hat{\sigma}^{\alpha}\hat{\sigma}^{\beta} = 2\delta_{\alpha,\beta}. \tag{10.12}$$

Also, with the aid of this result,

$$\begin{aligned}\mathrm{Tr}\,\hat{\sigma}^{\alpha}\hat{\sigma}^{\beta}\hat{\sigma}^{\gamma} &= \mathrm{i}\sum_{\gamma'}\epsilon^{\alpha\beta\gamma'}\,\mathrm{Tr}\,\hat{\sigma}^{\gamma'}\hat{\sigma}^{\gamma}\\ &= 2\mathrm{i}\epsilon^{\alpha\beta\gamma}.\end{aligned} \tag{10.36}$$

With the aid of these formulae we find for (10.34) the result,

$$\operatorname{Tr}\rho\hat{v}^{+}\hat{v}=\hat{\boldsymbol{\alpha}}^{+}\cdot\hat{\boldsymbol{\alpha}}+\hat{\beta}^{+}\hat{\beta}+\hat{\beta}^{+}(\hat{\boldsymbol{\alpha}}\cdot\mathbf{P})+(\hat{\boldsymbol{\alpha}}^{+}\cdot\mathbf{P})\hat{\beta}+\mathrm{i}\mathbf{P}\cdot(\hat{\boldsymbol{\alpha}}^{+}\times\hat{\boldsymbol{\alpha}}). \tag{10.37}$$

When it is adequate to consider only nuclear and magnetic scattering, an examination of the corresponding interaction potentials, (10.21) and (10.23), shows that $\hat{\beta}$ contains only nuclear terms and $\hat{\boldsymbol{\alpha}}$ both nuclear and magnetic scattering. Thus the third and fourth terms in (10.37) result in interference between nuclear and magnetic scattering. This interference has been utilized to perform very accurate measurements of magnetic structure factors. Bearing in mind that the cross-section must be averaged over the orientations of the nuclei, it can be seen that for randomly orientated nuclei the last term on the right-hand side of (10.37) is purely magnetic.

In order to evaluate

$$\operatorname{Tr}\hat{\rho}\hat{v}^{+}\hat{\boldsymbol{\sigma}}\hat{v} \tag{10.38}$$

as required in the formula for $\mathbf{P}'$ (eqn (10.32)) we need, in addition to (10.12) and (10.36), the identity

$$\operatorname{Tr}\hat{\sigma}^{\alpha}\hat{\sigma}^{\beta}\hat{\sigma}^{\gamma}\hat{\sigma}^{\delta}=2(\delta_{\alpha\beta}\,\delta_{\gamma\delta}-\delta_{\alpha\gamma}\,\delta_{\beta\delta}+\delta_{\alpha\delta}\,\delta_{\beta\gamma}), \tag{10.39}$$

which can be derived from (10.35). Also note that, if $\mathbf{A}$ and $\mathbf{B}$ are arbitrary vectors,

$$(\mathbf{A}\times\mathbf{B})_{\alpha}=\sum_{\beta,\gamma}\epsilon^{\alpha\beta\gamma}A_{\beta}B_{\gamma}. \tag{10.40}$$

With these results,

$$\begin{aligned}\operatorname{Tr}\hat{\rho}\hat{v}^{+}\hat{\boldsymbol{\sigma}}\hat{v}=\;&\hat{\beta}^{+}\hat{\boldsymbol{\alpha}}+\hat{\boldsymbol{\alpha}}^{+}\hat{\beta}+\hat{\beta}^{+}\hat{\beta}\mathbf{P}+\hat{\boldsymbol{\alpha}}^{+}(\hat{\boldsymbol{\alpha}}\cdot\mathbf{P})+(\hat{\boldsymbol{\alpha}}^{+}\cdot\mathbf{P})\hat{\boldsymbol{\alpha}}\\&-\mathbf{P}(\hat{\boldsymbol{\alpha}}^{+}\cdot\hat{\boldsymbol{\alpha}})-\mathrm{i}\hat{\boldsymbol{\alpha}}^{+}\times\hat{\boldsymbol{\alpha}}+\mathrm{i}\hat{\beta}^{+}(\hat{\boldsymbol{\alpha}}\times\mathbf{P})+\mathrm{i}(\mathbf{P}\times\hat{\boldsymbol{\alpha}}^{+})\hat{\beta}.\end{aligned} \tag{10.41}$$

If we again consider just nuclear and magnetic scattering, we see that in the limit of an unpolarized incident beam of neutrons there is a creation of polarization by a purely magnetic term (for randomly oriented nuclei), $-\mathrm{i}\hat{\boldsymbol{\alpha}}^{+}\times\hat{\boldsymbol{\alpha}}$.

To sum up, we use (10.37) in conjunction with (10.31) to give the cross-sections and we use (10.41) in conjunction with (10.33) to give the polarization $\mathbf{P}'$ of the scattered neutrons. We shall discuss how to do this first for elastic scattering and then for inelastic magnetic scattering. We begin our discussion of elastic scattering by examining the scattering of neutrons by the electrostatic fields created by the nuclei and atomic electrons in a solid.

10.3. Scattering by electrostatic field

If the nucleus at the site **d** in a unit cell has a charge eZ_d then the total charge density at position **r** due to all the nuclei in the solid is

$$\sum_{\mathbf{l},\mathbf{d}} eZ_d\, \delta(\mathbf{r}-\mathbf{R}_{ld}). \qquad (10.42)$$

Since the charge density at **r** due to the atomic electrons is simply

$$-e\sum_i \delta(\mathbf{r}-\mathbf{r}_i),$$

the total charge density $n(\mathbf{r})$ is

$$n(\mathbf{r}) = e\left\{\sum_{\mathbf{l},\mathbf{d}} Z_d\, \delta(\mathbf{r}-\mathbf{R}_{ld}) - \sum_i \delta(\mathbf{r}-\mathbf{r}_i)\right\}. \qquad (10.43)$$

Because the crystal is electrostatically neutral we must have

$$\int \mathrm{d}\mathbf{r}\, n(\mathbf{r}) = 0. \qquad (10.44)$$

The electrostatic potential $\phi(\mathbf{r})$ of the charge density $n(\mathbf{r})$ is determined by

$$\nabla^2\phi(\mathbf{r}) = -4\pi n(\mathbf{r}) \qquad (10.45)$$

and the electric field **E** satisfies

$$\mathbf{E} = -\boldsymbol{\nabla}\phi(\mathbf{r}). \qquad (10.46)$$

The interaction potential between a neutron and the charge distribution $n(\mathbf{r})$ consists of two terms, namely

$$\hat{V}_{\mathrm{E}}(\mathbf{r}) = (-\gamma\mu_{\mathrm{N}}/mc)\hat{\boldsymbol{\sigma}}\cdot\mathbf{E}\times\hat{\mathbf{p}} + (-\hbar\gamma\mu_{\mathrm{N}}/2mc)\boldsymbol{\nabla}\cdot\mathbf{E}. \qquad (10.47)$$

The first term on the right-hand side is usually referred to as the spin–orbit interaction while the second term is the neutron analogue of the Darwin interaction for electrons. The existence of this term for neutrons in an electrostatic field was first pointed out by Foldy.

For the contribution to the cross-section from the spin–orbit term we need

$$(-\gamma\mu_{\mathrm{N}}/mc)\int \mathrm{d}\mathbf{r}\, \exp(-\mathrm{i}\mathbf{k}'\cdot\mathbf{r})\hat{\boldsymbol{\sigma}}\cdot\mathbf{E}\times\hat{\mathbf{p}}\exp(\mathrm{i}\mathbf{k}\cdot\mathbf{r})$$
$$= (-\gamma\mu_{\mathrm{N}}/mc)\hat{\boldsymbol{\sigma}}\cdot\int \mathrm{d}\mathbf{r}\, \exp(\mathrm{i}\boldsymbol{\kappa}\cdot\mathbf{r})\mathbf{E}\times\hbar\mathbf{k}. \qquad (10.48)$$

The Fourier integral of the electric field in (10.48) is readily effected by the following argument. From (10.45) the potential $\phi(\mathbf{r})$ is given by

$$\phi(\mathbf{r}) = \int \mathrm{d}\mathbf{r}' \frac{n(\mathbf{r}')}{|\mathbf{r}-\mathbf{r}'|}. \tag{10.49}$$

If $n(\boldsymbol{\kappa})$ is the Fourier transform of the charge density, i.e.

$$n(\boldsymbol{\kappa}) = \int \mathrm{d}\mathbf{r} \exp(-\mathrm{i}\boldsymbol{\kappa}\cdot\mathbf{r}) n(\mathbf{r}), \tag{10.50}$$

then (10.49) shows that

$$\phi(\mathbf{r}) = \frac{1}{2\pi^2} \int \mathrm{d}\boldsymbol{\kappa} \exp(\mathrm{i}\boldsymbol{\kappa}\cdot\mathbf{r}) \frac{n(\boldsymbol{\kappa})}{|\boldsymbol{\kappa}|^2}. \tag{10.51}$$

In obtaining (10.51) from (10.49) it is convenient to use the integral

$$\int \mathrm{d}\mathbf{r}' \frac{\exp(\mathrm{i}\boldsymbol{\kappa}\cdot\mathbf{r}')}{|\mathbf{r}-\mathbf{r}'|} = \frac{4\pi}{|\boldsymbol{\kappa}|^2} \exp(\mathrm{i}\boldsymbol{\kappa}\cdot\mathbf{r}).$$

Hence we have now

$$\begin{aligned}\int \mathrm{d}\mathbf{r} \exp(\mathrm{i}\boldsymbol{\kappa}\cdot\mathbf{r})\mathbf{E}\times\hbar\mathbf{k} &= \int \mathrm{d}\mathbf{r} \exp(\mathrm{i}\boldsymbol{\kappa}\cdot\mathbf{r})\{-\boldsymbol{\nabla}\phi(\mathbf{r})\}\times\hbar\mathbf{k} \\ &= \frac{\mathrm{i}(2\pi)^3}{2\pi^2} \frac{n(-\boldsymbol{\kappa})}{|\boldsymbol{\kappa}|^2} (\boldsymbol{\kappa}\times\hbar\mathbf{k}) \\ &= -\frac{4\pi\mathrm{i}\hbar}{|\boldsymbol{\kappa}|^2} n(-\boldsymbol{\kappa})(\mathbf{k}'\times\mathbf{k}). \end{aligned} \tag{10.52}$$

Noting that for elastic scattering $|\boldsymbol{\kappa}| = 2k \sin\frac{1}{2}\theta$, (10.48) reads

$$\left(\frac{2\pi\hbar^2}{m}\right)\left(\frac{\gamma e}{2mc^2}\right) n(-\boldsymbol{\kappa})\mathrm{i}\cot(\tfrac{1}{2}\theta)\tilde{\mathbf{n}}\cdot\hat{\boldsymbol{\sigma}} \tag{10.53}$$

where the unit vector $\tilde{\mathbf{n}}$ perpendicular to the scattering plane is defined in (10.25). Observe that (10.52) depends on $\mathbf{k}$ and $\mathbf{k}'$ and not just $\boldsymbol{\kappa}$.

The evaluation of the contribution to the cross-section from the Foldy interaction is straightforward. From the definition of $n(\boldsymbol{\kappa})$,

$$\begin{aligned}\int \mathrm{d}\mathbf{r} \exp(\mathrm{i}\boldsymbol{\kappa}\cdot\mathbf{r})\boldsymbol{\nabla}\cdot\mathbf{E} &= \int \mathrm{d}\mathbf{r} \exp(\mathrm{i}\boldsymbol{\kappa}\cdot\mathbf{r})\{4\pi n(\mathbf{r})\} \\ &= 4\pi n(-\boldsymbol{\kappa}). \end{aligned} \tag{10.54}$$

Therefore from (10.53) and (10.54)

$$\begin{aligned}\hat{V}_{\mathrm{E}}(\boldsymbol{\kappa}) &= \left(\frac{m}{2\pi\hbar^2}\right)\int \mathrm{d}\mathbf{r} \exp(-\mathrm{i}\mathbf{k}'\cdot\mathbf{r})\hat{V}_{\mathrm{E}}(\mathbf{r})\exp(\mathrm{i}\mathbf{k}\cdot\mathbf{r}) \\ &= \left(\frac{m_{\mathrm{e}}}{2m}\right) r_0 n(-\boldsymbol{\kappa})\{\mathrm{i}\cot(\tfrac{1}{2}\theta)\tilde{\mathbf{n}}\cdot\hat{\boldsymbol{\sigma}} - 1\}. \end{aligned} \tag{10.55}$$

By analogy with the elastic scattering of neutrons by the unpaired electrons of an ion, we write for the matrix element

$$\langle\lambda|\, n(-\boldsymbol{\kappa})\,|\lambda\rangle = e\sum_{\mathbf{l},\mathbf{d}} \exp(\mathrm{i}\boldsymbol{\kappa}\cdot\mathbf{R}_{ld})\{Z_d - f_d(\boldsymbol{\kappa})\}$$

$$= e\sum_{\mathbf{l}} \exp(\mathrm{i}\boldsymbol{\kappa}\cdot\mathbf{l})F_{\mathrm{E}}(\boldsymbol{\kappa}) \tag{10.56}$$

where $f_d(\boldsymbol{\kappa})$ is the atomic form factor of the *charge density* associated with the atom at $\mathbf{R}_{ld}$ and $F_{\mathrm{E}}(\boldsymbol{\kappa})$ is the *electrostatic unit-cell structure factor.*

The form factor entering in (10.56) is that encountered in X-ray scattering. Note that in view of (10.44) we have $f_d(0) = Z_d$.

In the notation of § 10.2, we see from (10.55) that

$$\langle\lambda|\,\hat{\boldsymbol{\alpha}}\,|\lambda\rangle \equiv \tilde{\mathbf{n}}\left(\frac{m_{\mathrm{e}}}{2m}\right) r_0\, \mathrm{i}\cot(\tfrac{1}{2}\theta)\sum_{\mathbf{l}} \exp(\mathrm{i}\boldsymbol{\kappa}\cdot\mathbf{l})F_{\mathrm{E}}(\boldsymbol{\kappa})$$

and

$$\langle\lambda|\,\hat{\beta}\,|\lambda\rangle \equiv -\left(\frac{m_{\mathrm{e}}}{2m}\right) r_0 \sum_{\mathbf{l}} \exp(\mathrm{i}\boldsymbol{\kappa}\cdot\mathbf{l})F_{\mathrm{E}}(\boldsymbol{\kappa}).$$

By using these results in our general expression for the neutron spin-dependent part of the elastic cross-section (eqn (10.37)) we obtain for the cross-section for the elastic scattering of polarized neutrons by the Coulomb field of the nuclei and atomic electrons in a solid the result

$$\left(\frac{\mathrm{d}\sigma}{\mathrm{d}\Omega}\right)_{\mathrm{EE}} = \left|\sum_{\mathbf{l}} \exp(\mathrm{i}\boldsymbol{\kappa}\cdot\mathbf{l})\right|^2 \left(\frac{m_{\mathrm{e}}}{2m}\right)^2 r_0^2(\sin\tfrac{1}{2}\theta)^{-2}\,|F_{\mathrm{E}}(\boldsymbol{\kappa})|^2 \tag{10.57}$$

and

$$\left|\sum_{\mathbf{l}} \exp(\mathrm{i}\boldsymbol{\kappa}\cdot\mathbf{l})\right|^2 = N\frac{(2\pi)^3}{v_0}\sum_{\boldsymbol{\tau}} \delta(\boldsymbol{\kappa}-\boldsymbol{\tau}).$$

The scattering is therefore confined to nuclear Bragg peaks ($\boldsymbol{\kappa} = \boldsymbol{\tau}$), a result that is simply a consequence of the fact that the neutrons are scattered by a (rigid) periodic array of potentials. That $(\mathrm{d}\sigma/\mathrm{d}\Omega)_{\mathrm{EE}}$ is independent of the polarization $\mathbf{P}$ of the incident beam is an example of the general principle stated at the beginning of the chapter, namely, that the cross-section is independent of $\mathbf{P}$ if the target system does not possess a preferred axis.

We do not pause to calculate the polarization of the scattered beam but go on to consider the scattering of polarized neutrons through their nuclear interaction with a crystal.

10.4. Nuclear scattering

The cross-section for purely nuclear scattering of a beam of polarized neutrons is independent of the polarization $\mathbf{P}$ because, unless the temperature is exceptionally low (Schermer and Blume 1968; Abragam and Bleaney 1983), the nuclei are randomly oriented and thus the crystal does not possess a preferred axis. However, to gain some familiarity with the formulae it is worthwhile to demonstrate this and to ask for the polarization of the scattered beam.

From § 10.2 we know that

$$\hat{V}_{\mathrm{N}}(\boldsymbol{\kappa}) = \sum_{\mathbf{l},\mathbf{d}} \exp(\mathrm{i}\boldsymbol{\kappa} \cdot \mathbf{R}_{ld})(A_{ld} + \tfrac{1}{2}B_{ld}\hat{\boldsymbol{\sigma}} \cdot \hat{\mathbf{i}}_{ld}), \tag{10.58}$$

so, from (10.26),

$$\hat{\boldsymbol{\alpha}} \equiv \sum_{\mathbf{l},\mathbf{d}} \exp(\mathrm{i}\boldsymbol{\kappa} \cdot \mathbf{R}_{ld})\tfrac{1}{2}B_{ld}\hat{\mathbf{i}}_{ld}$$

and

$$\hat{\beta} \equiv \sum_{\mathbf{l},\mathbf{d}} \exp(\mathrm{i}\boldsymbol{\kappa} \cdot \mathbf{R}_{ld})A_{ld}. \tag{10.59}$$

Now suppose (10.59) is inserted into (10.37) and (10.37) in turn inserted into (10.31). Then in (10.31) one operation we must perform is to average over nuclear spin orientations and isotope distributions. Because we assume the energy and all other properties of the target are independent of these factors we can do this immediately and it is convenient to do this step on (10.37) directly (rather than inserting it into (10.31) and then averaging).

For a discussion of polarization effects it is useful to have an explicit notation for averaging over nuclear spin orientations separately from averaging over both orientations and isotope distributions. We shall denote $\mathcal{O}$ = *average over nuclear spin orientations.*

Now consider this averaging process over (10.37), i.e. consider

$$\mathcal{O}\,\mathrm{Tr}\{\hat{\rho}\hat{v}^{+}\hat{v}\}.$$

We notice that (10.37) contains a term $\hat{\beta}^{+}(\hat{\boldsymbol{\alpha}} \cdot \mathbf{P})$ and from (10.59) this is linear in the nuclear spins $\hat{\mathbf{i}}_{ld}$. Hence

$$\mathcal{O}\{\hat{\beta}^{+}(\hat{\boldsymbol{\alpha}} \cdot \mathbf{P})\} = 0. \tag{10.60a}$$

Similarly

$$\mathcal{O}\{(\hat{\boldsymbol{\alpha}}^{+} \cdot \mathbf{P})\hat{\beta}\} = 0. \tag{10.60b}$$

We also notice that (10.37) contains a term $\mathrm{i}\mathbf{P} \cdot \hat{\boldsymbol{\alpha}}^{+} \times \hat{\boldsymbol{\alpha}}$ and from (10.59) we see this involves expressions like

$$\hat{\mathbf{i}}_{ld} \times \hat{\mathbf{i}}_{l'd'} = \delta_{\mathbf{l},\mathbf{l}'}\,\delta_{\mathbf{d},\mathbf{d}'}\,\mathrm{i}\hat{\mathbf{i}}_{ld}.$$

Since this also is linear in the nuclear spins

$$\mathcal{O}\{\mathrm{i}\mathbf{P} \cdot \hat{\boldsymbol{\alpha}}^{+} \times \hat{\boldsymbol{\alpha}}\} = 0. \tag{10.60c}$$

We therefore see that the three terms in (10.37) which involve the polarization all vanish. It follows that the cross-section itself is independent of **P**, as we had anticipated. We are left only with the terms $\hat{\boldsymbol{\alpha}}^+ \cdot \hat{\boldsymbol{\alpha}}$ and $\hat{\beta}^+\hat{\beta}$. The latter is independent of the $\mathcal{O}$ operation, hence

$$\mathcal{O}\ \mathrm{Tr}\{\hat{\rho}\hat{v}^+\hat{v}\} = \hat{\beta}^+\hat{\beta} + \mathcal{O}\{\hat{\boldsymbol{\alpha}}^+ \cdot \hat{\boldsymbol{\alpha}}\}$$

$$= \sum_{\mathbf{l,d,l',d'}} \exp\{i\boldsymbol{\kappa}\cdot(\mathbf{R}_{ld} - \mathbf{R}_{l'd'})\}\{A^*_{l'd'}A_{ld} + \tfrac{1}{4}\mathcal{O}(B^*_{l'd'}B_{ld}\hat{\mathbf{i}}_{l'd'}\cdot\hat{\mathbf{i}}_{ld})\}$$

$$= \sum_{\mathbf{l,d,l',d'}} \exp\{i\boldsymbol{\kappa}\cdot(\mathbf{R}_{ld} - \mathbf{R}_{l'd'})\}[\,|\bar{A}_d|^2 + \delta_{\mathbf{l,l'}}\,\delta_{\mathbf{d,d'}}\{\overline{|A_d|^2} - |\bar{A}_d|^2 + \tfrac{1}{4}\overline{|B_d|^2\, i_d(i_d+1)}\}] \tag{10.61}$$

where in the last step we have also averaged over isotope distributions.

We now suppose the incident beam is polarized and calculate the final polarization using (10.33) and (10.41). We first evaluate (10.41),

$$\mathcal{O}\{\mathrm{Tr}\ \hat{\rho}\hat{v}^+\hat{\boldsymbol{\sigma}}\hat{v}\}$$

and, as in the previous example, note that all the terms linear in $\hat{\boldsymbol{\alpha}}$ vanish and so also does the term $i\hat{\boldsymbol{\alpha}}^+\times\hat{\boldsymbol{\alpha}}$. Furthermore,

$$\mathcal{O}\{\hat{\boldsymbol{\alpha}}^+(\hat{\boldsymbol{\alpha}}\cdot\mathbf{P})\} = \mathcal{O}\{(\hat{\boldsymbol{\alpha}}^+\cdot\mathbf{P})\hat{\boldsymbol{\alpha}}\} = \tfrac{1}{3}\mathbf{P}\mathcal{O}\{\hat{\boldsymbol{\alpha}}^+\cdot\hat{\boldsymbol{\alpha}}\}.$$

Hence

$$\mathcal{O}\{\mathrm{Tr}\ \hat{\rho}\hat{v}^+\hat{\boldsymbol{\sigma}}\hat{v}\} = \mathbf{P}\{\hat{\beta}^+\hat{\beta} - \tfrac{1}{3}\mathcal{O}(\hat{\boldsymbol{\alpha}}^+\cdot\hat{\boldsymbol{\alpha}})\}$$

and by comparison with the derivation of (10.61) this gives, after averaging over isotope distributions,

$$\sum_{\mathbf{l,d,l',d'}} \exp\{i\boldsymbol{\kappa}\cdot(\mathbf{R}_{ld} - \mathbf{R}_{l'd'})\}$$

$$\times\mathbf{P}[\,|\bar{A}_d|^2 + \delta_{\mathbf{l,l'}}\,\delta_{\mathbf{d,d'}}\{\overline{|A_d|^2} - |\bar{A}_d|^2 - \tfrac{1}{12}\overline{|B_d|^2\, i_d(i_d+1)}\}]. \tag{10.62}$$

Comparing (10.62) with (10.61) term by term we note that: (a) the nuclear coherent scattering has a final polarization equal to the initial polarization **P**; (b) the nuclear incoherent scattering due to the random isotope distributions also has a final polarization **P**; (c) the nuclear incoherent scattering due to the random nuclear spin orientations has a polarization $-\tfrac{1}{3}\mathbf{P}$. Physically these results are easy to understand. Neither (a) nor (b) can possibly involve a neutron spin-flip and therefore the polarization is unchanged. However, for (c) one-third of the interaction, namely $\hat{\sigma}_z\hat{i}_z$, scatters the neutron without spin-flip whereas two-thirds, namely $\hat{\sigma}_x\hat{i}_x + \hat{\sigma}_y\hat{i}_y$, scatters with spin-flip and hence gives a final polarization of opposite sign and one-third the magnitude of the initial polarization.

In general we have incoherent scattering due to both random isotope distributions and nuclear spin orientations and therefore the final polarization of the incoherent scattering lies between the limits of $\mathbf{P}$ and $-\frac{1}{3}\mathbf{P}$. From (10.61) and (10.62) we see that

$$\mathbf{P}'_{\text{coh}} = \mathbf{P} \tag{10.63}$$

and

$$\mathbf{P}'_{\text{incoh}} = \mathbf{P} \frac{\sum_{\mathbf{d}} \{\overline{|A_d|^2} - |\bar{A}_d|^2 - \frac{1}{12}\overline{|B_d|^2\, i_d(i_d+1)}\}}{\sum_{\mathbf{d}} \{\overline{|A_d|^2} - |\bar{A}_d|^2 + \frac{1}{4}\overline{|B_d|^2 i_d(i_d+1)}\}}. \tag{10.64}$$

We now remember, that

$$\bar{A}_d = \bar{b}_d \tag{10.65}$$

and

$$\overline{|A_d|^2} + \frac{1}{4}\overline{|B_d|^2\, i_d(i_d+1)} = \overline{|b_d|^2} \tag{10.66}$$

and rewrite (10.64) as

$$\mathbf{P}'_{\text{incoh}} = \mathbf{P} \frac{\sum_{\mathbf{d}} \{\frac{4}{3}\overline{|A_d|^2} - |\bar{b}_d|^2 - \frac{1}{3}\overline{|b_d|^2}\}}{\sum_{\mathbf{d}} \{\overline{|b_d|^2} - |\bar{b}_d|^2\}} \tag{10.67}$$

where the quantities $\bar{b}_d$ and $\overline{|b_d|^2}$ are defined in Chapter 1 and, from the definition in this section,

$$\overline{|A|^2} = \sum_{\xi} c_{\xi}\, |A_{\xi}|^2. \tag{10.68}$$

An experimental arrangement used in polarization analysis work is illustrated in Fig. 10.1. The actual layout shown is that used with the High Flux Isotope Reactor at Oak Ridge National Laboratory. It consists of a triple-axis spectrometer with a polarizer and an analyser. With this scheme an energy analysis of the scattered neutrons can be made, as with any triple-axis instrument, and, in addition, production of polarized neutrons and selection of a particular polarization state of the scattered beam. The polarization of the incident and scattered beams can be reversed by activating the r.f. coils, or flippers. The electromagnet around the target and the guide fields 2 and 3 can be rotated so that the polarization is either perpendicular or parallel to the scattering vector $\boldsymbol{\kappa}$. In detail the experimental scheme is as follows.

With both flippers off, the beam of neutrons incident on the target system is polarized in a direction perpendicular to the scattering plane (↑) and only those scattered neutrons in the same polarization state are reflected by the analysing crystal. With the first flipper on the neutrons

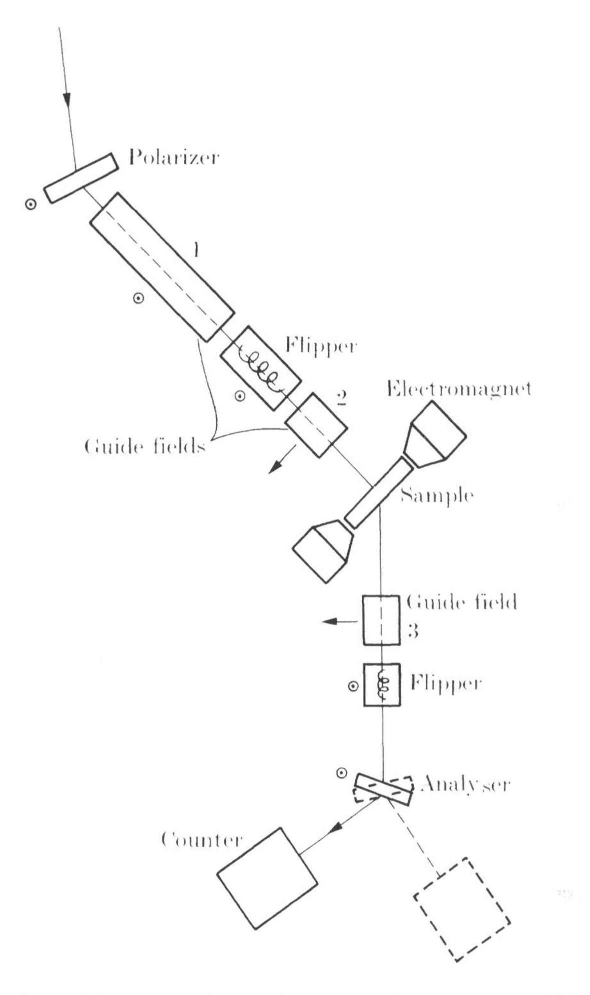

FIG. 10.1. Experimental layout used in polarization analysis experiments with HFIR at Oak Ridge National Laboratory. (Figs. 10.1–10.3 and 10.5–10.9 are from a lecture by R. M. Moon at Kjeller, Netherlands—Norwegian Reactor School and from Moon, Riste, and Koehler (1969).)

incident on the target are in the ↓ state and the analyser affects only those scattered neutrons whose polarization is oppositely orientated to that of the incident neutrons. Similarly, if just the second flipper is on, the measured cross-section is that for the ↑–↓ transition. With both flippers on the measured cross-section corresponds to ↓–↓ transitions.

An example of isotopic incoherent scattering is shown in Fig. 10.2. This gives the number of neutrons scattered from polycrystalline nickel for the two cases where neither flipper was activated and where one

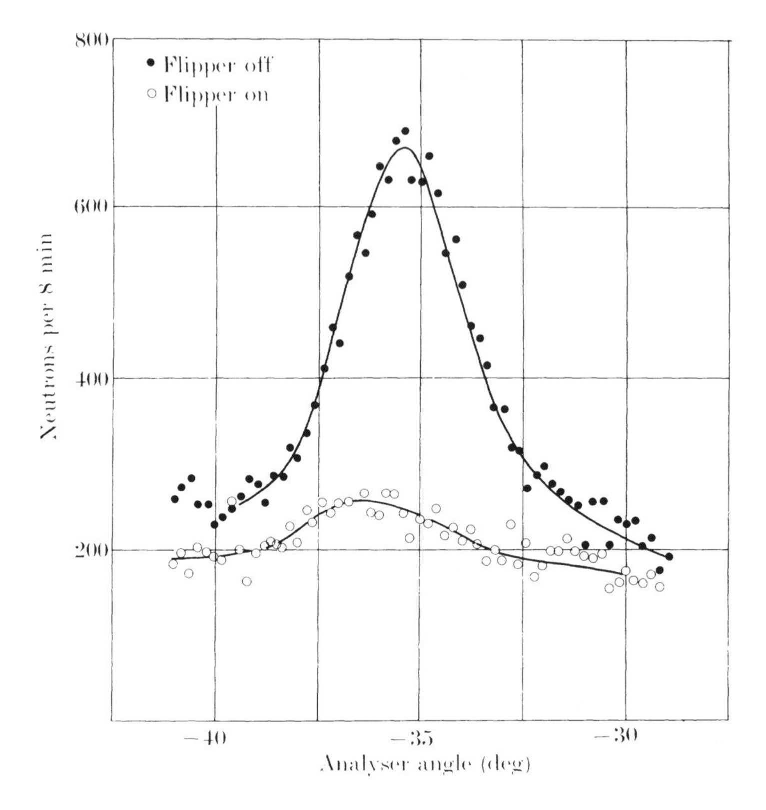

FIG. 10.2. Isotopic incoherent scattering from polycrystalline Ni.

flipper was activated. All nickel isotopes have zero spin so our discussion given above tells us that the incoherent nuclear scattering has the same magnitude as for the scattering of unpolarized neutrons and the polarization of the scattered beam is parallel to that of the incident beam. These facts are clearly borne out by the experiment. The small peak in the flipper on measurements is magnon scattering.

Because there is only one vanadium isotope we expect $\frac{1}{3}$ of the incoherent scattering to be without spin-flip and $\frac{2}{3}$ with spin-flip. The measurements shown in Fig. 10.3 verify that this is so, independent of whether the incident beam is polarized parallel or perpendicular to $\boldsymbol{\kappa}$.

In the previous section we considered the electrostatic scattering of polarized neutrons, so let us now ask for the cross-section when we have both electrostatic and nuclear elastic scattering. For this case,

$$\langle\lambda|\,\hat{V}(\boldsymbol{\kappa})\,|\lambda\rangle=\langle\lambda|\,\hat{V}_{\mathrm{E}}(\boldsymbol{\kappa})+\hat{V}_{\mathrm{N}}(\boldsymbol{\kappa})\,|\lambda\rangle$$

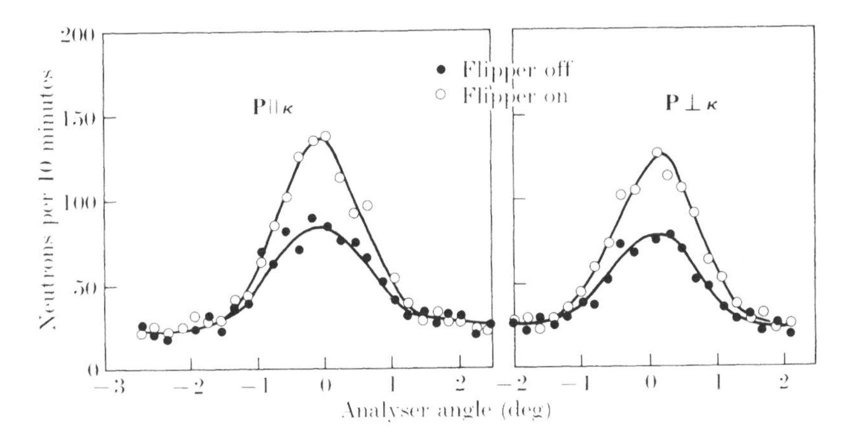

FIG. 10.3. Nuclear spin incoherent scattering from vanadium.

and so, from (10.55) and (10.58),

$$\langle\lambda|\,\hat{\boldsymbol{\alpha}}\,|\lambda\rangle \equiv \sum_{\mathbf{l}} \exp(\mathrm{i}\boldsymbol{\kappa}\cdot\mathbf{l})\left\{\sum_{\mathbf{d}} \exp(\mathrm{i}\boldsymbol{\kappa}\cdot\mathbf{d})\tfrac{1}{2}B_{ld}\hat{\mathbf{i}}_{ld} + \tfrac{1}{2}r_0\left(\frac{m_e}{m}\right)\mathrm{i}\cot(\tfrac{1}{2}\theta)F_{\mathrm{E}}(\boldsymbol{\kappa})\tilde{\mathbf{n}}\right\} \tag{10.69}$$

and

$$\langle\lambda|\,\hat{\beta}\,|\lambda\rangle \equiv \sum_{\mathbf{l}} \exp(\mathrm{i}\boldsymbol{\kappa}\cdot\mathbf{l})\left\{\sum_{\mathbf{d}} \exp(\mathrm{i}\boldsymbol{\kappa}\cdot\mathbf{d})A_{ld} - \tfrac{1}{2}r_0\left(\frac{m_e}{m}\right)F_{\mathrm{E}}(\boldsymbol{\kappa})\right\}.$$

When we use the result (10.69) in (10.37) we shall, of course, recover the nuclear-coherent, nuclear-incoherent, and Coulomb cross-sections separately. In addition we now look for interference effects. No interference effects can come from the term $\hat{\boldsymbol{\alpha}}^{+}\cdot\hat{\boldsymbol{\alpha}}$ because any such term would be linear in the nuclear spins $\hat{\mathbf{i}}_{ld}$ and therefore vanish with the averaging operating $\mathcal{O}$. For a similar reason no contribution comes from the term $\mathrm{i}\mathbf{P}\cdot\hat{\boldsymbol{\alpha}}^{+}\times\hat{\boldsymbol{\alpha}}$. But the other terms in (10.37) do give an interference effect which, remembering to apply the averaging operation $\mathcal{O}$, adds to the previous results to give in total

$$\frac{\mathrm{d}\sigma}{\mathrm{d}\Omega} = \left(\frac{\mathrm{d}\sigma}{\mathrm{d}\Omega}\right)_{\mathrm{NN}} + \left(\frac{\mathrm{d}\sigma}{\mathrm{d}\Omega}\right)_{\mathrm{EE}} + \left|\sum_{\mathbf{l}} \exp(\mathrm{i}\boldsymbol{\kappa}\cdot\mathbf{l})\right|^2\left(\frac{m_e}{m}\right)r_0\{\cot(\tfrac{1}{2}\theta)\mathbf{P}\cdot\tilde{\mathbf{n}}\,\mathrm{Im}\,F_{\mathrm{N}}(\boldsymbol{\kappa})F_{\mathrm{E}}^{*}(\boldsymbol{\kappa}) - \mathrm{Re}\,F_{\mathrm{N}}(\boldsymbol{\kappa})F_{\mathrm{E}}^{*}(\boldsymbol{\kappa})\} \tag{10.70}$$

where the unit-cell structure factor

$$F_{\mathrm{N}}(\boldsymbol{\kappa}) = \sum_{\mathbf{d}} \exp(\mathrm{i}\boldsymbol{\kappa}\cdot\mathbf{d})\bar{b}_d. \tag{10.71}$$

Thus we have, in addition to the pure nuclear and pure electrostatic cross-sections, two interference terms that contribute at Bragg reflections, one of which depends on the polarizations. The polarization-independent term arises from the Foldy part of the electrostatic interaction potential while the other term arises from the spin–orbit part of the electrostatic interaction. As a consequence of this latter term, containing the factor $\cot \frac{1}{2}\theta$, the Bragg cross-section is asymmetric, i.e. the cross-section differs for scattering to the right and to the left of the incident beam. The presence of the asymmetric term in the scattering from vanadium crystals has been demonstrated by Shull (1963). He measured the flipping ratios for the (110) vanadium reflection for θ and $-\theta$ and found that, for the former setting, the ratio was greater than unity and, for the latter, less than unity, the magnitude of the deviation from unity being the same in the two cases. Shull's results are given in Fig. 10.4.

Because it depends upon the real part of the scattering length, the polarization-independent interference term in (10.70) may well be an order of magnitude greater than the asymmetric term; nevertheless its presence need not be significant in an experiment undertaken to observe the asymmetric contribution. If the coherent nuclear scattering gives the

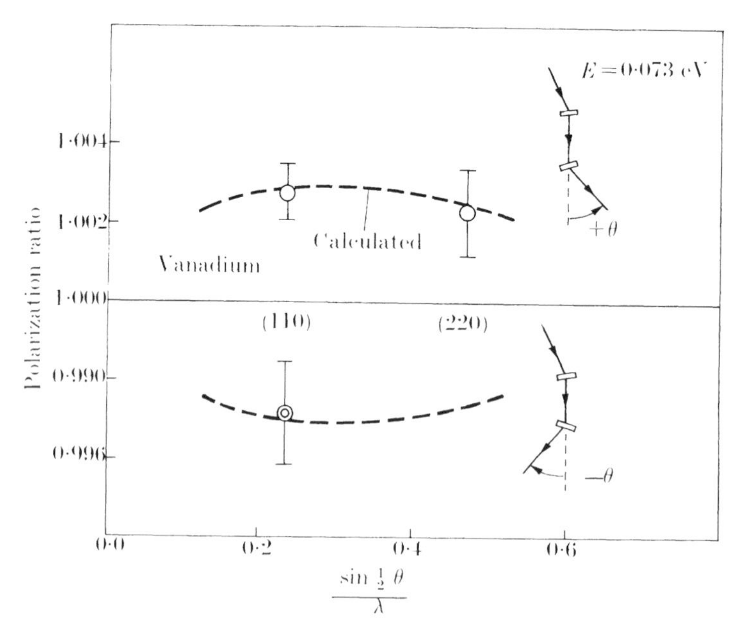

Fig. 10.4. Slow neutron spin–orbit asymmetry in the scattering by vanadium with zero magnetic field (Shull 1963).

major part of the measured intensity, the flipping ratio can be written

$$R = \frac{1-\delta+\delta'}{1-\delta-\delta'}$$

where

$$\delta = (m_e r_0/m)\mathrm{Re}\{F_N(\boldsymbol{\tau})F_E^*(\boldsymbol{\tau})\}/|F_N(\boldsymbol{\tau})|^2$$

and

$$\delta' = (m_e r_0/m)\cot(\tfrac{1}{2}\theta)\mathrm{Im}\{F_N(\boldsymbol{\tau})F_E^*(\boldsymbol{\tau})\}/|F_N(\boldsymbol{\tau})|^2.$$

Assuming that both δ and δ' are much less than unity,

$$R \simeq 1 + 2\delta' - \delta^2.$$

Thus if $2\delta'$ is about the limit of what can be measured with any certainty the contribution δ^2 can certainly be neglected.

In most cases the contributions to the cross-section coming from the nuclear–electrostatic interference terms are completely negligible. However, Shull pointed out that the presence of the asymmetric term means that all polarized-beam spectrometers are inherently left- and right-handed.

If the incident neutrons are unpolarized, it is very easy to calculate the polarization of the scattered neutrons, for we have only to consider the terms

$$\hat{\beta}^+\hat{\boldsymbol{\alpha}} + \hat{\boldsymbol{\alpha}}^+\hat{\beta} - \mathrm{i}\hat{\boldsymbol{\alpha}}^+ \times \hat{\boldsymbol{\alpha}} \tag{10.72}$$

in (10.41), and all these occur in the cross-section (10.70). Thus we have immediately (for a Bragg reflection)

$$\mathbf{P}'_{\mathrm{NE}} = \frac{(m_e/m)r_0\cot(\tfrac{1}{2}\theta)\tilde{\mathbf{n}}\,\mathrm{Im}\,F_N(\boldsymbol{\tau})F_E^*(\boldsymbol{\tau})}{\{|F_N(\boldsymbol{\tau})|^2 + (m_e r_0/2m)^2(\sin\tfrac{1}{2}\theta)^{-2}|F_E(\boldsymbol{\tau})|^2 - (m_e r_0/m)\mathrm{Re}\,F_N(\boldsymbol{\tau})F_E^*(\boldsymbol{\tau})\}}. \tag{10.73}$$

Thus the polarization arising from the interference of the nuclear electrostatic scattering $\mathbf{P}'_{\mathrm{NE}}$ is perpendicular to the scattering plane. The magnitude of $\mathbf{P}'_{\mathrm{NE}}$ is probably largest at the first Bragg peak owing to the factor $\cot(\tfrac{1}{2}\theta)$.

10.5. Magnetic elastic scattering

10.5.1. Perfect paramagnet

Because such a system does not possess a preferred axis, the cross-section for scattering of polarized neutrons is independent of $\mathbf{P}$ and is given by (7.33). The polarization of the scattered beam $\mathbf{P}'$ is of interest and readily calculated.

For purely magnetic scattering the interaction potential $\hat{V}_{\mathrm{M}}(\boldsymbol{\kappa})$ is given by eqn (10.23). From this equation we see that $\hat{\beta} \equiv 0$. Hence, from the general expression (10.41) for $\mathbf{P}'$, we have only to calculate the terms corresponding to

$$\hat{\boldsymbol{\alpha}}^{+}(\hat{\boldsymbol{\alpha}} \cdot \mathbf{P}) + (\hat{\boldsymbol{\alpha}}^{+} \cdot \mathbf{P})\hat{\boldsymbol{\alpha}} - \mathbf{P}(\hat{\boldsymbol{\alpha}}^{+} \cdot \hat{\boldsymbol{\alpha}}) - \mathrm{i}\hat{\boldsymbol{\alpha}}^{+} \times \hat{\boldsymbol{\alpha}}. \tag{10.74}$$

In the dipole approximation (cf. § 7.3)

$$\hat{\boldsymbol{\alpha}} \equiv r_0 \sum_{\mathbf{l},\mathbf{d}} \exp(\mathrm{i}\boldsymbol{\kappa} \cdot \mathbf{R}_{ld}) \tfrac{1}{2} g_d F_d(\boldsymbol{\kappa}) \{\tilde{\boldsymbol{\kappa}} \times (\hat{\mathbf{S}}_{ld} \times \tilde{\boldsymbol{\kappa}})\}. \tag{10.75}$$

To calculate the contribution to the cross-section coming from the first term in (10.74) we need to evaluate

$$\sum_{\mathbf{l},\mathbf{d}} \sum_{\mathbf{l}',\mathbf{d}'} \exp\{\mathrm{i}\boldsymbol{\kappa} \cdot (\mathbf{R}_{ld} - \mathbf{R}_{l'd'})\} g_d g_{d'} F_d(\boldsymbol{\kappa}) F^{*}_{d'}(\boldsymbol{\kappa})$$
$$\times \sum_{\lambda} p_{\lambda} \langle \lambda | \{\tilde{\boldsymbol{\kappa}} \times (\hat{\mathbf{S}}_{l'd'} \times \tilde{\boldsymbol{\kappa}})\}\{\mathbf{P} \cdot \tilde{\boldsymbol{\kappa}} \times (\hat{\mathbf{S}}_{ld} \times \tilde{\boldsymbol{\kappa}})\} | \lambda \rangle. \tag{10.76}$$

Now

$$\sum_{\lambda} p_{\lambda} \langle \lambda | \{\tilde{\boldsymbol{\kappa}} \times (\hat{\mathbf{S}}_{l'd'} \times \tilde{\boldsymbol{\kappa}})\}\{\mathbf{P} \cdot \tilde{\boldsymbol{\kappa}} \times (\hat{\mathbf{S}}_{ld} \times \tilde{\boldsymbol{\kappa}})\} | \lambda \rangle$$
$$= \sum_{\alpha,\beta} (\tilde{\boldsymbol{\alpha}} - \tilde{\boldsymbol{\kappa}}\tilde{\kappa}_{\alpha})\{P^{\beta} - (\tilde{\boldsymbol{\kappa}} \cdot \mathbf{P})\tilde{\kappa}_{\beta}\}\langle \hat{S}^{\alpha}_{l'd'} \hat{S}^{\beta}_{ld} \rangle$$
$$= \sum_{\alpha,\beta} (\tilde{\boldsymbol{\alpha}} - \tilde{\boldsymbol{\kappa}}\tilde{\kappa}_{\alpha})\{P^{\beta} - (\tilde{\boldsymbol{\kappa}} \cdot \mathbf{P})\tilde{\kappa}_{\beta}\}\, \delta_{\mathbf{l},\mathbf{l}'}\, \delta_{\mathbf{d},\mathbf{d}'}\, \delta_{\alpha,\beta} \tfrac{1}{3} S_d(S_d + 1) \tag{10.77}$$

where to obtain the last line we have used (7.32). We therefore obtain for (10.76)

$$\tfrac{1}{3} N \sum_{\mathbf{d}} |g_d F_d(\boldsymbol{\kappa})|^2\, S_d(S_d + 1)\mathbf{P}_{\perp} \tag{10.78}$$

where

$$\mathbf{P}_{\perp} = \tilde{\boldsymbol{\kappa}} \times (\mathbf{P} \times \tilde{\boldsymbol{\kappa}}) \tag{10.79}$$

is the component of $\mathbf{P}$ perpendicular to $\boldsymbol{\kappa}$.

The second term of (10.74) gives an identical result and the third term is easily evaluated as (10.78) with $-2\mathbf{P}$ replacing $\mathbf{P}_{\perp}$. Hence the first three terms of (10.74) give

$$r_0^2 \tfrac{2}{3} N \sum_{\mathbf{d}} |\tfrac{1}{2} g_d F_d(\boldsymbol{\kappa})|^2\, S_d(S_d + 1)\{\mathbf{P}_{\perp} - \mathbf{P}\}$$
$$= r_0^2 \tfrac{2}{3} N \sum_{\mathbf{d}} |\tfrac{1}{2} g_d F_d(\boldsymbol{\kappa})|^2\, S_d(S_d + 1)\{-\tilde{\boldsymbol{\kappa}}(\tilde{\boldsymbol{\kappa}} \cdot \mathbf{P})\}. \tag{10.80}$$

We have now to evaluate the $\hat{\boldsymbol{\alpha}}^{+} \times \hat{\boldsymbol{\alpha}}$ term in (10.74). This requires

$$\sum_{\lambda} p_{\lambda} \langle \lambda | \{\tilde{\boldsymbol{\kappa}} \times (\hat{\mathbf{S}}_{l'd'} \times \tilde{\boldsymbol{\kappa}})\} \times \{\tilde{\boldsymbol{\kappa}} \times (\hat{\mathbf{S}}_{ld} \times \tilde{\boldsymbol{\kappa}})\} | \lambda \rangle$$
$$= \langle \hat{\mathbf{S}}_{l'd'} \times \hat{\mathbf{S}}_{ld} \rangle + \langle (\tilde{\boldsymbol{\kappa}} \times \hat{\mathbf{S}}_{l'd'})(\tilde{\boldsymbol{\kappa}} \cdot \hat{\mathbf{S}}_{ld}) - (\tilde{\boldsymbol{\kappa}} \cdot \hat{\mathbf{S}}_{l'd'})(\tilde{\boldsymbol{\kappa}} \times \hat{\mathbf{S}}_{ld}) \rangle. \tag{10.81}$$

The first term in (10.81) is zero by an argument analogous to that used in deriving (10.60c). Further, each of the remaining two terms is zero. For

$$\langle(\tilde{\boldsymbol{\kappa}}\times\hat{\mathbf{S}}_{l'd'})_\alpha(\tilde{\boldsymbol{\kappa}}\cdot\hat{\mathbf{S}}_{ld})\rangle = \sum_{\beta,\gamma,\mu}\epsilon^{\alpha\beta\gamma}\tilde{\kappa}_\beta\tilde{\kappa}_\mu\langle\hat{S}^\gamma_{l'd'}\hat{S}^\mu_{ld}\rangle$$

$$= \sum_{\beta,\gamma,\mu}\epsilon^{\alpha\beta\gamma}\tilde{\kappa}_\beta\tilde{\kappa}_\mu\,\delta_{\gamma,\mu}\,\delta_{\mathbf{l},\mathbf{l}'}\,\delta_{\mathbf{d},\mathbf{d}'}\tfrac{1}{3}S_d(S_d+1) \tag{10.82}$$

and

$$\sum_{\beta,\gamma,\mu}\epsilon^{\alpha\beta\gamma}\tilde{\kappa}_\beta\tilde{\kappa}_\mu\,\delta_{\gamma,\mu} = \sum_{\beta,\gamma}\epsilon^{\alpha\beta\gamma}\tilde{\kappa}_\beta\tilde{\kappa}_\gamma = 0. \tag{10.83}$$

Thus (10.80) represents the only non-zero part of (10.74). The final result for $\mathbf{P}'$ is, therefore,

$$\mathbf{P}' = -\tilde{\boldsymbol{\kappa}}(\tilde{\boldsymbol{\kappa}}\cdot\mathbf{P}). \tag{10.84}$$

The polarization of the scattered beam is seen to be zero when the incident beam is unpolarized, and when the incident beam is polarized the scattered beam is polarized in the direction of the scattering vector.

Result (10.84) for the polarization of neutrons scattered by a paramagnet is illustrated in Fig. 10.5. This shows the paramagnetic scattering from MnF_2 with $\mathbf{P}$ parallel to the scattering vector and with $\mathbf{P}$ perpendicular to the scattering vector. In the latter case no difference is observed in the scattered neutrons with flipper on or off, as expected from the above result. For $\mathbf{P}$ parallel to the scattering vector, the scattered neutrons are only reflected by the analyser when one of the flippers is activated, in accord with the negative sign in the result (10.84). The small

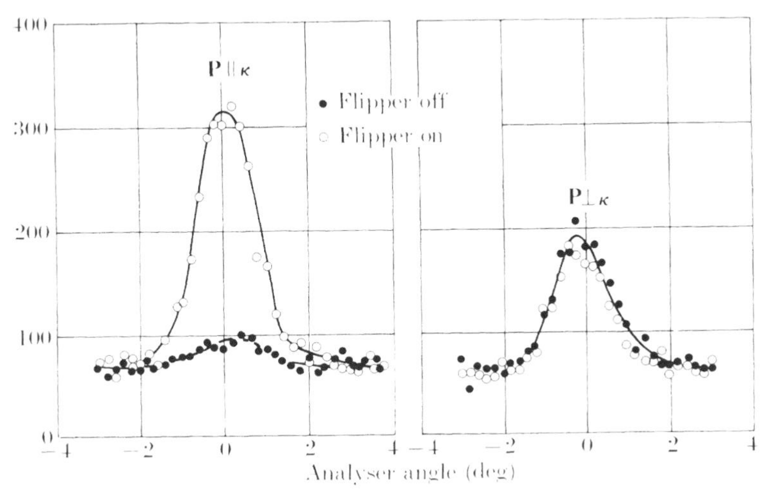

FIG. 10.5. Paramagnetic scattering from MnF_2.

peak observed in the **P** parallel to $\boldsymbol{\kappa}$ case with the flipper off, is due to multiple Bragg scattering.

Paramagnetic scattering studies afford an area of research where polarization analysis can be of great value. For in the usual diffuse scattering experiment where the total cross-section is measured, several sources of scattering additional to the paramagnetic scattering can be present, e.g. isotopic disorder, multiple Bragg, Bragg, and nuclear spin-coherent. If the spin-flip scattering is measured only the nuclear spin and paramagnetic scattering processes contribute, thus greatly simplifying the task of extracting the purely paramagnetic scattering (Steinsvoll *et al.* 1983).

10.5.2. Ordered structures (Tofield 1975; Brown 1979; Dachs 1978)

One of the most rewarding applications of polarized neutrons has been in the field of magnetic spin density and form factor measurements in ferromagnetic metals, a subject discussed in Chapter 12. These experiments utilize the polarization-dependent interference term between nuclear and magnetic scattering to obtain accurate measurements of the magnetic scattering amplitude (Menzinger and Sacchetti 1979). The actual quantity measured is the ratio of the scattered intensity (at a Bragg reflection) for one setting of the polarization to the intensity with polarization direction reversed. Since the interference term is linear in the magnetic scattering amplitude, this technique enables the sign of the magnetic amplitude to be determined as well. We do nto discuss elastic scattering of polarized neutrons specifically for ferromagnets but choose to discuss a far more general case. First we calculate the cross-section for the scattering of polarized neutrons through their nuclear and magnetic interactions with a solid, and then go on to evaluate the polarization of the scattered beam. Because of the importance of the polarized-beam technique in gaining detailed information on the magnetic properties of crystals we present the calculations in detail. The general structure of the expressions for the coherent elastic cross-section and final polarization are reviewed in detail in § 12.2.

For coherent elastic scattering by ions with only spin angular momentum,

$$\langle\lambda|\,\hat{\boldsymbol{\alpha}}\,|\lambda\rangle \Rightarrow \langle\hat{\boldsymbol{\alpha}}\rangle = r_0 \sum_{\mathbf{l},\mathbf{d}} \exp(\mathrm{i}\boldsymbol{\kappa}\cdot\mathbf{R}_{ld})\tfrac{1}{2}g_d F_d(\boldsymbol{\kappa})\langle\hat{S}_{ld}\rangle[\tilde{\boldsymbol{\kappa}}\times\{\tilde{\boldsymbol{\eta}}_{ld}(\boldsymbol{\kappa})\times\tilde{\boldsymbol{\kappa}}\}] \tag{10.85}$$

where $\tilde{\boldsymbol{\eta}}_{ld}(\boldsymbol{\kappa})$ is a unit vector in the direction of the spin at site $\mathbf{R}_{ld}$ and for non-collinear spin systems it is a function of the scattering vector. For brevity of notation we do not show this dependence in the course of the following calculations.

We must add to (10.85) the expression for $\hat{\boldsymbol{\alpha}}$ for purely nuclear scattering. This is given in (10.59) together with the required expression for $\hat{\beta}$. Because we assume the nuclear spins to be randomly orientated any term linear in $\hat{\mathbf{i}}_{ld}$ can be dropped. Similarly terms containing $\hat{\mathbf{i}}_{ld} \times \hat{\mathbf{i}}_{l'd'}$ can be dropped. Consider the term corresponding to $\hat{\boldsymbol{\alpha}}^{+} \cdot \hat{\boldsymbol{\alpha}}$ in expression (10.37).

We require

$$\begin{aligned}
&\sum_{\mathbf{l,d}} \sum_{\mathbf{l',d'}} \exp\{\mathrm{i}\boldsymbol{\kappa} \cdot (\mathbf{R}_{ld} - \mathbf{R}_{l'd'})\} \\
&\quad \times \mathcal{O}\Big[\Big\{\tfrac{1}{2}B^{*}_{l'd'}\hat{\mathbf{i}}_{l'd'} + r_0\,\tfrac{1}{2}g_{d'}F^{*}_{d'}\langle \hat{S}_{l'd'}\rangle[\tilde{\boldsymbol{\kappa}} \times (\tilde{\boldsymbol{\eta}}_{l'd'} \times \tilde{\boldsymbol{\kappa}})]\Big\} \cdot \\
&\quad \cdot \Big\{\tfrac{1}{2}B_{ld}\hat{\mathbf{i}}_{ld} + r_0\,\tfrac{1}{2}g_d F_d\langle \hat{S}_{ld}\rangle[\tilde{\boldsymbol{\kappa}} \times (\tilde{\boldsymbol{\eta}}_{ld} \times \tilde{\boldsymbol{\kappa}})]\Big\}\Big] \\
&= \sum_{\mathbf{l,d}} \sum_{\mathbf{l',d'}} \exp\{\mathrm{i}\boldsymbol{\kappa} \cdot (\mathbf{R}_{ld} - \mathbf{R}_{l'd'})\}\Big[\tfrac{1}{4}\delta_{\mathbf{l,l'}}\,\delta_{\mathbf{d,d'}}\{|B_{ld}|^2\, i_d(i_d+1)\} \qquad (10.86) \\
&\quad + r_0^2\,\tfrac{1}{2}g_d F_d\langle \hat{S}_{ld}\rangle \tfrac{1}{2}g_{d'}F^{*}_{d'}\langle \hat{S}_{l'd'}\rangle[\tilde{\boldsymbol{\kappa}} \times (\tilde{\boldsymbol{\eta}}_{ld} \times \tilde{\boldsymbol{\kappa}})] \cdot [\tilde{\boldsymbol{\kappa}} \times (\tilde{\boldsymbol{\eta}}_{l'd'} \times \tilde{\boldsymbol{\kappa}})]\Big].
\end{aligned}$$

It remains to average the first term on the right-hand side over the (random) distribution of nuclear isotopes. This same result was given in eqn (10.61).

The calculation of the $\hat{\beta}^{+}\hat{\beta}$ term in (10.37) is straightforward. We have only to average

$$\sum_{\mathbf{l,d}} \sum_{\mathbf{l',d'}} \exp\{\mathrm{i}\boldsymbol{\kappa} \cdot (\mathbf{R}_{ld} - \mathbf{R}_{l'd'})\} A_{ld} A^{*}_{l'd'} \qquad (10.87)$$

over the distribution of isotopes and find using (10.71),

$$\Big|\sum_{\mathbf{l}} \exp(\mathrm{i}\boldsymbol{\kappa} \cdot \mathbf{l})\Big|^2 |F_{\mathrm{N}}(\boldsymbol{\kappa})|^2 + N\sum_{\mathbf{d}}\{\overline{|A_d|^2} - |\bar{A}_d|^2\}. \qquad (10.88)$$

Combining this with the first part of (10.86), and using (10.65) and (10.66), the total purely nuclear contribution to the cross-section is,

$$N\sum_{\mathbf{d}}\{\overline{|b_d|^2} - |\bar{b}_d|^2\} + \Big|\sum_{\mathbf{l}} \exp(\mathrm{i}\boldsymbol{\kappa} \cdot \mathbf{l})\Big|^2 |F_{\mathrm{N}}(\boldsymbol{\kappa})|^2 = \left(\frac{\mathrm{d}\sigma}{\mathrm{d}\Omega}\right)_{\mathrm{NN}}. \qquad (10.89)$$

This result is, of course, identical to that derived in Chapter 2; the first term is the incoherent nuclear cross-section and the second the coherent nuclear cross-section.

The evaluation of the terms $\hat{\beta}^{+}(\hat{\boldsymbol{\alpha}} \cdot \mathbf{P})$ and $(\hat{\boldsymbol{\alpha}}^{+} \cdot \mathbf{P})\hat{\beta}$ in (10.37) proceeds as follows. Because we need take only the magnetic part of $\hat{\boldsymbol{\alpha}}$, we have for the first of the two terms,

$$r_0 \sum_{\mathbf{l},\mathbf{d}} \sum_{\mathbf{l}',\mathbf{d}'} \exp\{-i\boldsymbol{\kappa}\cdot(\mathbf{R}_{ld}-\mathbf{R}_{l'd'})\}\bar{b}_d^* \tfrac{1}{2}g_{d'}F_{d'}(\boldsymbol{\kappa})\langle\hat{S}_{l'd'}\rangle\mathbf{P}\cdot[\tilde{\boldsymbol{\kappa}}\times(\tilde{\boldsymbol{\eta}}_{l'd'}\times\tilde{\boldsymbol{\kappa}})] \tag{10.90}$$

where we have averaged over the isotope distribution. The term $(\hat{\boldsymbol{\alpha}}^+\cdot\mathbf{P})\hat{\beta}$ is similar, the difference being that we take the complex conjugate of the magnetic part instead of the nuclear. Note that

$$\mathbf{P}\cdot\tilde{\boldsymbol{\kappa}}\times(\tilde{\boldsymbol{\eta}}_{ld}\times\tilde{\boldsymbol{\kappa}})=\tilde{\boldsymbol{\eta}}_{ld}\cdot\tilde{\boldsymbol{\kappa}}\times(\mathbf{P}\times\tilde{\boldsymbol{\kappa}})\equiv\tilde{\boldsymbol{\eta}}_{ld}\cdot\mathbf{P}_\perp, \tag{10.91}$$

which shows that the interference term is zero for any spins that are orientated parallel to the scattering vector.

We have now only to calculate the $\hat{\boldsymbol{\alpha}}^+\times\hat{\boldsymbol{\alpha}}$ term in (10.37). For randomly orientated nuclear spins this is seen to be purely magnetic

$$r_0^2 \sum_{\mathbf{l},\mathbf{d}} \sum_{\mathbf{l}',\mathbf{d}'} \exp\{i\boldsymbol{\kappa}\cdot(\mathbf{R}_{l'd'}-\mathbf{R}_{ld})\}\tfrac{1}{2}g_{d'}F_{d'}(\boldsymbol{\kappa})\langle\hat{S}_{l'd'}\rangle$$
$$\times\tfrac{1}{2}g_dF_d^*(\boldsymbol{\kappa})\langle\hat{S}_{ld}\rangle[\tilde{\boldsymbol{\kappa}}\times(\tilde{\boldsymbol{\eta}}_{l'd'}\times\tilde{\boldsymbol{\kappa}})]\times[\tilde{\boldsymbol{\kappa}}\times(\tilde{\boldsymbol{\eta}}_{ld}\times\tilde{\boldsymbol{\kappa}})]. \tag{10.92}$$

We find for the cross-product in (10.92)

$$[\tilde{\boldsymbol{\kappa}}\times(\tilde{\boldsymbol{\eta}}_{l'd'}\times\tilde{\boldsymbol{\kappa}})]\times[\tilde{\boldsymbol{\kappa}}\times(\tilde{\boldsymbol{\eta}}_{ld}\times\tilde{\boldsymbol{\kappa}})]$$
$$=\tilde{\boldsymbol{\eta}}_{l'd'}\times\tilde{\boldsymbol{\eta}}_{ld}+(\tilde{\boldsymbol{\kappa}}\times\tilde{\boldsymbol{\eta}}_{l'd'})(\tilde{\boldsymbol{\kappa}}\cdot\tilde{\boldsymbol{\eta}}_{ld})-(\tilde{\boldsymbol{\kappa}}\times\tilde{\boldsymbol{\eta}}_{ld})(\tilde{\boldsymbol{\kappa}}\cdot\tilde{\boldsymbol{\eta}}_{l'd'}). \tag{10.93}$$

For certain magnetic structures this term is zero. Two important cases are (a) a ferromagnet in which all the spins are aligned parallel to a single direction and (b) an antiferromagnet in which the spins are either all parallel or antiparallel to a single direction.

We have now all the terms in the cross-section for the elastic scattering of polarized neutrons from a magnetic crystal. Combining (10.86), (10.89), (10.90), and (10.92) the result is

$$\frac{d\sigma}{d\Omega}=\left(\frac{d\sigma}{d\Omega}\right)_{\mathrm{NN}}+\left(\frac{d\sigma}{d\Omega}\right)_{\mathrm{MM}}+r_0\sum_{\mathbf{l},\mathbf{d}}\sum_{\mathbf{l}',\mathbf{d}'}\exp\{i\boldsymbol{\kappa}\cdot(\mathbf{R}_{ld}-\mathbf{R}_{l'd'})\}$$
$$+\Big[\{\bar{b}_{d'}\tfrac{1}{2}g_dF_d(\boldsymbol{\kappa})\langle\hat{S}_{ld}\rangle\tilde{\boldsymbol{\eta}}_{ld}\cdot\mathbf{P}_\perp+\bar{b}_d\tfrac{1}{2}g_{d'}F_{d'}(\boldsymbol{\kappa})\langle\hat{S}_{l'd'}\rangle\tilde{\boldsymbol{\eta}}_{l'd'}\cdot\mathbf{P}_\perp\}$$
$$+r_0\tfrac{1}{2}g_{d'}F_{d'}^*(\boldsymbol{\kappa})\langle\hat{S}_{l'd'}\rangle\tfrac{1}{2}g_dF_d(\boldsymbol{\kappa})\langle\hat{S}_{ld}\rangle i\mathbf{P}\cdot[\tilde{\boldsymbol{\kappa}}\times(\tilde{\boldsymbol{\eta}}_{l'd'}\times\tilde{\boldsymbol{\kappa}})]\times[\tilde{\boldsymbol{\kappa}}\times(\tilde{\boldsymbol{\eta}}_{ld}\times\tilde{\boldsymbol{\kappa}})]\Big]. \tag{10.94}$$

In this expression $(d\sigma/d\Omega)_{\mathrm{MM}}$ is the purely magnetic cross-section given by the second part of the right-hand side of (10.86).

We see from the expression (10.94) that the use of polarized neutrons introduces into the cross-section two additional contributions. First there is a contribution arising from interference between nuclear and magnetic scattering. This term will be zero for a given Bragg reflection for two sets of conditions: (a) if the spins concerned in the magnetic scattering are aligned parallel to the scattering vector, and (b) if both nuclear and magnetic scattering do not occur at the Bragg reflection in question. The latter case is, of course, encountered in scattering from antiferromagnets. For instance, in a rutile antiferromagnet like MnF_2 the $(0, t_2, t_3)$ reflections with t_2+t_3 odd are purely magnetic; reflections with $t_1+t_2+t_3$ even are, in the absence of covalency, purely nuclear; while all remaining reflections with $t_1+t_2+t_3$ odd are mixed nuclear and magnetic with the nuclear scattering coming solely from the nuclei of the anions. Thus in this type of antiferromagnet only the reflections with $t_1+t_2+t_3$ odd can be polarization-dependent (unless covalency effects are present).

Besides the nuclear and magnetic interference terms the cross-section (10.94) contains an additional, purely magnetic term. As we have seen in (10.93), this term vanishes for a simple magnetic system in which all the spins are either parallel or antiparallel to a single direction. However, it can contribute in scattering from complicated spin configurations. An example of this is given in § 10.5.3 when we calculate the cross-section for scattering of polarized neutrons by the antiferromagnetic MnO_2, which possesses a helical spin configuration.

A particular case of (10.94) that is of interest is that where all the spins in the crystal are either parallel or antiparallel to a single direction $\tilde{\boldsymbol{\eta}}$. Taking the Bragg part and writing the purely nuclear and magnetic contributions explicitly gives

$$\frac{\mathrm{d}\sigma}{\mathrm{d}\Omega}=\left|\sum_{l}\exp(\mathrm{i}\boldsymbol{\kappa}\cdot\mathbf{l})\right|^{2}$$
$$\times\left[|F_{\mathrm{N}}(\boldsymbol{\kappa})|^{2}+r_{0}^{2}\{1-(\tilde{\boldsymbol{\kappa}}\cdot\tilde{\boldsymbol{\eta}})^{2}\}\,|F_{\mathrm{M}}(\tilde{\boldsymbol{\kappa}})|^{2}+2r_{0}\tilde{\boldsymbol{\eta}}\cdot\mathbf{P}_{\perp}\ \mathrm{Re}\ F_{\mathrm{N}}(\boldsymbol{\kappa})F_{\mathrm{M}}^{*}(\boldsymbol{\kappa})\right] \quad (10.95)$$

where, as usual, cf. eqn (12.31),

$$F_{\mathrm{M}}(\boldsymbol{\kappa})=\sum_{\mathbf{d}}\exp(\mathrm{i}\boldsymbol{\kappa}\cdot\mathbf{d})\tfrac{1}{2}g_{d}F_{d}(\boldsymbol{\kappa})\langle\hat{S}_{d}\rangle\sigma_{d}$$

and

$$\left|\sum_{l}\exp(\mathrm{i}\boldsymbol{\kappa}\cdot\mathbf{l})\right|^{2}=N\frac{(2\pi)^{3}}{v_{0}}\sum_{\boldsymbol{\tau}}\delta(\boldsymbol{\kappa}-\boldsymbol{\tau}).$$

Consider now a simple ferromagnet with just one atom per unit cell, i.e. Bravais lattice structure. Because each ion in such a system is at a centre of symmetry the atomic magnetic form factor $F(\boldsymbol{\kappa})$ for the ions is purely real. The cross-section for Bragg scattering is therefore

$$\frac{\mathrm{d}\sigma}{\mathrm{d}\Omega}=\left|\sum_{\mathbf{l}}\exp(\mathrm{i}\boldsymbol{\kappa}\cdot\mathbf{l})\right|^{2}\left[\bar{b}^{2}+2r_{0}\tfrac{1}{2}gF(\boldsymbol{\kappa})\langle\hat{S}\rangle\bar{b}\ \tilde{\boldsymbol{\eta}}\cdot\mathbf{P}\right.$$
$$\left.+r_{0}^{2}\{1-(\tilde{\boldsymbol{\kappa}}\cdot\tilde{\boldsymbol{\eta}})^{2}\}\{\tfrac{1}{2}gF(\boldsymbol{\kappa})\langle\hat{S}\rangle\}^{2}\right]. \quad (10.96)$$

If the magnetization of the ferromagnet is aligned by an external field perpendicular to the scattering plane,

$$\tilde{\boldsymbol{\eta}} \cdot \mathbf{P}_{\perp} = \tilde{\boldsymbol{\eta}} \cdot \mathbf{P} \quad \text{and} \quad \tilde{\boldsymbol{\kappa}} \cdot \tilde{\boldsymbol{\eta}} = 0.$$

If now the polarization $\mathbf{P}$ is made to coincide with $\tilde{\boldsymbol{\eta}}$,

$$\frac{\mathrm{d}\sigma}{\mathrm{d}\Omega} = \left|\sum_{\mathbf{l}} \exp(\mathrm{i}\boldsymbol{\kappa} \cdot \mathbf{l})\right|^2 \left\{\bar{b} \pm r_0 \tfrac{1}{2} g F(\boldsymbol{\kappa}) \langle \hat{S} \rangle\right\}^2, \tag{10.97}$$

where the $\pm$ signs are taken according to

$$\tilde{\boldsymbol{\eta}} \cdot \mathbf{P} = \pm 1. \tag{10.98}$$

The ratio of the Bragg intensities, or the flipping ratio, is therefore given by

$$R = \left\{\frac{1 + r_0 \frac{1}{2} g F(\boldsymbol{\tau}) \langle \hat{S} \rangle / \bar{b}}{1 - r_0 \frac{1}{2} g F(\boldsymbol{\tau}) \langle \hat{S} \rangle / \bar{b}}\right\}^2. \tag{10.99}$$

Let us consider the value of R for scattering from nickel. The first Bragg reflection is (111) and the corresponding value of $F(\boldsymbol{\tau})$ is 0.79. Thus with

$$\tfrac{1}{2} g \langle \hat{S} \rangle = 0.3$$

at room temperature, the magnetic scattering amplitude is

$$|r_0 \tfrac{1}{2} g \langle \hat{S} \rangle F(\boldsymbol{\tau})| = 0.54 \times 0.3 \times 0.79 \times 10^{-12}\ \mathrm{cm} = 0.128 \times 10^{-12}\ \mathrm{cm}.$$

For Ni, $\bar{b} = 1.03 \times 10^{-12}$ cm, so the ratio of the magnetic to nuclear scattering at the (111) reflection is

$$\frac{0.128}{1.03} = 0.124 \quad \text{and} \quad R = \left(\frac{1 + 0.124}{1 - 0.124}\right)^2 = 1.646.$$

At the (440) reflection $F \simeq 0.06$ and the ratio of the magnetic and nuclear scattering is 0.0097 with

$$R = 1.028.$$

10.5.3. *Helical structures*

Several examples of complicated helical spin ordering have been determined by neutron diffraction (Coqblin 1977). The use of polarized neutrons enables more information to be gained about these structures than would otherwise be possible. Yoshimori (1959) was the first to point out the existence of helical spin orderings and demonstrated that such an ordering was consistent with the neutron diffraction pattern obtained by Erickson for MnO_2. This is discussed in § 12.6. Here we evaluate the cross-section (10.94) for a helical spin structure and choose MnO_2 as the example.

MnO_2 is a rutile structure for which Yoshimori proposed that the unit vector $\tilde{\boldsymbol{\eta}}_l$ is given by

$$\tilde{\boldsymbol{\eta}}_l = \exp(\mathrm{i}\mathbf{w}\cdot\mathbf{l})\{\tilde{\mathbf{x}}\cos\mathbf{Q}\cdot\mathbf{l}+\tilde{\mathbf{y}}\sin\mathbf{Q}\cdot\mathbf{l}\}. \tag{10.100}$$

The reciprocal vector $\mathbf{w}$ is such that the factor $\exp(\mathrm{i}\mathbf{w}\cdot\mathbf{l})$ has the value $+1$ when l_1, l_2, and l_3 are all even and -1 when they are all odd. The vector $\mathbf{Q}$ in (10.100) is

$$\mathbf{Q}=\frac{2\pi}{c}(0, 0, \tfrac{2}{7}).$$

Thus (10.100) describes an antiferromagnetic spin configuration in which the spins screw along the z-axis with pitch $7c/2$.

In § 12.6 we show that the elastic magnetic cross-section for the scattering of unpolarized neutrons by MnO_2 is

$$\frac{\mathrm{d}\sigma}{\mathrm{d}\Omega}=r_0^2|F(\boldsymbol{\kappa})|^2\tfrac{1}{4}\langle\hat{S}\rangle^2(1+\tilde{\kappa}_z^2)$$
$$\times\left\{\left|\sum_{\mathbf{l}}\exp\{\mathrm{i}\mathbf{l}\cdot(\boldsymbol{\kappa}+\mathbf{w}+\mathbf{Q})\}\right|^2+\left|\sum_{\mathbf{l}}\exp\{\mathrm{i}\mathbf{l}\cdot(\boldsymbol{\kappa}+\mathbf{w}-\mathbf{Q})\}\right|^2\right\}, \tag{10.101}$$

i.e. the Bragg scattering occurs in satellite reflections equally displaced from $\boldsymbol{\kappa}+\mathbf{w}=\boldsymbol{\tau}$ positions. This means that the magnetic and nuclear scattering occur at different positions and cannot therefore interfere with one another. Hence it is sufficient to consider only the purely magnetic part of the cross-section (10.94). This entails the evaluation of the contribution

$$r_0^2|F(\boldsymbol{\kappa})|^2\langle\hat{S}\rangle^2\sum_{\mathbf{l},\mathbf{l}'}\exp\{\mathrm{i}\boldsymbol{\kappa}\cdot(\mathbf{l}-\mathbf{l}')\}\mathrm{i}\mathbf{P}\cdot[\tilde{\boldsymbol{\kappa}}\times(\tilde{\boldsymbol{\eta}}_{l'}\times\tilde{\boldsymbol{\kappa}})]\times[\tilde{\boldsymbol{\kappa}}\times(\tilde{\boldsymbol{\eta}}_l\times\tilde{\boldsymbol{\kappa}})]. \tag{10.102}$$

By using (10.93),

$$\mathbf{P}\cdot[\tilde{\boldsymbol{\kappa}}\times(\tilde{\boldsymbol{\eta}}_{l'}\times\tilde{\boldsymbol{\kappa}})]\times[\tilde{\boldsymbol{\kappa}}\times(\tilde{\boldsymbol{\eta}}_l\times\tilde{\boldsymbol{\kappa}})]=\mathbf{P}\cdot\tilde{\boldsymbol{\eta}}_{l'}\times\tilde{\boldsymbol{\eta}}_l+\sum_{\alpha,\beta}(\mathbf{P}\times\tilde{\boldsymbol{\kappa}})_\alpha\tilde{\kappa}_\beta(\tilde{\eta}_{l'}^\alpha\tilde{\eta}_l^\beta-\tilde{\eta}_{l'}^\beta\tilde{\eta}_l^\alpha)$$

and, since we have only α, $\beta=x$ or y, this reduces to

$$[P_z+(\mathbf{P}\times\tilde{\boldsymbol{\kappa}})_x\tilde{\kappa}_y-(\mathbf{P}\times\tilde{\boldsymbol{\kappa}})_y\tilde{\kappa}_x](\tilde{\eta}_{l'}^x\tilde{\eta}_l^y-\tilde{\eta}_{l'}^y\tilde{\eta}_l^x)=\tilde{\kappa}_z(\mathbf{P}\cdot\tilde{\boldsymbol{\kappa}})(\tilde{\eta}_{l'}^x\tilde{\eta}_l^y-\tilde{\eta}_{l'}^y\tilde{\eta}_l^x). \tag{10.103}$$

From (10.100)

$$\tilde{\eta}_{l'}^x\tilde{\eta}_l^y-\tilde{\eta}_{l'}^y\tilde{\eta}_l^x\equiv\exp\{\mathrm{i}\mathbf{w}\cdot(\mathbf{l}-\mathbf{l}')\}\sin\{\mathbf{Q}\cdot(\mathbf{l}-\mathbf{l}')\}. \tag{10.104}$$

Thus with the results (10.103) and (10.104) we have for (10.102),

$$r_0^2|F(\boldsymbol{\kappa})|^2\tfrac{1}{2}\langle\hat{S}\rangle^2\tilde{\kappa}_z(\mathbf{P}\cdot\tilde{\boldsymbol{\kappa}})$$
$$\times\left\{\left|\sum_{\mathbf{l}}\exp\{\mathrm{i}\mathbf{l}\cdot(\boldsymbol{\kappa}+\mathbf{w}+\mathbf{Q})\}\right|^2-\left|\sum_{\mathbf{l}}\exp\{\mathrm{i}\mathbf{l}\cdot(\boldsymbol{\kappa}+\mathbf{w}-\mathbf{Q})\}\right|^2\right\}.$$

Adding this result to (10.101) as required in the cross-section (10.94), we have for the elastic magnetic cross-section for the scattering of polarized neutrons by the helical spin structure MnO_2 the result

$$\frac{d\sigma}{d\Omega}=r_0^2|F(\boldsymbol{\kappa})|^2\tfrac{1}{4}\langle\hat{S}\rangle^2\Big[\Big|\sum_{\mathbf{l}}\exp\{i\mathbf{l}\cdot(\boldsymbol{\kappa}+\mathbf{w}+\mathbf{Q})\}\Big|^2\{1+\tilde{\kappa}_z^2+2\tilde{\kappa}_z(\mathbf{P}\cdot\tilde{\boldsymbol{\kappa}})\}$$
$$+\Big|\sum_{\mathbf{l}}\exp\{i\mathbf{l}\cdot(\boldsymbol{\kappa}+\mathbf{w}-\mathbf{Q})\}\Big|^2\{1+\tilde{\kappa}_z^2-2\tilde{\kappa}_z(\mathbf{P}\cdot\tilde{\boldsymbol{\kappa}})\}\Big]. \quad (10.105)$$

From this we see that the polarization dependence is different for the two satellites, the difference arising from the

$$i\mathbf{P}\cdot[\tilde{\boldsymbol{\kappa}}\times(\tilde{\boldsymbol{\eta}}_{l'}\times\tilde{\boldsymbol{\kappa}})]\times[\tilde{\boldsymbol{\kappa}}\times(\tilde{\boldsymbol{\eta}}_l\times\tilde{\boldsymbol{\kappa}})]$$

term in the cross-section (Izyumov and Ozerov 1970).

If both the scattering vector $\boldsymbol{\kappa}$ and the polarization $\mathbf{P}$ are made parallel to the z-axis, the intensity of the reflection at $\boldsymbol{\kappa}=\boldsymbol{\tau}+\mathbf{Q}-\mathbf{w}$ is seen to be reduced to zero and the total intensity concentrated in the

$$\boldsymbol{\kappa}=\boldsymbol{\tau}-\mathbf{Q}-\mathbf{w}$$

reflection. If $\boldsymbol{\kappa}$ is maintained in the same direction but $\mathbf{P}$ altered to make it antiparallel, the situation is reversed. Finally, when $\mathbf{P}$ is perpendicular to the scattering plane the cross-section is independent of the polarization and the cross-section (10.105) reduces to that for the scattering of unpolarized neutrons.

10.5.4. *Polarization of scattered beam*

We turn now to the calculation of the polarization of the scattered beam for elastic magnetic scattering in the dipole approximation.

First note that there are several terms in (10.41) that we evaluated above when calculating the cross-section: thus, the first two terms are given in (10.90), the third is (10.88), the sixth is (10.86), and the seventh is (10.92). Of the remaining terms to be calculated we start with $\hat{\boldsymbol{\alpha}}^+(\hat{\boldsymbol{\alpha}}\cdot\mathbf{P})$ and $(\hat{\boldsymbol{\alpha}}^+\cdot\mathbf{P})\hat{\boldsymbol{\alpha}}$.

For the first of these we need

$$\sum_{\mathbf{l,d}}\sum_{\mathbf{l',d'}}\exp\{i\boldsymbol{\kappa}\cdot(\mathbf{R}_{ld}-\mathbf{R}_{l'd'})\}$$
$$\times\mathcal{O}[\{\tfrac{1}{2}B^*_{l'd'}\hat{\mathbf{i}}_{l'd'}+r_0\tfrac{1}{2}g_{d'}F^*_{d'}\langle\hat{S}_{l'd'}\rangle[\tilde{\boldsymbol{\kappa}}\times(\tilde{\boldsymbol{\eta}}_{l'd'}\times\tilde{\boldsymbol{\kappa}})]\}$$
$$\times\{\tfrac{1}{2}B_{ld}\mathbf{P}\cdot\hat{\mathbf{i}}_{ld}+r_0\tfrac{1}{2}g_dF_d\langle\hat{S}_{ld}\rangle\tilde{\boldsymbol{\eta}}_{ld}\cdot\mathbf{P}_\perp\}]$$
$$=\sum_{\mathbf{l,d}}\sum_{\mathbf{l',d'}}\exp\{i\boldsymbol{\kappa}\cdot(\mathbf{R}_{ld}-\mathbf{R}_{l'd'})\}[\tfrac{1}{4}\mathbf{P}\delta_{\mathbf{l,l'}}\,\delta_{\mathbf{d,d'}}\{i_d(i_d+1)\tfrac{1}{3}|B_d|^2\}$$
$$+r_0^2\tfrac{1}{2}g_{d'}F^*_{d'}\langle\hat{S}_{l'd'}\rangle\{\tilde{\boldsymbol{\kappa}}\times(\tilde{\boldsymbol{\eta}}_{l'd'}\times\tilde{\boldsymbol{\kappa}})\}\tfrac{1}{2}g_dF_d\langle\hat{S}_{ld}\rangle\tilde{\boldsymbol{\eta}}_{ld}\cdot\mathbf{P}_\perp]. \quad (10.106)$$

Here we have employed the same argument as that following (10.61).

The calculation of the term corresponding to $(\hat{\boldsymbol{\alpha}}^{+}\cdot\mathbf{P})\hat{\boldsymbol{\alpha}}$ is similar to that just given for $\hat{\boldsymbol{\alpha}}^{+}(\hat{\boldsymbol{\alpha}}\cdot\mathbf{P})$.

The only terms in (10.41) that remain to be evaluated are the last two. Consider $\mathrm{i}\hat{\beta}^{+}(\hat{\boldsymbol{\alpha}}\times\mathbf{P})$; we need

$$\mathrm{i}r_0\sum_{\mathbf{l},\mathbf{d}}\sum_{\mathbf{l}',\mathbf{d}'}\exp\{\mathrm{i}\boldsymbol{\kappa}\cdot(\mathbf{R}_{ld}-\mathbf{R}_{l'd'})\}A^{*}_{l'd'}\tfrac{1}{2}g_dF_d\langle\hat{S}_{ld}\rangle\{\tilde{\boldsymbol{\kappa}}\times(\tilde{\boldsymbol{\eta}}_{ld}\times\tilde{\boldsymbol{\kappa}})\}\times\mathbf{P} \tag{10.107}$$

the nuclear part of $\hat{\boldsymbol{\alpha}}$ not entering because the nuclear spins are randomly orientated. The term corresponding to $\mathrm{i}(\mathbf{P}\times\hat{\boldsymbol{\alpha}})\hat{\beta}$ follows immediately from (10.107).

Combining the results (10.106) and (10.107) with those evaluated previously for the cross-section, the polarization of the scattered beam is given by the expression

$$\begin{aligned}\mathbf{P}'\left(\frac{\mathrm{d}\sigma}{\mathrm{d}\Omega}\right)&=\mathbf{P}\left|\sum_{\mathbf{l}}\exp(\mathrm{i}\boldsymbol{\kappa}\cdot\mathbf{l})\right|^2|F_{\mathrm{N}}(\boldsymbol{\kappa})|^2-\mathbf{P}N\sum_{\mathbf{d}}\{|\bar{b}_d|^2+\tfrac{1}{3}\overline{|b_d|^2}-\tfrac{4}{3}\overline{|A_d|^2}\}\\&+r_0^2\sum_{\mathbf{l},\mathbf{d}}\sum_{\mathbf{l}',\mathbf{d}'}\exp\{\mathrm{i}\boldsymbol{\kappa}\cdot(\mathbf{R}_{ld}-\mathbf{R}_{l'd'})\}\tfrac{1}{2}g_{d'}F^{*}_{d'}(\boldsymbol{\kappa})\langle\hat{S}_{l'd'}\rangle\tfrac{1}{2}g_dF_d(\boldsymbol{\kappa})\langle\hat{S}_{ld}\rangle\\&\times\{-\mathbf{P}[\tilde{\boldsymbol{\kappa}}\times(\tilde{\boldsymbol{\eta}}_{l'd'}\times\tilde{\boldsymbol{\kappa}})]\cdot[\tilde{\boldsymbol{\kappa}}\times(\tilde{\boldsymbol{\eta}}_{ld}\times\tilde{\boldsymbol{\kappa}})]+[\tilde{\boldsymbol{\kappa}}\times(\tilde{\boldsymbol{\eta}}_{l'd'}\times\tilde{\boldsymbol{\kappa}})]\tilde{\boldsymbol{\eta}}_{ld}\cdot\mathbf{P}_{\perp}\\&+\tilde{\boldsymbol{\eta}}_{l'd'}\cdot\mathbf{P}_{\perp}[\tilde{\boldsymbol{\kappa}}\times(\tilde{\boldsymbol{\eta}}_{ld}\times\tilde{\boldsymbol{\kappa}})]-\mathrm{i}[\tilde{\boldsymbol{\kappa}}\times(\tilde{\boldsymbol{\eta}}_{l'd'}\times\tilde{\boldsymbol{\kappa}})]\times[\tilde{\boldsymbol{\kappa}}\times(\tilde{\boldsymbol{\eta}}_{ld}\times\tilde{\boldsymbol{\kappa}})]\}\\&+r_0\sum_{\mathbf{l},\mathbf{d}}\sum_{\mathbf{l}',\mathbf{d}'}\exp\{\mathrm{i}\boldsymbol{\kappa}\cdot(\mathbf{R}_{ld}-\mathbf{R}_{l'd'})\}\{\bar{b}^{*}_{d'}\tfrac{1}{2}g_dF_d(\boldsymbol{\kappa})\langle\hat{S}_{ld}\rangle[\tilde{\boldsymbol{\kappa}}\times(\tilde{\boldsymbol{\eta}}_{ld}\times\tilde{\boldsymbol{\kappa}})]\\&+\bar{b}_d\tfrac{1}{2}g_{d'}F_{d'}(\boldsymbol{\kappa})\langle\hat{S}_{l'd'}\rangle[\tilde{\boldsymbol{\kappa}}\times(\tilde{\boldsymbol{\eta}}_{l'd'}\times\tilde{\boldsymbol{\kappa}})]+\mathrm{i}\bar{b}^{*}_{d'}\tfrac{1}{2}g_dF_d(\boldsymbol{\kappa})\langle\hat{S}_{ld}\rangle[\tilde{\boldsymbol{\kappa}}\times(\tilde{\boldsymbol{\eta}}_{ld}\times\tilde{\boldsymbol{\kappa}})]\times\mathbf{P}\\&+\mathrm{i}\bar{b}^{*}_{d'}\tfrac{1}{2}g_{d'}F_{d'}(\boldsymbol{\kappa})\langle\hat{S}_{l'd'}\rangle\mathbf{P}\times[\tilde{\boldsymbol{\kappa}}\times(\tilde{\boldsymbol{\eta}}_{l'd'}\times\tilde{\boldsymbol{\kappa}})]\}.\end{aligned} \tag{10.108}$$

The first two terms in (10.108) are the purely nuclear contributions to $\mathbf{P}'$ and are identical to the results derived in § 10.4. The second set of terms in (10.108), with the coefficient r_0^2, are the purely magnetic contributions to $\mathbf{P}'$ and the third set, with coefficient r_0, are the contributions arising from the interference between the nuclear and magnetic scattering. When we evaluate $\mathbf{P}'$ for a Bragg reflection the nuclear incoherent contribution can be neglected. Note that two contributions in (10.108) create polarization, one of them being purely magnetic and the other arising from the interference between nuclear and magnetic scattering. If, for the Bragg reflection being examined, there is not both nuclear and magnetic scattering, then the latter cannot contribute to the polarization of an initially unpolarized beam.

If all the spins in the crystal are parallel or antiparallel to a single direction $\tilde{\boldsymbol{\eta}}$, the expression (10.108) for $\mathbf{P}'$ simplifies considerably. Taking just the Bragg part,

$$\mathbf{P}'\left(\frac{\mathrm{d}\sigma}{\mathrm{d}\Omega}\right)=\left|\sum_{\mathbf{l}}\exp(\mathrm{i}\boldsymbol{\kappa}\cdot\mathbf{l})\right|^2 \times\Big\{\mathbf{P}\,|F_{\mathrm{N}}(\boldsymbol{\kappa})|^2+r_0^2[-\mathbf{P}\{1-(\tilde{\boldsymbol{\kappa}}\cdot\tilde{\boldsymbol{\eta}})^2\}+2\tilde{\boldsymbol{\eta}}\cdot\mathbf{P}_{\perp}\{\tilde{\boldsymbol{\kappa}}\times(\tilde{\boldsymbol{\eta}}\times\tilde{\boldsymbol{\kappa}})\}]\,|F_{\mathrm{M}}(\boldsymbol{\kappa})|^2 + 2r_0[\tilde{\boldsymbol{\kappa}}\times(\tilde{\boldsymbol{\eta}}\times\tilde{\boldsymbol{\kappa}})]\mathrm{Re}\,F_{\mathrm{N}}(\boldsymbol{\kappa})F_{\mathrm{M}}^*(\boldsymbol{\kappa})-2r_0\mathbf{P}\times[\tilde{\boldsymbol{\kappa}}\times(\tilde{\boldsymbol{\eta}}\times\tilde{\boldsymbol{\kappa}})]\mathrm{Im}\,F_{\mathrm{N}}(\boldsymbol{\kappa})F_{\mathrm{M}}^*(\boldsymbol{\kappa})\Big\} \tag{10.109}$$

where $(\mathrm{d}\sigma/\mathrm{d}\Omega)$ appearing in this expression is given by (10.95). Notice that the last term in this expression for $\mathbf{P}'$ involves the imaginary part of the magnetic unit-cell structure factor. If the magnetic crystal structure is Bravais, then each ion is at a centre of inversion symmetry and the atomic magnetic form factor purely real. However, in non-centrosymmetric magnetic structures, $F(\boldsymbol{\kappa})$ can possess an imaginary part and this may be detected by observing the polarization of the scattered beam. For, if it is arranged to have $\mathbf{P}$ perpendicular to $\tilde{\boldsymbol{\kappa}}\times(\tilde{\boldsymbol{\eta}}\times\tilde{\boldsymbol{\kappa}})$, the term containing $\mathrm{Im}\,F(\boldsymbol{\kappa})$ contributes a component to $\mathbf{P}'$ that is perpendicular to both vectors whereas all other contributions coincide with either $\mathbf{P}$ or

$$\tilde{\boldsymbol{\kappa}}\times(\tilde{\boldsymbol{\eta}}\times\tilde{\boldsymbol{\kappa}}).$$

Fig. 10.6 shows the various vectors involved in (10.109) with the purpose of clarifying the relations amongst them.

For the case of a ferromagnet with just one atom per unit cell the expression (10.109) takes the very simple form

$$\mathbf{P}'\left(\frac{\mathrm{d}\sigma}{\mathrm{d}\Omega}\right)=\left|\sum_{\mathbf{l}}\exp(\mathrm{i}\boldsymbol{\kappa}\cdot\mathbf{l})\right|^2\Big\{\mathbf{P}\bar{b}^2+r_0^2[-\mathbf{P}\{1-(\tilde{\boldsymbol{\kappa}}\cdot\tilde{\boldsymbol{\eta}})^2\} + 2(\tilde{\boldsymbol{\eta}}\cdot\mathbf{P}_{\perp})\{\tilde{\boldsymbol{\kappa}}\times(\tilde{\boldsymbol{\eta}}\times\tilde{\boldsymbol{\kappa}})\}]\{\tfrac{1}{2}gF(\boldsymbol{\kappa})\langle\hat{S}\rangle\}^2+2r_0[\tilde{\boldsymbol{\kappa}}\times(\tilde{\boldsymbol{\eta}}\times\tilde{\boldsymbol{\kappa}})]\tfrac{1}{2}gF(\boldsymbol{\kappa})\langle\hat{S}\rangle\bar{b}\Big\} \tag{10.110}$$

with $(\mathrm{d}\sigma/\mathrm{d}\Omega)$ given by (10.96).

A ferromagnet can be used to produce a polarized beam of neutrons. If the incident beam is unpolarized, the expression (10.110) shows that the polarization of the scattered beam is given by

$$\mathbf{P}'=\frac{2r_0\{\tilde{\boldsymbol{\tau}}\times(\tilde{\boldsymbol{\eta}}\times\tilde{\boldsymbol{\tau}})\}\tfrac{1}{2}gF(\boldsymbol{\tau})\langle\hat{S}\rangle\bar{b}}{\bar{b}^2+r_0^2\{1-(\tilde{\boldsymbol{\tau}}\cdot\tilde{\boldsymbol{\eta}})^2\}\{\tfrac{1}{2}gF(\boldsymbol{\tau})\langle\hat{S}\rangle\}^2}. \tag{10.111}$$

Hence if a reflection exists where

$$r_0\{1-(\tilde{\boldsymbol{\tau}}\cdot\tilde{\boldsymbol{\eta}})^2\}^{1/2}\tfrac{1}{2}gF(\boldsymbol{\tau})\langle\hat{S}\rangle=\pm\bar{b},$$

the scattered Bragg beam is seen to be completely polarized.

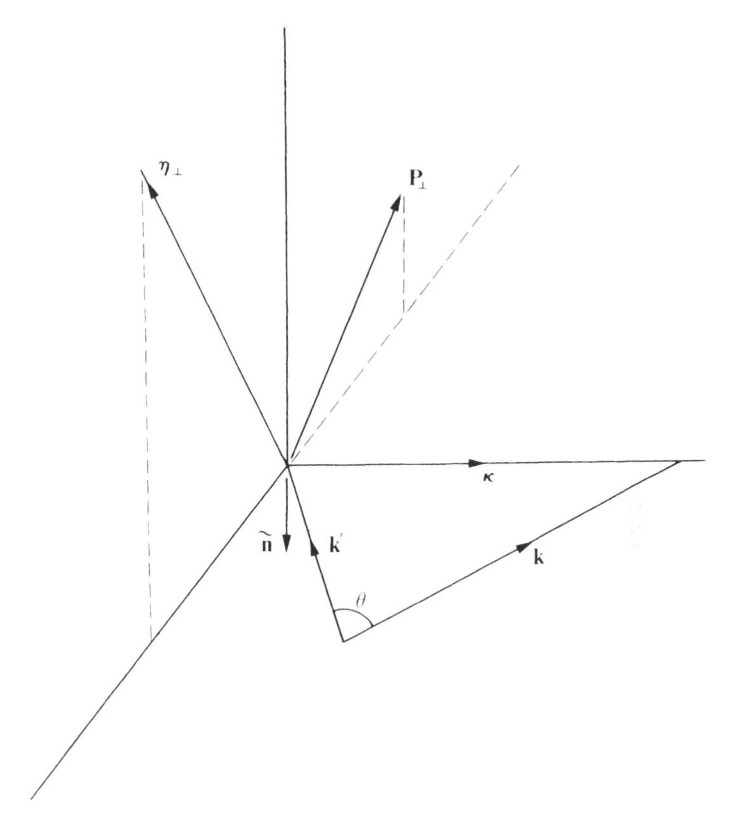

FIG. 10.6. Vectors in elastic neutron cross-section and expression for $\mathbf{P}_\perp = \tilde{\boldsymbol{\kappa}}\times(\mathbf{P}\times\tilde{\boldsymbol{\kappa}})$ and similarly for $\tilde{\mathbf{n}} = \mathbf{k}'\times\mathbf{k}/k^2 \sin\theta$.

We calculate the polarization of neutrons scattered by a helical spin configuration only for the special case in which the incident beam is unpolarized. We choose as our example antiferromagnetic MnO_2. The spin configuration for this system is defined by $\tilde{\boldsymbol{\eta}}_l$ given in (10.100). Because the nuclear and magnetic scattering from MnO_2 do not coincide, the created polarization comes solely from the term

$$r_0^2\,|F(\boldsymbol{\kappa})|^2\,\langle\hat{S}\rangle^2\sum_{\mathbf{l},\mathbf{l}'}\exp\{\mathrm{i}\boldsymbol{\kappa}\cdot(\mathbf{l}-\mathbf{l}')\}(-\mathrm{i})[\tilde{\boldsymbol{\kappa}}\times(\tilde{\boldsymbol{\eta}}_{l'}\times\tilde{\boldsymbol{\kappa}})]\times[\tilde{\boldsymbol{\kappa}}\times(\tilde{\boldsymbol{\eta}}_l\times\tilde{\boldsymbol{\kappa}})] \tag{10.112}$$

in the general expression (10.108).

By using the results (10.103) and (10.104) we find that (10.112) is equal to

$$r_0^2\,|F(\boldsymbol{\kappa})|^2\,\tfrac{1}{2}\langle\hat{S}\rangle^2\tilde{\kappa}_z\tilde{\boldsymbol{\kappa}}\left\{\left|\sum_{\mathbf{l}}\exp\{\mathrm{i}\mathbf{l}\cdot(\boldsymbol{\kappa}+\mathbf{w}+\mathbf{Q})\}\right|^2-\left|\sum_{\mathbf{l}}\exp\{\mathrm{i}\mathbf{l}\cdot(\boldsymbol{\kappa}+\mathbf{w}-\mathbf{Q})\}\right|^2\right\}. \tag{10.113}$$

This result shows that the polarization created in the two Bragg peaks $\boldsymbol{\kappa} = \boldsymbol{\tau} - \mathbf{w} \pm \mathbf{Q}$ is oppositely directed; in one it is parallel to the scattering vector $\boldsymbol{\kappa}$ and in the other it is antiparallel to $\boldsymbol{\kappa}$. We also note that if the scattering vector is arranged perpendicular to the screw axis, in this instance the z-axis, the polarization is not observed.

10.5.5. Cross-section and created polarization with magnetic, nuclear, and electrostatic scattering

As a final topic in elastic scattering we briefly discuss the cross-section and polarization of an initially unpolarized beam for scattering by magnetic, nuclear, and electrostatic interactions. Since we have considered pure nuclear, magnetic, and electrostatic scattering and also the nuclear–electrostatic and nuclear–magnetic interference terms, we have only to consider the magnetic and electrostatic interference terms. We simplify our discussion by restricting attention to magnetic systems where the spins are either parallel or antiparallel to the direction defined by the unit vector $\tilde{\boldsymbol{\eta}}$. The interaction potential for the magnetic and electrostatic scattering of neutrons is obtained from the sum of (10.55) and (10.85) and is, for the particular system under discussion,

$$\langle\lambda|\,\hat{V}(\boldsymbol{\kappa})\,|\lambda\rangle = r_0 \sum_{\mathbf{l}} \exp(\mathrm{i}\boldsymbol{\kappa}\cdot\mathbf{l})$$
$$\times\left\{-\frac{1}{2}\frac{m_e}{m}F_E(\boldsymbol{\kappa}) + \hat{\boldsymbol{\sigma}}\cdot\left[\tfrac{1}{2}\mathrm{i}\frac{m_e}{m}\cot(\tfrac{1}{2}\theta)F_E(\boldsymbol{\kappa})\tilde{\mathbf{n}} + F_M(\boldsymbol{\kappa})\{\tilde{\boldsymbol{\kappa}}\times(\tilde{\boldsymbol{\eta}}\times\tilde{\boldsymbol{\kappa}})\}\right]\right\}. \tag{10.114}$$

A straightforward calculation gives for the cross-section the result

$$\frac{\mathrm{d}\sigma}{\mathrm{d}\Omega} = \left(\frac{\mathrm{d}\sigma}{\mathrm{d}\Omega}\right)_{\mathrm{EE}} + \left(\frac{\mathrm{d}\sigma}{\mathrm{d}\Omega}\right)_{\mathrm{MM}} + \left|\sum_{\mathbf{l}}\exp(\mathrm{i}\boldsymbol{\kappa}\cdot\mathbf{l})\right|^2 \frac{m_e}{m} r_0^2$$
$$\times[\tilde{\boldsymbol{\eta}}\cdot\tilde{\mathbf{n}}\cot(\tfrac{1}{2}\theta)\mathrm{Im}\,F_M(\boldsymbol{\kappa})F_E^*(\boldsymbol{\kappa}) + \tilde{\boldsymbol{\eta}}\cdot\{\cot(\tfrac{1}{2}\theta)(\tilde{\boldsymbol{\kappa}}\cdot\mathbf{P})(\tilde{\boldsymbol{\kappa}}\times\tilde{\mathbf{n}}) - \mathbf{P}_\perp\}$$
$$\times\,\mathrm{Re}\,F_M(\boldsymbol{\kappa})F_E^*(\boldsymbol{\kappa})]. \tag{10.115}$$

We see that there is a polarization-independent interference term between magnetic and electrostatic scattering. Since this term involves the imaginary parts of the magnetic and electrostatic unit-cell structure factors, it can be neglected in the great majority of cases. For the other interference term we note that it involves the scalar product of $\tilde{\boldsymbol{\eta}}$ with two vectors both of which are perpendicular to the scattering vector. Thus this whole term is zero, if $\tilde{\boldsymbol{\eta}}$ is parallel to $\boldsymbol{\kappa}$, and the cross-section is then independent of the polarization $\mathbf{P}$. That $\tilde{\boldsymbol{\eta}}$ be parallel to $\boldsymbol{\kappa}$ is also the condition for the nuclear–magnetic interference term to be zero. In view of the identity

$$\tilde{\boldsymbol{\eta}}\cdot\{\cot(\tfrac{1}{2}\theta)(\tilde{\boldsymbol{\kappa}}\cdot\mathbf{P})(\tilde{\boldsymbol{\kappa}}\times\tilde{\mathbf{n}})-\mathbf{P}_{\perp}\}=\tilde{\boldsymbol{\eta}}\cdot\left\{\frac{2\mathbf{k}(\tilde{\boldsymbol{\kappa}}\cdot\mathbf{P})}{|\boldsymbol{\kappa}|}-\mathbf{P}\right\}, \qquad (10.116)$$

we observe that the polarization-dependent interference term has the dependence $-\tilde{\boldsymbol{\eta}}\cdot\mathbf{P}$ if $\tilde{\boldsymbol{\eta}}$ is arranged to be perpendicular to the incident beam. Clearly the term is zero, if, in addition to this condition, we have also $\mathbf{P}$ perpendicular to the magnetization.

For the polarization of the scattered beam when the incident beam is unpolarized we find

$$\mathbf{P}'\cdot\tilde{\mathbf{R}}\propto-\tilde{\mathbf{R}}\cdot[\{\tilde{\boldsymbol{\kappa}}\times(\tilde{\boldsymbol{\eta}}\times\tilde{\boldsymbol{\kappa}})\}+\cot(\tfrac{1}{2}\theta)\tilde{\boldsymbol{\kappa}}\{\tilde{\boldsymbol{\eta}}\cdot\tilde{\boldsymbol{\kappa}}\times\tilde{\mathbf{n}}\}]\mathrm{Re}\,F_{\mathrm{M}}(\boldsymbol{\kappa})F_{\mathrm{E}}^{*}(\boldsymbol{\kappa}) \qquad (10.117)$$

where $\tilde{\mathbf{R}}$ defines the direction relative to which the polarization is observed. Since

$$\tilde{\mathbf{R}}\cdot[\{\tilde{\boldsymbol{\kappa}}\times(\tilde{\boldsymbol{\eta}}\times\tilde{\boldsymbol{\kappa}})\}+\cot(\tfrac{1}{2}\theta)\tilde{\boldsymbol{\kappa}}\{\tilde{\boldsymbol{\eta}}\cdot\tilde{\boldsymbol{\kappa}}\times\tilde{\mathbf{n}}\}]=\tilde{\mathbf{R}}\cdot\left\{\tilde{\boldsymbol{\eta}}+\frac{2\tilde{\boldsymbol{\kappa}}(\mathbf{k}'\cdot\tilde{\boldsymbol{\eta}})}{|\boldsymbol{\kappa}|}\right\}, \qquad (10.118)$$

it is readily shown that the created polarization is perpendicular to the wave vector of the scattered beam $\mathbf{k}'$. Furthermore, (10.118) allows us to write the polarization in terms of the vectors $\tilde{\mathbf{n}}$ and $\tilde{\mathbf{n}}\times\mathbf{k}'$; namely

$$\mathbf{P}'\cdot\tilde{\mathbf{R}}\propto\tilde{\mathbf{R}}\cdot\left\{(\tilde{\boldsymbol{\eta}}\cdot\tilde{\mathbf{n}})\tilde{\mathbf{n}}+\frac{2\tilde{\boldsymbol{\eta}}\cdot(\tilde{\boldsymbol{\kappa}}\times\tilde{\mathbf{n}})}{|\boldsymbol{\kappa}|}(\tilde{\boldsymbol{\eta}}\times\mathbf{k}')\right\}\mathrm{Re}\,F_{\mathrm{M}}(\boldsymbol{\kappa})F_{\mathrm{E}}^{*}(\boldsymbol{\kappa}). \qquad (10.119)$$

If we combine the results for magnetic and electrostatic scattering with those derived earlier for nuclear and electrostatic scattering, and nuclear and magnetic scattering, the polarization of an initially unpolarized beam of neutrons is given by

$$\begin{aligned}\mathbf{P}'\cdot\tilde{\mathbf{R}}=r_0\Bigg[&\tilde{\mathbf{R}}\cdot\{\tilde{\boldsymbol{\tau}}\times(\tilde{\boldsymbol{\eta}}\times\tilde{\boldsymbol{\tau}})\}2\mathrm{Re}\,F_{\mathrm{N}}^{*}(\boldsymbol{\tau})F_{\mathrm{M}}(\boldsymbol{\tau})\\&+\left(\frac{m_{\mathrm{e}}}{m}\right)\cot(\tfrac{1}{2}\theta)\tilde{\mathbf{R}}\cdot\tilde{\mathbf{n}}\,\mathrm{Im}\,F_{\mathrm{N}}(\boldsymbol{\tau})F_{\mathrm{E}}^{*}(\boldsymbol{\tau})\\&-\left(\frac{m_{\mathrm{e}}}{m}\right)r_0\tilde{\mathbf{R}}\cdot\left\{\tilde{\boldsymbol{\eta}}+\frac{2\tilde{\boldsymbol{\tau}}(\mathbf{k}'\cdot\tilde{\boldsymbol{\eta}})}{|\boldsymbol{\tau}|}\right\}\mathrm{Re}\,F_{\mathrm{M}}(\boldsymbol{\tau})F_{\mathrm{E}}^{*}(\boldsymbol{\tau})\Bigg]\\&\times\Bigg[|F_{\mathrm{N}}(\boldsymbol{\tau})|^2+r_0^2\{1-(\tilde{\boldsymbol{\tau}}\cdot\tilde{\boldsymbol{\eta}})^2\}\,|F_{\mathrm{M}}(\boldsymbol{\tau})|^2\\&+\left(\frac{m_{\mathrm{e}}}{2m}\right)^2r_0^2(\sin\tfrac{1}{2}\theta)^{-2}\,|F_{\mathrm{E}}(\boldsymbol{\tau})|^2-\left(\frac{m_{\mathrm{e}}}{m}\right)r_0\,\mathrm{Re}\,F_{\mathrm{N}}(\boldsymbol{\tau})F_{\mathrm{E}}^{*}(\boldsymbol{\tau})\Bigg]^{-1}. \qquad (10.120)\end{aligned}$$

Only under special circumstances will the components in $\mathbf{P}'$ involving the electrostatic scattering be significant because normally the nuclear and magnetic contributions will be dominant.

10.6. Inelastic magnetic scattering

To calculate the cross-section for inelastic magnetic scattering of polarized neutrons and, also, the polarization of the scattered beam, we take in our general expressions (10.37) and (10.41), $\hat{\beta}\equiv 0$ and for $\hat{\boldsymbol{\alpha}}$ expression (10.75). Thus from (10.37) the full expression for the required cross-section is

$$\frac{\mathrm{d}^2\sigma}{\mathrm{d}\Omega\,\mathrm{d}E'}=\frac{k'}{k}\sum_{\lambda,\lambda'}p_\lambda \times\{\langle\lambda|\,\hat{\boldsymbol{\alpha}}^+\,|\lambda'\rangle\cdot\langle\lambda'|\,\hat{\boldsymbol{\alpha}}\,|\lambda\rangle+\mathrm{i}\mathbf{P}\cdot\langle\lambda|\,\hat{\boldsymbol{\alpha}}^+\,|\lambda'\rangle\times\langle\lambda'|\,\hat{\boldsymbol{\alpha}}\,|\lambda\rangle\}\,\delta(\hbar\omega+E_\lambda-E_{\lambda'}). \tag{10.121}$$

For the polarization of the scattered beam we have from (10.41)

$$\mathbf{P}'\left(\frac{\mathrm{d}^2\sigma}{\mathrm{d}\Omega\,\mathrm{d}E'}\right)=\frac{k'}{k}\sum_{\lambda,\lambda'}p_\lambda\{\langle\lambda|\,\hat{\boldsymbol{\alpha}}^+\,|\lambda'\rangle\langle\lambda'|\,\hat{\boldsymbol{\alpha}}\cdot\mathbf{P}\,|\lambda\rangle+\langle\lambda|\,\hat{\boldsymbol{\alpha}}^+\cdot\mathbf{P}\,|\lambda'\rangle\langle\lambda'|\,\hat{\boldsymbol{\alpha}}\,|\lambda\rangle -\mathbf{P}\langle\lambda|\,\hat{\boldsymbol{\alpha}}^+\,|\lambda'\rangle\cdot\langle\lambda'|\,\hat{\boldsymbol{\alpha}}\,|\lambda\rangle-\mathrm{i}\langle\lambda|\,\hat{\boldsymbol{\alpha}}^+\,|\lambda'\rangle\times\langle\lambda'|\,\hat{\boldsymbol{\alpha}}\,|\lambda\rangle\}\,\delta(\hbar\omega+E_\lambda-E_{\lambda'}). \tag{10.122}$$

We have previously shown that the cross-section for magnetic scattering of unpolarized neutrons can be expressed in terms of correlation functions of operators that belong to the target system (cf. Chapter 8). Such a programme can also be undertaken for the additional term in the cross-section (10.121) that arises because we now have polarized incident neutrons. Clearly then, each term in the expression on the right-hand side of (10.122) for $\mathbf{P}'$ can be written in terms of correlation functions. If we define

$$\hat{\mathbf{S}}^{(\perp)}_{ld}(t)=\tilde{\boldsymbol{\kappa}}\times\{\hat{\mathbf{S}}_{ld}(t)\times\tilde{\boldsymbol{\kappa}}\}, \tag{10.123}$$

the cross-section (10.121) for the magnetic scattering of polarized neutrons

$$\frac{\mathrm{d}^2\sigma}{\mathrm{d}\Omega\,\mathrm{d}E'}=r_0^2\frac{k'}{k}\sum_{\mathbf{l,d}}\sum_{\mathbf{l'd'}}\exp\{\mathrm{i}\boldsymbol{\kappa}\cdot(\mathbf{R}_{l'd'}-\mathbf{R}_{ld})\}\tfrac{1}{2}g_dF^*_d(\boldsymbol{\kappa})\tfrac{1}{2}g_{d'}F_{d'}(\boldsymbol{\kappa}) \times\frac{1}{2\pi\hbar}\int_{-\infty}^{\infty}\mathrm{d}t\,\exp(-\mathrm{i}\omega t)\{\langle\hat{\mathbf{S}}^{(\perp)}_{ld}\cdot\hat{\mathbf{S}}^{(\perp)}_{l'd'}(t)\rangle+\mathrm{i}\mathbf{P}\cdot\langle\hat{\mathbf{S}}^{(\perp)}_{ld}\times\hat{\mathbf{S}}^{(\perp)}_{l'd'}(t)\rangle\}. \tag{10.124}$$

By inspecting (10.122) we see that the polarization of the scattered neutrons can be written

$$\mathbf{P}'\left(\frac{\mathrm{d}^2\sigma}{\mathrm{d}\Omega\,\mathrm{d}E'}\right)=r_0^2\frac{k'}{k}\sum_{\mathbf{l},\mathbf{d}}\sum_{\mathbf{l}',\mathbf{d}'}\exp\{\mathrm{i}\boldsymbol{\kappa}\cdot(\mathbf{R}_{l'd'}-\mathbf{R}_{ld})\}\tfrac{1}{2}g_dF_d^*(\boldsymbol{\kappa})\tfrac{1}{2}g_{d'}F_{d'}(\boldsymbol{\kappa})$$
$$\times\frac{1}{2\pi\hbar}\int_{-\infty}^{\infty}\mathrm{d}t\,\exp(-\mathrm{i}\omega t)\{\langle\hat{\mathbf{S}}_{ld}^{(\perp)}\{\mathbf{P}\cdot\hat{\mathbf{S}}_{l'd'}^{(\perp)}(t)\}+\{\mathbf{P}\cdot\hat{\mathbf{S}}_{ld}^{(\perp)}\}\hat{\mathbf{S}}_{l'd'}^{(\perp)}(t)\rangle$$
$$-\mathbf{P}\langle\hat{\mathbf{S}}_{ld}^{(\perp)}\cdot\hat{\mathbf{S}}_{l'd'}^{(\perp)}(t)\rangle-\mathrm{i}\langle\hat{\mathbf{S}}_{ld}^{(\perp)}\times\hat{\mathbf{S}}_{l'd'}^{(\perp)}(t)\rangle\}. \qquad (10.125)$$

In (10.124) and (10.125), $\hat{\mathbf{S}}^{(\perp)}\equiv\hat{\mathbf{S}}^{(\perp)}(0)$. (10.124) and (10.125) are the key equations for discussing polarization effects in inelastic magnetic scattering (Kakurai *et al.* 1984).

In the particular case when the total z-component of spin is a constant of motion, eqns (10.124) and (10.125) are greatly simplified. The reason for this is that if

$$[\hat{S}^z_{\text{tot}},\mathscr{H}]=0, \qquad (10.126)$$

then we have the identity

$$\langle\hat{S}^{\pm}_{ld}\hat{S}^{\pm}_{l'd'}(t)\rangle=0 \qquad (10.127)$$

and as a result many of the terms in (10.124) and (10.125) are identically zero.

In investigating the consequences of (10.126) in eqns (10.124) and (10.125) we assume a Bravais-crystal structure and take $\tilde{\boldsymbol{\eta}}$ to be a unit vector in the direction of quantization, in this instance the z-axis.

First look at

$$\langle\hat{\mathbf{S}}_l^{(\perp)}\cdot\hat{\mathbf{S}}_{l'}^{(\perp)}(t)\rangle=\sum_{\alpha,\beta}(\delta_{\alpha\beta}-\tilde{\kappa}_\alpha\tilde{\kappa}_\beta)\langle\hat{S}_l^\alpha\hat{S}_{l'}^\beta(t)\rangle, \qquad (10.128)$$

which appears in both (10.124) and (10.125). The terms in this expression with coefficient $-\tilde{\kappa}_x\tilde{\kappa}_y$ are zero. For

$$\langle\hat{S}_l^x\hat{S}_{l'}^y(t)+\hat{S}_l^y\hat{S}_{l'}^x(t)\rangle=\frac{1}{2\mathrm{i}}\langle\hat{S}_l^+\hat{S}_{l'}^+(t)-\hat{S}_l^-\hat{S}_{l'}^-(t)\rangle=0.$$

Furthermore, correlation functions of the form $\langle\hat{S}_l^z\hat{S}_{l'}^\alpha(t)\rangle$ with $\alpha,\ \beta=x$ or y must be identically zero. In addition, $\langle\hat{S}_l^z\hat{S}_{l'}^z(t)\rangle$ does not contribute in one-magnon processes. As a consequence of these results we have for (10.143)

$$\langle\hat{\mathbf{S}}_l^{(\perp)}\cdot\hat{\mathbf{S}}_{l'}^{(\perp)}(t)\rangle=(1-\tilde{\kappa}_x^2)\langle\hat{S}_l^x\hat{S}_{l'}^x(t)\rangle+(1-\tilde{\kappa}_y^2)\langle\hat{S}_l^y\hat{S}_{l'}^y(t)\rangle$$
$$=\tfrac{1}{4}(1+\tilde{\kappa}_z^2)\langle\hat{S}_l^+\hat{S}_{l'}^-(t)+\hat{S}_l^-\hat{S}_{l'}^+(t)\rangle. \qquad (10.129)$$

The reduction of the second part of the cross-section (10.124) proceeds in much the same way. With the definition (10.123) of the operator $\hat{\mathbf{S}}_l^{(\perp)}(t)$,

$$\mathbf{P}\cdot\langle\hat{\mathbf{S}}_l^{(\perp)}\times\hat{\mathbf{S}}_{l'}^{(\perp)}(t)\rangle$$

$$=\langle\mathbf{P}\cdot\hat{\mathbf{S}}_l\times\hat{\mathbf{S}}_{l'}(t)+\hat{\mathbf{S}}_l\cdot(\mathbf{P}\times\tilde{\boldsymbol{\kappa}})\{\tilde{\boldsymbol{\kappa}}\cdot\hat{\mathbf{S}}_{l'}(t)\}-\{\tilde{\boldsymbol{\kappa}}\cdot\hat{\mathbf{S}}_l(t)\}\hat{\mathbf{S}}_{l'}(t)\cdot(\mathbf{P}\times\tilde{\boldsymbol{\kappa}})\rangle$$

$$=\langle\mathbf{P}\cdot\hat{\mathbf{S}}_l\times\hat{\mathbf{S}}_{l'}(t)+\sum_{\alpha,\beta}(\mathbf{P}\times\tilde{\boldsymbol{\kappa}})_\alpha\tilde{\kappa}_\beta\{\hat{S}_l^\alpha\hat{S}_{l'}^\beta(t)-\hat{S}_l^\beta\hat{S}_{l'}^\alpha(t)\}\rangle.$$

Because we only have to keep those terms that contain the products of $\hat{S}^x$ and $\hat{S}^y$, this reduces to

$$\mathbf{P}\cdot\langle\hat{\mathbf{S}}_l^{(\perp)}\times\hat{\mathbf{S}}_{l'}^{(\perp)}(t)\rangle$$

$$=\{P_z+(\mathbf{P}\times\tilde{\boldsymbol{\kappa}})_x\tilde{\kappa}_y-(\mathbf{P}\times\tilde{\boldsymbol{\kappa}})_y\tilde{\kappa}_x\}\langle\hat{S}_l^x\hat{S}_{l'}^y(t)-\hat{S}_l^y\hat{S}_{l'}^x(t)\rangle$$

$$=(\mathbf{P}\cdot\tilde{\boldsymbol{\kappa}})(\tilde{\boldsymbol{\kappa}}\cdot\tilde{\boldsymbol{\eta}})\frac{1}{2\mathrm{i}}\langle\hat{S}_l^-\hat{S}_{l'}^+(t)-\hat{S}_l^+\hat{S}_{l'}^-(t)\rangle. \tag{10.130}$$

On combining this result with (10.129) we have the following expression for the transverse cross-section for inelastic magnetic scattering of polarized neutrons from a target whose total component of spin in the direction defined by the unit vector $\tilde{\boldsymbol{\eta}}$ is a constant of motion (Lowde *et al.* 1983)

$$\left(\frac{\mathrm{d}^2\sigma}{\mathrm{d}\Omega\,\mathrm{d}E'}\right)_{\text{inel}}^{\text{trans}}=r_0^2\frac{k'}{k}\{\tfrac{1}{2}gF(\boldsymbol{\kappa})\}^2\sum_{\mathbf{l},\mathbf{l}'}\exp\{\mathrm{i}\boldsymbol{\kappa}\cdot(\mathbf{l}'-\mathbf{l})\}\frac{1}{2\pi\hbar}\int_{-\infty}^{\infty}\mathrm{d}t\exp(-\mathrm{i}\omega t)$$
$$\times\tfrac{1}{4}[\{1+(\tilde{\boldsymbol{\kappa}}\cdot\tilde{\boldsymbol{\eta}})^2+2(\mathbf{P}\cdot\tilde{\boldsymbol{\kappa}})(\tilde{\boldsymbol{\kappa}}\cdot\tilde{\boldsymbol{\eta}})\}\langle\hat{S}_l^-\hat{S}_{l'}^+(t)\rangle$$
$$+\{1+(\tilde{\boldsymbol{\kappa}}\cdot\tilde{\boldsymbol{\eta}})^2-2(\mathbf{P}\cdot\tilde{\boldsymbol{\kappa}})(\tilde{\boldsymbol{\kappa}}\cdot\tilde{\boldsymbol{\eta}})\}\langle\hat{S}_l^+\hat{S}_{l'}^-(t)\rangle]. \tag{10.131}$$

Note that this expression consists of the sum of two terms, one containing the correlation function $\langle\hat{S}_l^+\hat{S}_{l'}(t)\rangle$ and the other the correlation function $\langle\hat{S}_l^+\hat{S}_{l'}^-(t)\rangle$. These correlation functions are related to each other via

$$\langle\hat{S}_l^+\hat{S}_{l'}^-(t)\rangle=\langle\hat{S}_{l'}^-\hat{S}_l^+(-t+\mathrm{i}\hbar\beta)\rangle. \tag{10.132}$$

Eqn (10.131) is the generalization of (8.80) to the case of a polarized incident beam.

If it is arranged to have the polarization and magnetization of the sample parallel to the scattering vector, then, clearly, the cross-section contains a contribution from only one of the two correlation functions. At low temperatures there is a very small density of magnetic excitations in the target system, so the cross-section involving the creation of an excitation will be much greater in this instance than that involving the annihilation of an excitation.

The correlation functions $\langle\hat{S}_l^\pm\hat{S}_{l'}^\mp(t)\rangle$ are evaluated for a Heisenberg ferromagnet and an antiferromagnet in §§ 9.2.2 and 9.6.3.

Using the results of this chapter, the cross-sections for spin-wave creation (upper sign) and spin-wave annihilation (lower sign) in the scattering of polarized neutrons from a Heisenberg ferromagnet are, from (10.131) (Holden and Stirling 1977)

$$\left(\frac{\mathrm{d}^2\sigma}{\mathrm{d}\Omega\,\mathrm{d}E'}\right)^{(\pm)} = r_0^2 \frac{k'}{k}\{\tfrac{1}{2}gF(\boldsymbol{\kappa})\}^2 \tfrac{1}{2}S \frac{(2\pi)^3}{v_0} \sum_{\mathbf{q},\boldsymbol{\tau}} (n_{\mathbf{q}} + \tfrac{1}{2} \pm \tfrac{1}{2})$$
$$\times\, \delta(\hbar\omega \mp \hbar\omega_{\mathbf{q}})\, \delta(\boldsymbol{\kappa} \mp \mathbf{q} - \boldsymbol{\tau})\{1 + (\tilde{\boldsymbol{\kappa}} \cdot \tilde{\boldsymbol{\eta}})^2 \mp 2(\mathbf{P} \cdot \tilde{\boldsymbol{\kappa}})(\tilde{\boldsymbol{\kappa}} \cdot \tilde{\boldsymbol{\eta}})\} \tag{10.133}$$

where $\hbar\omega_{\mathbf{q}}$ is the energy of a spin-wave with wave vector $\mathbf{q}$ and

$$n_{\mathbf{q}} = \{\exp(\hbar\omega_{\mathbf{q}}\beta) - 1\}^{-1}.$$

For a two-sublattice Heisenberg antiferromagnet in an external magnetic field H applied parallel to the axis of quantization, there are two spin-wave branches with energies

$$\hbar\omega_{\mathbf{q},a} = (-1)^a g\mu_{\mathrm{B}}H + \mathscr{E}_{\mathbf{q}} \quad (a = 0 \text{ or } 1). \tag{10.134}$$

The function $\mathscr{E}_{\mathbf{q}}$ depends upon the various exchange parameters and is given explicitly in § 9.6. The cross-sections for spin-wave creation and annihilation for the scattering of polarized neutrons by a Heisenberg two-sublattice antiferromagnet can be shown with the results of § 9.6 to be given by

$$\left(\frac{\mathrm{d}^2\sigma}{\mathrm{d}\Omega\,\mathrm{d}E'}\right)^{(\pm)} = \sum_{a=0,1} \left(\frac{\mathrm{d}^2\sigma}{\mathrm{d}\Omega\,\mathrm{d}E'}\right)_a^{(\pm)} \tag{10.135}$$

where

$$\left(\frac{\mathrm{d}^2\sigma}{\mathrm{d}\Omega\,\mathrm{d}E'}\right)_a^{(\pm)} = r_0^2 \frac{k'}{k}\{\tfrac{1}{2}gF(\boldsymbol{\kappa})\}^2 \frac{1}{4N} \frac{(2\pi)^3}{v_0} \sum_{\mathbf{q},\boldsymbol{\tau}} (n_{\mathbf{q},a} + \tfrac{1}{2} \pm \tfrac{1}{2})$$
$$\times\, \delta(\hbar\omega \mp \hbar\omega_{\mathbf{q},a})\, \delta(\boldsymbol{\kappa} \mp \mathbf{q} - \boldsymbol{\tau})(u_{\mathbf{q}}^2 + v_{\mathbf{q}}^2 + 2u_{\mathbf{q}}v_{\mathbf{q}} \cos \boldsymbol{\tau} \cdot \boldsymbol{\rho})$$
$$\times \{1 + (\tilde{\boldsymbol{\kappa}} \cdot \tilde{\boldsymbol{\eta}})^2 \mp (-1)^a 2(\mathbf{P} \cdot \tilde{\boldsymbol{\kappa}})(\tilde{\boldsymbol{\kappa}} \cdot \tilde{\boldsymbol{\eta}})\}. \tag{10.136}$$

The functions $u_{\mathbf{q}}$ and $v_{\mathbf{q}}$ appearing in (10.136) are given in § 9.6. $\boldsymbol{\rho}$ is the vector joining nearest-neighbour ions on opposite sublattices.

If no external field H is applied to the antiferromagnet, we see from (10.134) that the two spin-wave branches become degenerate, each having energy $\mathscr{E}_{\mathbf{q}}$. In this instance the cross-sections $(\mathrm{d}^2\sigma/\mathrm{d}\Omega\,\mathrm{d}E')^{(\pm)}$ become independent of the polarization; in (10.136) $n_{\mathbf{q},a} = n_{\mathbf{q}}$ and $\omega_{\mathbf{q},a} = \mathscr{E}_{\mathbf{q}}$. This result could have been anticipated because it is merely an example of the general principle stated in the introduction to this chapter, namely that the cross-section is independent of $\mathbf{P}$ if the target system has no preferred axis. In the present case, the two sublattices are identical in the absence of an external magnetic field and as a consequence the two spin-wave branches are degenerate.

Let us now seek the expression for the polarization of the scattered neutrons that corresponds to the cross-section (10.131). From the general expression for $\mathbf{P}'$ (eqn (10.125)) we see that we have only to consider the first correlation function appearing on the right-hand side because the remaining two have already been considered. For this correlation function we have

$$\begin{aligned}\langle \hat{\mathbf{S}}_l^{(\perp)}\{\mathbf{P}\cdot\hat{\mathbf{S}}_{l'}^{(\perp)}(t)\}+\{\mathbf{P}\cdot\hat{\mathbf{S}}_l^{(\perp)}\}\hat{\mathbf{S}}_{l'}^{(\perp)}(t)\rangle \\ =\sum_{\alpha\beta}\{\tilde{\boldsymbol{\alpha}}P_\beta+\tilde{\boldsymbol{\beta}}P_\alpha-\tilde{\boldsymbol{\kappa}}(\tilde{\kappa}_\alpha P_\beta+\tilde{\kappa}_\beta P_\alpha)-\mathbf{P}\cdot\tilde{\boldsymbol{\kappa}}(\tilde{\boldsymbol{\alpha}}\tilde{\kappa}_\beta+\tilde{\boldsymbol{\beta}}\tilde{\kappa}_\alpha) \\ +2\tilde{\boldsymbol{\kappa}}(\mathbf{P}\cdot\tilde{\boldsymbol{\kappa}})(\tilde{\kappa}_\alpha\tilde{\kappa}_\beta)\}\langle\hat{S}_l^\alpha\hat{S}_{l'}^\beta(t)\rangle.\end{aligned}$$

Taking $\alpha=\beta=x$ or y, as required for the particular case under consideration, this reduces to

$$\begin{aligned}\tfrac{1}{2}\{\mathbf{P}_x+\mathbf{P}_y-\mathbf{P}\cdot\tilde{\boldsymbol{\kappa}}(\tilde{\boldsymbol{\kappa}}_x+\tilde{\boldsymbol{\kappa}}_y)+\tilde{\boldsymbol{\kappa}}(\tilde{\boldsymbol{\kappa}}\cdot\tilde{\boldsymbol{\eta}})\tilde{\boldsymbol{\eta}}\cdot\mathbf{P}_\perp\}\langle\hat{S}_l^+\hat{S}_{l'}^-(t)+\hat{S}_l^-\hat{S}_{l'}^+(t)\rangle \\ =\tfrac{1}{2}\{\mathbf{P}_\perp-\boldsymbol{\eta}_\perp(\tilde{\boldsymbol{\eta}}\cdot\mathbf{P}_\perp)\}\langle\hat{S}_l^+\hat{S}_{l'}^-(t)+\hat{S}_l^-\hat{S}_{l'}^+(t)\rangle\end{aligned}$$

where, in the usual notation, $\boldsymbol{\eta}_\perp=\tilde{\boldsymbol{\kappa}}\times(\tilde{\boldsymbol{\eta}}\times\tilde{\boldsymbol{\kappa}})$.

By combining this result with the results (10.129) and (10.130) we obtain from the general expression (10.140) for $\mathbf{P}'$ the following result, valid under the same conditions as applied to eqn (10.131) for the cross-section

$$\mathbf{P}'\left(\frac{\mathrm{d}^2\sigma}{\mathrm{d}\Omega\,\mathrm{d}E'}\right)=r_0^2\frac{k'}{k}\{\tfrac{1}{2}gF(\boldsymbol{\kappa})\}^2\sum_{\mathbf{l},\mathbf{l}'}\exp\{\mathrm{i}\boldsymbol{\kappa}\cdot(\mathbf{l}'-\mathbf{l})\}\frac{1}{2\pi\hbar}\int_{-\infty}^{\infty}\mathrm{d}t\,\exp(-\mathrm{i}\omega t)$$
$$\times\tfrac{1}{4}\{\mathbf{Q}^{(+)}\langle\hat{S}_l^+\hat{S}_{l'}^-(t)\rangle+\mathbf{Q}^{(-)}\langle\hat{S}_l^-\hat{S}_{l'}^+(t)\rangle\} \quad (10.137)$$

with

$$\mathbf{Q}^{(\pm)}=\mathbf{P}\{1-(\tilde{\boldsymbol{\kappa}}\cdot\tilde{\boldsymbol{\eta}})^2\}+2\tilde{\boldsymbol{\kappa}}\{\pm(\tilde{\boldsymbol{\kappa}}\cdot\tilde{\boldsymbol{\eta}})-(\tilde{\boldsymbol{\kappa}}\cdot\mathbf{P})+(\tilde{\boldsymbol{\kappa}}\cdot\tilde{\boldsymbol{\eta}})(\tilde{\boldsymbol{\eta}}\cdot\mathbf{P}_\perp)\}-2\tilde{\boldsymbol{\eta}}(\tilde{\boldsymbol{\eta}}\cdot\mathbf{P}_\perp). \quad (10.138)$$

For the particular case of the Heisenberg ferromagnet considered before, the polarization of the scattered neutrons is

$$\mathbf{P}'_{(\pm)}=\mathbf{Q}^{(\pm)}\{1+(\tilde{\boldsymbol{\kappa}}\cdot\tilde{\boldsymbol{\eta}})^2\mp2(\tilde{\boldsymbol{\kappa}}\cdot\mathbf{P})(\tilde{\boldsymbol{\kappa}}\cdot\tilde{\boldsymbol{\eta}})\}^{-1} \quad (10.139)$$

where the upper sign refers to the creation process and the lower to the annihilation process. If the initial beam is unpolarized ($\mathbf{P}=0$), (10.139) reduces to

$$\mathbf{P}'_{(\pm)}=\frac{\pm2\tilde{\boldsymbol{\kappa}}(\tilde{\boldsymbol{\kappa}}\cdot\tilde{\boldsymbol{\eta}})}{1+(\tilde{\boldsymbol{\kappa}}\cdot\tilde{\boldsymbol{\eta}})^2}. \quad (10.140)$$

Thus the polarization created in the two scattering processes is oppositely aligned; for the creation of a single spin-wave the created polarization is parallel to the scattering vector and for the annihilation process the created polarization is antiparallel to the scattering vector. Complete polarization is achieved if the magnetization of the ferromagnet is aligned parallel to the scattering vector, i.e. if $-\tilde{\boldsymbol{\eta}}$ is made parallel to $\boldsymbol{\kappa}$.

The various results derived so far in this section for ferromagnets are illustrated in Figs. 10.7, 10.8, and 10.9 for scattering of both polarized and unpolarized neutrons from lithium ferrite. If the polarization and magnetization are made to coincide with the scattering vector, then either the creation or annihilation cross-sections are zero and correspondingly the polarization of the scattered neutrons is zero for one process. Fig. 10.7(a) shows the number of neutrons scattered from Li ferrite with $\mathbf{P}$, $\tilde{\boldsymbol{\eta}}$, and $\boldsymbol{\kappa}$ all parallel, and it is seen that neutrons are reflected by the analyser only if one flipper is activated. Thus the allowed scattering processes involve spin-flip, i.e. are either $\uparrow$–$\downarrow$ or $\downarrow$–$\uparrow$ in agreement with (10.139). The small peak in the measurements shown in Fig. 10.7 with the flipper off is due to nuclear disorder scattering. If the polarization and magnetization coincide and are perpendicular, the result (10.139) shows that neutrons will be reflected by the analyser only if one flipper is activated, in accord with the results shown in Fig. 10.7(b). Also (10.139) shows that

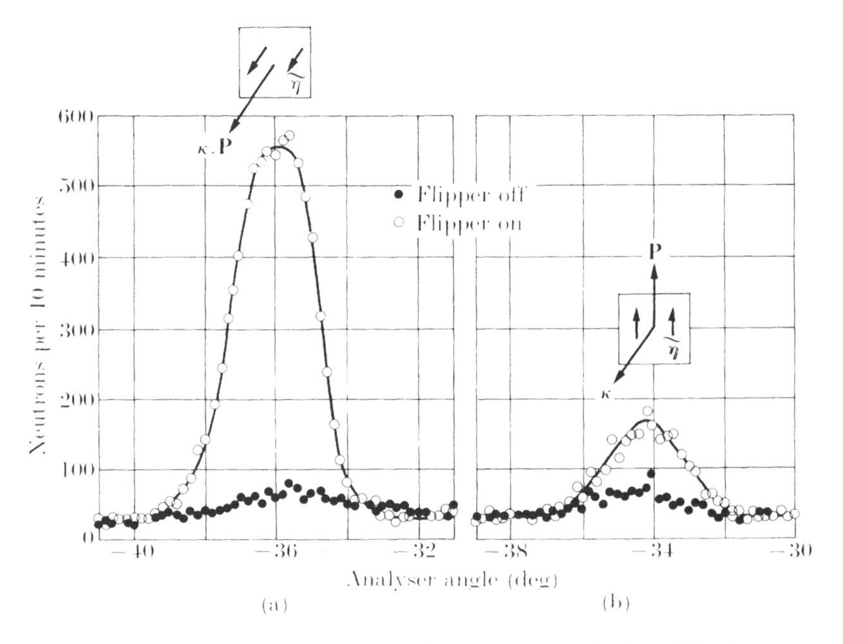

FIG. 10.7. Polarization reversal in magnon scattering from Li ferrite.

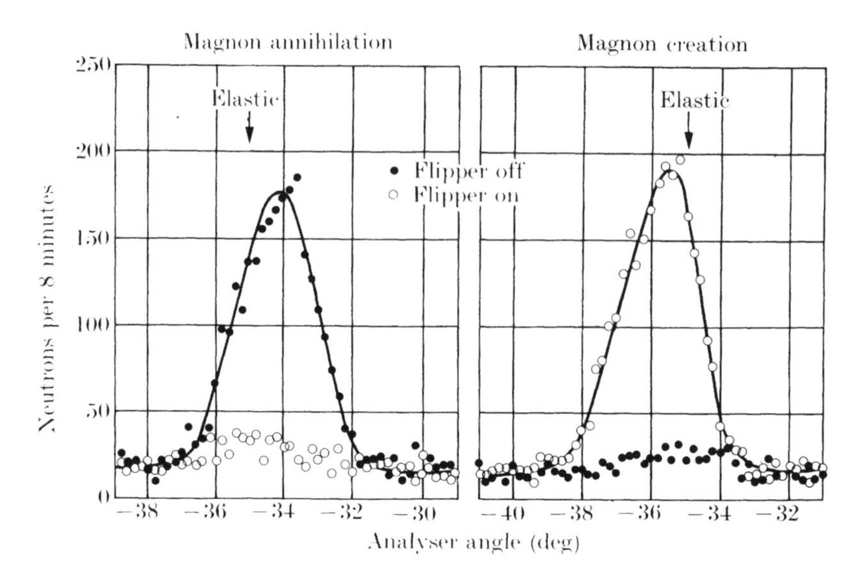

FIG. 10.8. Polarization creation in magnon scattering from Li ferrite: $\tilde{\boldsymbol{\kappa}} \cdot \tilde{\boldsymbol{\eta}} = -1$ and the incident beam is unpolarized with the flipper after the target.

the cross-section of the scattered neutrons is only $\frac{1}{4}$ of its value in this instance compared to the previous case where **P**, $\tilde{\boldsymbol{\eta}}$, and $\boldsymbol{\kappa}$ were all praallel. This too is borne out by Fig. 10.7(a) and (b).

If the incident neutron beam is unpolarized, the polarization of the scattered beam is oppositely directed for spin-wave creation and annihilation scattering. This point is illustrated in Fig. 10.8. The activated flipper was placed after the sample and the magnetization and the scattering vector made parallel, i.e. $\tilde{\boldsymbol{\kappa}} \cdot \tilde{\boldsymbol{\eta}} = -1$ in eqn (10.140). Fig. 10.9 shows the number of neutrons inelastically scattered from Li ferrite with unpolarized incident neutrons and the magnetization perpendicular to the scattering vector. It is seen that there is no polarization dependence of the cross-section in accord with the theory given above.

The expression corresponding to (10.139) for a simple two-sublattice antiferromagnet is easily shown to be

$$\mathbf{P}'_{a,(\pm)} = \mathbf{Q}_a^{(\pm)}\{1 + (\tilde{\boldsymbol{\kappa}} \cdot \tilde{\boldsymbol{\eta}})^2 \mp (-1)^a 2(\tilde{\boldsymbol{\kappa}} \cdot \mathbf{P})(\tilde{\boldsymbol{\kappa}} \cdot \tilde{\boldsymbol{\eta}})\}^{-1} \tag{10.141}$$

with

$$\mathbf{Q}_a^{(\pm)} = \mathbf{P}\{1 - (\tilde{\boldsymbol{\kappa}} \cdot \tilde{\boldsymbol{\eta}})^2\} + 2\tilde{\boldsymbol{\kappa}}\{\pm(-1)^a(\tilde{\boldsymbol{\kappa}} \cdot \tilde{\boldsymbol{\eta}}) - (\tilde{\boldsymbol{\kappa}} \cdot \mathbf{P}) + (\tilde{\boldsymbol{\kappa}} \cdot \tilde{\boldsymbol{\eta}})(\tilde{\boldsymbol{\eta}} \cdot \mathbf{P}_\perp)\} - 2\tilde{\boldsymbol{\eta}}(\tilde{\boldsymbol{\eta}} \cdot \mathbf{P}_\perp). \tag{10.142}$$

Thus the created polarization for a given scattering process is seen to lie in opposite directions for the two spin-wave modes.

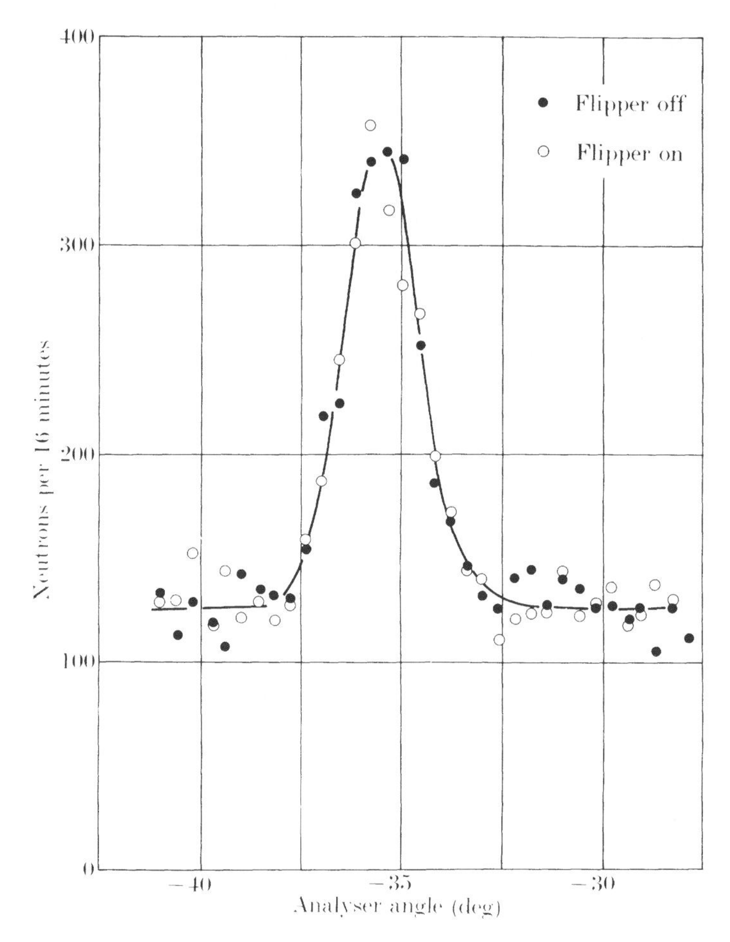

FIG. 10.9. Absence of polarization dependence in magnon scattering from Li ferrite with $\tilde{\boldsymbol{\kappa}} \cdot \tilde{\boldsymbol{\eta}} = 0$. Incident beam unpolarized and flipper after target.

If there is no external field applied to the antiferromagnet so that the two sublattices are identical to one another, then in place of (10.141) we have the result

$$\mathbf{P}' = -\mathbf{P} + 2\{\mathbf{P}_\perp - (\tilde{\boldsymbol{\eta}} \cdot \mathbf{P}_\perp)\boldsymbol{\eta}_\perp\}/\{1 + (\tilde{\boldsymbol{\kappa}} \cdot \tilde{\boldsymbol{\eta}})^2\} \tag{10.143}$$

From this we see that there is no created polarization nor any dependence of the polarization of the scattered beam on the scattering process involved. These conclusions are just what we expect for scattering from a target system with no preferred axis of quantization.

Polarization analysis is a valuable method for distinguishing between scattering from phonons and spin-waves. For hybridized modes, of the type considered in § 9.8, it affords a direct measure of the magnetic content of a mode (Lovesey 1974).

10.7. Inelastic nuclear scattering

We express the nuclear cross-section and final polarization in terms of correlation functions using the results derived in § 10.4 for a sample with randomly oriented nuclei. For the cross-section we require the terms

$$\hat{\beta}^+\hat{\beta} + \mathcal{O}\{\hat{\boldsymbol{\alpha}}^+ \cdot \hat{\boldsymbol{\alpha}}\} \tag{10.144}$$

where the operators $\hat{\boldsymbol{\alpha}}$ and $\hat{\beta}$ are defined in (10.59), and $\mathcal{O}$ indicates an average over nuclear spin orientations. The corresponding terms in the final polarization are

$$\mathbf{P}[\hat{\beta}^+\hat{\beta} - \tfrac{1}{3}\mathcal{O}\{\hat{\boldsymbol{\alpha}}^+ \cdot \hat{\boldsymbol{\alpha}}\}]. \tag{10.145}$$

From these results we deduce that the same correlation functions appear in the cross-section and final polarization.

The correlation function which arises from the term $\hat{\beta}^+\hat{\beta}$ is

$$Y_{jj'}(\boldsymbol{\kappa}, t) = \langle \exp(-i\boldsymbol{\kappa} \cdot \hat{\mathbf{R}}_j) \exp\{i\boldsymbol{\kappa} \cdot \hat{\mathbf{R}}_{j'}(t)\}\rangle \tag{10.146}$$

where $\mathbf{R}_j$ is the position of the jth nucleus; thus,

$$\langle \hat{\beta}^+\hat{\beta}(t)\rangle = \sum_{jj'} A_j^* A_{j'} Y_{jj'}(\boldsymbol{\kappa}, t). \tag{10.147}$$

The correlation function which arises from $\hat{\boldsymbol{\alpha}}^+ \cdot \hat{\boldsymbol{\alpha}}$ is,

$$\langle \boldsymbol{\alpha}^+ \cdot \hat{\boldsymbol{\alpha}}(t)\rangle = \tfrac{1}{4}\sum_{jj'} B_j^* B_{j'} \langle \exp(-i\boldsymbol{\kappa} \cdot \hat{\mathbf{R}}_j)\hat{\mathbf{i}}_j \cdot \hat{\mathbf{i}}_{j'}(t) \exp\{i\boldsymbol{\kappa} \cdot \hat{\mathbf{R}}_{j'}(t)\}\rangle. \tag{10.148}$$

The nuclear spin operators are included in the time-dependent correlation function because, in general, their states are correlated with the motion of the nuclei through quantum mechanical exchange forces. However, for many situations of interest the quantum mechanical effect can be safely neglected, in which case, for random spin orientations,

$$\mathcal{O}\langle \hat{\boldsymbol{\alpha}}^+ \cdot \hat{\boldsymbol{\alpha}}(t)\rangle = \tfrac{1}{4}\sum_j |B_j|^2\, i_j(i_j+1) Y_{jj}(\boldsymbol{\kappa}, t). \tag{10.149}$$

Hence, both correlation functions are expressed in terms of $Y_{jj'}(\boldsymbol{\kappa}, t)$.

The corresponding partial differential cross-section is

$$\frac{d^2\sigma}{d\Omega\, dE'} = \frac{k'}{k}\frac{1}{2\pi\hbar}\int_{-\infty}^{\infty} dt \exp(-i\omega t)$$
$$\times \overline{\left\{\sum_{jj'} A_j^* A_{j'} Y_{jj'}(\boldsymbol{\kappa}, t) + \tfrac{1}{4}\sum_j |B_j|^2\, i_j(i_j+1) Y_{jj}(\boldsymbol{\kappa}, t)\right\}}. \tag{10.150}$$

Here, the horizontal bar refers to an average over the distribution of atomic species and isotopes. The corresponding expression for the polarization of the scattered beam is

$$\mathbf{P}'\left(\frac{\mathrm{d}^2\sigma}{\mathrm{d}\Omega\,\mathrm{d}E'}\right)=\mathbf{P}\frac{k'}{k}\frac{1}{2\pi\hbar}\int_{-\infty}^{\infty}\mathrm{d}t\exp(-\mathrm{i}\omega t)$$
$$\times\overline{\left\{\sum_{ii'}A_i^*A_{i'}Y_{ii'}(\boldsymbol{\kappa},t)-\tfrac{1}{12}\sum_i|B_i|^2\,i_i(i_i+1)Y_{ii}(\boldsymbol{\kappa},t)\right\}}.\quad(10.151)$$

Eqns (10.150) and (10.151) are general expressions given that the nuclei are randomly oriented and quantum mechanical exchange forces are negligible.

Consider the special case of a monatomic system. We will assume that the dynamic properties of the nuclei are independent of the isotope distribution and therefore remove the correlation functions from the average. Values of A_i for different nuclei are quite independent of each other, so that

$$\overline{A_i^*A_{i'}}=|\bar{A}|^2+\delta_{ii'}\{\overline{|A|^2}-|\bar{A}|^2\}.\quad(10.152)$$

The single-atom coherent and total cross-sections are defined to be

$$\sigma_\mathrm{c}=4\pi\,|\bar{A}|^2$$

and

$$\sigma=4\pi\{\overline{|A|^2}+\tfrac{1}{4}i(i+1)\overline{|B|^2}\},$$

and the incoherent cross-section

$$\sigma_\mathrm{i}=\sigma-\sigma_\mathrm{c}.\quad(10.153)$$

Using these definitions in (10.150),

$$\frac{\mathrm{d}^2\sigma}{\mathrm{d}\Omega\,\mathrm{d}E'}=\frac{1}{4\pi}\frac{k'}{k}\frac{1}{2\pi\hbar}\int_{-\infty}^{\infty}\mathrm{d}t\exp(-\mathrm{i}\omega t)\left\{\sigma_\mathrm{c}\sum_{ii'}Y_{ii'}(\boldsymbol{\kappa},t)+\sigma_\mathrm{i}\sum_i Y_{ii}(\boldsymbol{\kappa},t)\right\}.\quad(10.154)$$

For the polarization of the scattered beam we obtain the expression

$$\mathbf{P}'\left[\int_{-\infty}^{\infty}\mathrm{d}t\exp(-\mathrm{i}\omega t)\left\{\sigma_\mathrm{c}\sum_{ii'}Y_{ii'}(\boldsymbol{\kappa},t)+\sigma_\mathrm{i}\sum_i Y_{ii}(\boldsymbol{\kappa},t)\right\}\right]$$
$$=\mathbf{P}\left[\int_{-\infty}^{\infty}\mathrm{d}t\exp(-\mathrm{i}\omega t)\left\{\sigma_\mathrm{c}\sum_{ii'}Y_{ii'}(\boldsymbol{\kappa},t)+(\sigma_\mathrm{i}-\tfrac{4}{3}\sigma+\tfrac{16}{3}\pi\,\overline{|A|^2})\sum_i Y_{ii}(\boldsymbol{\kappa},t)\right\}\right].$$
$$(10.155)$$

If there is just one type of isotope,

$$\sigma_{\mathrm{i}} - \tfrac{4}{3}\sigma + \tfrac{16}{3}\pi\,\overline{|A|^2} = -\tfrac{1}{3}\sigma_{\mathrm{i}}.$$

These results are of interest in the study of single-particle and collective motions in solids and liquids. The cross-section (10.154) is a weighted sum of the two types of motion. Measurements of the cross-sections for non-spin-flip and spin-flip scattering, using a polarized incident beam, together with a knowledge of the various scattering lengths, enable one (at least in principle) to separate the single-particle and collective motions.

10.8. Magnetic and nuclear scattering

As a final topic in this chapter, we consider the form of the cross-section for scattering polarized neutrons, expressed in terms of correlation functions, and including both magnetic and nuclear interactions. Thus, $\hat{\beta}$ is given by (10.59) and $\hat{\boldsymbol{\alpha}}$ is the sum of the magnetic and nuclear interaction operators in (10.59) and (10.75); in using the latter we employ the dipole approximation for magnetic scattering. We will assume that the nuclei are randomly oriented, as in § 10.4.

Let the scattering ions be at positions defined by $\mathbf{R}_j$. After averaging over the orientations of the nuclei, and using the notation (10.146) for the purely nuclear correlation function, we arrive at the expression

$$\begin{aligned}\frac{\mathrm{d}^2\sigma}{\mathrm{d}\Omega\,\mathrm{d}E'} = \frac{k'}{k}\frac{1}{2\pi\hbar}\int_{-\infty}^{\infty} \mathrm{d}t\,\exp(-\mathrm{i}\omega t)\sum_{ii'}\{\overline{[A_i^* A_{i'} + \tfrac{1}{4}|B_i|^2 i_i(i_i+1)\,\delta_{ii'}]}\\
\times Y_{ii'}(\boldsymbol{\kappa},t) + (\tfrac{1}{2}gr_0)^2 F_i^*(\boldsymbol{\kappa})F_{i'}(\boldsymbol{\kappa})\langle\{\hat{\mathbf{S}}_i^{(\perp)}\exp(-\mathrm{i}\boldsymbol{\kappa}\cdot\hat{\mathbf{R}}_i)\}\cdot\{\exp\{\mathrm{i}\boldsymbol{\kappa}\cdot\hat{\mathbf{R}}_{i'}(t)\}\hat{\mathbf{S}}_{i'}^{(\perp)}(t)\}\rangle\\
+\mathrm{i}(\tfrac{1}{2}gr_0)^2 F_i^*(\boldsymbol{\kappa})F_{i'}(\boldsymbol{\kappa})\mathbf{P}\cdot\langle\{\hat{\mathbf{S}}_i^{(\perp)}\exp(-\mathrm{i}\boldsymbol{\kappa}\cdot\hat{\mathbf{R}}_i)\}\times\{\exp\{\mathrm{i}\boldsymbol{\kappa}\cdot\hat{\mathbf{R}}_{i'}(t)\}\hat{\mathbf{S}}_{i'}^{(\perp)}(t)\}\rangle\\
+(\tfrac{1}{2}gr_0)[\bar{A}_i^* F_{i'}(\boldsymbol{\kappa})\langle\exp(-\mathrm{i}\boldsymbol{\kappa}\cdot\hat{\mathbf{R}}_i)\exp\{\mathrm{i}\boldsymbol{\kappa}\cdot\hat{\mathbf{R}}_{i'}(t)\}\mathbf{P}\cdot\hat{\mathbf{S}}_{i'}^{(\perp)}(t)\rangle\\
+F_i^*(\boldsymbol{\kappa})\bar{A}_{i'}\langle\mathbf{P}\cdot\hat{\mathbf{S}}_i^{(\perp)}\exp(-\mathrm{i}\boldsymbol{\kappa}\cdot\hat{\mathbf{R}}_i)\exp\{-\mathrm{i}\boldsymbol{\kappa}\cdot\hat{\mathbf{R}}_{i'}(t)\}\rangle]\}.\end{aligned} \tag{10.156}$$

Here, we have assumed that quantum mechanical exchange forces are negligible and, therefore, (10.156) is the analogue of (10.150) for purely nuclear scattering.

If we assume, furthermore, that the spin and vibrational motions are uncorrelated, the expression (10.156) reduces to

$$\begin{aligned}\frac{\mathrm{d}^2\sigma}{\mathrm{d}\Omega\,\mathrm{d}E'} = \frac{k'}{k}\frac{1}{2\pi\hbar}\int_{-\infty}^{\infty} \mathrm{d}t\,\exp(-\mathrm{i}\omega t)\sum_{ii'}\{\overline{[A_i^* A_{i'} + \tfrac{1}{4}|B_i|^2\, i_i(i_i+1)\,\delta_{ii'}]}\\
+(\tfrac{1}{2}gr_0)^2 F_i^*(\boldsymbol{\kappa})F_{i'}(\boldsymbol{\kappa})\langle\hat{\mathbf{S}}_i^{(\perp)}\cdot\hat{\mathbf{S}}_{i'}^{(\perp)}(t)\rangle + \mathrm{i}(\tfrac{1}{2}gr_0)^2 F_i^*(\boldsymbol{\kappa})F_{i'}(\boldsymbol{\kappa})\mathbf{P}\cdot\langle\hat{\mathbf{S}}_i^{(\perp)}\times\hat{\mathbf{S}}_{i'}^{(\perp)}(t)\rangle\\
+(\tfrac{1}{2}gr_0)[\bar{A}_i^* F_{i'}(\boldsymbol{\kappa})\langle\mathbf{P}\cdot\hat{\mathbf{S}}_{i'}^{(\perp)}\rangle + F_i^*(\boldsymbol{\kappa})\bar{A}_{i'}\langle\mathbf{P}\cdot\hat{\mathbf{S}}_i^{(\perp)}\rangle]\}\,Y_{ii'}(\boldsymbol{\kappa},t).\end{aligned} \tag{10.157}$$

Note that, for a stationary system, $\langle \hat{\mathbf{S}}^{(\perp)}(t)\rangle = \langle \hat{\mathbf{S}}^{(\perp)}(0)\rangle \equiv \langle \hat{\mathbf{S}}^{(\perp)}\rangle$, where the final equality arises from our use of a shorthand notation.

We focus attention on that part of (10.157) which is elastic in the magnetic scattering, from which we can deduce the magnetovibrational cross-section for scattering polarized neutrons (cf. § 8.1). Now, for a bulk sample and ordered spins,

$$\lim_{t\to\infty} \langle \hat{\mathbf{S}}_j^{(\perp)} \cdot \hat{\mathbf{S}}_{j'}^{(\perp)}(t)\rangle = \langle \hat{\mathbf{S}}_j^{(\perp)}\rangle \cdot \langle \hat{\mathbf{S}}_{j'}^{(\perp)}\rangle$$

$$= \langle \hat{S}\rangle^2 \boldsymbol{\eta}_{j,\perp} \cdot \boldsymbol{\eta}_{j',\perp} \tag{10.158}$$

and

$$\lim_{t\to\infty} \langle \hat{\mathbf{S}}_j^{(\perp)} \times \hat{\mathbf{S}}_{j'}^{(\perp)}(t)\rangle = \langle \hat{S}\rangle^2 \{\boldsymbol{\eta}_{j,\perp} \times \boldsymbol{\eta}_{j',\perp}\} \tag{10.159}$$

where, in the usual notation, $\boldsymbol{\eta}_\perp = \tilde{\boldsymbol{\kappa}} \times (\tilde{\boldsymbol{\eta}} \times \tilde{\boldsymbol{\kappa}})$ and $\tilde{\boldsymbol{\eta}}$ is a unit vector in ordering direction. The magnetovibrational scattering is inelastic with respect to the vibrational motion of the ions. Writing

$$Y_{jj'}(\boldsymbol{\kappa}, t) = \{Y_{jj'}(\boldsymbol{\kappa}, t) - Y_{jj'}(\boldsymbol{\kappa}, \infty)\} + Y_{jj'}(\boldsymbol{\kappa}, \infty),$$

the cross-section for magnetovibrational scattering of polarized neutrons is (Steinsvoll *et al.* 1981).

$$\left(\frac{\mathrm{d}^2\sigma}{\mathrm{d}\Omega\,\mathrm{d}E'}\right)_{\mathrm{m.v.}} = \frac{k'}{k}(\tfrac{1}{2}gr_0\langle \hat{S}\rangle)\frac{1}{2\pi\hbar}\int_{-\infty}^{\infty} \mathrm{d}t\, \exp(-\mathrm{i}\omega t)$$
$$\times \sum_{jj'} \{(\tfrac{1}{2}gr_0\langle \hat{S}\rangle) F_j^*(\boldsymbol{\kappa}) F_{j'}(\boldsymbol{\kappa}) \boldsymbol{\eta}_{j,\perp} \cdot \boldsymbol{\eta}_{j',\perp}$$
$$+ \mathrm{i}(\tfrac{1}{2}gr_0\langle \hat{S}\rangle) F_j^*(\boldsymbol{\kappa}) F_{j'}(\boldsymbol{\kappa}) \mathbf{P} \cdot (\boldsymbol{\eta}_{j,\perp} \times \boldsymbol{\eta}_{j',\perp})$$
$$+ \mathbf{P}_\perp \cdot [\bar{A}_j^* F_{j'}(\boldsymbol{\kappa}) \tilde{\boldsymbol{\eta}}_{j'} + F_j^*(\boldsymbol{\kappa}) \bar{A}_{j'} \tilde{\boldsymbol{\eta}}_j]\}\{Y_{jj'}(\boldsymbol{\kappa}, t) - Y_{jj'}(\boldsymbol{\kappa}, \infty)\}. \tag{10.160}$$

The interference term is expressed in terms of $\mathbf{P}_\perp$ by using the identity (10.91). An expression for $\boldsymbol{\eta}_{j,\perp} \times \boldsymbol{\eta}_{j',\perp}$ is given in (10.93), followed by an explanation of its features. It is zero for simple magnetic structures, in which case the polarization dependence arises from the interference between the magnetic and nuclear amplitudes.

The nuclear correlation functions for harmonic lattice vibrations are summarized in § 8.1. On setting $\mathbf{P} = 0$ in (10.160), we recover (8.14) when the magnetic ions are identical.

10.9. Crystal field states (Lovesey and Gunn 1984)

We calculate the polarization of an initially unpolarized beam following inelastic scattering by a magnetic ion subject to a crystal field. The cross-section for this process is discussed in § 11.7. The result given here is based on the dipole approximation for the interaction operator given in eqn (11.45). From (10.41) we find the polarization created in the transition $\Gamma_n \to \Gamma_{n'}$ is

$$\mathbf{P}'\left(\frac{d^2\sigma}{d\Omega\, dE'}\right) = 2r_0^2 \frac{k'}{k} \{\tfrac{1}{2}gF(\kappa)\}^2 \sum_{v,v'} p_n\, \delta(\hbar\omega + E_n - E_{n'})$$
$$\times \operatorname{Im}\{(\langle\Gamma_n v|\, \hat{J}^x\, |\Gamma_{n'}v'\rangle\langle\Gamma_n v|\, \hat{J}^y\, |\Gamma_{n'}v'\rangle^* \tilde{\mathbf{z}} + \text{permutations in } x, y, z)$$
$$+ \langle\Gamma_n v|\, \tilde{\boldsymbol{\kappa}} \times \hat{\mathbf{J}}\, |\Gamma_{n'}v'\rangle\langle\Gamma_n v|\, \tilde{\boldsymbol{\kappa}} \cdot \hat{\mathbf{J}}\, |\Gamma_{n'}v'\rangle^*\}. \qquad (10.161)$$

Here, the cross-section on the left-hand side is given by (11.144), and Im denotes the imaginary part of the following expression. In deriving (10.161) we have used the identity

$$\langle\lambda|\, \hat{J}^\alpha\, |\lambda'\rangle^* = \langle\lambda'|\, \hat{J}^\alpha\, |\lambda\rangle, \qquad (10.162)$$

which follows on noting that $\hat{\mathbf{J}}$ is a Hermitean operator. The final polarization is seen to be zero if the matrix elements of the angular momentum operator are all purely real, or imaginary. The wave function for a non-degenerate state can always be taken to be real. If the z-axis is the axis of quantization, then for such states $\mathbf{P}'$ is different from zero only if the matrix element of

$$\hat{J}^y = (\hat{J}^+ - \hat{J}^-)/2i$$

is non-zero, i.e. the transverse matrix elements must be finite for $\mathbf{P}' \neq 0$. We illustrate the use of (10.161) by two examples. First, we consider the simple, non-degenerate states

$$|v\rangle = |J, M = 0\rangle$$
$$|v'\rangle = |J, M = 1\rangle$$

and find

$$\langle v|\, \hat{J}^z\, |v'\rangle = 0,$$
$$\langle v|\, \hat{J}^x\, |v'\rangle = -i\langle v|\, \hat{J}^y\, |v'\rangle = \tfrac{1}{2}\{J(J+1)\}^{1/2}.$$

Using these results in (10.161) and (11.144) we obtain,

$$\mathbf{P}' = -2\tilde{\kappa}_z \tilde{\boldsymbol{\kappa}}/(1 + \tilde{\kappa}_z^2),$$

which agrees with the result for the polarization created in the annihilation of a ferromagnetic spin wave, eqn (10.140), as should be expected.

As a second example, we consider the polarization created in the transition $\Gamma_1 \to \Gamma_4$ illustrated in the level scheme given in Fig. 11.3. In the absence of perturbations to remove the degeneracy of the Γ_4 state we find $\mathbf{P}' = 0$. This result arises because for full cubic symmetry there is no preferred axis for the states of the ion.

For the calculation we use the following wave functions ($|M\rangle$ is shorthand for $|J, M\rangle$)

$$|\Gamma_1\rangle = (\tfrac{7}{12})^{1/2}\,|0\rangle + (\tfrac{5}{24})^{1/2}\{|4\rangle + |-4\rangle\},$$

and, taking $v' = a$, b, and c,

$$|\Gamma_4, a\rangle = (\tfrac{1}{2})^{1/2}\{|4\rangle - |-4\rangle\}$$
$$|\Gamma_4, b\rangle = (\tfrac{1}{8})^{1/2}\{\sqrt{7}\,|-1\rangle + |3\rangle\}$$
$$|\Gamma_4, c\rangle = (\tfrac{1}{8})^{1/2}\{\sqrt{7}\,|1\rangle + |-3\rangle\}$$

The matrix elements of $\hat{\mathbf{J}}$ are given in the table in terms of the parameter $\alpha = (10/3)^{1/2}$

	J^x	J^y	J^z
a	0	0	$\alpha\sqrt{2}$
b	α	$-i\alpha$	0
c	α	$i\alpha$	0

We note that all matrix elements $\langle\Gamma_1|\,\hat{J}^\alpha\,|a\rangle$ are real (namely, 0, 0, and $\alpha\sqrt{2}$). Hence, no polarization is created in scattering from the ground state to $|\Gamma_4, a\rangle$. For the remaining two states we obtain

$$\mathbf{P}'_b = -\mathbf{P}'_c = 2\tilde{\kappa}_z\tilde{\boldsymbol{\kappa}}/(1 + \tilde{\kappa}_z^2),$$

and if these states are degenerate $\mathbf{P}'$ is zero.

REFERENCES

Abragam, A. and Bleaney, B. (1983). *Proc. R. Soc.* **A387,** 221.

—— and Goldman, M. (1982). *Nuclear magnetism; order and disorder.* Oxford University Press, Oxford.

Brown, P. J. (1979). *Treatise on materials science and technology*, Vol. 15 (ed. G. Kostorz). Academic Press, New York.

Coqblin, B. (1977). *The electronic structure of rare-earth metals and alloys; the magnetic heavy rare-earths.* Academic Press, New York.

Dachs, H. (1978). *Topics in current physics*, Vol. 6. Springer-Verlag, Heidelberg.

Fano, U. (1983). *Rev. mod. Phys.* **55,** 855.

Holden, T. M. and Stirling, W. G. (1977). *J. Phys.* **F7,** 1901.

Izyumov, Y. A. and Ozerov, R. P. (1970). *Magnetic neutron diffraction.* Plenum Press, New York.

Joachain, C. J. (1983). *Quantum Collision Theory.* North-Holland, Amsterdam.

Kakurai, K., Pynn, R., Dorner, B. and Steiner, M. (1984). *J. Phys.* **C17,** L123.

Lovesey, S. W. (1974). *J. Phys.* **C7,** 2039.

—— Gunn, J. M. F. (1984). *Z. Phys.* **B57,** 191.

Lowde, R. D., Moon, R. M., Pagonis, B., Perry, C. H., Sokoloff, J. B., Vaughan-Watkins, R. S., Wiltshire, M. C. K., and Crangle, J. (1983). *J. Phys.* **F13,** 249.

Menzinger, F. and Sacchetti, F. (1979). *Nukleonika* **24,** 737.
Moon, R. M., Riste, T., and Koehler, W. C. (1969). *Phys. Rev.* **181,** 920.
Schermer, R. I. and Blume, M. (1968). *Phys. Rev.* **166,** 554.
Shull, C. G. (1963). *Phys. Rev. Lett.* **10,** 297.
Steinsvoll, O., Moon, R. M., Koehler, W. C. and Windsor, C. G. (1981). *Phys. Rev.* **B24,** 4031.
—— Majkrazak, C. F., Shirane, G., and Wicksted, J. (1983). *Phys. Rev. Lett.* **51,** 300.
Tofield, B. C. (1975). *Structure and bonding,* Vol. 21. Springer-Verlag, Heidelberg.
Yoshimori, A. (1959). *J. Phys. Soc. Japan* **14,** 807.

11

ATOMIC ELECTRONS

The theory of magnetic scattering by ions is required for the interpretation of several types of elastic and inelastic scattering experiments. These experiments may yield information obtainable in no other way. Unfortunately, the theory for ions with more than one unpaired electron is probably quite complicated, at first sight, for readers not familiar with the theory of atomic or nuclear structure. For these reasons it seems wise to devote an entire chapter to requisite theoretical methods, together with some illustrative examples.

Developments in the theory of magnetic scattering from ions have, to a large extent, followed developments in the theory of atomic and nuclear structure. The original work by Trammell (1953) exploited the theory of atomic spectra set out by Condon and Shortley (1935). Trammell's theory has been applied, extensively, to scattering from magnetic molecules (Kleiner 1955; Meier and Helmholdt 1984) and magnetic salts (Blume 1961; Brown and Forsyth 1981). Later work showed that there was a certain redundancy in Trammell's theory and reformulated it in terms of Racah algebra (Condon and Odabaşi 1980: Lovesey and Rimmer 1969). This development was timely since it cast the theory into a form amenable to computation. Applications of the theory, in terms of Racah algebra, have included the calculation of cross-sections for inelastic scattering by crystal-field levels in rare earths (Balcar and Lovesey 1970; Balcar, Lovesey, and Wedgwood 1970) and impurity electron states (Lovesey 1978) and elastic scattering from magnetic salts (Balcar, Lovesey, and Wedgwood 1973; Balcar 1975). A theory suitable for relativistic calculations of atomic states has also been developed (Stassis and Deckman 1975, 1976*a,b*) and applied to metallic magnets (Stassis 1979). The authors state that in the nonrelativistic limit their theory contains contributions absent in the theory given here, but they are mistaken and the two theories are identical as should be expected. While most theories are directed toward a calculation of the neutron cross-section, and thus the Fourier transform of the magnetization density, it is useful, and in some instances highly desirable, to calculate the magnetization density profile directly (Brown and Forsyth 1981; Balcar 1975).

A full review of the theories mentioned in the preceding paragraph would take more than one chapter of reasonable length, and such a review would be somewhat out of line with the remainder of the book. The theory developed here uses Racah algebra and dovetails the spin and orbital contributions of the magnetic scattering amplitude. Use of the final formulae is illustrated in applications to several simple examples. Mindful of the reader's time and patience, the final, working formulae are gathered together in § 11.4, which can be consulted without reference to the lengthy derivations given in § 11.3. We will not discuss neutron scattering from magnetic molecules, such as O_2 (Meier and Helmholdt 1984) and the molecular anion O_2^- (Lines and Bösch 1983).

11.1. Introduction

It is clear from the outset that the magnetic neutron cross-section must depend on all the factors that determine the electronic state of an ion in a crystal, i.e. it must be sensitive to the crystal-field environment of the ion, the degree of covalency, and exchange interactions (Harrison 1980). The two latter effects may well be small. In nearly all cases of interest the unpaired electrons of an ion in a crystal couple according to Hund's rule (the cyanides form an example of salts where Hund's rule breaks down) and the neutrons are unlikely to have enough energy to break this coupling.

For transition-metal ions, which are characterized by having an incomplete 3d shell, the crystal field is a small perturbation compared with the intra-atomic electrostatic repulsion but large compared with the spin–orbit coupling (intermediate crystal-field case). Experiments show that the main crystal-field splittings in the transition-metal salts are of the order of 1 eV, whereas the spin–orbit coupling parameter is at most 0.1 eV. The ion is therefore described first in terms of its total orbital angular momentum L in the crystal field. If this field removes completely the orbital degeneracy of the ground-state wave function, then the expectation value of the orbital angular momentum operator is zero; the crystal field is said to quench the orbital moment. When the spin–orbit coupling is introduced, however, it mixes higher energy states into the ground state, with the result that the g value of the ion differs from its spin-only value $g = 2$. Physically, the spin, acting through the spin–orbit coupling, induces an orbital moment. For example, the orbital angular momentum of Ni^{2+} is completely quenched by a cubic crystal field but the measured value of g is 2.2, so that spin–orbit coupling gives rise to a 10 per cent orbital moment.

The behaviour of the rare earths in a crystal is somewhat simpler (Coqblin 1977). The mean radius of the 4f shell is smaller than the overall size of the ion and the perturbation of the crystal field is smaller than both the intra-atomic electrostatic repulsion between the electrons and the spin–orbit coupling (weak-field case). This means that the 'finished' ion, with L and S coupled to give a total angular momentum J, is examined in the environment of the crystal field, which has the effect of partially or totally removing the $(2J+1)$-fold degeneracy. The total spread in energy of these levels is found to be of the order 5–80 meV. If the incident neutron energy is much greater than this, the cross-section will be independent of the detailed structure of the crystal-field splittings. If, on the other hand, less energetic neutrons are used and an energy analysis is performed on the scattered neutrons, then it is possible to determine the energy-level splittings.

The amplitude of the magnetic scattering by an ion with both spin and orbital angular momentum is proportional to the Fourier transform of the total magnetization density. To prove this we need to express

$$\hat{\mathbf{Q}}_{\perp} = \sum_i \exp(\mathrm{i}\boldsymbol{\kappa}\cdot\mathbf{r}_i)\left\{\tilde{\boldsymbol{\kappa}}\times(\hat{\mathbf{s}}_i\times\tilde{\boldsymbol{\kappa}}) - \frac{\mathrm{i}}{\hbar\,|\boldsymbol{\kappa}|}\tilde{\boldsymbol{\kappa}}\times\hat{\mathbf{p}}_i\right\} = \tilde{\boldsymbol{\kappa}}\times(\hat{\mathbf{Q}}\times\tilde{\boldsymbol{\kappa}}) \quad (11.1)$$

(cf. (7.10)) in terms of the spin and orbital angular momentum density operators. The sum in (11.1) is over all unpaired electrons in the target sample.

The spin magnetization density operator $\hat{\mathbf{M}}_s(\mathbf{r})$ is defined through the relationship

$$\sum_i \exp(\mathrm{i}\boldsymbol{\kappa}\cdot\mathbf{r}_i)\hat{\mathbf{s}}_i = \int \mathrm{d}\mathbf{r}\exp(\mathrm{i}\boldsymbol{\kappa}\cdot\mathbf{r})\sum_i \hat{\mathbf{s}}_i\,\delta(\mathbf{r}-\mathbf{r}_i)$$

$$= -\frac{1}{2\mu_{\mathrm{B}}}\int \mathrm{d}\mathbf{r}\exp(\mathrm{i}\boldsymbol{\kappa}\cdot\mathbf{r})\hat{\mathbf{M}}_s(\mathbf{r}),$$

i.e.

$$\hat{\mathbf{M}}_s(\mathbf{r}) = -2\mu_{\mathrm{B}}\sum_i \hat{\mathbf{s}}_i\,\delta(\mathbf{r}-\mathbf{r}_i). \quad (11.2)$$

The orbital part can be expressed in terms of the orbital current density operator $\hat{\mathbf{j}}(\mathbf{r})$, where

$$\hat{\mathbf{j}}(\mathrm{r}) = -\frac{e}{2m_e}\sum_i \hat{\mathbf{p}}_i\,\delta(\mathbf{r}-\mathbf{r}_i) + \delta(\mathbf{r}-\mathbf{r}_i)\hat{\mathbf{p}}_i. \quad (11.3)$$

For

$$-\frac{m_e}{e}\int \mathrm{d}\mathbf{r}\exp(\mathrm{i}\boldsymbol{\kappa}\cdot\mathbf{r})\hat{\mathbf{j}}(\mathbf{r}) = \tfrac{1}{2}\int \mathrm{d}\mathbf{r}\exp(\mathrm{i}\boldsymbol{\kappa}\cdot\mathbf{r})\sum_i \hat{\mathbf{p}}_i\delta(\mathbf{r}-\mathbf{r}_i) + \delta(\mathbf{r}-\mathbf{r}_i)\hat{\mathbf{p}}_i$$

$$= \tfrac{1}{2}\sum_i \hat{\mathbf{p}}_i\exp(\mathrm{i}\boldsymbol{\kappa}\cdot\mathbf{r}_i) + \exp(\mathrm{i}\boldsymbol{\kappa}\cdot\mathbf{r}_i)\hat{\mathbf{p}}_i,$$

so that

$$-\frac{\mathrm{i}}{\hbar\,|\boldsymbol{\kappa}|}\tilde{\boldsymbol{\kappa}}\times\tfrac{1}{2}\sum_i \exp(\mathrm{i}\boldsymbol{\kappa}\cdot\mathbf{r}_i)\hat{\mathbf{p}}_i + \hat{\mathbf{p}}_i\exp(\mathrm{i}\boldsymbol{\kappa}\cdot\mathbf{r}_i)$$

$$= \frac{im_e}{\hbar e\,|\boldsymbol{\kappa}|}\tilde{\boldsymbol{\kappa}}\times\int \mathrm{d}\mathbf{r}\exp(\mathrm{i}\boldsymbol{\kappa}\cdot\mathbf{r})\hat{\mathbf{j}}(\mathbf{r}). \quad (11.4)$$

The classical mean current density $\mathbf{j}$ in a conductor carrying a constant (conduction) current density $\mathbf{j}'$ is given (Balcar 1975) by

$$\mathbf{j} = c \operatorname{curl} \mathbf{M} + \mathbf{j}' \tag{11.5}$$

where $\mathbf{M}$ is the magnetic moment density. The conduction–current density is proportional to the electric field and can, therefore, be expressed as the gradient of a scalar function $\phi(\mathbf{r})$, say. Clearly

$$\tilde{\boldsymbol{\kappa}} \times \int d\mathbf{r} \exp(i\boldsymbol{\kappa} \cdot \mathbf{r}) \boldsymbol{\nabla} \phi(\mathbf{r}) = 0,$$

so that the second term in (11.5) cannot contribute to the right-hand side of (11.4). We introduce an orbital magnetic moment density operator $\hat{\mathbf{M}}_{\mathrm{L}}$ by analogy with $\mathbf{M}$ in eqn (11.5), in terms of which

$$\begin{aligned}\int d\mathbf{r} \exp(i\boldsymbol{\kappa} \cdot \mathbf{r}) \hat{\mathbf{j}}(\mathbf{r}) &= c \int d\mathbf{r} \exp(i\boldsymbol{\kappa} \cdot \mathbf{r}) \operatorname{curl} \hat{\mathbf{M}}_{\mathrm{L}}(\mathbf{r}) \\ &= c \int d\mathbf{r} \operatorname{curl}\{\exp(i\boldsymbol{\kappa} \cdot \mathbf{r}) \hat{\mathbf{M}}_{\mathrm{L}}(\mathbf{r})\} \\ &\quad - c \int d\mathbf{r} \operatorname{grad}\{\exp(i\boldsymbol{\kappa} \cdot \mathbf{r})\} \times \hat{\mathbf{M}}_{\mathrm{L}}(\mathbf{r}).\end{aligned}$$

The first term integrates to zero, hence

$$\int d\mathbf{r} \exp(i\boldsymbol{\kappa} \cdot \mathbf{r}) \hat{\mathbf{j}}(\mathbf{r}) = -ic\boldsymbol{\kappa} \times \int d\mathbf{r} \exp(i\boldsymbol{\kappa} \cdot \mathbf{r}) \hat{\mathbf{M}}_{\mathrm{L}}(\mathbf{r}). \tag{11.6}$$

Whence, using (11.2), (11.4), and (11.6), we have

$$\begin{aligned}\hat{\mathbf{Q}}_{\perp} &= \sum_i \exp(i\boldsymbol{\kappa} \cdot \mathbf{r}_i) \left\{ \tilde{\boldsymbol{\kappa}} \times (\hat{\mathbf{s}}_i \times \tilde{\boldsymbol{\kappa}}) - \frac{i}{\hbar |\boldsymbol{\kappa}|} \tilde{\boldsymbol{\kappa}} \times \hat{\mathbf{p}}_i \right\} \\ &= \int d\mathbf{r} \exp(i\boldsymbol{\kappa} \cdot \mathbf{r}) \left\{ -\frac{1}{2\mu_{\mathrm{B}}} \tilde{\boldsymbol{\kappa}} \times \{\hat{\mathbf{M}}_{\mathrm{s}}(\mathbf{r}) \times \tilde{\boldsymbol{\kappa}}\} + \frac{im_{\mathrm{e}}}{\hbar e |\boldsymbol{\kappa}|} \tilde{\boldsymbol{\kappa}} \times (-ic)\{\boldsymbol{\kappa} \times \hat{\mathbf{M}}_{\mathrm{L}}(\mathbf{r})\} \right\} \\ &= -\frac{1}{2\mu_{\mathrm{B}}} \int d\mathbf{r} \exp(i\mathbf{k} \cdot \mathbf{r}) \tilde{\boldsymbol{\kappa}} \times \{\hat{\mathbf{M}}(\mathbf{r}) \times \tilde{\boldsymbol{\kappa}}\}\end{aligned} \tag{11.7}$$

where $\hat{\mathbf{M}}(\mathbf{r}) = \hat{\mathbf{M}}_{\mathrm{L}}(\mathbf{r}) + \hat{\mathbf{M}}_{\mathrm{s}}(\mathbf{r})$. This is the required result.

This result might have been anticipated since the neutrons are scattered through the interaction of their magnetic moment with the magnetic field arising from the unpaired electrons and this, in turn, is determined by their total magnetization. The fact that the scattering amplitude contains the Fourier transform of the magnetization density is a consequence of the Born approximation, for in this approximation the scattering amplitude is given by the Fourier transform of the interaction potential.

Let us now examine the precise form of the matrix elements of $\hat{\mathbf{Q}}_\perp$ taken between the target states $|\lambda\rangle$ and $|\lambda'\rangle$ that occur in the cross-section.

The allowed states of the unpaired electrons associated with the jth ion ($j = 1, 2, \ldots, rN$) are defined by wave functions $\phi_j, \phi'_j, \ldots$. The functions ϕ_i and ϕ_j are assumed orthogonal.

In the Hartree–Fock model the configurational part of the wave function is constructed from a simple product of the individual wave functions ϕ_j. In the simple version of this model it is assumed that these configurational wave functions are unaffected by the immediate environment of spin ordering. It follows immediately that the quantity we need for Bragg scattering

$$\sum_\lambda p_\lambda \langle\lambda|\, \hat{\mathbf{Q}}_\perp \,|\lambda\rangle \quad ,$$

involves simply the sum of single-atom contributions. Thus

$$\sum_\lambda p_\lambda \langle\lambda|\, \hat{\mathbf{Q}}_\perp \,|\lambda\rangle = \sum_l \exp(i\boldsymbol{\kappa}\cdot\mathbf{l}) \sum_d \exp(i\boldsymbol{\kappa}\cdot\mathbf{d}) \langle\phi_d|\, \hat{\mathbf{Q}}_\perp(d) \,|\phi_d\rangle_{\text{av}} \tag{11.8}$$

where the subscript av means that the average is taken over the occupied wave functions ϕ_d at the appropriate temperature. Since the configurational part of ϕ_d is always the same in this model, the averaging process is, in most cases, trivial. Thus for a ferromagnet the matrix element can be computed for the ground state and the averaging process simply introduces a factor $\langle\hat{\mathbf{S}}\rangle/S$.

For an antiferromagnet of simple type we need to compute

$$\langle\phi_d|\, \hat{\mathbf{Q}}_\perp(d) \,|\phi_d\rangle,$$

first assuming the spin on the atom is fully aligned and then multiplying by $\langle\hat{\mathbf{S}}\rangle/S$—which for an antiferromagnet is not unity even at absolute zero. For more complex spin-ordering patterns, helical arrangements for example, the precise nature of the averaging process must be thought out carefully; particularly when orbital moments are contributing to $\hat{\mathbf{Q}}_\perp$. Nevertheless in all cases the calculation follows two steps: first calculate the individual atom matrix elements; secondly average these in some way dictated by the thermodynamics of the total spin system of the target.

In the following sections we shall be concerned with the evaluation of the matrix elements $\langle\phi_d|\, \hat{\mathbf{Q}}_\perp(d) \,|\phi_d\rangle$. For the sake of brevity we will drop the explicit reference to the dth ion within a unit cell except where the omission is likely to cause confusion.

11.2. Small $|\boldsymbol{\kappa}|$ approximation

The evaluation of the matrix elements of $\hat{\mathbf{Q}}_\perp$ in the general case is complicated. However, we can obtain a simple result for small $|\boldsymbol{\kappa}|$, which is a meaningful approximation to examine since $|\boldsymbol{\kappa}|^{-1}$ is often much greater than the mean radius of the orbital wave function of the unpaired electrons.

For small $|\boldsymbol{\kappa}|$ and elastic scattering we can take

$$\exp(\mathrm{i}\boldsymbol{\kappa}\cdot\mathbf{r})\hat{\mathbf{p}}+\hat{\mathbf{p}}\exp(\mathrm{i}\boldsymbol{\kappa}\cdot\mathbf{r})\simeq \mathrm{i}\hbar(\hat{\mathbf{l}}\times\boldsymbol{\kappa}) \tag{11.9}$$

where $\hat{\mathbf{l}}$ is the orbital angular momentum operator of the electron with respect to the position of the parent nucleus. The proof of (11.9) is as follows.

On expanding the exponentials in the left-hand side of (11.9) we require to evaluate

$$(\boldsymbol{\kappa}\cdot\mathbf{r})\hat{\mathbf{p}}+\hat{\mathbf{p}}(\boldsymbol{\kappa}\cdot\mathbf{r}), \tag{11.10}$$

which we divide into two parts by both adding and subtracting

$$\tfrac{1}{2}\{(\boldsymbol{\kappa}\cdot\hat{\mathbf{p}})\mathbf{r}+\mathbf{r}(\boldsymbol{\kappa}\cdot\hat{\mathbf{p}})\}$$

to give

$$\tfrac{1}{2}\{(\boldsymbol{\kappa}\cdot\mathbf{r})\hat{\mathbf{p}}+\hat{\mathbf{p}}(\boldsymbol{\kappa}\cdot\mathbf{r})\}+\tfrac{1}{2}\{(\boldsymbol{\kappa}\cdot\mathbf{r})\hat{\mathbf{p}}+\hat{\mathbf{p}}(\boldsymbol{\kappa}\cdot\mathbf{r})\}$$
$$+\tfrac{1}{2}\{(\boldsymbol{\kappa}\cdot\hat{\mathbf{p}})\mathbf{r}+\mathbf{r}(\boldsymbol{\kappa}\cdot\hat{\mathbf{p}})\}-\tfrac{1}{2}\{(\boldsymbol{\kappa}\cdot\hat{\mathbf{p}})\mathbf{r}+\mathbf{r}(\boldsymbol{\kappa}\cdot\hat{\mathbf{p}})\}. \tag{11.11a}$$

We now choose to group the first and third terms together and the second and fourth terms together to give

$$\tfrac{1}{2}\{(\boldsymbol{\kappa}\cdot\mathbf{r})\hat{\mathbf{p}}+(\boldsymbol{\kappa}\cdot\hat{\mathbf{p}})\mathbf{r}+\hat{\mathbf{p}}(\boldsymbol{\kappa}\cdot\mathbf{r})+\mathbf{r}(\boldsymbol{\kappa}\cdot\hat{\mathbf{p}})\}$$
$$+\tfrac{1}{2}\{(\boldsymbol{\kappa}\cdot\mathbf{r})\hat{\mathbf{p}}+\hat{\mathbf{p}}(\boldsymbol{\kappa}\cdot\mathbf{r})-(\boldsymbol{\kappa}\cdot\hat{\mathbf{p}})\mathbf{r}-\mathbf{r}(\boldsymbol{\kappa}\cdot\hat{\mathbf{p}})\}. \tag{11.11b}$$

In the first line we write $\hat{\mathbf{p}}=m_e(\partial/\partial t)\mathbf{r}$ to get

$$m_e\,\partial_t\{(\boldsymbol{\kappa}\cdot\mathbf{r})\mathbf{r}\} \tag{11.11c}$$

and the second line becomes

$$\begin{aligned}(\boldsymbol{\kappa}\cdot\mathbf{r})\hat{\mathbf{p}}-\mathbf{r}(\boldsymbol{\kappa}\cdot\hat{\mathbf{p}})&=-\boldsymbol{\kappa}\times(\mathbf{r}\times\hat{\mathbf{p}})\\&=-\hbar\boldsymbol{\kappa}\times\hat{\mathbf{l}}.\end{aligned} \tag{11.11d}$$

Hence, inserting these results into the expansion of the left-hand side of (11.9) we get

$$\begin{aligned}\exp(\mathrm{i}\boldsymbol{\kappa}\cdot\mathbf{r})\hat{\mathbf{p}}+\hat{\mathbf{p}}\exp(\mathrm{i}\boldsymbol{\kappa}\cdot\mathbf{r})&\simeq m_e\,\partial_t\{2\mathbf{r}+\mathrm{i}(\boldsymbol{\kappa}\cdot\mathbf{r})\mathbf{r}\}+\mathrm{i}\hbar\hat{\mathbf{l}}\times\boldsymbol{\kappa}\\&=\mathrm{i}\hbar\hat{\mathbf{l}}\times\boldsymbol{\kappa}+\frac{m_e}{\mathrm{i}\hbar}[(2+\mathrm{i}\boldsymbol{\kappa}\cdot\mathbf{r})\mathbf{r},\hat{\mathscr{H}}]\end{aligned} \tag{11.12}$$

where the equality follows from using the equation-of-motion for a variable described by a Hamiltonian $\hat{\mathscr{H}}$.

For elastic scattering the second term is zero and it is often feasible to neglect it also for inelastic scattering (Lovesey 1978). We thus arrive at eqn (11.9).

With the use of this result, valid for $\boldsymbol{\kappa} \to 0$,

$$\hat{\mathbf{Q}}_\perp = \tilde{\boldsymbol{\kappa}} \times (\hat{\mathbf{s}} \times \tilde{\boldsymbol{\kappa}}) - \frac{\mathrm{i}}{2\hbar\,|\boldsymbol{\kappa}|} \tilde{\boldsymbol{\kappa}} \times (\mathrm{i}\hbar\hat{\mathbf{l}} \times \boldsymbol{\kappa}) = \tfrac{1}{2}\tilde{\boldsymbol{\kappa}} \times \{(\hat{\mathbf{l}} + 2\hat{\mathbf{s}}) \times \tilde{\boldsymbol{\kappa}}\} \tag{11.13}$$

and

$$2\hat{\mathbf{Q}} = \hat{\mathbf{l}} + 2\hat{\mathbf{s}}.$$

Thus, if the unpaired electrons of the ion couple to form a total spin $\hat{\mathbf{S}}$ and orbital angular momentum $\hat{\mathbf{L}}$, the elastic magnetic cross-section (per ion) is approximately given, for small $\boldsymbol{\kappa}$, by

$$\frac{\mathrm{d}\sigma}{\mathrm{d}\Omega} = (\tfrac{1}{2}r_0)^2 \sum_{\alpha,\beta} (\delta_{\alpha\beta} - \tilde{\kappa}_\alpha \tilde{\kappa}_\beta) \sum_\lambda p_\lambda \langle\lambda|\,(\hat{\mathbf{L}} + 2\hat{\mathbf{S}})_\alpha\,|\lambda\rangle\langle\lambda|\,(\hat{\mathbf{L}} + 2\hat{\mathbf{S}})_\beta\,|\lambda\rangle. \tag{11.14}$$

The evaluation of the matrix elements of $(\hat{\mathbf{L}} + 2\hat{\mathbf{S}})_\alpha$ that occur in (11.14) for the case of scattering by transition-metal ions requires, in general, a detailed evaluation of the wave functions of the ion in the presence of the crystal field. For those ions in which $\hat{\mathbf{S}}$ is approximately a constant of motion (e.g. Ni^{2+} and Cr^{3+} normally belong to this class), $\hat{\mathbf{L}}$ can be replaced by $(g-2)\hat{\mathbf{S}}$, by definition of the g value of the ion.

To discuss the rare earths we recall that, in the ground-state manifold,

$$\hat{\mathbf{L}} + 2\hat{\mathbf{S}} = g\hat{\mathbf{J}}, \tag{11.15}$$

where g is the Landé splitting factor

$$g = 1 + \frac{J(J+1) - L(L+1) + S(S+1)}{2J(J+1)} \tag{11.16}$$

Under the condition that the incident neutron energy is much greater than the spread in energy of the $2J+1$ levels, the static approximation is valid and the scattering is elastic. For a single, free ion the cross-section around the forward direction ($\boldsymbol{\kappa}$ small) is

$$\frac{\mathrm{d}\sigma}{\mathrm{d}\Omega} = (\tfrac{1}{2}r_0)^2 g^2 \overline{\hat{\mathbf{J}}_\perp \cdot \hat{\mathbf{J}}_\perp} \tag{11.17}$$

and we can usually take

$$\overline{\hat{\mathbf{J}}_\perp \cdot \hat{\mathbf{J}}_\perp} = \overline{\hat{\mathbf{J}}^2 - (\tilde{\boldsymbol{\kappa}} \cdot \hat{\mathbf{J}})^2} = \mathbf{J}^2 - \tfrac{1}{3}\mathbf{J}^2 = \tfrac{2}{3}J(J+1). \tag{11.18}$$

The cross-section (11.17) is in this instance entirely independent of crystal-field effects and also the temperature.

11.3. Matrix elements

We turn now to a development of expressions for the matrix elements of the spin and orbital contributions to the magnetic interaction operator, (11.1). For a single unpaired electron and small scattering vectors, the calculations involved are simple, and the required expressions can almost be written down by inspection. Electrons in s and p shells, and even two electrons in d and f shells, can be treated by a straightforward application of the methods set out by Condon and Shortley (1935).

However, magnetic neutron scattering is applied to the properties of 3d transition and rare-earth ions with many electrons, and for these cases the Condon and Shortley methods become very cumbersome. In consequence, we appeal to the more powerful methods of Racah algebra; a review of Racah methods germane to the following development is given in Chapter 5 of Condon and Odabaşi (1980) and many specific results of interest are derived in Judd (1963). Results for a single electron are obtained as a special case of our general expressions, which are summarized in § 11.4. It is assumed throughout this section that the many electrons belong to a single atomic shell, and share the same angular momentum, l. The more general case, in which electrons are promoted between different atomic shells in the scattering process, is discussed in Lovesey (1978) using the methods applied here.

Although we choose to present the developments for spin and orbital scattering separately, they are ultimately dovetailed together. In this respect, the development is fundamentally different from Trammell's, although, of course, matrix elements of observed variables (operators) are equivalent (Lovesey and Rimmer 1969), for there is a simple redundancy in Trammell's theory. Notwithstanding, Trammell's approach is very appealing; it is cast in Racah algebra methods in Balcar (1975). The latter reference also gives expressions for the matrix elements in real space, i.e. the magnetization density profile.

Our expressions for the matrix elements of the spin and orbital contributions are derived for the Russell–Saunders case, or LS-coupling scheme. In this case, the orbital angular momentum of the electrons combine to give the total orbital angular momentum L of the unfilled shell, and their spins to give the total spin S. The wave functions are eigenfunctions of the operators $\hat{\mathbf{L}}$, $\hat{\mathbf{S}}$, $\hat{\mathbf{J}}$, and $\hat{J}_z$. In a homogeneous magnetic field, each energy level splits into $(2J+1)$ Zeeman components with magnetic energies $\mu_B gM$ where g is the Landé splitting factor (eqn (11.16)) and M is the (magnetic) quantum number associated with $\hat{J}_z$.

The Russell–Saunders coupling scheme is adequate when relativistic effects in the Hamiltonian for the electrons are small. In practice, the scheme is a good approximation for the ground-state multiplet of rare-earth ions, and it forms the basic states for further approximations in which no restriction is placed on the magnitude of the spin–orbit interaction (Edelstein 1982).

11.3.1. Orbital term

We assume that all n electrons in the unfilled shell of orbital angular momentum l have the same radial wave function $f(r)$. Thus, the wave functions of the individual electrons are

$$|lm\rangle = Y^l_m(\tilde{\mathbf{r}})f(r), \tag{11.19}$$

where $Y^l_m(\tilde{\mathbf{r}})$ is a spherical harmonic of rank l and order m (Condon and Odabaşi 1980).

It is convenient to work in terms of the spherical components of a vector $\mathbf{r}$ (instead of the cartesian components) and, furthermore, it is convenient to define these components so that they look exactly like the spherical harmonics $Y^1_q(\tilde{\mathbf{r}})$. Hence we define

$$r_{\pm 1} = \mp \frac{1}{\sqrt{2}}(x \pm \mathrm{i}y), \qquad r_0 = z. \tag{11.20a}$$

Then we have

$$\tilde{r}_q = r_q/|\mathbf{r}| = \left(\frac{4\pi}{3}\right)^{1/2} Y^1_q(\tilde{\mathbf{r}}), \quad \text{for} \quad q = -1, 0, +1. \tag{11.20b}$$

It is well known that product eigenstates, say of orbital and spin angular momentum, labelled by L and S respectively, can be constructed using Clebsch–Gordan coefficients. Thus L and S can be combined to give a total state of angular momentum J, where J ranges from $|L-S|$ to $L+S$. By exact analogy to this we can combine together tensors of rank k_1 and k_2, say, to give one of rank K, using these same Clebsch–Gordan coefficients. The formula for doing this is simply

$$T^K_Q = \sum_{q_1, q_2} z^{k_1}_{q_1} y^{k_2}_{q_2} (k_1 q_1 k_2 q_2 \,|\, KQ) \tag{11.21}$$

where the spherical tensors are z^{k_1} and y^{k_2}, which combine to give the spherical tensor T^K_Q.

In actual calculations it is convenient to express the Clebsch–Gordan coefficient $(k_1 q_1 k_2 q_2 \,|\, KQ)$ in terms of the $3j$ symbol $\begin{pmatrix} k_1 & k_2 & K \\ q_1 & q_2 & -Q \end{pmatrix}$, which possesses a number of extremely useful symmetry properties. In terms of the $3j$ symbol the Clebsch–Gordan coefficient is

$$(k_1 q_1 k_2 q_2 \,|\, K, -Q) = (-1)^{k_2 - k_1 + Q}(2K+1)^{1/2} \begin{pmatrix} k_1 & k_2 & K \\ q_1 & q_2 & Q \end{pmatrix}. \tag{11.22}$$

The $3j$ symbol automatically equals zero unless both $q_1+q_2+Q=0$ and k_1, k_2, K satisfy the triangle conditions

$$k_1+k_2-K\geqslant 0, \qquad k_1-k_2+K\geqslant 0, \qquad -k_1+k_2+K\geqslant 0.$$

The sum k_1+k_2+K must be an integer.

The $3j$ symbol is invariant under an even permutation of the columns and is multiplied by $(-1)^{k_1+k_2+K}$ for an odd permutation. Further,

$$\begin{pmatrix} k_1 & k_2 & K \\ q_1 & q_2 & Q \end{pmatrix} = (-1)^{k_1+k_2+K}\begin{pmatrix} k_1 & k_2 & K \\ -q_1 & -q_2 & -Q \end{pmatrix} \tag{11.23}$$

and from this it follows as a special case that

$$\begin{pmatrix} k_1 & k_2 & K \\ 0 & 0 & 0 \end{pmatrix}$$

is zero unless k_1+k_2+K is an even integer.

For a single electron our immediate task is to find the matrix element

$$\langle lm|\exp(\mathrm{i}\boldsymbol{\kappa}\cdot\mathbf{r})(\tilde{\boldsymbol{\kappa}}\times\boldsymbol{\nabla})_q\,|lm'\rangle. \tag{11.24}$$

As an example of the above notation we obtain an alternative form for $\tilde{\boldsymbol{\kappa}}\times\boldsymbol{\nabla}$. We know this is a vector, i.e. a tensor of rank one, and hence by analogy with (11.21),

$$(\tilde{\boldsymbol{\kappa}}\times\boldsymbol{\nabla})_q = c\sum_{q_1,q_2}\tilde{\kappa}_{q_1}\nabla_{q_2}(1q_11q_2\,|\,1q).$$

To determine the constant of proportionality c we put $q=0$ and rewrite the left-hand side as

$$(\tilde{\boldsymbol{\kappa}}\times\boldsymbol{\nabla})_0 = (\tilde{\boldsymbol{\kappa}}\times\boldsymbol{\nabla})_z = \tilde{\kappa}_x\nabla_y - \tilde{\kappa}_y\nabla_x = -\mathrm{i}(\tilde{\kappa}_{-1}\nabla_{+1}-\tilde{\kappa}_{+1}\nabla_{-1}).$$

But the right-hand side becomes

$$c\{\tilde{\kappa}_{-1}\nabla_{+1}(1,-111\,|\,10)+\tilde{\kappa}_0\nabla_0(1010\,|\,10)+\tilde{\kappa}_{+1}\nabla_{-1}(111,-1\,|\,10) \\ = \frac{c}{\sqrt{2}}(\tilde{\kappa}_{-1}\nabla_{+1}-\tilde{\kappa}_{+1}\nabla_{-1}).$$

Hence, $c=-\mathrm{i}\sqrt{2}$ and we get

$$(\tilde{\boldsymbol{\kappa}}\times\boldsymbol{\nabla})_q = \mathrm{i}(-1)^{1+q}\sqrt{6}\sum_{q_1q_2}\tilde{\kappa}_{q_1}\boldsymbol{\nabla}_{q_2}\begin{pmatrix} 1 & 1 & 1 \\ q_1 & q_2 & -q \end{pmatrix}. \tag{11.25}$$

Before using this result in (11.24) we rearrange the latter as follows.

If we make use of the fact that

$$\int \mathrm{d}\mathbf{r}\,\boldsymbol{\nabla}\{fY^{*l}_m\exp(\mathrm{i}\boldsymbol{\kappa}\cdot\mathbf{r})fY^l_{m'}\}=0$$

and also

$$(\boldsymbol{\kappa}\times\boldsymbol{\nabla})\exp(\mathrm{i}\boldsymbol{\kappa}\cdot\mathbf{r})=0,$$

the matrix element (11.24) can be written

$$\langle lm|\exp(\mathrm{i}\boldsymbol{\kappa}\cdot\mathbf{r})(\tilde{\boldsymbol{\kappa}}\times\boldsymbol{\nabla})\,|lm'\rangle$$
$$=\tfrac{1}{2}\int\mathrm{d}\mathbf{r}\,\exp(\mathrm{i}\boldsymbol{\kappa}\cdot\mathbf{r})[fY^{*l}_{m}\{(\tilde{\boldsymbol{\kappa}}\times\boldsymbol{\nabla})fY^{l}_{m'}\}-fY^{l}_{m'}\{(\tilde{\boldsymbol{\kappa}}\times\boldsymbol{\nabla})fY^{*l}_{m}\}]$$
$$=\tfrac{1}{2}\int\mathrm{d}\mathbf{r}\,\exp(\mathrm{i}\boldsymbol{\kappa}\cdot\mathbf{r})f^{2}[Y^{*l}_{m}\{(\tilde{\boldsymbol{\kappa}}\times\boldsymbol{\nabla})\,Y^{l}_{m'}\}-Y^{l}_{m'}\{(\tilde{\boldsymbol{\kappa}}\times\boldsymbol{\nabla})\,Y^{*l}_{m}\}]. \qquad (11.26)$$

To proceed further we note that

$$\boldsymbol{\nabla}=\tfrac{1}{2}[\nabla^{2},\mathbf{r}] \qquad (11.27)$$

and

$$\nabla^{2}\{Y^{K}_{Q}(\tilde{\mathbf{r}})f(r)\}=Y^{K}_{Q}(\tilde{\mathbf{r}})\left\{\frac{1}{r}\frac{\mathrm{d}^{2}}{\mathrm{d}r^{2}}r-\frac{K(K+1)}{r^{2}}\right\}f(r), \qquad (11.28)$$

which enables us to rewrite (11.26) in the form

$$\langle lm|\exp(\mathrm{i}\boldsymbol{\kappa}\cdot\mathbf{r})(\tilde{\boldsymbol{\kappa}}\times\boldsymbol{\nabla})\,|lm'\rangle$$
$$=\tfrac{1}{4}\tilde{\boldsymbol{\kappa}}\times\int\mathrm{d}\mathbf{r}\,f^{2}\exp(\mathrm{i}\boldsymbol{\kappa}\cdot\mathbf{r})\{Y^{*l}_{m}\nabla^{2}(\mathbf{r}Y^{l}_{m'})-Y^{l}_{m'}\nabla^{2}(\mathbf{r}Y^{*l}_{m})\}. \qquad (11.29)$$

Now, from (11.20b),

$$r_{q}=r\tilde{r}_{q}=r\left(\frac{4\pi}{3}\right)^{1/2}Y^{1}_{q}(\tilde{\mathbf{r}}).$$

The product of two spherical harmonics with the same argument can be expressed as an expansion in single spherical harmonics, namely,

$$Y^{K}_{Q}(\tilde{\mathbf{r}})\,Y^{K'}_{Q'}(\tilde{\mathbf{r}})=\sum_{\bar{K}\bar{Q}}\left\{\frac{(2K+1)(2K'+1)(2\bar{K}+1)}{4\pi}\right\}^{1/2}$$
$$\times\begin{pmatrix}K & K' & \bar{K}\\ Q & Q' & \bar{Q}\end{pmatrix}Y^{*\bar{K}}_{\bar{Q}}(\tilde{\mathbf{r}})\begin{pmatrix}K & K' & \bar{K}\\ 0 & 0 & 0\end{pmatrix}. \qquad (11.30)$$

From this we obtain

$$\tilde{r}_{q}Y^{l}_{m'}(\tilde{\mathbf{r}})$$
$$=\sum_{Q}(-1)^{l+Q}\left\{l^{1/2}Y^{l-1}_{Q}(\tilde{\mathbf{r}})\begin{pmatrix}1 & l & l-1\\ q & m' & -Q\end{pmatrix}-(l+1)^{1/2}Y^{l+1}_{Q}(\tilde{\mathbf{r}})\begin{pmatrix}1 & l & l+1\\ q & m' & -Q\end{pmatrix}\right\}. \qquad (11.31)$$

Note that, with our phase convention,

$$Y_Q^{*K}(\tilde{\mathbf{r}}) = (-1)^Q Y_{-Q}^K(\tilde{\mathbf{r}}). \tag{11.32}$$

If we use (11.31) in conjunction with (11.28)

$$\nabla^2(r_q Y_{m'}^l) = \frac{1}{r} \sum_Q (-1)^{l+Q+1} \left\{ l^{1/2}(l+1)(l-2) Y_Q^{l-1}(\tilde{\mathbf{r}}) \begin{pmatrix} 1 & l & l-1 \\ q & m' & -Q \end{pmatrix} \right.$$
$$\left. - (l+1)^{1/2} l(l+3) Y_Q^{l+1}(\tilde{\mathbf{r}}) \begin{pmatrix} 1 & l & l+1 \\ q & m' & -Q \end{pmatrix} \right\}. \tag{11.33}$$

Hence,

$$Y_m^{*l} \nabla^2(r_q Y_{m'}^l) - Y_{m'}^l \nabla^2(r_q Y_m^{*l}),$$

as is required in (11.29),

$$= \frac{1}{r}(-1)^{l+1+m} \sum_Q (-1)^Q \left[Y_{-m}^l \left\{ l^{1/2}(l+1)(l-2) Y_Q^{l-1} \begin{pmatrix} 1 & l & l-1 \\ q & m' & -Q \end{pmatrix} \right.\right.$$
$$\left. - (l+1)^{1/2} l(l+3) Y_Q^{l+1} \begin{pmatrix} 1 & l & l+1 \\ q & m' & -Q \end{pmatrix} \right\}$$
$$- Y_{m'}^l \left\{ l^{1/2}(l+1)(l-2) Y_Q^{l-1} \begin{pmatrix} 1 & l & l-1 \\ q & -m & -Q \end{pmatrix} \right.$$
$$\left.\left. - (l+1)^{1/2} l(l+3) Y_Q^{l+1} \begin{pmatrix} 1 & l & l+1 \\ q & -m & -Q \end{pmatrix} \right\} \right]. \tag{11.34}$$

The next step is to multiply (11.34) by $f^2 \exp(\mathrm{i}\boldsymbol{\kappa} \cdot \mathbf{r})$ and integrate over $\mathbf{r}$. To separate the radial and angular parts of the integral, $\exp(\mathrm{i}\boldsymbol{\kappa} \cdot \mathbf{r})$ is expressed as

$$\exp(\mathrm{i}\boldsymbol{\kappa} \cdot \mathbf{r}) = 4\pi \sum_{K'=0}^{\infty} \sum_{Q'=-K'}^{K'} \mathrm{i}^{K'} j_{K'}(\kappa r) Y_{Q'}^{K'}(\tilde{\mathbf{r}}) Y_{Q'}^{*K'}(\tilde{\boldsymbol{\kappa}}). \tag{11.35}$$

Here $j_{K'}(x)$ is a spherical Bessel function of order K', and we note the relation

$$j_{K'}(x) = \frac{x}{(2K'+1)} \{ j_{K'-1}(x) + j_{K'+1}(x) \}. \tag{11.36}$$

From (11.29), (11.34), (11.35), and (11.36) the radial part of the integration is seen to be

$$\int_0^{\infty} r^2 \, \mathrm{d}r \frac{1}{r} f^2 j_{K'}(\kappa r) = \frac{|\boldsymbol{\kappa}|}{(2K'+1)} \{ \bar{j}_{K'-1}(\kappa) + \bar{j}_{K'+1}(\kappa) \}, \tag{11.37}$$

where the radial integrals $\bar{j}_K(\kappa)$ are defined as

$$\bar{j}_K(\kappa) = \int_0^{\infty} r^2 f^2 j_K(\kappa r) \, \mathrm{d}r. \tag{11.38}$$

Turning now to the angular part of the integration over $\mathbf{r}$ it is evident from (11.34) and (11.35) that this involves the integral of the product of three spherical harmonics. For this we have the result

$$\int \mathrm{d}\tilde{\mathbf{r}}\; Y^K_Q(\tilde{\mathbf{r}})\, Y^{K'}_{Q'}(\tilde{\mathbf{r}})\, Y^{K''}_{Q''}(\tilde{\mathbf{r}})$$

$$= \left\{\frac{(2K+1)(2K'+1)(2K''+1)}{4\pi}\right\}^{1/2} \begin{pmatrix} K & K' & K'' \\ 0 & 0 & 0 \end{pmatrix} \begin{pmatrix} K & K' & K'' \\ Q & Q' & Q'' \end{pmatrix}. \tag{11.39}$$

Thus, on multiplying (11.34) by $Y^{K'}_{Q'}(\tilde{\mathbf{r}})$, which occurs in (11.35), and using (11.39),

$$\int \mathrm{d}\tilde{\mathbf{r}}\; Y^{K'}_{Q'}\{Y^{*l}_{m}\nabla^2(r_q Y^l_{m'}) - Y^l_{m'}\nabla^2(r_q Y^{*l}_{m})\}$$

$$= \frac{1}{r}(-1)^{l+1+m}\left\{\frac{(2K'+1)(2l+1)}{4\pi}\right\}^{1/2} \sum_Q (-1)^Q \Bigg[(l+1)(l-2)\{l(2l-1)\}^{1/2}$$

$$\times \begin{pmatrix} K' & l & l-1 \\ 0 & 0 & 0 \end{pmatrix} \Bigg\{ \begin{pmatrix} 1 & l & l-1 \\ q & m' & Q \end{pmatrix} \begin{pmatrix} K' & l & l-1 \\ Q' & -m & -Q \end{pmatrix}$$

$$- \begin{pmatrix} 1 & l & l-1 \\ q & -m & Q \end{pmatrix} \begin{pmatrix} K' & l & l-1 \\ Q' & m' & -Q \end{pmatrix} \Bigg\} - l(l+3)\{(l+1)(2l+3)\}^{1/2}$$

$$\times \begin{pmatrix} K' & l & l+1 \\ 0 & 0 & 0 \end{pmatrix} \Bigg\{ \begin{pmatrix} 1 & l & l+1 \\ q & m' & Q \end{pmatrix} \begin{pmatrix} K' & l & l+1 \\ Q' & -m & -Q \end{pmatrix}$$

$$- \begin{pmatrix} 1 & l & l+1 \\ q & -m & Q \end{pmatrix} \begin{pmatrix} K' & l & l+1 \\ Q' & m' & -Q \end{pmatrix} \Bigg\}\Bigg]. \tag{11.40}$$

We now note that (11.40) can be greatly simplified by using the relation

$$\sum_{m_3} \begin{pmatrix} j_1 & j_2 & j_3 \\ m_1 & m_2 & m_3 \end{pmatrix} \begin{pmatrix} l_1 & l_2 & j_3 \\ n_1 & n_2 & -m_3 \end{pmatrix}$$

$$= \sum_{l_3 n_3} (-1)^{l_3+l_3+m_1+n_1}(2l_3+1) \begin{Bmatrix} j_1 & j_2 & j_3 \\ l_1 & l_2 & l_3 \end{Bmatrix} \begin{pmatrix} l_1 & j_2 & l_3 \\ n_1 & m_2 & n_3 \end{pmatrix} \begin{pmatrix} j_1 & l_2 & l_3 \\ m_1 & n_2 & -n_3 \end{pmatrix}, \tag{11.41}$$

where $\begin{Bmatrix} j_1 & j_2 & j_3 \\ l_1 & l_2 & l_3 \end{Bmatrix}$ is a $6j$ symbol. The latter is automatically zero unless j_1, j_2, j_3, l_1, l_2, and l_3 satisfy the four triangle conditions represented by

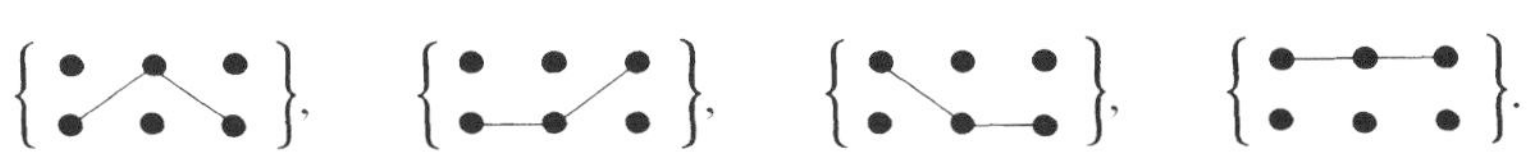

$$\begin{Bmatrix} \bullet & \bullet & \bullet \\ \bullet & \bullet & \bullet \end{Bmatrix}, \quad \begin{Bmatrix} \bullet & \bullet & \bullet \\ \bullet & \bullet & \bullet \end{Bmatrix}, \quad \begin{Bmatrix} \bullet & \bullet & \bullet \\ \bullet & \bullet & \bullet \end{Bmatrix}, \quad \begin{Bmatrix} \bullet & \bullet & \bullet \\ \bullet & \bullet & \bullet \end{Bmatrix}.$$

Further, the $6j$ symbol is invariant under either an even or odd permutation of its columns. It also possesses the symmetry

$$\begin{Bmatrix} j_1 & j_2 & j_3 \\ l_1 & l_2 & l_3 \end{Bmatrix} = \begin{Bmatrix} j_1 & l_2 & l_3 \\ l_1 & j_2 & j_3 \end{Bmatrix}.$$

On making use of (11.41) to perform the sum over Q in (11.40) we obtain for the latter

$$\int \mathrm{d}\tilde{\mathbf{r}}\; Y^{K'}_{Q'}\{Y^{*l}_{m}\nabla^2(r_q Y^l_{m'}) - Y^l_{m'}\nabla^2(r_q Y^{*l}_{m})\}$$
$$= \frac{1}{r}\left\{\frac{(2K'+1)}{4\pi}\right\}^{1/2}(2l+1)\sum_{KQ}(2K+1)(-1)^{K+m'}$$
$$\times\begin{pmatrix} K' & 1 & K \\ Q' & q & Q \end{pmatrix}\left\{\begin{pmatrix} l & l & K \\ m' & -m & -Q \end{pmatrix} - \begin{pmatrix} l & l & K \\ -m & m' & -Q \end{pmatrix}\right\}$$
$$\times\{(l+1)(l-2)(-1)^l B - l(l+3)(-1)^{l-1}A\} \tag{11.42}$$

where

$$A(K, K', l) = (-1)^{l-1}\left\{\frac{(l+1)(2l+3)}{(2l+1)}\right\}^{1/2}\begin{pmatrix} l & K' & l+1 \\ 0 & 0 & 0 \end{pmatrix}\begin{Bmatrix} l & 1 & l+1 \\ K' & l & K \end{Bmatrix} \tag{11.43}$$

and

$$B(K, K', l) = (-1)^l\left\{\frac{l(2l-1)}{(2l+1)}\right\}^{1/2}\begin{pmatrix} l & K' & l-1 \\ 0 & 0 & 0 \end{pmatrix}\begin{Bmatrix} l & 1 & l-1 \\ K' & l & K \end{Bmatrix}.$$

Because A and B contain the $3j$ symbols

$$\begin{pmatrix} l & K' & l\pm 1 \\ 0 & 0 & 0 \end{pmatrix},$$

they are both non-zero only if K' is an odd integer. The triangle condition on K and K' requires

$$K = K' \quad \text{or} \quad K = K' \pm 1.$$

However,

$$\begin{pmatrix} l & l & K \\ m' & -m & -Q \end{pmatrix} - \begin{pmatrix} l & l & K \\ -m & m' & -Q \end{pmatrix} = \begin{pmatrix} l & l & K \\ m' & -m & -Q \end{pmatrix}\{1-(-1)^{2l+K}\}$$

and this is zero unless K is odd. Hence $K = K'$.

A direct calculation shows that

$$A(K', K', l) = -B(K', K', l)$$

with

$$A(K', K', l) = (-1)^{(K'-1)/2} \frac{1}{2(2l+1)}$$

$$\times \left\{ \frac{2l+1-K'}{2K'+1} \right\}^{1/2} \frac{\{\frac{1}{2}(2l+1+K')\}!}{\{\frac{1}{2}(K'-1)\}!\,\{\frac{1}{2}(K'+1)\}!\,\{\frac{1}{2}(2l+1-K')\}!}$$

$$\times \left\{ \frac{(K'-1)!\,(K'+1)!\,(2l+1-K')!}{(2l+1+K')!} \right\}^{1/2}. \tag{11.44}$$

Thus we obtain from (11.42) the result

$$\int \mathrm{d}\tilde{\mathbf{r}}\, Y^{K'}_{Q'}\{Y^{*l}_{m}\nabla^2(r_q Y^{l}_{m'}) - Y^{l}_{m'}\nabla^2(r_q Y^{*l}_{m})\}$$

$$= \frac{4}{r}\left(\frac{1}{4\pi}\right)^{1/2} \{(2K'+1)(2l+1)\}^{3/2} A(K', K', l)$$

$$\times \sum_{Q} (-1)^{Q} (K'Qlm' \mid lm) \begin{pmatrix} K' & 1 & K' \\ Q' & q & -Q \end{pmatrix}. \tag{11.45}$$

From (11.25), (11.29), (11.35), and (11.37), together with (11.45), we obtain

$$\langle lm|\exp(\mathrm{i}\boldsymbol{\kappa}\cdot\mathbf{r})(\tilde{\boldsymbol{\kappa}}\times\boldsymbol{\nabla})_q\,|lm'\rangle$$

$$= \tfrac{1}{4}|\boldsymbol{\kappa}|\,(4\pi)\sqrt{6}(-1)^{1+q} \sum_{q_1 q_2} \begin{pmatrix} 1 & 1 & 1 \\ q_1 & q_2 & -q \end{pmatrix} \sum_{K'Q'} \mathrm{i}^{K'+1} \frac{1}{(2K'+1)}$$

$$\times (\bar{j}_{K'-1} + \bar{j}_{K'+1})\tilde{\kappa}_{q_1} Y^{*K'}_{Q'}(\tilde{\boldsymbol{\kappa}}) \int \mathrm{d}\tilde{\mathbf{r}}\, Y^{K'}_{Q'}\{Y^{*l}_{m}\nabla^2(r_{q_2} Y^{l}_{m'}) - Y^{l}_{m'}\nabla^2(r_{q_2} Y^{*l}_{m})\}$$

$$= |\boldsymbol{\kappa}|\,(4\pi)^{1/2}\sqrt{6}(-1)^{1+q}(2l+1)^{3/2} \sum_{K'Q} \sum_{\bar{K}\bar{Q}} \mathrm{i}^{K'+1}(\bar{j}_{K'-1} + \bar{j}_{K'+1})(2K'+1)$$

$$\times (-1)^{Q+\bar{Q}} A(K', K', l)(K'Qlm' \mid lm)(2\bar{K}+1)^{1/2} Y^{\bar{K}}_{\bar{Q}}(\tilde{\boldsymbol{\kappa}}) \begin{pmatrix} 1 & K' & \bar{K} \\ 0 & 0 & 0 \end{pmatrix}$$

$$\times \sum_{q_1 q_2 Q'} (-1)^{Q'} \begin{pmatrix} 1 & 1 & 1 \\ q_1 & q_2 & -q \end{pmatrix} \begin{pmatrix} K' & 1 & K' \\ Q' & q_2 & -Q \end{pmatrix} \begin{pmatrix} 1 & K' & \bar{K} \\ q_1 & -Q' & \bar{Q} \end{pmatrix} \tag{11.46}$$

where in obtaining the last line we have made use of (11.20b) and (11.30).

The sums over q_1, q_2, and Q' that involve the product of the last three $3j$ symbols on the right-hand side of (11.46) can be replaced by the product of a $3j$ and a $6j$ symbol through the relation

$$\begin{pmatrix} j_1 & j_2 & j_3 \\ m_1 & m_2 & m_3 \end{pmatrix}\begin{Bmatrix} j_1 & j_2 & j_3 \\ l_1 & l_2 & l_3 \end{Bmatrix} = \sum_{\text{all } n} (-1)^{l_1+l_2+l_3+n_1+n_2+n_3}$$

$$\times \begin{pmatrix} j_1 & l_2 & l_3 \\ m_1 & n_2 & -n_3 \end{pmatrix}\begin{pmatrix} l_1 & j_2 & l_3 \\ -n_1 & m_2 & n_3 \end{pmatrix}\begin{pmatrix} l_1 & l_2 & j_3 \\ n_1 & -n_2 & m_3 \end{pmatrix}. \quad (11.47)$$

If in addition to using (11.47) we make the change in notation $Q \to Q'$ and $\bar{K}, \bar{Q} \to K, Q$, respectively, then we have finally from (11.46) the result

$$\langle lm|\exp(\mathrm{i}\boldsymbol{\kappa}\cdot\mathbf{r})(\tilde{\boldsymbol{\kappa}}\times\boldsymbol{\nabla})_q|lm'\rangle$$

$$= -|\boldsymbol{\kappa}|(8\pi)^{1/2}(2l+1)^{3/2}\sum_{KQ}\sum_{K'Q'}\mathrm{i}^{K'+1}Y^K_Q(\tilde{\boldsymbol{\kappa}})(\bar{j}_{K'-1}+\bar{j}_{K'+1})$$

$$\times(2K'+1)(2K+1)^{1/2}A(K',K',l)\begin{pmatrix} 1 & K & K' \\ 0 & 0 & 0 \end{pmatrix}\begin{Bmatrix} 1 & 1 & 1 \\ K' & K & K' \end{Bmatrix}$$

$$\times(K'Q'lm'\mid lm)(KQK'Q'\mid 1q). \quad (11.48)$$

We recall that the integer $K' = 1, 3, \ldots, (2l-1)$. The $3j$ symbol

$$\begin{pmatrix} 1 & K & K' \\ 0 & 0 & 0 \end{pmatrix}$$

is non-zero only if K is even and therefore $K = K' \pm 1$, i.e. $K = 0, 2, \ldots, 2l$.

The result (11.48) applies to the case of a single electron, but it is not difficult to generalize it to the configuration of l^n, the n electrons coupling to give a total spin S and orbital angular momentum L. Since Hund's rule for the coupling is assumed to apply, $2S$ is either the number of electrons or the number of holes, $2(2l+1)-n$, whichever is the smaller.

In general, if the operator $\hat{F}$ is the sum of single-particle operators $\hat{f}$, then

$$\langle l^n vSM_{\mathrm{S}}LM_{\mathrm{L}}|\hat{F}|l^n v'S'M'_{\mathrm{S}}L'M'_{\mathrm{L}}\rangle$$

$$= n\sum_{\bar{\theta}}(\theta\{|\bar{\theta})(\theta'\{|\bar{\theta})\sum_{m,\bar{M}_{\mathrm{L}},m'}(\bar{L}\bar{M}_{\mathrm{L}}lm\mid LM_{\mathrm{L}})(\bar{L}\bar{M}_{\mathrm{L}}lm'\mid L'M'_{\mathrm{L}})$$

$$\times\sum_{m_{\mathrm{s}},\bar{M}_{\mathrm{S}},m'_{\mathrm{s}}}(\bar{S}\bar{M}_{\mathrm{S}}sm_{\mathrm{s}}\mid SM_{\mathrm{S}})(\bar{S}\bar{M}_{\mathrm{S}}sm'_{\mathrm{s}}\mid S'M'_{\mathrm{S}})\langle sm_{\mathrm{s}}lm|\hat{f}|sm'_{\mathrm{s}}lm'\rangle. \quad (11.49)$$

The notation used in eqn (11.49) is as follows: the symbol v stands for any other quantum numbers that are needed when the set $SM_{\mathrm{S}}LM_{\mathrm{L}}$ fails to define the state uniquely; θ, θ', and $\bar{\theta}$ are shorthand for

$$\theta \equiv vSL, \qquad \theta' \equiv v'S'L', \qquad \bar{\theta} \equiv \bar{v}\bar{S}\bar{L},$$

and it is to be understood that θ and θ' define terms of l^n, while $\bar{\theta}$ defines a term of l^{n-1}. $(\theta\{|\bar{\theta})$ are Racah's coefficients of fractional parentage and are purely real (Condon and Odabaşi 1980; Nielson and Koster 1963).

In (11.48) the dependence of $\langle lm|\exp(\mathrm{i}\boldsymbol{\kappa}\cdot\mathbf{r})(\tilde{\boldsymbol{\kappa}}\times\boldsymbol{\nabla})_q|lm'\rangle$ on the quantum numbers m and m' is contained solely in the Clebsch–Gordan coefficient $(K'Q'lm'|lm)$. If we make use of (11.49), then

$$\left\langle vSM_sLM_L\right|\sum_{\text{electrons}}\exp(\mathrm{i}\boldsymbol{\kappa}\cdot\mathbf{r})(\tilde{\boldsymbol{\kappa}}\times\boldsymbol{\nabla})_q\left|v'S'M_s'L'M_L'\right\rangle \tag{11.50}$$

is obtained from (11.48) merely by replacing $(K'Q'lm'|lm)$ by

$$\sqrt{\{(2l+1)(2L+1)(2L'+1)\}}\,\delta_{S,S'}\,\delta_{M_s,M_s'}(-)^{l-K'+Q'-L+L'-M_L'}$$
$$\times\begin{pmatrix}L' & K' & L\\ M_L' & Q' & -M_L\end{pmatrix} n\sum_{\bar{\theta}}(-)^{\bar{L}}(\theta\{|\bar{\theta})(\theta'\{|\bar{\theta})\begin{Bmatrix}L' & K' & L\\ l & \bar{L} & l\end{Bmatrix}. \tag{11.51}$$

If $J=L\pm S$ is a good quantum number, then we need to calculate

$$\left\langle\theta JM\right|\sum_{\text{electrons}}\exp(\mathrm{i}\boldsymbol{\kappa}\cdot\mathbf{r})(\tilde{\boldsymbol{\kappa}}\times\boldsymbol{\nabla})_q\left|\theta'J'M'\right\rangle. \tag{11.52}$$

To do this it is only necessary to multiply (11.50) by $(SM_sLM_L|JM)$ and $(S'M_s'L'M_L'|J'M')$ and sum over M_s, M_s', M_L, and M_L'. The result is that (11.52) is obtained from (11.48) by replacing $(K'Q'lm'|lm)$ by

$$\sqrt{\{(2l+1)(2L+1)(2L'+1)(2J'+1)\}}\,\delta_{S,S'}(-)^{S+l+J'+L+L'}$$
$$\times\begin{Bmatrix}J' & K' & J\\ L & S & L'\end{Bmatrix} n\sum_{\bar{\theta}}(-)^{\bar{L}}(\theta\{|\bar{\theta})(\theta'\{|\bar{\theta})\begin{Bmatrix}L' & K' & L\\ l & \bar{L} & l\end{Bmatrix}(K'Q'J'M'|JM). \tag{11.53}$$

Now combine (11.48) and (11.53) and write (11.52) in the form

$$\left\langle\theta JM\right|\sum_{\text{electrons}}\exp(\mathrm{i}\boldsymbol{\kappa}\cdot\mathbf{r})(\tilde{\boldsymbol{\kappa}}\times\nabla)_q\left|\theta'J'M'\right\rangle$$
$$=-(4\pi)^{1/2}|\boldsymbol{\kappa}|\sum_{K,K'}A(K,K')\sum_{Q,Q'}Y_Q^K(\tilde{\boldsymbol{\kappa}})(K'Q'J'M'|JM)(KQK'Q'|1q). \tag{11.54}$$

In eqn (11.54),

$$A(K,K')=(2l+1)^2\sqrt{\{2(2L+1)(2L'+1)(2J'+1)\}}\,\delta_{S,S'}(-)^{S+l+J'+L+L'}$$
$$\times\mathrm{i}^{K'+1}(2K'+1)(2K+1)^{1/2}\begin{pmatrix}1 & K & K'\\ 0 & 0 & 0\end{pmatrix}\begin{Bmatrix}1 & 1 & 1\\ K' & K & K'\end{Bmatrix}$$
$$\times A(K',K',l)\{\bar{j}_{K'-1}+\bar{j}_{K'+1}\}\begin{Bmatrix}J' & K' & J\\ L & S & L'\end{Bmatrix}$$
$$\times n\sum_{\bar{\theta}}(-)^{\bar{L}}(\theta\{|\bar{\theta})(\theta'\{|\bar{\theta})\begin{Bmatrix}L' & K' & L\\ l & \bar{L} & l\end{Bmatrix} \tag{11.55}$$

and it is a simple matter to show that

$$\frac{A(K'-1, K')}{A(K'+1, K')} = \left(\frac{K'+1}{K'}\right)^{1/2}. \tag{11.56}$$

11.3.2. Spin term

If we write

$$\hat{\mathbf{Q}} = \sum_i \exp(\mathrm{i}\boldsymbol{\kappa}\cdot\mathbf{r}_i)\left(\hat{\mathbf{s}}_i - \frac{1}{|\boldsymbol{\kappa}|}\tilde{\boldsymbol{\kappa}}\times\boldsymbol{\nabla}_i\right) + (\text{any function proportional to } \tilde{\boldsymbol{\kappa}}), \tag{11.57}$$

then the interaction operator

$$\hat{\mathbf{Q}}_\perp = \tilde{\boldsymbol{\kappa}}\times(\hat{\mathbf{Q}}\times\tilde{\boldsymbol{\kappa}})$$

and

$$\hat{\mathbf{Q}}_\perp \cdot \hat{\mathbf{Q}}_\perp = \sum_{\alpha,\beta} (\delta_{\alpha,\beta} - \tilde{\kappa}_\alpha\tilde{\kappa}_\beta)\hat{Q}^\alpha\hat{Q}^\beta, \tag{11.58}$$

with $\alpha, \beta = x, y,$ or z.

In the previous section we evaluated the matrix elements of the orbital contribution to $\hat{\mathbf{Q}}$ and we turn now to the evaluation of the spin part

$$\sum_i \exp(\mathrm{i}\boldsymbol{\kappa}\cdot\mathbf{r}_i)\hat{\mathbf{s}}_i.$$

We emphasize that $\hat{\mathbf{Q}}$ is an intermediate operator inasmuch as the cross-section involves $\hat{\mathbf{Q}}_\perp$. However, $\hat{\mathbf{Q}}$ has the conceptual value that it enables us to retain the same form for the cross-section in the general case as in the spin-only case and, in addition, it is easier to use in calculations. Note that it is defined only to within any function that is proportional to the scattering vector, a property that we shall later use to simplify the algebra.

To evaluate the matrix elements of $\sum \exp(\mathrm{i}\boldsymbol{\kappa}\cdot\mathbf{r})\hat{\mathbf{s}}$ we again make use of (11.35) to expand $\exp(\mathrm{i}\boldsymbol{\kappa}\cdot\mathbf{r})$ and obtain

$$\begin{aligned}\exp(\mathrm{i}\boldsymbol{\kappa}\cdot\mathbf{r})\hat{s}_q &= 4\pi \sum_{K,Q} \mathrm{i}^K j_K(\kappa r) Y_Q^{*K}(\tilde{\boldsymbol{\kappa}}) Y_Q^K(\tilde{\mathbf{r}})\hat{s}_q \\ &= (4\pi)^{1/2} \sum_{K,Q}\sum_{K',Q'} \mathrm{i}^K Y_Q^{*K}(\tilde{\boldsymbol{\kappa}}) f_{Q'}^{K'}(1qKQ \mid K'Q')\end{aligned} \tag{11.59}$$

where the tensor $f_{Q'}^{K'}$ is defined by

$$f_{Q'}^{K'} = (4\pi)^{1/2} j_K(\kappa r) \sum_{q,\bar{Q}} \hat{s}_{\bar{q}} Y_{\bar{Q}}^K(\tilde{\mathbf{r}})(1\bar{q}K\bar{Q} \mid K'Q'). \tag{11.60}$$

In rewriting the spin term in this way we have utilized the relation

$$\sum_{K',Q'} (2K'+1)\begin{pmatrix} 1 & K & K' \\ \bar{q} & \bar{Q} & -Q' \end{pmatrix}\begin{pmatrix} 1 & K & K' \\ q & Q & -Q' \end{pmatrix} = \delta_{q,\bar{q}}\,\delta_{Q,\bar{Q}}. \quad (11.61)$$

Since $f^{K'}_{Q'}$ is a spherical tensor, of rank K', the Wigner-Eckart theorem applies to its matrix elements, i.e.

$$\langle \theta JM|\sum f^{K'}_{Q'}|\theta' J'M'\rangle = (-1)^{J-M}\begin{pmatrix} J & K' & J' \\ -M & Q' & M' \end{pmatrix}[\theta J\,\|\textstyle\sum f^{K'}\|\,\theta' J']. \quad (11.62)$$

Here $[\theta J\,\|\sum f^{K'}\|\,\theta' J']$ is called a reduced matrix element and it is independent of the quantum numbers M and M'.

In evaluating this reduced matrix element it is convenient to obtain some general results for the matrix elements of a mixed tensor, $z^{k_1}_{q_1}y^{k_2}_{q_2}$, where z acts only on the spin part of the wave function and y on the orbital part.

The Wigner–Eckart theorem for double tensors is

$$\langle vSM_SLM_L|\, z^{k_1}_{q_1}y^{k_2}_{q_2}\,|v'S'M'_SL'M'_L\rangle$$

$$= (-)^{S-M_S+L-M_L}\begin{pmatrix} S & k_1 & S' \\ -M_S & q_1 & M'_S \end{pmatrix}\begin{pmatrix} L & k_2 & L' \\ -M_L & q_2 & M'_L \end{pmatrix}[\theta\,\|z^{k_1}y^{k_2}\|\,\theta']. \quad (11.63)$$

From (11.47) and (11.49),

$$\sum_{\text{electrons}} [\theta\,\|z^{k_1}y^{k_2}\|\,\theta']$$

$$= (-)^{l+s+L+S+k_1+k_2}\sqrt{\{(2S+1)(2S'+1)(2L+1)(2L'+1)\}}[s\,\|z^{k_1}\|\,s][l\,\|y^{k_2}\|\,l]$$

$$\times n\sum_{\bar{\theta}}(\theta\{|\bar{\theta})(\theta'\{|\bar{\theta})(-)^{\bar{S}+\bar{L}}\begin{Bmatrix} S' & k_1 & S \\ s & \bar{S} & s \end{Bmatrix}\begin{Bmatrix} L' & k_2 & L \\ l & \bar{L} & l \end{Bmatrix}. \quad (11.64)$$

If

$$F^{K'}_{Q'} = \sum_{\text{electrons}}\sum_{q_1,q_2} z^{k_1}_{q_1}y^{k_2}_{q_2}(k_1q_1k_2q_2\,|\,K'Q'), \quad (11.65)$$

then (cf. eqn (11.62))

$$\langle \theta JM|\, F^{K'}_{Q'}\,|\theta' J'M'\rangle = (-)^{J-M}\begin{pmatrix} J & K' & J' \\ -M & Q' & M' \end{pmatrix}[\theta J\,\|F^{K'}\|\,\theta' J'] \quad (11.66)$$

by the Wigner-Eckart theorem. Hence, on using (11.64),

$$[\theta J \| F^{K'} \| \theta' J'] = \sum_{M,M',Q'} (-)^{M-J} \begin{pmatrix} J & K' & J' \\ -M & Q' & M' \end{pmatrix}$$

$$\times \sum_{M_s,M_L} \sum_{M'_s,M'_L} (SM_s LM_L \mid JM)(S'M'_s L'M'_L \mid J'M')$$

$$\times \sum_{\text{electrons}} \sum_{q_1,q_2} (k_1 q_1 k_2 q_2 \mid K'Q') \langle vSM_s LM_L | z^{k_1}_{q_1} y^{k_2}_{q_2} | v'S'M'_s L'M'_L \rangle$$

$$= \sqrt{\{(2K'+1)(2S+1)(2S'+1)(2L+1)(2L'+1)(2J+1)(2J'+1)\}}$$

$$\times (-)^{L+k_1+k_2+l+s+S} [s \| z^{k_1} \| s][l \| y^{k_2} \| l]$$

$$\times n \sum_{\bar{\theta}} (\theta \{ | \bar{\theta})(\theta' \{ | \bar{\theta})(-)^{\bar{S}+\bar{L}} \begin{Bmatrix} S' & k_1 & S \\ s & \bar{S} & s \end{Bmatrix} \begin{Bmatrix} L' & k_2 & L \\ l & \bar{L} & l \end{Bmatrix}$$

$$\times \begin{Bmatrix} S & S' & k_1 \\ L & L' & k_2 \\ J & J' & K' \end{Bmatrix}. \tag{11.67}$$

In arriving at the last line of (11.67), we have used the relation

$$\begin{Bmatrix} j_{11} & j_{12} & j_{13} \\ j_{21} & j_{22} & j_{23} \\ j_{31} & j_{32} & j_{33} \end{Bmatrix} = \sum_{\text{all } m} \begin{pmatrix} j_{11} & j_{12} & j_{13} \\ m_{11} & m_{12} & m_{13} \end{pmatrix} \begin{pmatrix} j_{21} & j_{22} & j_{23} \\ m_{21} & m_{22} & m_{23} \end{pmatrix}$$
$$\times \begin{pmatrix} j_{31} & j_{32} & j_{33} \\ m_{31} & m_{32} & m_{33} \end{pmatrix} \begin{pmatrix} j_{11} & j_{21} & j_{31} \\ m_{11} & m_{21} & m_{31} \end{pmatrix}$$
$$\times \begin{pmatrix} j_{12} & j_{22} & j_{32} \\ m_{12} & m_{22} & m_{32} \end{pmatrix} \begin{pmatrix} j_{13} & j_{23} & j_{33} \\ m_{13} & m_{23} & m_{33} \end{pmatrix} \tag{11.68}$$

to reduce the sum over the six $3j$ symbols to a $9j$ symbol

$$\begin{Bmatrix} S & S' & k_1 \\ L & L' & k_2 \\ J & J' & K' \end{Bmatrix}.$$

From (11.68) it is clear that the arguments of each row and column of a $9j$ symbol must satisfy a triangle condition. The $9j$ symbol is invariant under an even permutation of either rows or columns and is multiplied by a factor $(-1)^N$, where N is the sum of the nine parameters, for an odd permutation. It is also invariant under reflection about either diagonal.

We now set

$$z^{k_1}_{q_1} = \hat{s}_{q_1} \quad (k_1 = 1)$$

and

$$y^{k_2}_{q_2} = Y^K_Q(\hat{\mathbf{r}}) \quad (k_2 = K, q_2 = Q),$$

so that the reduced matrix element required in (11.62) is

$$[\theta J \,\|\textstyle\sum f^{K'}\|\, \theta' J'] = (4\pi)^{1/2} \bar{j}_K(\kappa)[\theta J \,\|F^{K'}\|\, \theta' J']. \tag{11.69}$$

Since

$$[s \,\|\hat{s}\|\, s] = \sqrt{(\tfrac{3}{2})}$$

and

$$[l \,\|Y^K\|\, l] = (-1)^l (2l+1)\left(\frac{2K+1}{4\pi}\right)^{1/2} \begin{pmatrix} l & K & l \\ 0 & 0 & 0 \end{pmatrix},$$

we obtain from (11.67) and (11.69) the result

$$\begin{aligned}
&[\theta J \,\|\textstyle\sum f^{K'}\|\, \theta' J'] \\
&= (-1)^{K'+J'-J+S'+L'}(2l+1)\bar{j}_K(\kappa)\{\tfrac{3}{2}(2S+1)(2S'+1)(2L+1)(2L'+1) \\
&\quad \times (2J+1)(2J'+1)(2K+1)(2K'+1)\}^{\frac{1}{2}} \begin{pmatrix} l & K & l \\ 0 & 0 & 0 \end{pmatrix} \begin{Bmatrix} 1 & K & K' \\ S' & L' & J' \\ S & L & J \end{Bmatrix} \\
&\quad \times n \sum_{\bar{\theta}} (\theta\{|\bar{\theta})(\theta'\{|\bar{\theta})(-1)^{\bar{S}+\bar{L}+\frac{1}{2}} \begin{Bmatrix} S & 1 & S' \\ \frac{1}{2} & \bar{S} & \frac{1}{2} \end{Bmatrix} \begin{Bmatrix} L & K & L' \\ l & \bar{L} & l \end{Bmatrix} \\
&= (-1)^{K'+J'-J}(2J+1)^{1/2} C(K, K').
\end{aligned} \tag{11.70}$$

The $3j$ symbol

$$\begin{pmatrix} l & K & l \\ 0 & o & 0 \end{pmatrix}$$

in $C(K, K')$ is non-zero only if K is an even integer, i.e. $K = 0, 2, \ldots, 2l$. In general $K' = K$ or $K \pm 1$. If, however, $S = S'$, $L = L'$, $J = J'$ interchange of the second and third row of the $9j$ symbol in $C(K, K')$ multiplies it by

$$(-1)^{1+K+K'}$$

while leaving the $9j$ symbol unchanged. Hence we conclude that the $9j$ is non-zero only if K' is odd, i.e. we have $K' = K \pm 1$.

If we combine (11.59) (11.62), and (11.70)

$$\begin{aligned}
&\langle \theta J M | \textstyle\sum \exp(\mathrm{i}\boldsymbol{\kappa} \cdot \mathbf{r})\hat{s}_q \,|\theta' J' M'\rangle \\
&= (4\pi)^{1/2} \sum_{K,Q} \sum_{K',Q'} \mathrm{i}^K Y^{*K}_Q(\tilde{\boldsymbol{\kappa}})(1qKQ \mid K'Q') \\
&\quad \times (-1)^{J-M} \begin{pmatrix} J & K' & J' \\ -M & Q' & M' \end{pmatrix} [\theta J \,\|\textstyle\sum f^{K'}\|\, \theta' J'] \\
&= (4\pi)^{1/2} \sum_{K,Q} \sum_{K',Q'} \mathrm{i}^K Y^K_Q(\tilde{\boldsymbol{\kappa}}) \left(\frac{2K'+1}{3}\right)^{1/2} C(K, K') \\
&\quad \times (K'Q'J'M' \mid JM)(KQK'Q' \mid 1q).
\end{aligned} \tag{11.71}$$

Finally, if we combine (11.54) and (11.71) the matrix elements of $\hat{\mathbf{Q}}$ in the *SLJ* basis are

$$\langle \theta JM | \hat{Q}_q | \theta' J' M' \rangle$$

$$= \left\langle \theta JM \right| \sum_{\text{electrons}} \exp(\mathrm{i}\boldsymbol{\kappa} \cdot \mathbf{r}) \left\{ \hat{s}_q - \frac{1}{|\boldsymbol{\kappa}|} (\tilde{\boldsymbol{\kappa}} \times \boldsymbol{\nabla})_q \right\} \left| \theta' J' M' \right\rangle$$

$$= (4\pi)^{1/2} \sum_{K,Q} \sum_{K',Q'} Y^K_Q(\tilde{\boldsymbol{\kappa}}) \left\{ \mathrm{i}^K \left(\frac{2K'+1}{3} \right)^{1/2} C(K, K') + A(K, K') \right\}$$

$$\times (K'Q'J'M' \mid JM)(KQK'Q' \mid 1q). \quad (11.72)$$

Let us now evaluate the matrix elements of $\hat{\mathbf{Q}}_\perp$ (eqn (11.58)). This merely entails the derivation of

$$\langle \theta JM | \sum \exp(\mathrm{i}\boldsymbol{\kappa} \cdot \mathbf{r}) \{ \tilde{\boldsymbol{\kappa}} \times (\hat{\mathbf{s}} \times \tilde{\boldsymbol{\kappa}}) \}_q | \theta' J' M' \rangle \quad (11.73)$$

from (11.71).

Since $\tilde{\boldsymbol{\kappa}} \times (\hat{\mathbf{s}} \times \tilde{\boldsymbol{\kappa}}) = \hat{\mathbf{s}} - \tilde{\boldsymbol{\kappa}}(\hat{\mathbf{s}} \cdot \tilde{\boldsymbol{\kappa}})$ we look first at

$$\tilde{\kappa}_q (\hat{\mathbf{s}} \cdot \tilde{\boldsymbol{\kappa}}) = \tilde{\kappa}_q \sum_{\bar{q}} (-1)^{\bar{q}} \tilde{\kappa}_{\bar{q}} \hat{s}_{-\bar{q}} = \left(\frac{4\pi}{3} \right) \sum_{\bar{q}} (-1)^{\bar{q}} \, Y^1_q(\tilde{\boldsymbol{\kappa}}) Y^1_{\bar{q}}(\tilde{\boldsymbol{\kappa}}) \hat{s}_{-\bar{q}}. \quad (11.74)$$

Since the dependence on q of $\langle \theta JM | \sum \hat{s}_q | \theta' J' M' \rangle$ is in the Clebsch–Gordan coefficient $(KQK'Q' \mid 1q)$, we need to evaluate

$$\left(\frac{4\pi}{3} \right) \sum_{\bar{q}} (-1)^{\bar{q}} Y^1_q(\tilde{\boldsymbol{\kappa}}) Y^1_{\bar{q}}(\tilde{\boldsymbol{\kappa}}) (KQK'Q' \mid 1 - \bar{q}). \quad (11.75)$$

We wish to combine this term with the matrix elements of

$$\sum \exp(\mathrm{i}\boldsymbol{\kappa} \cdot \mathbf{r}) \hat{s}_q.$$

Therefore we use (11.30), to put the $\boldsymbol{\kappa}$-dependence of (11.75) into a single spherical harmonic, and obtain for the latter the result

$$(12\pi)^{1/2} (-1)^{K+K'} \sum_{\bar{K},\bar{Q}} (2\bar{K}+1)^{1/2} \, Y^{*\bar{K}}_{\bar{Q}}(\tilde{\boldsymbol{\kappa}}) \begin{pmatrix} 1 & 1 & \bar{K} \\ 0 & 0 & 0 \end{pmatrix}$$

$$\times \sum_{\bar{q}} \begin{pmatrix} 1 & 1 & \bar{K} \\ q & \bar{q} & \bar{Q} \end{pmatrix} \begin{pmatrix} K & K' & 1 \\ Q & Q' & \bar{q} \end{pmatrix}.$$

If this result is used in (11.73) then we must use (11.30) again to combine $Y^K_Q(\tilde{\boldsymbol{\kappa}})$ and $Y^{*\bar{K}}_{\bar{Q}}(\tilde{\boldsymbol{\kappa}})$. On rearranging the sum to produce an expansion in $Y^K_Q(\tilde{\boldsymbol{\kappa}})$ we find, on using (11.71),

$$\langle\theta JM|\sum \exp(\mathrm{i}\boldsymbol{\kappa}\cdot\mathbf{r})\{\tilde{\boldsymbol{\kappa}}\times(\hat{\mathbf{s}}\times\tilde{\boldsymbol{\kappa}})\}_q\,|\theta'J'M'\rangle$$

$$=(4\pi)^{1/2}\sum_{K,Q}\sum_{K',Q'}Y^K_Q(\tilde{\boldsymbol{\kappa}})\left(\frac{2K'+1}{3}\right)^{1/2}(K'Q'J'M'\mid JM)$$

$$\times\Bigg[\mathrm{i}^K C(K,K')(KQK'Q'\mid 1q)-\sqrt{3}\sum_{\bar{K},\bar{Q}}\sum_{K'',Q''}(-1)^{\bar{Q}+K'+Q}$$

$$\times\,\mathrm{i}^{K''}(2\bar{K}+1)C(K'',K')\{(2K+1)(2K''+1)\}^{1/2}$$

$$\times\begin{pmatrix}1 & 1 & \bar{K}\\ 0 & 0 & 0\end{pmatrix}\begin{pmatrix}K'' & \bar{K} & K\\ Q'' & -\bar{Q} & -Q\end{pmatrix}\begin{pmatrix}K'' & \bar{K} & K\\ 0 & 0 & 0\end{pmatrix}$$

$$\times\sum_{\bar{q}}\begin{pmatrix}1 & 1 & \bar{K}\\ q & \bar{q} & \bar{Q}\end{pmatrix}\begin{pmatrix}K'' & K' & 1\\ Q'' & Q' & \bar{q}\end{pmatrix}\Bigg].\tag{11.76}$$

The sums over $\bar{q}$, $\bar{Q}$, and Q'' in the second part of (11.76) can be performed with (11.47). This part then becomes

$$(-1)^{K'+K+1}\sum_{K''}\mathrm{i}^{K''}C(K'',K')\{(2K+1)(2K''+1)\}^{1/2}(KQK'Q'\mid 1q)$$

$$\times\sum_{\bar{K}}(2\bar{K}+1)\begin{pmatrix}1 & 1 & \bar{K}\\ 0 & 0 & 0\end{pmatrix}\begin{pmatrix}K'' & \bar{K} & K\\ 0 & 0 & 0\end{pmatrix}\begin{Bmatrix}1 & K & K'\\ K'' & 1 & \bar{K}\end{Bmatrix}.\tag{11.77}$$

The triangle condition on $\bar{K}$ allows the values $\bar{K}=0, 1, 2$, but since $\bar{K}$ must be even for a non-zero contribution $\bar{K}=0$ and 2.

For $\bar{K}=0$, $K''=K$ and

$$\begin{pmatrix}1 & 1 & 0\\ 0 & 0 & 0\end{pmatrix}\begin{pmatrix}K & K & 0\\ 0 & 0 & 0\end{pmatrix}\begin{Bmatrix}1 & K & K'\\ K & 1 & 0\end{Bmatrix}=\frac{(-1)^{K'}}{3(2K+1)},$$

so the $\bar{K}=0$ term in (11.77) is simply

$$-\tfrac{1}{3}\mathrm{i}^K C(K,K')(KQK'Q'\mid 1q).$$

For the contribution from $\bar{K}=2$ we use

$$(2\bar{K}+1)\begin{pmatrix}1 & 1 & \bar{K}\\ 0 & 0 & 0\end{pmatrix}=\left(\frac{10}{3}\right)^{1/2}$$

and thus finally obtain from (11.76) the result

$$\langle\theta JM|\sum\exp(\mathrm{i}\boldsymbol{\kappa}\cdot\mathbf{r})\{\tilde{\boldsymbol{\kappa}}\times(\hat{\mathbf{s}}\times\tilde{\boldsymbol{\kappa}})\}_q\,|\theta'J'M'\rangle$$

$$=(4\pi)^{1/2}\sum_{K,Q}\sum_{K',Q'}Y^K_Q(\tilde{\boldsymbol{\kappa}})B(K,K')(K'Q'J'M'\mid JM)(KQK'Q'\mid 1q)\tag{11.78}$$

where

$$B(K,K')=\left(\frac{2K'+1}{3}\right)^{1/2}\left[\tfrac{2}{3}\mathrm{i}^{K}C(K,K')-(-1)^{K'}\{\tfrac{10}{3}(2K+1)\}^{1/2}\right.$$

$$\left.\times\sum_{\bar{K}}\mathrm{i}^{\bar{K}}(2\bar{K}+1)^{1/2}C(\bar{K},K')\begin{pmatrix}2 & \bar{K} & K\\ 0 & 0 & 0\end{pmatrix}\begin{Bmatrix}K' & \bar{K} & 1\\ 2 & 1 & K\end{Bmatrix}\right]. \quad (11.79)$$

It is a simple matter to show that

$$B(K',K')=\mathrm{i}^{K'}\{\tfrac{1}{3}(2K'+1)\}^{1/2}C(K',K'), \quad (11.80)$$

$$B(K'-1,K')$$
$$=\mathrm{i}^{K'-1}\frac{(K'+1)}{\{3(2K'+1)\}^{1/2}}\left\{C(K'-1,K')-\left(\frac{K'}{K'+1}\right)^{1/2}C(K'+1,K')\right\}, \quad (11.81)$$

and

$$\frac{B(K'-1,K')}{B(K'+1,K')}=\left(\frac{K'+1}{K'}\right)^{1/2}.$$

If we combine the result (11.54) for the orbital part with (11.78) for the spin part of the neutron-ion interaction we obtain the expression,

$$\langle\theta JM|\,\hat{Q}_{\perp,q}\,|\theta'J'M'\rangle$$
$$=\left\langle l^n vSLJM\right|\sum_{\text{electrons}}\exp(\mathrm{i}\boldsymbol{\kappa}\cdot\mathbf{r})\left\{\tilde{\boldsymbol{\kappa}}\times(\hat{\mathbf{s}}\times\tilde{\boldsymbol{\kappa}})-\frac{\mathrm{i}}{\hbar\,|\boldsymbol{\kappa}|}\tilde{\boldsymbol{\kappa}}\times\hat{\mathbf{p}}\right\}_q\left|l^n v'S'L'J'M'\right\rangle$$
$$=(4\pi)^{1/2}\sum_{K,Q}\sum_{K',Q'}Y^K_Q(\tilde{\boldsymbol{\kappa}})\{A(K,K')+B(K,K')\}(K'Q'J'M'\,|\,JM)$$
$$\times(KQK'Q'\,|\,1q). \quad (11.82)$$

Let us now return to and simplify eqn (11.72) for the matrix elements of $\hat{\mathbf{Q}}$. For $C(K,K')$, $K=K'$ and $K'\pm1$, and for $A(K,K')$, $K=K'\pm1$. We can add to (11.72) a function proportional to $\tilde{\boldsymbol{\kappa}}$ (which leaves $\hat{\mathbf{Q}}_\perp$ unchanged) and choose this function so that it cancels the contributions in (11.72) from $K=K'+1$.

Consider first the calculation for the orbital term. Let us add the function

$$\tilde{\kappa}_q\sum_{K',Q'}Y^{*K'}_{Q'}(\tilde{\boldsymbol{\kappa}})F^{K'}_{Q'}$$
$$=\sum_{\substack{K',Q'\\Q}}(-1)^{K'+Q'+Q}F^{K'}_{Q'}\left\{(K')^{\frac{1}{2}}Y^{K'-1}_{Q}\begin{pmatrix}1 & K' & K'-1\\ q & -Q' & -Q\end{pmatrix}\right.$$
$$\left.-(K'+1)^{\frac{1}{2}}Y^{K'+1}_{Q}\begin{pmatrix}1 & K' & K'+1\\ q & -Q' & -Q\end{pmatrix}\right\}. \quad (11.83)$$

In arriving at the second line we have used (11.20b) and (11.31).

We find that, if we set

$$F_{Q'}^{K'}=(-1)^{1+q+K'+Q'+Q}(4\pi)^{1/2}\frac{(3K')^{1/2}}{(K'+1)}A(K'+1,K')(K'Q'J'M'\mid JM),$$

then

$$(4\pi)^{1/2}\sum_{K,Q}\sum_{K',Q'}Y_Q^K(\tilde{\boldsymbol{\kappa}})A(K,K')(K'Q'J'M'\mid JM)$$
$$\times(KQK'Q'\mid 1q)+\tilde{\kappa}_q\sum_{K',Q'}Y_{Q'}^{*K'}(\tilde{\boldsymbol{\kappa}})F_{Q'}^{K'}$$
$$=(4\pi)^{1/2}\sum_{\substack{K',Q'\\Q}}Y_Q^{K'-1}(\tilde{\boldsymbol{\kappa}})\left(\frac{2K'+1}{K'+1}\right)A(K'-1,K')(K'Q'J'M'\mid JM)$$
$$\times(K'-1QK'Q'\mid 1q). \tag{11.84}$$

Similarly, if for the spin contribution to (11.72) we set

$$F_{Q'}^{K'}=(-1)^{1+q+K'+Q'+Q}(4\pi)^{1/2}\left(\frac{3}{K'+1}\right)^{1/2}\mathrm{i}^{K'+1}C(K'+1,K')(K'Q'J'M'\mid JM),$$

then

$$(4\pi)^{1/2}\sum_{K,Q}\sum_{K',Q'}Y_Q^K(\tilde{\boldsymbol{\kappa}})\mathrm{i}^K\left(\frac{2K'+1}{3}\right)^{1/2}C(K,K')(K'Q'J'M'\mid JM)$$
$$\times(KQK'Q'\mid 1q)+\tilde{\kappa}_q\sum_{K',Q'}Y_{Q'}^{*K'}(\tilde{\boldsymbol{\kappa}})F_{Q'}^{K'}$$
$$=(4\pi)^{1/2}\sum_{\substack{K',Q'\\Q}}\left\{Y_Q^{K'-1}(\tilde{\boldsymbol{\kappa}})\left(\frac{2K'+1}{K'+1}\right)B(K'-1,K')(K'-1QK'Q'\mid 1q)\right.$$
$$\left.+Y_Q^{K'}(\tilde{\boldsymbol{\kappa}})B(K',K')(K'QK'Q'\mid 1q)\right\}(K'Q'J'M'\mid JM). \tag{11.85}$$

On combining (11.84) and (11.85) we have the desired expression for the matrix elements of $\hat{Q}_q$ in the $SLJM$ basis, namely,

$$\langle\theta JM|\,\hat{Q}_q\,|\theta'J'M'\rangle$$
$$=(4\pi)^{1/2}\sum_{\substack{K',Q'\\Q}}\left[Y_Q^{K'-1}(\tilde{\boldsymbol{\kappa}})\left(\frac{2K'+1}{K'+1}\right)\{A(K'-1,K')+B(K'-1,K')\}\right.$$
$$\left.\times(K'-1QK'Q'\mid 1q)+Y_Q^{K'}(\tilde{\boldsymbol{\kappa}})B(K',K')(K'QK'Q'\mid 1q)\right](K'Q'J'M'\mid JM). \tag{11.86}$$

We recall the fact that if $S=S'$, $L=L'$, $J=J'$, then $B(K',K')=0$.

11.4. Summary of formulae. *JM*- and *LSM*-quantization

Since the algebra encountered in § 11.3 has been heavy we gather together the principal results derived therein.

For a particular ion in a unit cell the matrix elements of the qth spherical component of the interaction operator $\hat{\mathbf{Q}}_\perp$ in the SLJ basis are

$$\langle l^n vSLJM|\, \hat{Q}_{\perp,q}\, |l^n v'S'L'J'M'\rangle = (4\pi)^{1/2} \sum_{KK'} \{A(K, K') + B(K, K')\}$$

$$\times \sum_{QQ'} Y^K_Q(\tilde{\boldsymbol{\kappa}})(K'Q'J'M' \mid JM)(KQK'Q' \mid 1q). \quad (11.87\text{a})$$

The coefficients $A(K, K')$ and $B(K, K')$ are as follows:

$$\begin{aligned} A(K, K') = {} & (-)^{l+L+L'+S+J'}\, \mathrm{i}^{K'+1}\, \delta_{S,S'}(2l+1)^2 \\ & \times \{(2L+1)(2L'+1)(2J'+1)/3\}^{1/2} \\ & \times \begin{Bmatrix} K' & L' & L \\ S & J & J' \end{Bmatrix} \Gamma(K, K') A(K', K', l)(\bar{\jmath}_{K'+1} + \bar{\jmath}_{K'-1}) \\ & \times n \sum_{\bar{\theta}} (-)^{\bar{L}} (\theta\{\, \bar{\theta})(\theta'\{|\, \bar{\theta}) \begin{Bmatrix} K' & l & l \\ \bar{L} & L & L' \end{Bmatrix}. \end{aligned} \quad (11.88)$$

The summation is over all $n-1$ electron states $\bar{\theta}$ that are common to $|\theta\rangle$ and $|\theta'\rangle$, and $(\{|)$ denotes the corresponding fractional parentage coefficients (Nielson and Koster 1963; Condon and Odabaşi 1980). $\Gamma(K, K')$ has the values

$$\begin{aligned} \Gamma(K, K') &= (K')^{1/2}, && \text{if} \quad K = K'+1, \\ &= (K'+1)^{1/2}, && \text{if} \quad K = K'-1, \\ &= 0, && \text{otherwise.} \end{aligned} \quad (11.89)$$

$A(K', K', l)$ is zero unless K' is one of the integers $1, 3, 5, \ldots, (2l-1)$. If this condition is satisfied and we set

$$K' = 2m + 1,$$

then

$$A(K', K', l) = \frac{(-)^m}{(2l+1)} \frac{(m+l+1)!}{m!\,(m+1)!\,(l-m)!} \times \left\{ \frac{(l-m)(2m)!\,(2m+2)!\,(2l-2m)!}{2(4m+3)(2l+2m+2)!} \right\}^{1/2}. \quad (11.90)$$

The requirement that K' be odd, together with eqn (11.89), implies that the integer K must be even and not greater than $2l$. Hence, for a given K', the only non-zero coefficients are $A(K' \pm 1, K')$, and these are related by

$$\frac{A(K'+1, K')}{A(K'-1, K')} = \left(\frac{K'}{K'+1}\right)^{1/2}. \tag{11.91}$$

The spin factor $B(K, K')$ is slightly more cumbersome, namely

$$B(K, K') = \tfrac{2}{3}\mathrm{i}^K\left(\frac{2K'+1}{3}\right)^{1/2} C(K, K') + \tfrac{1}{3}(-)^{K'+1}\{10(2K+1)(2K'+1)\}^{1/2}$$
$$\times \sum_{\bar{K}} \mathrm{i}^{\bar{K}}(2\bar{K}+1)^{1/2}\begin{pmatrix} 2 & \bar{K} & K \\ 0 & 0 & 0 \end{pmatrix}\begin{Bmatrix} K' & \bar{K} & 1 \\ 2 & 1 & K \end{Bmatrix} C(\bar{K}, K'), \tag{11.92}$$

where

$$C(\dot{K}, K') = (-1)^{L'+S'}$$
$$\times\{\tfrac{3}{2}(2\dot{K}+1)(2K'+1)(2S+1)(2S'+1)(2L+1)(2L'+1)(2J'+1)\}^{1/2}$$
$$\times(2l+1)\begin{pmatrix} l & \dot{K} & l \\ 0 & 0 & 0 \end{pmatrix}\begin{Bmatrix} 1 & \dot{K} & K' \\ S' & L' & J' \\ S & L & J \end{Bmatrix} \bar{j}_{\dot{K}}(\kappa)$$
$$\times n\sum_{\bar{\theta}}(-1)^{\frac{1}{2}+\bar{S}+\bar{L}}(\theta\{|\bar{\theta})(\theta'\{|\bar{\theta})\begin{Bmatrix} S & 1 & S' \\ \frac{1}{2} & \bar{S} & \frac{1}{2} \end{Bmatrix}\begin{Bmatrix} L & \dot{K} & L' \\ l & \bar{L} & l \end{Bmatrix}. \tag{11.93}$$

The $3j$ symbol in (11.93) makes $\dot{K}$, and hence $\bar{K}$ and K in (11.92), even for a non-zero $B(K, K')$. The maximum value of $\dot{K}$ is $2l$. Hence triangular conditions ensure that the upper limit of K' is $2l+1$ and of K, $2l+2$. If K' is even, the only non-zero coefficient occurs when $K = K'$, in which case

$$B(K', K') = \mathrm{i}^{K'}\left(\frac{2K'+1}{3}\right)^{1/2} C(K', K'). \tag{11.94}$$

If K' is odd, then $K = K' \pm 1$. For $K = K'-1$,

$$B(K'-1, K')$$
$$= \mathrm{i}^{K'-1}[(K'+1)C(K'-1, K') - \{K'(K'+1)\}^{1/2}C(K'+1, K')]\{3(2K'+1)\}^{-1/2} \tag{11.95}$$

and the coefficient $B(K'+1, K')$ is got from this through

$$B(K'+1, K') = \{K'/(K'+1)\}^{1/2}B(K'-1, K'). \tag{11.96}$$

For the particular case $S = S'$, $L = L'$, and $J = J'$ the $9j$ coefficient in $C(\dot{K}, K')$ requires that $B(K, K')$ is non-zero for $K = K' \pm 1$ with $K' = 1, 3, \ldots, (2l+1)$. Thus in this instance the conditions on K and K' for non-zero $B(K, K')$ coincide with those for $A(K, K')$, except for the fact that for $B(K, K')$ the maximum value of K' is $(2l+1)$ whereas for $A(K, K')$ it is $(2l-1)$.

In some instances it proves more convenient to calculate the cross-section in terms of the matrix elements of the operator $\hat{\mathbf{Q}}$, which is related to $\hat{\mathbf{Q}}_{\perp}$ through $\hat{\mathbf{Q}}_{\perp} = \tilde{\boldsymbol{\kappa}} \times (\hat{\mathbf{Q}} \times \tilde{\boldsymbol{\kappa}})$ and has matrix elements

$$\langle l^n vSLJM|\, \hat{Q}_q \,|l^n v'S'L'J'M'\rangle = (4\pi)^{1/2} \sum_{\substack{K',Q'\\Q}} \left[Y_Q^{K'-1}(\tilde{\boldsymbol{\kappa}})\left(\frac{2K'+1}{K'+1}\right) \times \{A(K'-1, K') + B(K'-1, K')\}(K'-1QK'Q' \mid 1q) + Y_Q^{K'}(\tilde{\boldsymbol{\kappa}})B(K', K')(K'QK'Q' \mid 1q)\right](K'Q'J'M' \mid JM). \quad (6.87b)$$

The operator $\hat{\mathbf{Q}}$ has the conceptual value of retaining the same form for the cross-section in the general case of both spin and orbital scattering as in the spin-only case.

The above formulae for the matrix elements $\hat{\mathbf{Q}}_{\perp}$ refer to the $|SLJM\rangle$ basis. However, we remarked in the introduction that often the magnetic state of an ion is appropriately described in terms of the $|SM_sLM_L\rangle$ basis, e.g. 3d transitional-metal ions. The matrix elements can easily be evaluated directly in this basis or through the transformation

$$|SM_sLM_L\rangle = \sum_{JM} (SM_sLM_L \mid JM)\,|SLJM\rangle,$$

noting that $A(K, K')$ and $C(\dot{K}, K')$ are functions of J.

The spin matrix element is

$$\left\langle SM_SLM_L \middle| \sum_{\text{electrons}} \exp(\mathrm{i}\boldsymbol{\kappa}\cdot\mathbf{r})\hat{s}_q \middle| S'M_S'L'M_L' \right\rangle$$
$$= (4\pi)^{1/2}(-1)^{3/2+M_S-M_L}(2l+1)[\tfrac{3}{2}(2S+1)(2S'+1)(2L+1)(2L'+1)]^{1/2}$$
$$\times \begin{pmatrix} S & 1 & S' \\ -M_S & q & M_S' \end{pmatrix} \sum_{KQ} \mathrm{i}^K \bar{j}_K Y_Q^{*K}(\tilde{\boldsymbol{\kappa}})(2K+1)^{1/2}$$
$$\times \begin{pmatrix} l & K & l \\ 0 & 0 & 0 \end{pmatrix}\begin{pmatrix} L & K & L' \\ -M_L & Q & M_L' \end{pmatrix}$$
$$\times n \sum_{\bar{\theta}} (\theta\{|\bar{\theta})(\theta'\{|\bar{\theta})(-1)^{\bar{S}+\bar{L}} \begin{Bmatrix} S' & 1 & S \\ \frac{1}{2} & \bar{S} & \frac{1}{2} \end{Bmatrix}\begin{Bmatrix} L' & K & L \\ l & L & l \end{Bmatrix} \quad (11.97a)$$

and for the orbital matrix we find, using (11.48) and (11.51),

$$\left\langle SM_SLM_L \left| \sum_{\text{electrons}} \exp(\mathrm{i}\boldsymbol{\kappa}\cdot\mathbf{r})(\tilde{\boldsymbol{\kappa}}\times\boldsymbol{\nabla})_q \right| S'M_S'L'M_L' \right\rangle$$
$$= |\boldsymbol{\kappa}|\,(4\pi)^{1/2}(2l+1)^2(-1)^{l+M_L+L+L'}[\tfrac{1}{3}(2L+1)(2L'+1)]^{1/2}$$
$$\times \sum_{K'Q'} \mathrm{i}^{K'+1} Y_Q^{K'-1}(\tilde{\boldsymbol{\kappa}})[\bar{\jmath}_{K'-1}+\bar{\jmath}_{K'+1}]\frac{(2K'+1)}{(K'+1)^{1/2}}A(K',K',l)$$
$$\times (K'-1QK'Q'\,|\,1q)\begin{pmatrix} L' & K' & L \\ M_L' & Q' & -M_L \end{pmatrix}$$
$$\times n\sum_{\bar{\theta}}(\theta\{|\bar{\theta})(\theta'\{|\bar{\theta})(-1)^{\bar{L}}\begin{Bmatrix} L & K' & L' \\ l & \bar{L} & l \end{Bmatrix}\delta_{S,S'}\delta_{M_S,M_S'}. \tag{11.97b}$$

Though the detailed structure of the matrix elements appears very complicated, involving summations over $3j$, $6j$, and $9j$ symbols, they are very amenable to computing. There is just one set of parameters, the radial integrals $\bar{\jmath}_K$, which are calculated from a given radial-wave function. By comparing the cross-section computed from (11.87a), (11.87b), or (11.97) with that measured, one can determine the detailed nature of the unpaired electrons in the ion, provided, of course, one is able to make proper allowance for covalency.

The relative significance of the terms in eqns (11.88) and (11.92) for $A(K, K')$ and $B(K, K')$ depends upon the relative size of the radial integrals

$$\bar{\jmath}_K(\kappa) = \int_0^\infty \mathrm{d}r\, r^2 f^2 j_K(\kappa r).$$

We note that in general more than one order K of $\bar{\jmath}_K(\kappa)$ appears in a given term in (11.87). Since the small argument expansion of the spherical Bessel function is

$$j_K(x) = \frac{x^K}{1\cdot 3\cdot 5\cdot\cdots\cdot(2K+1)}, \tag{11.98}$$

it is often sufficient to consider only the first few terms in the sum over K. The *dipole approximation* $K'=1$, $K=0, 2$, is considered in detail in the next section.

11.5. Dipole approximation

In this approximation ($K'=1$, $K=0, 2$), we assume that $|\boldsymbol{\kappa}|^{-1}$ is much larger than the mean radius of the wave function of the unpaired electrons.

First, consider the cross-section for the scattering of unpolarized neutrons by a system of free ions. In this instance

$$|\lambda\rangle = |JM\rangle, \qquad |\lambda'\rangle = |JM'\rangle \tag{11.99}$$

and we must average over all the initial orientations of the ion and sum over all possible final orientations, i.e.

$$\sum_{\lambda,\lambda} p_\lambda \equiv \sum_{M,M'} (2J+1)^{-1} \tag{11.100}$$

and so, from eqn (11.87a),

$$\begin{aligned}
&\sum_q \sum_{\lambda\lambda'} p_\lambda \, |\langle\lambda'|\, \hat{Q}_{\perp,q}\, |\lambda\rangle|^2 \\
&= \sum_{MM'} (2J+1)^{-1} \sum_{\substack{qK,K' \\ K_1,K_1'}} \{A(K,K')+B(K,K')\}\{A(K_1,K_1')+B(K_1,K_1')\} \\
&\quad \times \sum_{QQ'} \sum_{Q_1Q_1'} 4\pi Y_Q^K(\tilde{\boldsymbol{\kappa}})\, Y_{Q_1}^{*K_1}(\tilde{\boldsymbol{\kappa}})(K'Q'JM' \mid JM)(K_1'Q_1'JM' \mid JM) \\
&\quad \times (KQK'Q' \mid 1q)(K_1Q_1K_1'Q_1' \mid 1q) \\
&= \sum_q \sum_{KK'} \sum_{QQ'} \frac{4\pi}{2K'+1} \, |Y_Q^K(\tilde{\boldsymbol{\kappa}})|^2 \, \{A(K,K')+B(K,K')\}^2 (KQK'Q' \mid 1q)^2.
\end{aligned} \tag{11.101}$$

Since

$$\sum_{qQ'} (KQK'Q' \mid 1q)^2 = \frac{3}{2K+1}$$

and

$$\sum_Q |Y_Q^K(\tilde{\boldsymbol{\kappa}})|^2 = \frac{2K+1}{4\pi},$$

the required cross-section is

$$\frac{d\sigma}{d\Omega} = r_0^2 \sum_{K,K'} \frac{3}{2K'+1} \{A(K,K')+B(K,K')\}^2. \tag{11.102}$$

Taking $K'=1$, $K=0, 2$,

$$\frac{d\sigma}{d\Omega} = r_0^2 \tfrac{3}{2}\{A(0,1)+B(0,1)\}^2 \tag{11.103}$$

on using (11.91) and (11.96).

To evaluate $A(0, 1)$ from (11.88) note that

$$n\sum_{\bar{\theta}}(\theta\{|\bar{\theta})^2(-)^{\bar{L}}\begin{Bmatrix} L & 1 & L \\ l & \bar{L} & l \end{Bmatrix} = (-)^{-l+1-L}\left\{\frac{L(L+1)}{(2L+1)l(l+1)(2l+1)}\right\}^{1/2} \tag{11.104a}$$

and

$$\begin{Bmatrix} L & L & 1 \\ J & J & S \end{Bmatrix} = \tfrac{1}{2}(-)^{J+L+S+1}\frac{J(J+1)+L(L+1)-S(S+1)}{\{J(J+1)(2J+1)L(L+1)(2L+1)\}^{1/2}}. \tag{11.104b}$$

Also from (6.90)

$$A(1, 1, l) = \frac{1}{2l+1}\left\{\frac{l(l+1)}{6(2l+1)}\right\}^{1/2}. \tag{11.104c}$$

With the results (11.104), the value of $A(0, 1)$ is found to be

$$A(0, 1) = -\tfrac{1}{6}(\bar{J}_0 + \bar{J}_2)\frac{1}{\sqrt{\{J(J+1)\}}}\{J(J+1)+L(L+1)-S(S+1)\}. \tag{11.105}$$

In evaluating $B(0, 1)$ from (11.92) and (11.93) we assume, for the present, that the coefficient of $\bar{J}_2$ in $C(2, 1)$ is small compared to the coefficient of $\bar{J}_2$ in $A(0, 1)$. From (11.93),

$$\begin{aligned} C(0, 1) = \bar{J}_{02}\tfrac{3}{2}\sqrt{(2)}(2S+1)(2L+1)\sqrt{\{(2J+1)(2l+1)\}} \\ \times(-)^{L+l+1/2+S}\begin{Bmatrix} S & S & 1 \\ L & L & 0 \\ J & J & 1 \end{Bmatrix} \\ \times n\sum_{\bar{\theta}}(\theta\{|\bar{\theta})^2(-)^{\bar{S}+\bar{L}}\begin{Bmatrix} S & 1 & S \\ \frac{1}{2} & \bar{S} & \frac{1}{2} \end{Bmatrix}\begin{Bmatrix} L & 0 & L \\ l & \bar{L} & l \end{Bmatrix}. \end{aligned} \tag{11.106}$$

The $9j$ symbol in (11.106) reduces to

$$\begin{Bmatrix} S & S & 1 \\ L & L & 0 \\ J & J & 1 \end{Bmatrix} = \frac{(-)^{S+L+J+1}}{\sqrt{\{3(2L+1)\}}}\begin{Bmatrix} S & S & 1 \\ J & J & L \end{Bmatrix}. \tag{11.107a}$$

and

$$\begin{Bmatrix} L & 0 & L \\ l & \bar{L} & l \end{Bmatrix} = \frac{(-)^{\bar{L}+l+L}}{\sqrt{\{(2l+1)(2L+1)\}}}. \tag{11.107b}$$

The sum over the coefficients of fractional parentage in (11.106) can be written down by analogy with (11.104a) and the $6j$ symbol in

(11.107a) is obtained from (11.104b). The result for $C(0,1)$ is then

$$C(0,1) = -\tfrac{1}{2}\bar{j}_0 \frac{1}{\sqrt{\{J(J+1)\}}}\{J(J+1)+S(S+1)-L(L+1)\} \tag{11.108a}$$

and we shall take

$$B(0,1) = \tfrac{2}{3}C(0,1) \tag{11.108b}$$

from (11.92).

On combining (11.105) and (11.108b) in (11.103) the neutron cross-section in the dipole approximation for the scattering from an isolated ion is

$$\frac{d\sigma}{d\Omega} = r_0^2 \frac{J(J+1)}{6} g^2 F^2(\kappa) \tag{11.109}$$

where g is the Landé splitting factor (11.16) and

$$F(\kappa) = \bar{j}_0 + \bar{j}_2 \left\{ \frac{J(J+1)+L(L+1)-S(S+1)}{3J(J+1)+S(S+1)-L(L+1)} \right\} \tag{11.110}$$

is the free-ion form factor in the dipole approximation. In the limit $|\boldsymbol{\kappa}| \to 0$, $F \to 1$, and (11.109) becomes identical with (11.17).

In deriving (11.109) we neglected $C(2,1)$, which contains $\bar{j}_2$. We now examine the validity of this step in more detail. From (11.93),

$$C(2,1) = -\bar{j}_2 \left[(2S+1)(2L+1)(2l+1) \begin{pmatrix} l & 2 & l \\ 0 & 0 & 0 \end{pmatrix} \times 3\{\tfrac{5}{2}(2J+1)\}^{1/2}(-)^{L+S-1/2} \begin{Bmatrix} S & S & 1 \\ L & L & 2 \\ J & J & 1 \end{Bmatrix} \right] \times n \sum_{\bar{\theta}} (\theta\{|\bar{\theta})^2(-)^{\bar{S}+\bar{L}} \begin{Bmatrix} S & 1 & S \\ \frac{1}{2} & \bar{S} & \frac{1}{2} \end{Bmatrix} \begin{Bmatrix} L & 2 & L \\ l & \bar{L} & l \end{Bmatrix}. \tag{11.111}$$

The $9j$ symbol occurring in this expression can easily be evaluated from

$$\begin{Bmatrix} S & S & 1 \\ L & L & 2 \\ J & J & 1 \end{Bmatrix} = \frac{1}{60}\left\{\frac{1}{4L(L+1)-3}\right\} \times \left\{ \frac{30(2L-1)(2L+3)}{S(S+1)(2S+1)L(L+1)(2L+1)J(J+1)(2J+1)} \right\}^{1/2} \times [L(L+1)\{L(L+1)+2J(J+1)+2S(S+1)\} - 3\{J(J+1)-S(S+1)\}^2]. \tag{11.112}$$

As an example we consider d^3. The ground state is ^4F, for which the term in square brackets in (6.111) has the value ($L = 3$, $S = J = 3/2$)

$$-\frac{4 \cdot 3^2}{5}\sqrt{2}. \qquad (11.113a)$$

The parent states $\bar{\theta}$ of d^3 are $\bar{S} = 1$, and $\bar{L} = 1$ and 3, for which the coefficients $(\theta\{|\bar{\theta})^2$ have the values 1/5 and 4/5, respectively. Thus

$$n\sum_{\bar{\theta}}(\theta\{|\bar{\theta})^2(-)^{\bar{S}+\bar{L}}\begin{Bmatrix} S & 1 & S \\ \frac{1}{2} & \bar{S} & \frac{1}{2}\end{Bmatrix}\begin{Bmatrix} L & 2 & L \\ 2 & \bar{L} & 2\end{Bmatrix} = -\frac{1}{14\sqrt{15}}. \qquad (11.113b)$$

With the results (11.113a) and (11.113b) we find, for d^3,

$$C(2, 1) = -\frac{2 \cdot 3^2}{5 \cdot 7}\sqrt{\left(\frac{2}{15}\right)}\bar{j}_2.$$

The coefficient of $C(2, 1)$ in $B(0, 1)$ is $-\frac{1}{3}\sqrt{2}$, so that in neglecting $C(2, 1)$ in this instance we have dropped a term

$$\frac{4 \cdot 3}{35}\frac{1}{\sqrt{15}}\bar{j}_2.$$

The coefficient of $\bar{j}_2$ in $A(0, 1)$ for d^3 is

$$-4\frac{1}{\sqrt{15}}$$

so that we conclude that, at least for this example, the error involved in neglecting $C(2, 1)$ in $B(0, 1)$ is indeed small, the ratio of the $\bar{j}_2$ term in $A(0, 1)$ to that in $B(0, 1)$ being 3/35; for d^7 the ratio is 4/105.

We now consider the consequences of the dipole approximation in more detail.

In writing $\hat{Q}_{\perp,q}$ in the form (11.87a) we have implicitly assumed that the wave function of the unpaired electrons in a crystal environment takes the form

$$|\lambda\rangle = \sum_M |JM\rangle\langle JM\,|\,\lambda\rangle \qquad (11.114a)$$

where the matrix elements $\langle JM\,|\,\lambda\rangle$ are determined by the crystal field.† Thus

$$(K'Q'JM'\,|\,JM) \Rightarrow \sum_{M,M'}\langle\lambda\,|\,JM\rangle(K'Q'JM'\,|\,JM)\langle JM'\,|\,\lambda\rangle. \qquad (11.114b)$$

† Alternatively, we could, of course, base our discussion on $\hat{\mathbf{Q}}$ (eqn (11.86b)).

In the dipole approximation we are concerned only with the $K' = 1$ term. With the formalism of § 11.3 it can be shown that

$$[\theta J \,\|\hat{\mathbf{L}}\|\, \theta J] = (-)^{S+L+J+1}(2J+1)\begin{Bmatrix} J & J & 1 \\ L & L & S \end{Bmatrix}[L \,\|\hat{\mathbf{L}}\|\, L]$$

$$= (-)^{S+L+J+1}(2J+1)\begin{Bmatrix} J & J & 1 \\ L & L & S \end{Bmatrix}\sqrt{\{L(L+1)(2L+1)\}}$$

$$= \frac{1}{2}\left\{\frac{(2J+1)}{J(J+1)}\right\}^{1/2}\{J(J+1)+L(L+1)-S(S+1)\}$$

and, similarly,

$$[\theta J \,\|\hat{\mathbf{S}}\|\, \theta J] = \frac{1}{2}\left\{\frac{(2J+1)}{J(J+1)}\right\}^{1/2}\{J(J+1)+S(S+1)-L(L+1)\}.$$

Hence,

$$A(0, 1) = -\frac{1}{3\sqrt{(2J+1)}}(\bar{\jmath}_0+\bar{\jmath}_2)[\theta J \,\|\hat{\mathbf{L}}\|\, \theta J]$$

and, neglecting $C(2, 1)$,

$$B(0, 1) \simeq -\frac{2}{3\sqrt{(2J+1)}}\,\bar{\jmath}_0[\theta J \,\|\hat{\mathbf{S}}\|\, \theta J].$$

The latter two results imply the relations

$$(1Q'JM' \mid JM)A(0, 1) \Rightarrow \tfrac{1}{3}(\bar{\jmath}_0+\bar{\jmath}_2)\langle\lambda|\, \hat{L}_{Q'}\, |\lambda\rangle$$

and

$$(1Q'JM' \mid JM)B(0, 1) \Rightarrow \tfrac{2}{3}\bar{\jmath}_0\langle\lambda|\, \hat{S}_{Q'}\, |\lambda\rangle,$$

i.e.

$$3(1Q'JM' \mid JM)\{A(0, 1)+B(0, 1)\} \Rightarrow \langle\lambda|\, \{(\bar{\jmath}_0+\bar{\jmath}_2)\hat{\mathbf{L}}+2\bar{\jmath}_0\hat{\mathbf{S}}\}_{Q'}\, |\lambda\rangle. \tag{11.115}$$

In the limit $|\boldsymbol{\kappa}| \to 0$, the right-hand side of (11.115) becomes

$$\langle\lambda|\, (\hat{\mathbf{L}}+2\hat{\mathbf{S}})_{Q'}\, |\lambda\rangle.$$

Returning now to the evaluation of the cross-section for ions with wave functions of the form (11.114a), we make the assumption that on the right-hand side we can take $Q' = 0$ (this means that only those states for which $M = M'$ are retained). Hence

$$\langle\lambda|\, \hat{\mathbf{Q}}_{\perp,q}\, |\lambda\rangle \simeq (4\pi)^{1/2}\tfrac{1}{3}\langle\lambda|\, \{(\bar{\jmath}_0+\bar{\jmath}_2)\hat{\mathbf{L}}+2\bar{\jmath}_0\hat{\mathbf{S}}\}_z\, |\lambda\rangle$$
$$\times\left\{Y_0^0(\tilde{\boldsymbol{\kappa}})(0010 \mid 10)\delta_{q,0}+\frac{1}{\sqrt{2}}\,Y_q^2(\tilde{\boldsymbol{\kappa}})(2q10 \mid 1q)\right\}.$$

On making use of the results

$$(0010 \mid 10) = 1,$$

$$(2q10 \mid 1q) = -\left(\frac{4-q^2}{10}\right)^{\frac{1}{2}},$$

$$Y_0^0(\tilde{\boldsymbol{\kappa}}) = 1/(4\pi)^{1/2},$$

$$Y_0^2(\tilde{\boldsymbol{\kappa}}) = \left(\frac{5}{4\pi}\right)^{1/2}(1-\tfrac{3}{2}\sin^2\theta),$$

and

$$Y_{\pm 1}^2(\tilde{\boldsymbol{\kappa}}) = \mp\left(\frac{15}{2\cdot 4\pi}\right)^{1/2}\sin\theta\cos\theta\exp(\pm i\phi),$$

we obtain

$$\sum_q |\langle\lambda|\,\hat{Q}_{1,q}\,|\lambda\rangle|^2 \simeq |\tfrac{1}{2}\sin\theta\,\langle\lambda|\,\{(\bar{j}_0+\bar{j}_2)\hat{\mathbf{L}}+2\bar{j}_0\hat{\mathbf{S}}\}_z\,|\lambda\rangle|^2, \quad (11.116)$$

where θ is the angle between $\tilde{\boldsymbol{\kappa}}$ and the direction of quantization $\tilde{\mathbf{z}}$.

From (11.116) we see that in the dipole approximation the elastic cross-section for unpolarized neutrons is

$$\frac{d\sigma}{d\Omega} \simeq (\tfrac{1}{2}r_0)^2 \sum_\lambda p_\lambda \langle\lambda|\,\{(\bar{j}_0+\bar{j}_2)\hat{\mathbf{L}}+2\bar{j}_0\hat{\mathbf{S}}\}_\perp\,|\lambda\rangle \cdot \langle\lambda|\,\{(\bar{j}_0+\bar{j}_2)\hat{\mathbf{L}}+2\bar{j}_0\hat{\mathbf{S}}\}_\perp\,|\lambda\rangle. \quad (11.117)$$

More generally we have shown that in the dipole approximation (which can be said to be valid when $\bar{j}_0$ is larger than $\bar{j}_2$) the operator $\hat{\mathbf{Q}}$ (cf. (11.58)) can be replaced by

$$\hat{\mathbf{Q}}^{(D)} = \tfrac{1}{2}\{(\bar{j}_0(\kappa)+\bar{j}_2(\kappa))\hat{\mathbf{L}}+2\bar{j}_0(\kappa)\hat{\mathbf{S}}\}. \quad (11.118)$$

Two cases are of particular importance.

(i) When $\hat{\mathbf{L}}$ can be replaced by $(g-2)\hat{\mathbf{S}}$,

$$\hat{\mathbf{Q}}^{(D)} = \tfrac{1}{2}gF(\kappa)\hat{\mathbf{S}} \quad (11.119)$$

where $F(\kappa)$ is the form factor of the ion and is given by

$$F(\kappa) = \bar{j}_0(\kappa) + (1-2/g)\bar{j}_2(\kappa). \quad (11.120)$$

(ii) When the ion is characterized by a total angular momentum J,

$$\hat{\mathbf{Q}}^{(D)} = \tfrac{1}{2}gF(\kappa)\hat{\mathbf{J}} \quad (11.121)$$

where g is the Landé splitting factor (11.16) and the form factor $F(\kappa)$ is given by eqn (11.110).

Equations (11.97a) and (11.97b) together give the required expression for the matrix element of the intermediate operator $\hat{\mathbf{Q}}$ in the SM_SLM_L basis. An important feature of the expression, which is a direct outcome of the simplification of the matrix element of $\hat{Q}_{\text{orbital}}$, is that the coefficient of $(4\pi)^{1/2}Y_0^0(\tilde{\boldsymbol{\kappa}})$ coincides with the dipole approximation (eqn (11.118)).

11.6. Elastic scattering by rare earths (Balcar *et al.* 1970)

In the rare-earth ions the contribution to the cross-section from the orbital angular momentum cannot in general be disregarded. Further, as we go to larger values of $|\boldsymbol{\kappa}|$ so the validity of the dipole approximation breaks down. We now give a detailed discussion of the application of the formalism derived in § 11.3 and summarized in § 11.4 to rare-earth ions, first when they are fully saturated ($J = M$) and secondly when the crystal field partially and wholly removes the $2J+1$ degeneracy. Our discussion in principle covers all ions with unpaired f electrons, both actinide and rare-earth elements. However for actinide elements and compounds a band or itinerant electron model is probably more appropriate (Brooks and Kelly 1983).

11.6.1. Saturated rare earths

We expect the state $|SLJJ\rangle$ to be a good description of the ions in a ferromagnetically or antiferromagnetically ordered rare-earth compound or for ions fully aligned by an external magnetic field. The ions are assumed aligned parallel to the z-axis.

Our starting point is eqn (11.87b) for $\langle SLJM|\,\hat{Q}_q\,|S'L'J'M'\rangle$. We set $S = S'$, $L = L'$, $J = J'$ and consider first $q = \pm 1$. Thus

$$\langle SLJM|\,\hat{Q}_{\pm1}\,|SLJM\rangle$$
$$= (-1)^{J-M+1}\{12\pi(2J+1)\}^{1/2}\sum_{K'} Y_{\pm1}^{K'-1}(\tilde{\boldsymbol{\kappa}})\left(\frac{2K'+1}{K'+1}\right)$$
$$\times\{A(K'-1,K')+B(K'-1,K')\}\begin{pmatrix}K' & J & J\\ 0 & M & -M\end{pmatrix}\begin{pmatrix}K'-1 & K' & 1\\ \pm1 & 0 & \mp1\end{pmatrix}. \tag{11.122}$$

Note that $K' \leqslant (2J-1)$ if J is an integer or $K' \leqslant 2J$ if J is half integer.

By definition of the spherical components of $\hat{\mathbf{Q}}$ (cf. (11.20)),

$$\hat{Q}_x = -\frac{1}{\sqrt{2}}(\hat{Q}_{+1}-\hat{Q}_{-1}), \qquad \hat{Q}_y = \frac{\mathrm{i}}{\sqrt{2}}(\hat{Q}_{+1}+\hat{Q}_{-1}). \tag{11.123}$$

If we write $\tilde{\boldsymbol{\kappa}} = (\sin\theta\cos\phi, \sin\theta\sin\phi, \cos\theta)$ and use the results

$$Y_1^{K'-1} - Y_{-1}^{K'-1} = 2\cos\phi e^{-i\phi} Y_1^{K'-1}$$

and

$$Y_1^{K'-1} + Y_{-1}^{K'-1} = 2i\sin\phi e^{-i\phi} Y_1^{K'-1}, \tag{11.124}$$

then, from (11.122),

$$\langle SLJJ|\, \hat{Q}_x \,|SLJJ\rangle = M(\theta)\cos\phi \tag{11.125a}$$

and

$$\langle SLJJ|\, \hat{Q}_y \,|SLJJ\rangle = M(\theta)\sin\phi. \tag{11.125b}$$

In eqns (11.125a) and (11.125b) the function $M(\theta)$ is defined by

$$M(\theta) = (4\pi)^{1/2} \sum_{K'} e^{-i\phi} Y_1^{K'-1}(\theta, \phi) \left(\frac{K'-1}{K'}\right)^{1/2} Z(K') \tag{11.126}$$

and

$$Z(K') = \{3(2J+1)\}^{1/2} \left(\frac{2K'+1}{K'+1}\right) \{A(K'-1, K') + B(K'-1, K')\}$$
$$\times \begin{pmatrix} K' & J & J \\ 0 & J & -J \end{pmatrix} \begin{pmatrix} K'-1 & K' & 1 \\ 0 & 0 & 0 \end{pmatrix}. \tag{11.127}$$

There are several properties of $M(\theta)$ that are worth noting at this point. The smallest value of K' in eqn (11.126) is 3, and it follows from the structure of $A(K'-1, K')$ and $B(K'-1, K)$ (eqns (11.88) and (11.95)) that the lowest possible order radial integral in $M(\theta)$ is $\bar{j}_2(\kappa)$. Since only $\bar{j}_0(\kappa)$ is finite for $|\boldsymbol{\kappa}| \to 0$, $M(\theta)$ must be zero for zero wave vector. It is also zero for $|\boldsymbol{\kappa}| \to \infty$. Consider now its dependence on θ, the angle between the axis of quantization and the direction of $\tilde{\boldsymbol{\kappa}}$. First, $M(0) = 0$ and a simple calculation shows that $M(\pi - \theta) = -M(\theta)$, so that $M(\frac{1}{2}\pi) = 0$. $M(\theta)$ for Pr^{3+} is shown in Fig. 11.1 as a function of $|\boldsymbol{\kappa}|$ for various values of θ.

With $L = 0$, $A(K'-1, K')$ is zero, of course, and the only contribution from $B(K'-1, K')$ is for $K' = 1$, and this gives $Z(K') = S\bar{j}_0(\kappa)$. On the other hand, if $S = 0$, then $B(K'-1, K') = 0$ but $A(K'-1, K')$ is in general non-zero for all K'. Hence $M(\theta)$ is a manifestation of the orbital moment of the rare-earth ions.

We can also express $\langle SLJJ|\, \hat{Q}_z \,|SLJJ\rangle$ in terms of the functions $Z(K')$. From (11.87b) and (11.127)

$$\langle SLJJ|\, \hat{Q}_z \,|SLJJ\rangle = (4\pi)^{1/2} \sum_{K'} Y_0^{K'-1}(\theta, \phi) Z(K'). \tag{11.128}$$

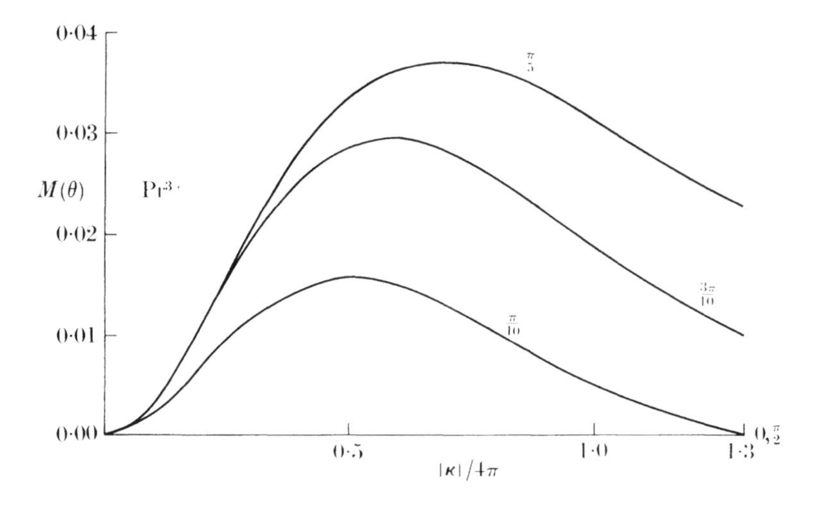

FIG. 11.1. The function $M(\theta)$ defined by eqns (11.126) and (11.127) is shown for the tripositive rare earth Pr^{3+} as a function of $|\boldsymbol{\kappa}|$ for various values of θ.

Clearly (11.128) is independent of ϕ. In contrast to $M(\theta)$, the smallest value of $K' = 1$ and hence $\langle SLJJ|\, \hat{Q}_z\, |SLJJ\rangle$ is non-zero in the limit $|\boldsymbol{\kappa}| \to 0$.

It is instructive to compare the above results for the diagonal matrix elements of $\hat{Q}_x$, $\hat{Q}_y$, and $\hat{Q}_z$ with those we would derive in the dipole approximation (eqn (11.121)). The latter case only involves the matrix elements of $\hat{J}_\alpha$, for which there are no matrix elements diagonal in M for $\alpha = x$ or y, i.e. both $\langle SLJJ|\, \hat{Q}_x\, |SLJJ\rangle$ and $\langle SLJJ|\, \hat{Q}_y\, |SLJJ\rangle$ are zero in the dipole approximation. Further, from (11.121),

$$\langle SLJJ|\, \hat{Q}_z^{(\mathrm{D})}\, |SLJJ\rangle = \tfrac{1}{2} g F(\kappa) J \qquad (11.129)$$

with $F(\kappa)$ given by eqn (11.110). On comparing (11.128) and (11.129) it is evident that the coefficient of $\bar{j}_0(\kappa)$ in $Z(1)$ must equal $\frac{1}{2}gJ$. (Note that $Y_0^0(\theta, \phi) = (4\pi)^{-1/2}$).

The functions $Z(K')$, which can be regarded as the 'characteristic' functions for the scattering from rare-earth ions, are tabulated in Table 11.1 for the tripositive rare-earth ions. It is of value to know the orbital contribution to these functions, and in Table 11.2 we give just this part of $Z(K')$, i.e. eqn (11.127) with $B(K'-1, K')$ set equal to zero. Inspection of eqn (11.88) for $A(K, K')$ shows that the orbital contribution to $Z(K')$ is of the form $c_{K'}^{(L)}(\bar{j}_{K'-1} + \bar{j}_{K'+1})$ and it is the coefficients $c_{K'}^{(L)}$ that are listed in Table 11.2. Further, the conditions on K and K' in $A(K, K')$ and $B(K, K')$ shows that $Z(7)$ comes entirely from $B(K, K')$ the spin part of the matrix elements. Note that for $L = 5$ there is no orbital contribution

Table 11.1

The coefficients $Z(K')$ defined by eqn (11.127) are given for the tripositive rare-earth ions. From (11.127) it follows that

$$Z(K') = c_{K'-1}\bar{j}_{K'-1}(\kappa) + c_{K'+1}\bar{j}_{K'+1}(\kappa)$$

and it is the coefficients $c_{K'\pm 1}$ that are tabulated

	$\bar{j}_0$	$\bar{j}_2$	$\bar{j}_4$	$\bar{j}_6$
		$Ce^{3+}f^1\ ^2F_{5/2}$		
$Z(1)$	1.07142857	1.71428571		
$Z(3)$		0.09583148	0.31943828	
$Z(5)$			0.00360750	0.04329004
		$Pr^{3+}f^2\ ^3H_4$		
$Z(1)$	1.60000000	2.63111111		
$Z(3)$		−0.09865803	−0.13453368	
$Z(5)$			−0.01836547	−0.11678557
$Z(7)$				0.00233133
		$Nd^{3+}f^3\ ^4I_{9/2}$		
$Z(1)$	1.63636364	2.95041322		
$Z(3)$		−0.20896503	−0.25329095	
$Z(5)$			0.03820789	0.14258681
$Z(7)$				−0.00614959
		$Pm^{3+}f^4\ ^5I_4$		
$Z(1)$	1.20000000	2.71515152		
$Z(3)$		−0.10866182	−0.07024602	
$Z(5)$			0.00509439	−0.06449156
$Z(7)$				0.00232318
		$Sm^{3+}f^5\ ^6H_{5/2}$		
$Z(1)$	0.35714286	1.93650794		
$Z(3)$		0.04614109	0.06291966	
$Z(5)$			−0.00426341	0.00335243
		$Tb^{3+}f^8\ ^7F_6$		
$Z(1)$	4.50000000	1.66666666		
$Z(3)$		0.00000000	0.28459047	
$Z(5)$			−0.05050505	0.03496503
$Z(7)$				−0.01131382
		$Dy^{3+}f^9\ ^6H_{15/2}$		
$Z(1)$	5.00000000	2.66666667		
$Z(3)$		−0.22360680	−0.08131156	
$Z(5)$			0.02525253	−0.12432012
$Z(7)$				0.05656908

Table 11.1 (*contd.*)

	$\bar{I}_0$	$\bar{I}_2$	$\bar{I}_4$	$\bar{I}_6$
		$Ho^{3+}f^{10}\ ^5I_8$		
$Z(1)$	5.00000000	3.066666667		
$Z(3)$		−0.31304952	−0.28459047	
$Z(5)$			0.12626263	0.14763015
$Z(7)$				−0.11313815
		$Er^{3+}f^{11}\ ^4I_{15/2}$		
$Z(1)$	4.50000000	2.93333333		
$Z(3)$		−0.13416408	−0.16262313	
$Z(5)$			−0.02525253	−0.04662005
$Z(7)$				0.11313815
		$Tm^{3+}f^{12}\ ^3H_6$		
$Z(1)$	3.50000000	2.33333333		
$Z(3)$		0.22360680	0.08131156	
$Z(5)$			−0.17676768	−0.02719503
$Z(7)$				−0.05656908
		$Yb^{3+}f^{13}\ ^2F_{7/2}$		
$Z(1)$	2.00000000	1.33333333		
$Z(3)$		0.44721360	0.16262313	
$Z(5)$			0.10101010	0.01554002
$Z(7)$				0.01131382

Table 11.2

The orbital contribution $Z_L(K') = c_{K'}^{(L)}(\bar{J}_{K'-1} + \bar{J}_{K'+1})$ *of the coefficients* $Z(K')$ *given in Table* 11.1

	$\bar{J}_0, \bar{J}_2$	$\bar{J}_2, \bar{J}_4$	$\bar{J}_4, \bar{J}_6$
		$Ce^{3+}f^1\ ^2F_{5/2}$	
$Z_L(1)$	1.42857143		
$Z_L(3)$		0.15971914	
$Z_L(5)$			0.00721501
		$Pr^{3+}f^2\ ^3H_4$	
$Z_L(1)$	2.40000000		
$Z_L(3)$		0.0	
$Z_L(5)$			−0.03673095

Table 11.2 (*contd.*)

	$\bar{J}_0, \bar{J}_2$	$\bar{J}_2, \bar{J}_4$	$\bar{J}_4, \bar{J}_6$
	$Nd^{3+}\ f^3\ {}^4I_{9/2}$		
$Z_L(1)$	2.86363636		
$Z_L(3)$		−0.16717202	
$Z_L(5)$			0.02292473
	$Pm^{3+}\ f^4\ {}^5I_4$		
$Z_L(1)$	2.80000000		
$Z_L(3)$		−0.14488242	
$Z_L(5)$			0.01528316
	$Sm^{3+}\ f^5\ {}^6H_{5/2}$		
$Z_L(1)$	2.14285714		
$Z_L(3)$		0.0	
$Z_L(5)$			−0.00284228
	$Tb^{3+}\ f^8\ {}^7F_6$		
$Z_L(1)$	1.50000000		
$Z_L(3)$		0.22360680	
$Z_L(5)$			0.02525253
	$Dy^{3+}\ f^9\ {}^6H_{15/2}$		
$Z_L(1)$	2.50000000		
$Z_L(3)$		0.0	
$Z_L(5)$			−0.07575758
	$Ho^{3+}\ f^{10}\ {}^5I_8$		
$Z_L(1)$	3.00000000		
$Z_L(3)$		−0.22360680	
$Z_L(5)$			0.05050505
	$Er^{3+}\ f^{11}\ {}^4I_{15/2}$		
$Z_L(1)$	3.00000000		
$Z_L(3)$		−0.22360680	
$Z_L(5)$			0.05050505
	$Tm^{3+}\ f^{12}\ {}^3H_6$		
$Z_L(1)$	2.50000000		
$Z_L(3)$		0.0	
$Z_L(5)$			−0.07575758
	$Yb^{3+}\ f^{13}\ {}^2F_{7/2}$		
$Z_L(1)$	1.50000000		
$Z_L(3)$		0.22360680	
$Z_L(5)$			0.02525253

to $Z(3)$, so that the leading-order part of $M(\theta)$ in these cases arises solely from the spin contribution to the scattering.

The cross-section for scattering of unpolarized neutrons is proportional to

$$\hat{\mathbf{Q}}_{\perp} \cdot \hat{\mathbf{Q}}_{\perp} = \sum_{\alpha,\beta} (\delta_{\alpha,\beta} - \tilde{\kappa}_{\alpha}\tilde{\kappa}_{\beta})\hat{Q}_{\alpha}\hat{Q}_{\beta}.$$

From (11.125) it follows that

$$\sum_{\alpha,\beta} (\delta_{\alpha,\beta} - \tilde{\kappa}_{\alpha}\tilde{\kappa}_{\beta})\langle SLJJ|\, \hat{Q}_{\alpha}\, |SLJJ\rangle\langle SLJJ|\, \hat{Q}_{\beta}\, |SLJJ\rangle$$
$$= \sin^2\theta[\langle SLJJ|\, \hat{Q}_z\, |SLJJ\rangle - \cot\theta\, M(\theta)]^2, \quad (11.130)$$

which is independent of ϕ, i.e. the cross-section has cylindrical symmetry.

In the limit of zero wave vector the argument of the square bracket on the right-hand side of (11.130) has the value $\frac{1}{2}gJ$. Thus we identify the function

$$F(\kappa, \theta) = \frac{2}{gJ}[\langle SLJJ|\, \hat{Q}_z\, |SLJJ\rangle - \cot\theta\, M(\theta)] \qquad (11.131)$$

as the atomic form factor.

Figure 11.2 shows $F(\kappa, \theta)$ for various values of θ as a function of $|\boldsymbol{\kappa}|/4\pi$ for the tripositive rare-earth ions Pr^{3+}, Sm^{3+}, and Er^{3+}. The unusual form factor for Sm^{3+} has been observed in the compound SmN at low temperatures (Moon and Koehler 1979). The large coefficient of $\bar{j}_2$ decreases with increasing temperature. For Pr^{3+} and Er^{3+} the atomic form factor contracts as θ decreases, the effect being more pronounced for Pr^{3+} (first half of series) than for Er^{3+} (second half). It is also evident that the atomic form factor for Er^{3+} at $\theta = \frac{1}{2}\pi$ is contracted with respect to that for Pr^{3+} for the same value of θ. This can be understood in terms of the dipole approximation.

There is probably little difference in the values of $\bar{j}_2(\kappa)$ between the rare earths, except perhaps at large values of $|\boldsymbol{\kappa}|$. Hence let us examine the coefficient of $\bar{j}_2(\kappa)$ as given by the dipole approximation (eqn (11.110)), namely,

$$\frac{J(J+1) + L(L+1) - S(S+1)}{3J(J+1) + S(S+1) - L(L+1)}.$$

In the first half of the series, spin–orbit interaction couples $\mathbf{L}$ and $\mathbf{S}$ so as to give $J = L - S$, for which the coefficient is

$$\{1 - 2S/(L+1)\}^{-1},$$

whereas in the second half, $J' = L' + S'$, say, and the coefficient is

$$\{1 + 2S'/L'\}^{-1}.$$

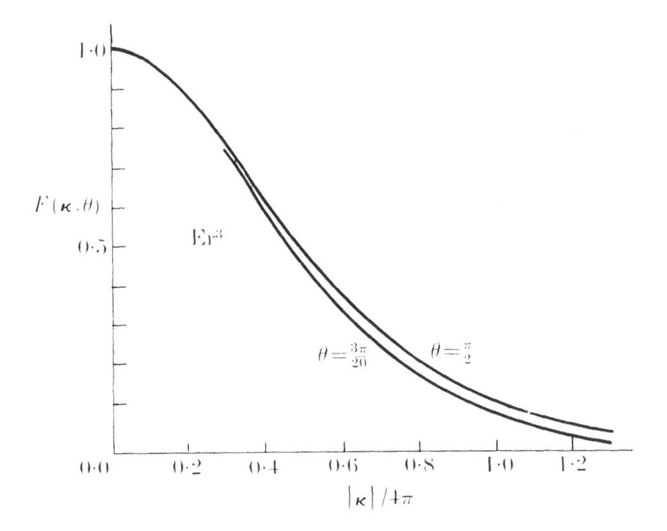

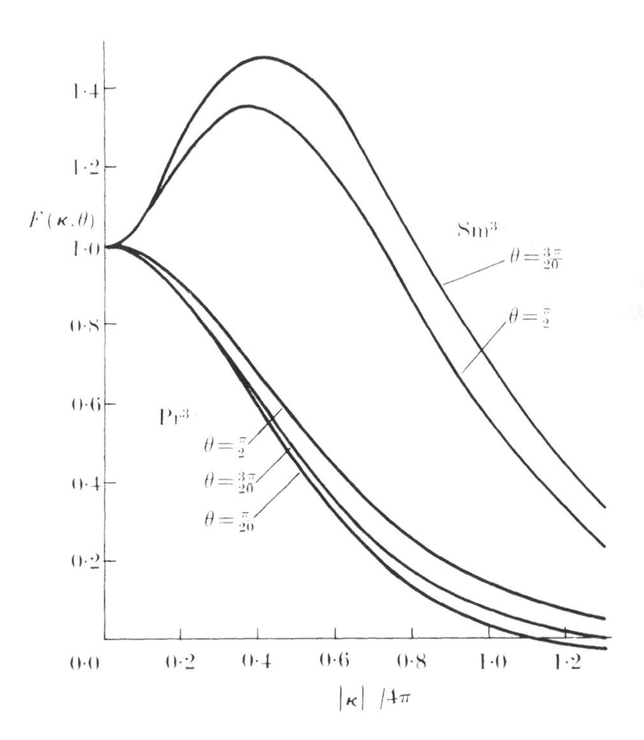

FIG. 11.2. The atomic form factor defined by eqn (11.131) is shown for the three rare-earth ions, Pr^{3+}, Sm^{3+}, and Er^{3+}, as a function of $|\boldsymbol{\kappa}|/4\pi$ for various values of θ.

Hence we see that, in the dipole approximation, the coefficient of $\bar{j}_2(\kappa)$ in $\langle SLJJ|\,\hat{Q}_z\,|SLJJ\rangle$ is always smaller in the second half of the rare earth series than in the first half, as shown in Fig. 11.2, for Pr^{3+} and Er^{3+}.

The calculation of $F(\kappa, \theta)$, as defined by (11.131), from the functions $Z(K')$ (eqn (11.127)) is greatly facilitated by making use of the following expression for

$$\sin\theta\langle SLJJ|\,\hat{Q}_z\,|SLJJ\rangle - \cos\theta\, M(\theta). \tag{11.132}$$

From (11.126) and (11.128) we see that (11.132) is equal to

$$(4\pi)^{1/2}\sum_{K'} Z(K')\mathrm{e}^{-\mathrm{i}\phi}\left\{\mathrm{e}^{\mathrm{i}\phi}\sin\theta\, Y_0^{K'-1}(\theta,\phi) - \cos\theta\left(\frac{K'-1}{K'}\right)^{1/2} Y_1^{K'-1}(\theta,\phi)\right\}.$$

Since

$$\mathrm{e}^{\mathrm{i}\phi}\sin\phi = -\left(\frac{8\pi}{3}\right)^{1/2} Y_1^1(\theta,\phi)$$

and

$$\cos\theta = \left(\frac{4\pi}{3}\right)^{1/2} Y_0^1(\theta,\phi),$$

we can use (11.30) to show that

$$\mathrm{e}^{\mathrm{i}\phi}\sin\theta\, Y_0^{K'-1}(\theta,\phi)$$
$$= -(2K'-1)^{-1/2}\left[\left\{\frac{K'(K'+1)}{(2K'+1)}\right\}^{1/2} Y_1^{K'}(\theta,\phi) - \left\{\frac{(K'-1)(K'-2)}{(2K'-3)}\right\}^{1/2} Y_1^{K'-2}(\theta,\phi)\right]$$

and

$$\cos\theta\, Y_1^{K'-1}(\theta,\phi)$$
$$= (2K'-1)^{-1/2}\left[\left\{\frac{(K'-1)(K'+1)}{(2K'+1)}\right\}^{1/2} Y_1^{K'}(\theta,\phi) + \left\{\frac{K'(K'-2)}{(2K'-3)}\right\}^{1/2} Y_1^{K'-2}(\theta,\phi)\right].$$

Hence

$$\mathrm{e}^{\mathrm{i}\phi}\sin\theta\, Y_0^{K'-1}(\theta,\phi) - \cos\theta\left(\frac{K'-1}{K'}\right)^{1/2} Y_1^{K'-1}(\theta,\phi)$$
$$= -\left\{\frac{(K'+1)(2K'-1)}{K'(2K'+1)}\right\}^{1/2} Y_1^{K'}(\theta,\phi),$$

and therefore

$$\sin\theta\langle SLJJ|\,\hat{Q}_z\,|SLJJ\rangle - \cos\theta\, M(\theta)$$
$$= -(4\pi)^{1/2}\mathrm{e}^{-\mathrm{i}\phi}\sum_{K'} Z(K')\left\{\frac{(K'+1)(2K'-1)}{K'(2K'+1)}\right\}^{1/2} Y_1^{K'}(\theta,\phi), \tag{11.133}$$

which is the desired relationship.

In those cases where it is appropriate to average the cross-section overall directions of the scattering vector, the relation

$$\int_0^{2\pi} \mathrm{d}\phi \int_0^{\pi} \sin\theta \, \mathrm{d}\theta \, Y_{Q'}^{*K'}(\theta, \phi) Y_{Q''}^{K''}(\theta, \phi) = \delta_{K',K''} \delta_{Q',Q''} \quad (11.134)$$

in conjunction with (11.133) gives the result

$$\frac{1}{4\pi} \int \mathrm{d}\tilde{\boldsymbol{\kappa}} \sum_{\alpha,\beta} (\delta_{\alpha,\beta} - \tilde{\kappa}_\alpha \tilde{\kappa}_\beta) \langle SLJJ | \hat{Q}_\alpha | SLJJ \rangle \langle SLJJ | \hat{Q}_\beta | SLJJ \rangle$$

$$= \sum_{K'} Z^2(K') \frac{(K'+1)(2K'-1)}{K'(2K'+1)}. \quad (11.135)$$

11.6.2. Crystal-field effects (*Boucherle, et al.* 1982)

When an ion is placed in a crystal it does, of course, experience many interactions that are absent for a free ion. For instance, there are electric and magnetic interactions with other ions in the crystal and to some degree a coupling to the crystal lattice. Even so, many of the properties of rare-earth ions have been well explained on the simple model of only taking into account the interaction arising from the electrostatic field created by neighbouring ions (Crow *et al.* 1980).

The crystal field, as it is called, possesses the point-group symmetry of the crystal lattice. Hence, to determine the allowed levels in the presence of a crystal field it is only necessary to determine what irreducible representations of the point group are contained in the reducible representation of the full rotation group of order J (Elliott and Dawber 1979). Consider, for example, a normal lattice site in a sodium chloride structure. Here the appropriate point group is the cubic group O whose irreducible representations we denote by Γ_n (Bethe's notation) with $n = 1$–8. Of these only Γ_1 and Γ_2 are non-degenerate. Γ_3, Γ_6, and Γ_7 are doubly degenerate, Γ_4 and Γ_5 are triply, and Γ_8 is fourfold degenerate. The decompositions for values of J appropriate to tripositive rare-earth ions are given in Table 11.3. From this table we see that for none of the tripositive rare-earth ions does a cubic crystal field completely remove the $2J+1$ degeneracy of the ground-state wave function.

Let us now turn to the evaluation of the elastic cross-section for rare earths in a crystal field. The ground-state wave functions $|\Gamma_n v\rangle$ are given by

$$|\Gamma_n v\rangle = \sum_M |SLJM\rangle \langle SLJM | \Gamma_n v\rangle, \quad (11.136)$$

where the label v distinguishes between the functions that belong to the representation Γ_n. In most cases the magnetic field experienced by the

Table 11.3

Decomposition of reducible representation of full rotation group of order J into irreducible representations of the cubic group O as appropriate for tripositive rare-earth ions in cubic crystals.

J	5/2	7/2	4	9/2
Decomposition into irred. rep.	$\Gamma_7+\Gamma_8$	$\Gamma_6+\Gamma_7+\Gamma_8$	$\Gamma_1+\Gamma_3+\Gamma_4+\Gamma_5$	$\Gamma_6+2\Gamma_8$
J	6		15/2	8
Decomposition into irred. rep.	$\Gamma_1+\Gamma_2+\Gamma_3+\Gamma_4+2\Gamma_5$		$\Gamma_6+\Gamma_7+3\Gamma_8$	$\Gamma_1+2\Gamma_3+2\Gamma_4+2\Gamma_5$

ions (here assumed small compared to the crystal field) will remove the degeneracy and make one of the $|\Gamma_n v\rangle$ the appropriate ground state.

The contribution to the cross-section is determined by

$$\langle\lambda|\,\hat{\mathbf{Q}}_\perp^+\,|\lambda\rangle\cdot\langle\lambda|\,\hat{\mathbf{Q}}_\perp\,|\lambda\rangle\equiv\langle\Gamma_n v|\,\hat{\mathbf{Q}}_\perp\,|\Gamma_n v\rangle\cdot\langle\Gamma_n v|\,\hat{\mathbf{Q}}_\perp\,|\Gamma_n v\rangle$$
$$=\sum_{\alpha,\beta}(\delta_{\alpha,\beta}-\tilde{\kappa}_\alpha\tilde{\kappa}_\beta)\langle\Gamma_n v|\,\hat{Q}_\alpha^+\,|\Gamma_n v\rangle\,\langle\Gamma_n v|\,\hat{Q}_\beta\,|\Gamma_n v\rangle. \tag{11.137}$$

Hence, in view of (11.136), we require the matrix elements

$$\langle SLJM|\,\hat{Q}_\alpha\,|SLJM'\rangle.$$

These can also be expressed in terms of the characteristic functions Z introduced in § 11.6.1.

From eqn (11.87b),

$$\langle SLJM|\,\hat{Q}_q\,|SLJM'\rangle$$
$$=(-1)^{J-M-q}\{12\pi(2J+1)\}^{1/2}\sum_{K',Q'}\left(\frac{2K'+1}{K'+1}\right)Y^{K'-1}_{Q'+q}(\tilde{\boldsymbol{\kappa}})$$
$$\times\{A(K'-1,K')+B(K'-1,K')\}\begin{pmatrix}K' & J & J\\ -Q' & M' & -M\end{pmatrix}\begin{pmatrix}K'-1 & K' & 1\\ Q'+q & -Q' & -q\end{pmatrix} \tag{11.138}$$

where $Q'=M'-M$. The matrix elements of $\hat{Q}_x$ and $\hat{Q}_y$ contain

$$Y^{K'-1}_{Q'+1}\begin{pmatrix}K'-1 & K' & 1\\ Q'+1 & -Q' & -1\end{pmatrix}\pm Y^{K'-1}_{Q'-1}\begin{pmatrix}K'-1 & K' & 1\\ Q'-1 & -Q' & 1\end{pmatrix}. \tag{11.139}$$

Only one of the $3j$ symbols need be evaluated since the other can be obtained merely by reversing the sign of Q'. We find

$$\begin{pmatrix} K'-1 & K' & 1 \\ Q'+1 & -Q' & -1 \end{pmatrix} = (-1)^{K'-Q'} \left\{ \frac{(K'-Q')(K'-Q'-1)}{(2K'-1)2K'(2K'+1)} \right\}^{1/2},$$

so that (11.139) becomes

$$(-1)^{K'-Q'}\{(2K'-1)2K'(2K'+1)\}^{-1/2}$$
$$\times [Y^{K'-1}_{Q'+1}\{(K'-Q')(K'-Q'-1)\}^{1/2} \pm Y^{K'-1}_{Q'-1}\{(K'+Q')(K'+Q'-1)\}^{1/2}].$$

Hence, for example,

$$\langle SLJM| \hat{Q}_x |SLJM'\rangle = \left(-\frac{1}{\sqrt{2}}\right)(4\pi)^{1/2} \sum_{K'Q'} Z(K')P(K', Q')$$
$$\times [Y^{K'-1}_{Q'+1}(\tilde{\boldsymbol{\kappa}})\{(K'-Q')(K'-Q'-1)\}^{1/2} - Y^{K'-1}_{Q'-1}(\tilde{\boldsymbol{\kappa}})\{(K'+Q')(K'+Q'-1)\}^{1/2}]$$
$$\times \{(2K'-1)2K'(2K'+1)\}^{-1/2} \begin{pmatrix} K'-1 & K' & 1 \\ 0 & 0 & 0 \end{pmatrix}^{-1},$$

where we have defined

$$P(K', Q') = (-1)^{J-M'} \begin{pmatrix} K' & J & J \\ -Q' & M' & -M \end{pmatrix} \begin{pmatrix} K' & J & J \\ 0 & J & -J \end{pmatrix}^{-1}. \tag{11.140}$$

Since

$$\{(2K'-1)2K'(2K'+1)\}^{-1/2} \begin{pmatrix} K'-1 & K' & 1 \\ 0 & 0 & 0 \end{pmatrix}^{-1} = -\frac{1}{K'\sqrt{2}},$$

we obtain for the matrix element of $\hat{Q}_x$ the result

$$\langle SLJM| \hat{Q}_x |SLJM'\rangle = \frac{i}{2}(4\pi)^{1/2} \sum_{K',Q'} \frac{Z(K')}{K'} P(K', Q')$$
$$\times [Y^{K'-1}_{Q'+1}(\tilde{\boldsymbol{\kappa}})\{(K'-Q')(K'-Q'-1)\}^{1/2} - Y^{K'-1}_{Q'-1}(\tilde{\boldsymbol{\kappa}})\{(K'+Q')(K'+Q'-1)\}^{1/2}] \tag{11.141}$$

and for $\hat{Q}_y$

$$\langle SLJM| \hat{Q}_y |SLJM'\rangle = -\tfrac{1}{2}(4\pi)^{1/2} \sum_{K',Q'} \frac{Z(K')}{K'} P(K', Q')$$
$$\times [Y^{K'-1}_{Q'+1}(\tilde{\boldsymbol{\kappa}})\{(K'-Q')(K'-Q'-1)\}^{1/2} + Y^{K'-1}_{Q'-1}(\tilde{\boldsymbol{\kappa}})\{(K'+Q')(K'+Q'-1)\}^{1/2}]. \tag{11.142}$$

Finally, for $\hat{Q}_0$, we note that

$$\begin{pmatrix} K'-1 & K' & 1 \\ Q' & -Q' & 0 \end{pmatrix} \begin{pmatrix} K'-1 & K' & 1 \\ 0 & 0 & 0 \end{pmatrix}^{-1} = \frac{(-1)^{Q'}}{K'}\{(K'-Q')(K'+Q')\}^{1/2}$$

and hence

$$\langle SLJM|\, \hat{Q}_z \,|SLJM'\rangle$$
$$= (4\pi)^{1/2} \sum_{K',Q'} \frac{Z(K')}{K'} P(K',Q') Y_{Q'}^{K'-1}(\tilde{\boldsymbol{\kappa}})\{(K'-Q')(K'+Q')\}^{1/2}. \quad (11.143)$$

The following property of the matrix elements (11.141), (11.142), and (11.143) are of help in actual calculations. If J is an integer,

$$\langle SLJM|\, \hat{Q}_\alpha \,|SLJM'\rangle = (-1)^{M'-M+1}\langle SLJ-M'|\, \hat{Q}_\alpha \,|SLJ-M\rangle.$$

Another property that it is worthwhile to recall is that if $|\lambda\rangle$ is a singlet state $\langle\lambda|\, \hat{Q}_\alpha \,|\lambda\rangle = 0$, i.e. the crystal field quenches out the magnetization.

11.7. Cross-section for transitions between crystal-field levels of a rare-earth ion (Balcar and Lovesey 1970; Coqblin 1977; Furrer 1977; Edelstein 1982)

As we mentioned in the preceding section, the effect of the crystal field on a rare-earth ion is to partially or totally remove the $2J+1$ degeneracy of the ground state.

Transitions between the crystal-field levels can be observed in neutron scattering. If the exchange interaction between the ions is weak, it is sufficient to consider the cross-section for scattering by a single isolated ion, in which case scattering will be present at all values of $|\boldsymbol{\kappa}|$, as in paramagnetic scattering. One condition that must be satisfied is that the total spread in energy of the levels must be comparable to the energy of the incident neutrons. If this is not so, only an average over the level structure will be observed. We do not consider the complications that occur in intermediate-valence compounds or heavy-fermion systems (Huber 1984; Sinha 1984; Aeppli *et al.* 1987).

In discussing the cross-section for transitions between crystal-field levels within the isolated ion model we shall pay particular attention to the case where the crystal field has cubic symmetry. The irreducible representations of the cubic group are here denoted by Γ_n (Bethe's notation) and the corresponding wave functions by $|\Gamma_n\, v\rangle$, where the label v distinguishes the degenerate wave functions. It follows from eqn (7.11b) that the partial differential cross-section per ion for the transition $\Gamma_n \rightarrow \Gamma_{n'}$ is

$$\frac{d^2\sigma}{d\Omega\, dE'} = r_0^2 \frac{k'}{k} \sum_{\alpha,\beta} (\delta_{\alpha,\beta} - \tilde{\kappa}_\alpha\tilde{\kappa}_\beta)$$
$$\times \sum_{v,v'} p_n \langle\Gamma_n v|\, \hat{Q}_\alpha^+ \,|\Gamma_{n'} v'\rangle\langle\Gamma_{n'} v'|\, \hat{Q}_\beta \,|\Gamma_n v\rangle\delta(\hbar\omega + E_n - E_{n'}).$$
$$(11.144)$$

Here p_n is the product of the probability that the Γ_n state is occupied and the appropriate degeneracy weighting factor, and E_n and $E_{n'}$ are, respectively, the energies of the Γ_n and $\Gamma_{n'}$ levels.

In general all the transitions will have a non-zero amplitude at some value of the scattering vector. However, just as in optical spectroscopy, there are strong and weak transitions. The strong transitions are those that occur for zero wave vector, that is to say, those allowed in the dipole approximation (dipole transitions). In this instance (cf. eqn (11.121)

$$\hat{\mathbf{Q}} \simeq \tfrac{1}{2} g F(\kappa) \hat{\mathbf{J}} \tag{11.145}$$

and we have therefore to determine for what levels Γ_n and $\Gamma_{n'}$ there is a non-zero matrix element $\langle \Gamma_n | \hat{J}^\alpha | \Gamma_{n'} \rangle$. This is a well-known question in optical spectroscopy and it is easily answered with a little group theory (Elliott and Dawber 1979).

In a cubic crystal field $\hat{\mathbf{J}}$ transforms as Γ_4. the matrix element $\langle \Gamma_n | \hat{J}^\alpha | \Gamma_{n'} \rangle$ is non-zero if the representation $\Gamma_n \times \Gamma_4 \times \Gamma_{n'}$ contains the unit representation Γ_1. To be specific, consider the simple case of Pr^{3+} ($f^2\ {}^3H_4$). Reference to Table 11.3 shows that the state $J = 4$ splits into four levels in a cubic crystal field, as illustrated in Fig. 11.3. To determine whether or not the $\Gamma_3 \leftrightarrow \Gamma_5$ transition is allowed or not in the dipole approximation we proceed as follows. The direct products of the representations $\Gamma_n \times \Gamma_{n'}$ are listed in Table 11.4, from which we see that $\Gamma_4 \times \Gamma_5 = \Gamma_2 + \Gamma_3 + \Gamma_4 + \Gamma_5$. Since $\Gamma_n \times \Gamma_n$ contains the unit representation the presence of Γ_3 in $\Gamma_4 \times \Gamma_5$ means that $\Gamma_3 \leftrightarrow \Gamma_5$ is indeed an allowed dipole transition. Similar calculations for the other possible transitions show that the dipole transitions for Pr^{3+} in a cubic crystal field are $\Gamma_1 \leftrightarrow \Gamma_4$, $\Gamma_4 \leftrightarrow \Gamma_5$, $\Gamma_3 \leftrightarrow \Gamma_4$, and $\Gamma_3 \leftrightarrow \Gamma_5$.

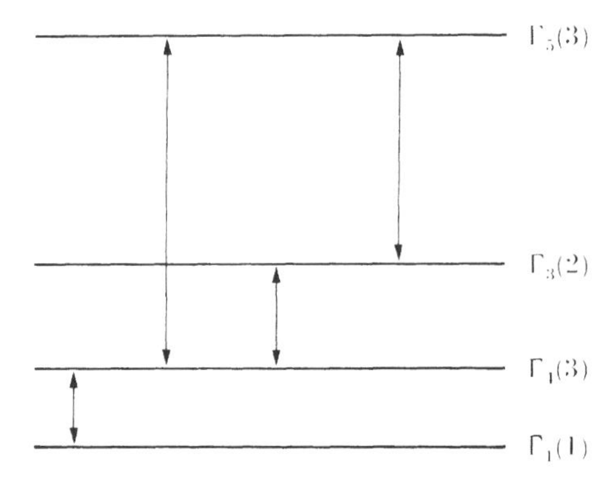

FIG. 11.3. Schematic diagram of crystal-field levels of Pr^{3+} in a cubic crystal field. The allowed dipole transitions are marked, together with the degenercies of the various crystal-field levels.

Table 11.4

Direct products $\Gamma_n \times \Gamma_{n'}$ of one- and two-valued representations of the cubic group O

	Γ_1	Γ_2	Γ_3	Γ_4	Γ_5	Γ_6	Γ_7	Γ_8
Γ_1	Γ_1	Γ_2	Γ_3	Γ_4	Γ_5	Γ_6	Γ_7	Γ_8
Γ_2	Γ_2	Γ_1	Γ_3	Γ_5	Γ_4	Γ_7	Γ_6	Γ_8
Γ_3	Γ_3	Γ_3	$\Gamma_1+\Gamma_2+\Gamma_3$	$\Gamma_4+\Gamma_5$	$\Gamma_4+\Gamma_5$	Γ_8	Γ_8	$\Gamma_6+\Gamma_7+\Gamma_8$
Γ_4	Γ_4	Γ_5	$\Gamma_4+\Gamma_5$	$\Gamma_1+\Gamma_3+\Gamma_4+\Gamma_5$	$\Gamma_2+\Gamma_3+\Gamma_4+\Gamma_5$	$\Gamma_6+\Gamma_8$	$\Gamma_7+\Gamma_8$	$\Gamma_6+\Gamma_7+2\Gamma_8$
Γ_5	Γ_5	Γ_4	$\Gamma_4+\Gamma_5$	$\Gamma_2+\Gamma_3+\Gamma_4+\Gamma_5$	$\Gamma_1+\Gamma_3+\Gamma_4+\Gamma_5$	$\Gamma_7+\Gamma_8$	$\Gamma_6+\Gamma_8$	$\Gamma_6+\Gamma_7+2\Gamma_8$
Γ_6	Γ_6	Γ_7	Γ_8	$\Gamma_6+\Gamma_8$	$\Gamma_7+\Gamma_8$	$\Gamma_1+\Gamma_4$	$\Gamma_2+\Gamma_5$	$\Gamma_3+\Gamma_4+\Gamma_5$
Γ_7	Γ_7	Γ_6	Γ_8	$\Gamma_7+\Gamma_8$	$\Gamma_6+\Gamma_8$	$\Gamma_2+\Gamma_5$	$\Gamma_1+\Gamma_4$	$\Gamma_3+\Gamma_4+\Gamma_5$
Γ_8	Γ_8	Γ_8	$\Gamma_6+\Gamma_7+\Gamma_8$	$\Gamma_6+\Gamma_7+2\Gamma_8$	$\Gamma_6+\Gamma_7+2\Gamma_8$	$\Gamma_3+\Gamma_4+\Gamma_5$	$\Gamma_3+\Gamma_4+\Gamma_5$	$\Gamma_1+\Gamma_2+\Gamma_3+2\Gamma_4+2\Gamma_5$

Table 11.5

The quantity defined by eqn (11.146), *which determines the cross-section for transitions between crystal-field levels, is tabulated for all six transitions appropriate to* Pr^{3+} *in a cubic crystal field*

n n'	$\bar{J}_0^2$	$\bar{J}_0\bar{J}_2$	$\bar{J}_2^2$	$\bar{J}_2\bar{J}_4$	$\bar{J}_4^2$	$\bar{J}_4\bar{J}_6$	$\bar{J}_6^2$
1—4	2.1333	7.0163	5.7805	0.0316	0.0272	0.0716	0.2292
3—5	1.2800	4.2098	3.4861	0.0673	0.0556	0.1232	0.3976
4—5	1.1200	3.8636	3.0813	0.1436	0.1067	0.1117	0.3659
4—3	2.9867	9.8228	8.1077	0.0849	0.0659	0.1021	0.3295
1—5	0.0	0.0	0.0248	0.0677	0.0464	0.0032	0.0144
1—3	0.0	0.0	0.0	0.0	0.0060	0.0758	0.2428

As far as the two remaining transitions, we know that they have zero amplitude at $|\boldsymbol{\kappa}|=0$ and also $|\boldsymbol{\kappa}|\to\infty$. In order to determine whether or not they are significant it is necessary to evaluate the cross-section in full. The results for Pr^{3+} are shown in Table 11.5. The actual quantity listed for each of the six transitions is

$$\frac{1}{4\pi}\int d\tilde{\boldsymbol{\kappa}} \sum_{\alpha\beta}(\delta_{\alpha,\beta}-\tilde{\kappa}_\alpha\tilde{\kappa}_\beta)\sum_{\upsilon,\upsilon'}\langle\Gamma_n\upsilon|\,\hat{Q}^+_\alpha\,|\Gamma_{n'}\upsilon'\rangle\langle\Gamma_{n'}\upsilon'|\,\hat{Q}_\beta\,|\Gamma_n\upsilon\rangle, \qquad (11.146)$$

which determines the cross-section averaged over all directions of the scattering vector. For each pair of levels (11.146) is in general the sum of products of two radial integrals $\bar{j}_K$, defined by eqn (11.38), and Table 11.5 contains the numerical coefficients associated with the various products. We note that for the two transitions not allowed in the dipole approximation, $\Gamma_1 \leftrightarrow \Gamma_5$ and $\Gamma_1 \leftrightarrow \Gamma_3$, there are no terms involving $\bar{j}_0$. Since this is the only radial integral that is non-zero for $|\boldsymbol{\kappa}|=0$, this result is in accord with our previous findings. A comparison of the coefficients of the remaining terms clearly shows that, in this particular case, the only transitions of importance are the dipole transitions shown in Fig. 11.3.

11.8. 3d transition-metal ions (Balcar *et al.* 1973)

Our discussion of 3d transition-metal ions is not quite so general as the preceding discussions of elastic and inelastic scattering by rare earths. The crystal-field perturbation is usually larger than the spin–orbit coupling and, therefore, wave functions generally have several components. Moreover, a range of ionicities is found for transition-metal ions. In consequence, determination of the wave functions is often a significant undertaking, and the form of the ground-state wave function might be ambiguous in the absence of pertinent experimental observations. Given the wave functions, the spin and orbital matrix elements in the cross-section are obtained from (11.97).

To illustrate the use of eqn (11.97) we consider a model for elastic scattering from $V^{3+}(3d^2)$. The calculated cross-section shows several features that are of some general interest.

The compound V_2O_3 is an antiferromagnetic insulator, below about 150 K, in which each vanadium ion possesses a moment of only $1.2\mu_B$ (Moon 1970). The Hund's rule ground-state configuration ^3F in an octahedral crystal field, together with a weak spin–orbit coupling, leads to a normalized wave function of the form,

$$|G\rangle=\{\epsilon\,|M_L=-3\rangle+(1-\epsilon^2)^{1/2}\,|M_L=1\rangle\}\,|M_S=1\rangle, \qquad (11.147)$$

where the parameter ϵ is to be determined, and

$$|M_L\rangle\equiv|L=3,\,M_L\rangle, \qquad |M_S\rangle\equiv|S=1,\,M_S\rangle.$$

We will not present arguments to support the validity of (11.147) as the ground-state wave function of the vanadium ions in V_2O_3, since we are primarily concerned here with the use of (11.97). Note, however, that the wave function (11.147) is unusually simple for a transition-metal ion because there is no explicit dependence on the spin–orbit coupling. By way of a contrast, the reader might wish to consult Brown and Forsyth (1981) which describes the wave function of another ^{3}F ion, $Ni^{2+}(3d^8)$, that is considerably more complicated.

The parameter ϵ in (11.147) is chosen to give a moment of $1.2\mu_B$, namely

$$\langle G|\,(\hat{\mathbf{L}}+2\hat{\mathbf{S}})_\alpha\,|G\rangle = \delta_{\alpha,z}\{-3\epsilon^2+(1-\epsilon^2)+2\}$$
$$= 1.2;\quad \epsilon^2 = 0.45 \quad \text{and} \quad \alpha = z. \qquad (11.148)$$

To obtain a first approximation to the elastic magnetic cross-section we can use the dipole approximation, (11.118). Thus

$$\langle G|\,\hat{Q}_\alpha^{(D)}\,|G\rangle = \tfrac{1}{2}\delta_{\alpha,z}\{(\bar{j}_0+\bar{j}_2)(1-4\epsilon^2)+2\bar{j}_0\} \qquad (11.149)$$

and, from (11.148), $1-4\epsilon^2 = -0.8$, so that the orbital moment leads to a contraction of the form factor, or equivalently an expansion of the magnetic wave function. The main effect of the orbital momentum on the form factor is seen at those values of $\boldsymbol{\kappa}$ at which $\bar{j}_2$ has its maximum value.

We turn now to the calculation of the complete elastic cross-section for the wave function (11.147). It is useful in the matrix elements derived from (11.97) to show the dependence on ϵ explicitly. There is only one parent state of d^2, namely $\bar{L}=2$, $\bar{S}=\frac{1}{2}$. All the matrix elements are diagonal in S, L, and M_S, with $L=L'=3$, $S=S'=1$, and $M_S=M'_S=1$. From this we conclude that there are no transverse components of the spin operator.

For the configuration appropriate to (11.147), the spin matrix element (11.97a) reduces to

$$\langle SM_SLM_L|\sum \exp(\mathrm{i}\boldsymbol{\kappa}\cdot\mathbf{r})\hat{s}_q\,|SM_SLM'_L\rangle$$
$$= (4\pi)^{1/2}\,\delta_{q,0}\,35(-1)^{M_L}\sum_{KQ}\mathrm{i}^K\bar{j}_K Y_Q^{*K}(\tilde{\boldsymbol{\kappa}})(2K+1)^{1/2}\begin{pmatrix}2 & K & 2\\ 0 & 0 & 0\end{pmatrix}$$
$$\times\begin{pmatrix}3 & K & 3\\ -M_L & Q & M'_L\end{pmatrix}\begin{Bmatrix}3 & K & 3\\ 2 & 2 & 2\end{Bmatrix}$$
$$= (4\pi)^{1/2}\,\delta_{q,0}(-1)^{M_L}\Bigg\{\bar{j}_0 Y_0^0\delta_{M_L,M'_L}(-1)^{M_L}$$
$$-\bar{j}_2 Y_Q^{*2}(3/7)^{1/2}\begin{pmatrix}3 & 2 & 3\\ -M_L & Q & M'_L\end{pmatrix}+\bar{j}_4 Y_Q^{*4}\,3(11/14)^{1/2}\begin{pmatrix}3 & 4 & 3\\ -M_L & Q & M'_L\end{pmatrix}\Bigg\}.$$
$$(11.150)$$

Eqn (11.97b) reduces to

$$\kappa^2(4\pi)^{1/2}(-1)^{M_L+1}\Big\{Y_0^0[\bar{j}_0+\bar{j}_2]\sqrt{21}\begin{pmatrix}3 & 1 & 3\\ M_L' & q & -M_L\end{pmatrix}$$

$$+Y_Q^2[\bar{j}_2+\bar{j}_4]\frac{3\sqrt{2}}{2}(-1)^q\begin{pmatrix}2 & 3 & 1\\ Q & Q' & -q\end{pmatrix}\begin{pmatrix}3 & 3 & 3\\ M_L' & Q' & -M_L\end{pmatrix}\Big\}. \quad (11.151)$$

Note that there is no contribution from a cross-term with $M_L=-3$, $M_L'=1$, and similarly for $M_L=1$, $M_L'=-3$. From the definition of the spherical components of $\hat{\mathbf{Q}}$, it follows from (11.147) and (11.151) that

$$\langle G|\,\hat{Q}_x\,|G\rangle = M(\theta)\cos\phi$$

and

$$\langle G|\,\hat{Q}_y\,|G\rangle = M(\theta)\sin\phi \quad (11.152)$$

with (note that M is independent of ϵ)

$$M(\theta)=(4\pi)^{1/2}\tfrac{1}{14}(\tfrac{6}{5})^{1/2}[\bar{j}_2+\bar{j}_4]e^{-i\phi}Y_1^2(\hat{\boldsymbol{\kappa}})$$

Here $\hat{\boldsymbol{\kappa}}=(\sin\theta\cos\phi,\ \sin\theta\sin\phi,\ \cos\theta)$.

Denote the $q=0$ matrix element of $\hat{\mathbf{Q}}$ by I, then from (11.152) we have,

$$\langle G|\,\hat{\mathbf{Q}}\,|G\rangle=(M\cos\phi,\ M\sin\phi,\ I)$$

and consequently (cf. (11.120)),

$$\langle G|\,\hat{\mathbf{Q}}_\perp^+\,|G\rangle\cdot\langle G|\,\hat{\mathbf{Q}}_\perp\,|G\rangle=\sin^2\theta[I(\theta,\phi)-M(\theta)\cot\theta]^2.$$

The argument of the square bracket for $\boldsymbol{\kappa}=0$ is $\frac{1}{2}(3-4\epsilon^2)$. We therefore define a form factor $F(\kappa)$ as

$$F(\boldsymbol{\kappa})=\left(\frac{2}{3-4\epsilon^2}\right)[I(\theta,\phi)-M(\theta)\cot\theta] \quad (11.153)$$

where, from (11.150) and (11.151),

$$I(\theta,\phi)=(4\pi)^{1/2}\Big\{\tfrac{1}{2}Y_0^0[\bar{j}_0(3-4\epsilon^2)+\bar{j}_2(1-4\epsilon^2)]$$

$$+Y_0^2\frac{1}{14\sqrt{5}}[8\epsilon^2\bar{j}_2+3\bar{j}_4]-Y_0^4\bar{j}_4\tfrac{1}{14}(2\epsilon^2+1)-\epsilon(1-\epsilon^2)^{1/2}(Y_4^4+Y_{-4}^4)\bar{j}_4(\tfrac{3}{14})^{1/2}\Big\}$$

$$(11.154)$$

and

$$-M(\theta)\cot\theta=\tfrac{3}{14}\cos^2\theta[\bar{j}_2+\bar{j}_4] \quad (11.155)$$

$$(Y_4^4+Y_{-4}^4)(\tfrac{3}{14})^{1/2}=\left(\frac{15}{\pi}\right)^{1/2}\tfrac{3}{16}\sin^4\theta\cos 4\phi$$

Several features of the result are worth explicit mention. First, the coefficient of $(4\pi)^{1/2} Y_0^0$ is identical to the dipole-approximation equation (11.149). This is in fact a general result as noted in § 11.4. Secondly, the transverse components of $\hat{\mathbf{Q}}$ eqn (11.152) are not observed with the conventional scattering geometry in which $\theta = \pi/2$ because $M(\theta)$ is proportional to $\sin(2\theta)$. Note also that the transverse components are independent of ϵ, and that their maximum contribution is at wave vectors where $\bar{\jmath}_2$ has its first maximum. Thirdly, the only dependence on ϕ is from the $Y_{\pm 4}^4$ terms in $\langle G| \hat{\mathbf{Q}}_{\text{spin}} |G\rangle$.

The fourth, and final, point we draw attention to is the ϵ-dependence of the form factor. For $\epsilon = \frac{1}{2}$, $\langle G| \hat{L}_z |G\rangle = 0$, and consequently there is no $\bar{\jmath}_0$ contribution to the form factor from $\hat{\mathbf{Q}}_{\text{orbital}}$. However, both I and M in eqn (11.153) for the form factor contain terms in $\bar{\jmath}_2$ and $\bar{\jmath}_4$ from $\hat{\mathbf{Q}}_{\text{orbital}}$ for this value of ϵ.

11.9. Intermultiplet transitions (Williams *et al.* 1987; Balcar and Lovesey 1986)

The cross-section for a transition between multiplets of a rare-earth ion is readily obtained from the general theory summarized in § 11.4. Because $B(K', K')$ is non zero in this instance full expressions for the magnetic amplitude are longer than for elastic scattering. The appropriate cross-section is akin to (11.144) and the intensity of dipole transitions, $J' = J \pm 1$, is finite for $\boldsymbol{\kappa} \to 0$. Using the approximate result (11.118) together with $\langle J|\hat{\mathbf{J}}|J'\rangle = 0$ we readily obtain the following estimate of the intensity of dipole-allowed transitions, valid for $J' = J + 1$ and $J = L - S$

$$(\bar{\jmath}_0 - \bar{\jmath}_2)^2(L + S + J + 2)(L + S - J)/24(J + 1). \tag{11.156}$$

In the second half of the rare-earth series, $J = L + S$, $J' = J - 1$ and the factor multiplying the radial integrals is

$$\{J^2 - (L - S)^2\}(2J + 1)/24J(2J - 1). \tag{11.157}$$

Observe that the κ-dependence of the intensity is different from the (elastic) atomic form factor; this feature is particularly striking for Sm^{3+}, Fig. 11.2, studied by Williams *et al.* (1987).

REFERENCES

Aeppli, G., Goldman, A., Shirane, G., Bucher, E., and Lux-Steiner, M.-Ch. (1987). *Phys. Rev. Lett.* **58,** 808.
Balcar, E. (1975). *J. Phys.* **C8,** 1581.
—— and Lovesey, S. W. (1970). *Phys. Lett.* **31A,** 67.
—— and Lovesey, S. W. (1986). *J. Phys.* **C19,** 4605.

——, ——, and Wedgwood, F. A. (1970). *J. Phys.* **C3,** 1292.
——, ——, and —— (1973). *J. Phys.* **C6,** 3746.
Blume, M. (1961). *Phys. Rev.* **124,** 96.
Boucherle, J. X., Givord, D., and Schweizer, J. (1982). *J. Physique* **C7,** 199.
Brooks, M. S. S. and Kelly, P. J. (1983). *Phys. Rev. Lett.* **51,** 1708.
Brown, P. J. and Forsyth, J. B. (1981). *J. Phys.* **C14,** 5171.
Condon, E. U. and Odabaşi, H. (1980). *Atomic structure.* Cambridge University Press, Cambridge.
—— and Shortley, G. H. (1935). *The theory of atomic spectra.* Cambridge University Press, Cambridge.
Coqblin, B. (1977). *The electronic structure of rare-earth metals and alloys: the magnetic heavy rare-earths.* Academic Press, New York.
Crow, J. E., Guertin, R. P., and Mihalisin, R. W. (Eds.) (1980). *Crystalline electric field and structural effects in f-electron systems.* Plenum Press, New York.
Edelstein, N. M. (Ed.) (1982). *Actinides in perspective.* Pergamon Press, Oxford.
Elliott, J. P. and Dawber, P. G. (1979). *Symmetry in physics,* Vols. 1 and 2. The Macmillan Press, London.
Furrer, A. (Ed.) (1977). *Crystal field effects in metals and alloys.* Plenum Press, New York.
Harrison, W. A. (1980). *Electronic structure and the properties of solids.* W. H. Freeman, New York.
Huber, D. L. (1984) *Phys. Rev.* **B29,** 456.
Judd, B. R. (1963). *Operator techniques in atomic spectroscopy.* McGraw-Hill, New York.
Kleiner, W. H. (1955). *Phys. Rev.* **97,** 411.
Lines, M. E. and Bösch, M. A. (1983). *Comments on Solid State Phys.* **XI,** 73.
Lovesey, S. W. (1978). *J. Phys.* **C11,** 3971.
—— and Rimmer, D. E. (1969). *Rep. Prog. Phys.* **32,** 333.
Meier, R. J. and Helmholdt, R. B. (1984). *Phys. Rev.* **B29,** 1387.
Moon, R. M. (1970). *Phys. Rev. Lett.* **25,** 527.
—— and Koehler, W. C. (1979). *J. Mag. Mag. Mat.* **14,** 265.
Nielson, C. W. and Koster, G. F. (1963). *Spectroscopic coefficients for the* p^n, d^n *and* f^n *configurations.* MIT Press, Cambridge, Massachusetts.
Sinha, S. K. (1984). *Condensed matter research using neutrons, today and tomorrow.* (Eds. S. W. Lovesey and R. Scherm) Plenum Press, New York.
Stassis, C. (1979). *Nukleonika* **24,** 765.
—— and Deckman, H. W. (1975). *Phys. Rev.* **12,** 1885.
—— and —— (1976*a*). *Phys. Rev.* **13,** 4943.
—— and —— (1976*b*). *J. Phys.* **C9,** 2241.
Trammell, G. T. (1953). *Phys. Rev.* **92,** 1387.
Williams, W. G., Boland, B. C., Bowden, Z. A., Taylor, A. D., Culverhouse, S., and Rainford, B. D. (1987). *J. Phys.* **F17,** L151.

12
ELASTIC MAGNETIC SCATTERING

12.1. Introduction

The value of neutron-scattering experiments to investigate matter on an atomic scale is amply demonstrated by magnetic-diffraction studies. Coherent elastic scattering from crystals (Bragg diffraction) determines the magnetic structure, i.e. the location and orientation of electrons in unfilled shells. Given sufficient, accurate, Bragg peak intensities a magnetization-density profile can be constructed by Fourier inversion. Such data provides a stringent test of theories of electronic structure. Moreover, a magnetization density can be interpreted in terms of covalency parameters which arise in the analysis of data obtained with other experimental techniques, e.g. nuclear magnetic resonance and Mössbauer studies.

Great benefits accrue from exploiting the dependence of magnetic scattering on the polarization state of the neutrons. In Chapter 10 we showed that the magnetic-scattering amplitude is obtained directly from a measurement of the magnetic–nuclear interference term. Here there are two benefits: first, increased sensitivity and accuracy and, second, the determination of the sign of the magnetic amplitude. Consequently, the polarized-beam technique is usually used to determine magnetization densities since this requires many accurate Bragg peak intensities, to minimize series termination noise in the Fourier transform, some of which are very weak at high scattering angles. Polarization analysis allows magnetic scattering to be separated from all nuclear scattering apart from the spin-incoherent contribution. This provides a method of measuring small diffuse cross-sections in mixed antiferromagnets, for example.

We derive general forms for the coherent elastic cross-section, and corresponding final polarization. These expressions complement results derived in Chapter 10 and reinforce the point that strictly elastic scattering measures the time-averaged moment of the electrons.

12.2. Coherent elastic cross-section and final polarization

We will derive an expression for the coherent elastic cross-section for magnetic and nuclear scattering of polarized neutrons. The neutron–matter interaction is expressed in terms of the operators $\hat{\boldsymbol{\alpha}}$ and $\hat{\beta}$ introduced in Chapter 10. For magnetic and nuclear scattering,

$$\hat{\boldsymbol{\alpha}} = r_0 \hat{\mathbf{Q}}_\perp + \tfrac{1}{2} \sum_i \exp(\mathrm{i}\boldsymbol{\kappa} \cdot \mathbf{R}_i) B_i \hat{\mathbf{i}}_i \qquad (12.1)$$

and

$$\hat{\beta} = \sum_{i} \exp(i\boldsymbol{\kappa} \cdot \mathbf{R}_i) A_i. \tag{12.2}$$

Here, $\hat{\mathbf{Q}}_{\perp}$ is the magnetic interaction operator (eqn (7.10)) and A_i and B_i are defined in terms of the scattering lengths of the nucleus at position $\mathbf{R}_i$ with angular momentum $\hat{\mathbf{i}}_i$ (cf. eqn (10.22)).

In order to minimize the length of the expressions involved, in the development of the elastic cross-section, we consider first the scattering of unpolarized neutrons. The partial differential cross-section is then

$$\frac{d^2\sigma}{d\Omega\, dE'} = \frac{k'}{k}\frac{1}{2\pi\hbar}\int_{-\infty}^{\infty} dt\, \exp(-i\omega t)\overline{\langle \hat{\boldsymbol{\alpha}}^{+} \cdot \hat{\boldsymbol{\alpha}}(t) + \hat{\beta}^{+}\hat{\beta}(t)\rangle}. \tag{12.3}$$

The additional correlation functions which arise in the cross-section for scattering polarized neutrons can be read off from eqn (10.37). The horizontal bar in (12.3) denotes, as usual, all relevant averages over and above the thermal average distinguished by angular brackets of the correlation function.

The elastic component of the cross-section is derived from the infinite time limit of the correlation function, namely,

$$\left(\frac{d^2\sigma}{d\Omega\, dE'}\right)^{\text{el}} = \frac{k'}{k}\,\delta(\hbar\omega)\overline{\langle \hat{\boldsymbol{\alpha}}^{+} \cdot \hat{\boldsymbol{\alpha}}(\infty) + \hat{\beta}^{+}\hat{\beta}(\infty)\rangle}.$$

Since $\omega = 0$, it follows that $k = k'$ and the elastic differential cross-section is

$$\left(\frac{d\sigma}{d\Omega}\right)^{\text{el}} = \overline{\langle \hat{\boldsymbol{\alpha}}^{+} \cdot \hat{\boldsymbol{\alpha}}(\infty) + \hat{\beta}^{+}\hat{\beta}(\infty)\rangle}. \tag{12.4}$$

The law of increase in entropy, or loss of information, requires that, for a bulk sample, there is no correlation between processes that are well separated in time. Consequently,

$$\begin{aligned}\langle \hat{\boldsymbol{\alpha}}^{+} \cdot \hat{\boldsymbol{\alpha}}(\infty)\rangle &= \langle \hat{\boldsymbol{\alpha}}^{+}\rangle \cdot \langle \hat{\boldsymbol{\alpha}}(\infty)\rangle \\ &= \langle \hat{\boldsymbol{\alpha}}^{+}\rangle \cdot \langle \hat{\boldsymbol{\alpha}}\rangle \end{aligned} \tag{12.5}$$

where the second equality is valid for stationary systems in which $\langle \hat{\boldsymbol{\alpha}}(t)\rangle = \langle \hat{\boldsymbol{\alpha}}(0)\rangle$ for any t. Inserting (12.5) in (12.4) together with the corresponding expression for $\langle \hat{\beta}^{+}\hat{\beta}(\infty)\rangle$, the elastic differential cross-section is

$$\left(\frac{d\sigma}{d\Omega}\right)^{\text{el}} = \left\{\overline{\langle \hat{\boldsymbol{\alpha}}^{+}\rangle \cdot \langle \hat{\boldsymbol{\alpha}}\rangle} + \overline{\langle \hat{\beta}^{+}\rangle\langle \hat{\beta}\rangle}\right\}. \tag{12.6}$$

For the more general case of scattering polarized neutrons,

$$\left(\frac{\mathrm{d}\sigma}{\mathrm{d}\Omega}\right)^{\mathrm{el}}=\left\{\overline{\langle\hat{\boldsymbol{\alpha}}^{+}\rangle\cdot\langle\hat{\boldsymbol{\alpha}}\rangle}+\overline{\langle\hat{\beta}^{+}\rangle\langle\hat{\beta}\rangle}+\overline{\langle\hat{\beta}^{+}\rangle\mathbf{P}\cdot\langle\hat{\boldsymbol{\alpha}}\rangle}+\overline{\mathbf{P}\cdot\langle\hat{\boldsymbol{\alpha}}^{+}\rangle\langle\hat{\beta}\rangle}+\overline{\mathrm{i}\mathbf{P}\cdot(\langle\hat{\boldsymbol{\alpha}}^{+}\rangle\times\langle\hat{\boldsymbol{\alpha}}\rangle)}\right\}\tag{12.7}$$

The polarization of the scattered beam is obtained from (10.41) and (12.7).

It is, perhaps, worthwhile to contrast the result (12.7) with the static approximation in which the scattering is also elastic. The static approximation is valid when the conditions of the experiment permit observation of the total cross-section, integrated over all energy transfers (cf. § 1.8). When this is accomplished, the observed quantity is related to the instantaneous value of a correlation function; in the notation of (12.3), the observed quantity will include $\langle\hat{\boldsymbol{\alpha}}^{+}\cdot\hat{\boldsymbol{\alpha}}\rangle$, for example. On the other hand, time-averaged variables are observed in strictly elastic scattering processes, since a basic principle in statistical physics is that statistical averaging, distinguished here by angular brackets, is completely equivalent to time averaging. The difference between instantaneous and time-averaged quantities

$$\{\langle\hat{\boldsymbol{\alpha}}^{+}\cdot\hat{\boldsymbol{\alpha}}\rangle-\langle\hat{\boldsymbol{\alpha}}^{+}\rangle\cdot\langle\hat{\boldsymbol{\alpha}}\rangle\}$$

is proportional to the corresponding isothermal susceptibility, as shown in eqn (8.47). In summary, a strictly elastic measurement differs from a measurement obtained when the static approximation is valid. The difference, being the mean-square thermal fluctuation in the observed variable, is generally small, except in the vicinity of a phase transition, when fluctuations take macroscopic values, and the variable is an order parameter.

For randomly oriented nuclei, $\langle\hat{\mathbf{i}}\rangle=0$, whereupon

$$\langle\hat{\boldsymbol{\alpha}}\rangle=r_0\langle\hat{\mathbf{Q}}_{\perp}\rangle;\qquad\langle\hat{\mathbf{i}}\rangle=0.\tag{12.8}$$

In this instance, $\langle\hat{\boldsymbol{\alpha}}\rangle$ and $\langle\hat{\beta}\rangle$ are purely magnetic and nuclear, respectively. If the electronic spins are randomly oriented, $\langle\hat{\boldsymbol{\alpha}}\rangle=0$ and (12.7) contains nuclear scattering only. The latter is coherent if the target sample is composed of a single isotope. For the more general case we write

$$\overline{\langle\hat{\beta}^{+}\rangle\langle\hat{\beta}\rangle}=\left|\overline{\langle\hat{\beta}\rangle}\right|^{2}+\left\{\overline{\langle\hat{\beta}^{+}\rangle\langle\hat{\beta}\rangle}-\left|\overline{\langle\hat{\beta}\rangle}\right|^{2}\right\},\tag{12.9}$$

and the second term represents diffuse scattering that arises from the distribution of isotopes; this diffuse contribution is usually referred to as isotopic incoherent scattering.

Thus, for randomly oriented nuclei, the coherent elastic differential cross-section is

$$\left(\frac{\mathrm{d}\sigma}{\mathrm{d}\Omega}\right)^{\mathrm{el}}_{\mathrm{coh}}=\left\{\langle\hat{\boldsymbol{\alpha}}^{+}\rangle\cdot\langle\hat{\boldsymbol{\alpha}}\rangle+\left|\overline{\langle\hat{\beta}\rangle}\right|^{2}+\overline{\langle\hat{\beta}^{+}\rangle}\mathbf{P}\cdot\langle\hat{\boldsymbol{\alpha}}\rangle+\mathbf{P}\cdot\langle\hat{\boldsymbol{\alpha}}\rangle\overline{\langle\hat{\beta}\rangle}+\mathrm{i}\mathbf{P}\cdot(\langle\hat{\boldsymbol{\alpha}}^{+}\rangle\times\langle\hat{\boldsymbol{\alpha}}\rangle)\right\}.\tag{12.10}$$

Here we assume that the target sample is magnetically pure. For, if there is more than one type of magnetic scattering centre, or unit, in the target sample, the departure from magnetic homogeneity generates magnetic diffuse scattering. The latter is discussed, for binary alloys, at the end of this chapter. Inelastic magnetic diffuse scattering is discussed in § 9.9.

The expression for the polarization of the scattered beam $\mathbf{P}'$ that corresponds to (12.10) is obtained from (10.41). We thus obtain a more general form of (10.109), namely,

$$\mathbf{P}'\left(\frac{d\sigma}{d\Omega}\right)^{el}_{coh} = \Big\{\overline{\langle\hat{\beta}^+\rangle}\langle\hat{\boldsymbol{\alpha}}\rangle + \langle\hat{\boldsymbol{\alpha}}^+\rangle\overline{\langle\hat{\beta}\rangle} + \mathbf{P}\left|\overline{\langle\hat{\beta}\rangle}\right|^2 + \langle\hat{\boldsymbol{\alpha}}^+\rangle\mathbf{P}\cdot\langle\hat{\boldsymbol{\alpha}}\rangle + \mathbf{P}\cdot\langle\hat{\boldsymbol{\alpha}}^+\rangle\langle\hat{\boldsymbol{\alpha}}\rangle - \mathbf{P}(\langle\hat{\boldsymbol{\alpha}}^+\rangle\cdot\langle\hat{\boldsymbol{\alpha}}\rangle) - \mathrm{i}(\langle\hat{\boldsymbol{\alpha}}^+\rangle\times\langle\hat{\boldsymbol{\alpha}}\rangle) - \mathrm{i}\overline{\langle\hat{\beta}^+\rangle}(\mathbf{P}\times\langle\hat{\boldsymbol{\alpha}}\rangle) + \mathrm{i}(\mathbf{P}\times\langle\hat{\boldsymbol{\alpha}}^+\rangle)\overline{\langle\hat{\beta}^+\rangle}\Big\}. \quad (12.11)$$

Expressions (12.10) and (12.11) are central in the interpretation of elastic scattering from magnetic crystals. In subsequent discussions of these expressions we will omit the subscript coh and superscript el.

The result (11.7) shows that the magnetic interaction operator $\hat{\mathbf{Q}}_\perp$ can be expressed in terms of the magnetic moment density operator $\hat{\mathbf{M}}(\mathbf{r})$. For a homogeneous magnetic crystal, $\langle\hat{\mathbf{M}}(\mathbf{r})\rangle$ is a periodic function that can be expressed as a Fourier series in reciprocal space. To explore the consequences of the crystal periodicity of $\langle\hat{\mathbf{M}}(\mathbf{r})\rangle$ we introduce a density function through the relation

$$\langle\hat{\mathbf{Q}}_\perp\rangle = -\frac{1}{2\mu_B}\int d\mathbf{r}\,\exp(\mathrm{i}\boldsymbol{\kappa}\cdot\mathbf{r})\tilde{\boldsymbol{\kappa}}\times\{\langle\hat{\mathbf{M}}(\mathbf{r})\rangle\times\tilde{\boldsymbol{\kappa}}\}$$
$$= \int d\mathbf{r}\,\exp(\mathrm{i}\boldsymbol{\kappa}\cdot\mathbf{r})\tilde{\boldsymbol{\kappa}}\times\{\mathscr{s}(\mathbf{r})\times\tilde{\boldsymbol{\kappa}}\}. \quad (12.12)$$

Recall that $\langle\hat{\mathbf{Q}}_\perp\rangle$ is dimensionless.

The position of a unit cell in the crystal is defined by a lattice vector $\mathbf{l}$ and the cell volume is denoted by v_0. An atom position $\mathbf{R}_j = \mathbf{l}+\mathbf{d}$, where $\mathbf{d}$ defines the position with a unit cell. The site $\mathbf{d}=0$ coincides with the corner of the unit cell, so there are $r-1$ non-null vectors $\mathbf{d}$. Vectors of the reciprocal lattice $\boldsymbol{\tau}$ satisfy

$$\exp(\mathrm{i}\boldsymbol{\tau}\cdot\mathbf{l}) = 1; \quad \text{for all } \mathbf{l}. \quad (12.13)$$

Because $\mathscr{s}(\mathbf{r})$ is periodic we can write

$$\int_V d\mathbf{r}\,\exp(\mathrm{i}\boldsymbol{\kappa}\cdot\mathbf{r})\mathscr{s}(\mathbf{r}) = \frac{(2\pi)^3}{v_0}\sum_{\boldsymbol{\tau}}\delta(\boldsymbol{\kappa}-\boldsymbol{\tau})\mathscr{F}(\boldsymbol{\tau}) \quad (12.14)$$

where $\mathscr{F}(\boldsymbol{\tau})$ is the *magnetic unit-cell vector structure factor.* Inverting

(12.14) we get

$$\mathscr{S}(\mathbf{r}) = \frac{1}{v_0} \sum_{\boldsymbol{\tau}} \exp(-\mathrm{i}\boldsymbol{\tau} \cdot \mathbf{r}) \mathscr{F}(\boldsymbol{\tau}). \tag{12.15}$$

Multiplying (12.15) by $\exp(\mathrm{i}\boldsymbol{\tau}' \cdot \mathbf{r})$ and integrating over a cell of volume v_0 gives

$$\int_{\text{cell}} \mathrm{d}\mathbf{r} \exp(\mathrm{i}\boldsymbol{\tau}' \cdot \mathbf{r}) \mathscr{S}(\mathbf{r}) = \frac{1}{v_0} \sum_{\boldsymbol{\tau}} \mathscr{F}(\boldsymbol{\tau}) \int_{\text{cell}} \mathrm{d}\mathbf{r} \exp\{\mathrm{i}(\boldsymbol{\tau}' - \boldsymbol{\tau}) \cdot \mathbf{r}\}.$$

The integral on the right-hand side vanishes unless $\boldsymbol{\tau}'$ and $\boldsymbol{\tau}$ coincide and, in that case, has value v_0. Hence

$$\mathscr{F}(\boldsymbol{\tau}) = \int_{\text{cell}} \mathrm{d}\mathbf{r} \exp(\mathrm{i}\boldsymbol{\tau} \cdot \mathbf{r}) \mathscr{S}(\mathbf{r}). \tag{12.16}$$

Finally we note that the magnetic cross-section involves

$$\int \mathrm{d}\mathbf{r} \exp(\mathrm{i}\boldsymbol{\kappa} \cdot \mathbf{r}) \mathscr{S}^{\alpha}(\mathbf{r}) \int \mathrm{d}\mathbf{r}' \exp(-\mathrm{i}\boldsymbol{\kappa} \cdot \mathbf{r}') \mathscr{S}^{\beta}(\mathbf{r}')$$

$$= \sum_{l,l'} \exp\{\mathrm{i}\boldsymbol{\kappa} \cdot (\mathbf{l} - \mathbf{l}')\} \int_{\text{cell}} \mathrm{d}\mathbf{r} \exp(\mathrm{i}\boldsymbol{\kappa} \cdot \mathbf{r}) \mathscr{S}^{\alpha}(\mathbf{r}) \int_{\text{cell}} \mathrm{d}\mathbf{r}' \exp(-\mathrm{i}\boldsymbol{\kappa} \cdot \mathbf{r}') \mathscr{S}^{\beta}(\mathbf{r}')$$

$$= N \frac{(2\pi)^3}{v_0} \sum_{\boldsymbol{\tau}} \delta(\boldsymbol{\kappa} - \boldsymbol{\tau}) \mathscr{F}^{\alpha}(\boldsymbol{\tau}) \mathscr{F}^{\beta}(\boldsymbol{\tau}).$$

Hence using (12.8), (12.12) and the preceding expression, the magnetic component of (12.10) with $\mathbf{P} = 0$ becomes

$$\frac{\mathrm{d}\sigma}{\mathrm{d}\Omega} = r_0^2 N \frac{(2\pi)^3}{v_0} \sum_{\boldsymbol{\tau}} \delta(\boldsymbol{\kappa} - \boldsymbol{\tau}) \, |\tilde{\boldsymbol{\kappa}} \times \{\mathscr{F}(\boldsymbol{\tau}) \times \tilde{\boldsymbol{\kappa}}\}|^2. \tag{12.17}$$

These results are to be contrasted with those obtained when we imagine $\mathscr{S}(\mathbf{r})$ to be made up of the superposition of individual atomic distributions. Thus, if

$$\mathscr{S}(\mathbf{r}) = \sum_{l,d} \mathbf{S}_d(\mathbf{r} - \mathbf{l} - \mathbf{d}), \tag{12.18}$$

then for each atom there exists a *vector form factor*

$$\mathscr{F}_d(\boldsymbol{\kappa}) = \int \mathrm{d}\mathbf{r} \exp(\mathrm{i}\boldsymbol{\kappa} \cdot \mathbf{r}) \mathbf{S}_d(\mathbf{r}) \tag{12.19}$$

with the inverse

$$\mathbf{S}_d(\mathbf{r}) = \frac{1}{(2\pi)^3} \int \mathrm{d}\boldsymbol{\kappa} \exp(-\mathrm{i}\boldsymbol{\kappa} \cdot \mathbf{r}) \mathscr{F}_d(\boldsymbol{\kappa}). \tag{12.20}$$

Note this is not the atomic form factor, which is introduced in (12.22). In terms of these quantities,

$$\mathscr{F}(\boldsymbol{\tau}) = \sum_{d} \exp(\mathrm{i}\boldsymbol{\tau} \cdot \mathbf{d})\mathscr{F}_d(\boldsymbol{\tau}). \tag{12.21}$$

From (12.20) we notice that to deduce $\mathbf{S}_d(\mathbf{r})$ we need to know $\mathscr{F}_d(\boldsymbol{\kappa})$ for all $\boldsymbol{\kappa}$. In principle this could be obtained from paramagnetic scattering, or indeed inelastic scattering, but it cannot be obtained from Bragg scattering, which gives only the form factors evaluated at the reciprocal lattice vectors $\boldsymbol{\tau}$.

On the other hand we notice from (12.15) that $\mathscr{s}(\mathbf{r})$ can be deduced rigorously from a knowledge of $\mathscr{F}(\boldsymbol{\tau})$ at the reciprocal lattice points only. In other words $\mathscr{s}(\mathbf{r})$ is a rigorously defined quantity, which can be determined from Bragg scattering; it tells us exactly what the spin density is at every point $\mathbf{r}$ without specifying which atom each part of the density 'belongs' to. In contrast the quantity $\mathbf{S}_d(\mathbf{r}-\mathbf{l}-\mathbf{d})$ in (12.18) gives us more information about the system, it describes exactly the spin density 'belonging' to the atom $\mathbf{l}$, $\mathbf{d}$ but it cannot be deduced from Bragg scattering because, by (12.20), this involves a full Fourier integral. But we must also recall that $\mathbf{S}_d(\mathbf{r})$ is defined only by a model of the systems and has a well-defined meaning only to the extent that the model faithfully represents the real system.

We conclude that Bragg scattering gives $\mathscr{F}(\boldsymbol{\tau})$ and that we can determine $\mathscr{s}(\mathbf{r})$ through (12.15). Now, as we have seen, $\mathscr{s}(\mathbf{r})$ is a rigorously defined quantity of great importance; it is not dependent on any model of the target system and the experiments to measure it can be performed with a high precision using polarized neutrons. It is very rare indeed in the study of materials, and especially in a study of transition metals, that such precise experimental results have such a direct and meaningful interpretation. There is only one small uncertainty in the interpretation and that is the contribution to the scattering arising from the orbital moment. The orbital moment is not quite quenched in these metals and to the extent that the scattering due to orbital moment has a different form factor from that due to the spin the interpretation is in doubt. However, we demonstrated in § 7.5 that for spins in a narrow energy band the orbital scattering is probably small, so that the error involved for the transition metals is not serious. We discuss the experimental determinations of $\mathscr{s}(\mathbf{r})$ in more detail in § 12.3.

In those materials where it is meaningful to speak of a spin density belonging to a particular ion in a unit cell we can rewrite (12.19) as

$$\mathscr{F}_d(\boldsymbol{\kappa}) = \tfrac{1}{2} g_d \langle \hat{S}_d \rangle F_d(\boldsymbol{\kappa}) \tilde{\boldsymbol{\eta}}_d(\boldsymbol{\kappa}). \tag{12.22}$$

Here $\tilde{\boldsymbol{\eta}}_d(\boldsymbol{\kappa})$ is a unit vector in the integral and $F_d(\boldsymbol{\kappa})$ is the *atomic form factor* for the ion and is defined such that $F_d(\boldsymbol{\kappa}=0)=1$. Further, $\langle\hat{S}_d\rangle$ is the average value of the spin associated with the ion and g_d the corresponding gyromagnetic ratio.

We note that for $\boldsymbol{\kappa}=0$, (12.22) becomes

$$\mathscr{F}_d(0)=\int \mathrm{d}\mathbf{r}\,\mathscr{s}(\mathbf{r})=\tfrac{1}{2}g_d\langle\hat{S}_d\rangle\tilde{\boldsymbol{\eta}}_d(0) \tag{12.23}$$

and from this it follows that $\tilde{\boldsymbol{\eta}}_d(0)$ gives the direction of the spin moment on the dth site. For most spin structures $\tilde{\boldsymbol{\eta}}_d(\boldsymbol{\kappa})$ is independent of $\boldsymbol{\kappa}$. This is the case for all the examples discussed below.

Using (12.22) in (12.21) and taking account of the vibration of the ions about their equilibrium positions we obtain for $\mathscr{F}(\boldsymbol{\tau})$

$$\mathscr{F}(\boldsymbol{\tau})=\tfrac{1}{2}\sum_d \exp(\mathrm{i}\boldsymbol{\tau}\cdot\mathbf{d})\exp\{-W_d(\boldsymbol{\tau})\}g_d\langle\hat{S}_d\rangle F_d(\boldsymbol{\tau})\tilde{\boldsymbol{\eta}}_d(\boldsymbol{\tau}) \tag{12.24}$$

where $\exp(-2W)$ is the usual Debye–Waller factor.

Hence the corresponding expression for the cross-section for coherent elastic magnetic scattering is

$$\frac{\mathrm{d}\sigma}{\mathrm{d}\Omega}=r_0^2N\frac{(2\pi)^3}{v_0}\sum_{\boldsymbol{\tau}}\delta(\boldsymbol{\kappa}-\boldsymbol{\tau})\times\left|\sum_d \exp(\mathrm{i}\boldsymbol{\tau}\cdot\mathbf{d})\exp\{-W_d(\boldsymbol{\tau})\}\,\tfrac{1}{2}g_d\langle\hat{S}_d\rangle F_d(\boldsymbol{\tau})[\tilde{\boldsymbol{\tau}}\times\{\tilde{\boldsymbol{\eta}}_d(\boldsymbol{\tau})\times\tilde{\boldsymbol{\tau}}\}]\right|^2. \tag{12.25}$$

For a simple ferromagnet, with one atom per unit cell, (12.25) reduces to

$$\frac{\mathrm{d}\sigma}{\mathrm{d}\Omega}=r_0^2\{\tfrac{1}{2}gF(\boldsymbol{\kappa})\}^2\exp\{-2W(\boldsymbol{\kappa})\}\langle\hat{S}\rangle^2N\frac{(2\pi)^3}{v_0}\sum_{\boldsymbol{\tau}}\delta(\boldsymbol{\kappa}-\boldsymbol{\tau})\{1-(\tilde{\boldsymbol{\tau}}\cdot\tilde{\boldsymbol{\eta}})^2\}_{\mathrm{av}}, \tag{12.26}$$

The subscript 'av' on the orientation factor in (12.26) signifies that an average over domain orientations $\tilde{\boldsymbol{\eta}}$ is to be taken.

We see immediately from the expression (12.26) for the cross-section for elastic magnetic scattering from a ferromagnet that the magnetic Bragg scattering occurs at exactly the same reciprocal lattice points as nuclear scattering, and, because the magnetic diffraction peaks exactly coincide with the nuclear diffraction peaks, difficulty arises in distinguishing one effect from the other. For a Bravais lattice structure the ratio of the magnetic to nuclear contributions for any given Bragg peak is

$$\frac{1}{|\bar{b}|^2}\,r_0^2\langle\hat{S}\rangle^2\,|F(\boldsymbol{\tau})|^2\,\{1-(\tilde{\boldsymbol{\tau}}\cdot\tilde{\boldsymbol{\eta}})^2\}_{\mathrm{av}}.$$

For iron $\langle\hat{S}\rangle$ at $T=0$ is 1.1 and this ratio is reasonable, but for nickel

$\langle \hat{S} \rangle_{T=0} = 0.3$ and the ratio is very small, certainly too small for the purely magnetic scattering to be easily seen in nickel. Magnetic scattering can be observed even for nickel by using polarized neutrons. Polarization phenomena in general are fully discussed in Chapter 10, though we shall consider the application of the polarized-beam technique in magnetic scattering on several occasions in the remaining part of this chapter.

Because the atomic form factor $F(\boldsymbol{\tau})$ in (12.26) decreases rapidly with increasing $\boldsymbol{\tau}$, the magnetic scattering is negligible for all except the first few Bragg peaks. It is possible to distinguish it from nuclear scattering in three ways.

(a) The magnetic scattering falls off rapidly when the temperature of the ferromagnetic target is raised near its Curie temperature T_c because the magnetic cross-section is proportional to $\langle \hat{\mathbf{S}} \rangle^2$. When comparing cross-sections at different temperatures it is important to make allowance for the small temperature dependence of the nuclear scattering. The latter results from the vibrations of the crystal lattice, as we saw in Chapter 4. The cross-section (12.26) contains the Debye–Waller factor $\exp\{-2W(\boldsymbol{\kappa})\}$.

(b) The magnetic scattering has a characteristic dependence on $\boldsymbol{\tau}$ through $F(\boldsymbol{\tau})$, whereas the nuclear scattering varies only slightly with $\boldsymbol{\tau}$ as a consequence of lattice vibrations.

(c) The magnetic scattering depends on the orientation of the spin $\langle \hat{\mathbf{S}} \rangle$ relative to the vector $\boldsymbol{\tau}$ through the last factor in (12.26).

In the absence of an external magnetic field a ferromagnet splits into domains and to obtain the cross-section we must average over the orientations of these domains, i.e. over the 'easy axes' for $\tilde{\boldsymbol{\eta}}$. If all directions in space were equally likely, then clearly $(\tilde{\boldsymbol{\tau}} \cdot \tilde{\boldsymbol{\eta}})^2$ would average to $\frac{1}{3}$ so that

$$\{1 - (\tilde{\boldsymbol{\tau}} \cdot \tilde{\boldsymbol{\eta}})^2\}_{\text{av}} = \tfrac{2}{3}. \tag{12.27}$$

This result is also correct for cubic symmetry. Thus when (12.27) is valid the cross-section (12.26) becomes

$$\frac{d\sigma}{d\Omega} = r_0^2\{\tfrac{1}{2}gF(\boldsymbol{\kappa})\}^2 \langle \hat{S} \rangle^2 N \frac{(2\pi)^3}{v_0} \tfrac{2}{3} \sum_{\tau} \delta(\boldsymbol{\kappa} - \boldsymbol{\tau}) \exp\{-2W(\boldsymbol{\tau})\}. \tag{12.28}$$

Now if an external magnetic field is applied in the direction of $\boldsymbol{\kappa}$, i.e. in the direction of $\boldsymbol{\tau}$ for a given Bragg peak, the ferromagnetic spins can be made parallel to $\boldsymbol{\tau}$ so that $\tilde{\boldsymbol{\tau}} \cdot \tilde{\boldsymbol{\eta}}$ is unity and thus the elastic magnetic cross-section vanishes. Hence the difference in cross-section for the external field off and the field on gives directly the magnetic scattering alone. This is a very simple and convenient technique for measuring the elastic magnetic scattering from ferromagnets.

It must be emphasized that the simple result (12.28) is strictly valid only for cubic crystals. For cobalt, for instance, the easy directions are along the hexagonal z-axis so there are just two 'easy' directions,

$$\tilde{\boldsymbol{\eta}} = (0, 0, \pm 1).$$

Thus in zero field the cross-section (12.26) becomes

$$\frac{d\sigma}{d\Omega} = r_0^2\{\tfrac{1}{2}gF(\boldsymbol{\kappa})\}^2\langle\hat{S}\rangle^2 N \frac{(2\pi)^3}{v_0} \sum_{\boldsymbol{\tau}} \delta(\boldsymbol{\kappa}-\boldsymbol{\tau})(1-\tilde{\tau}_z^2)\exp\{-2W(\boldsymbol{\tau})\}. \tag{12.29}$$

Of course, it is also possible to make this cross-section zero by applying a magnetic field but, because of the larger anisotropy energy in a noncubic crystal, it would have to be a larger field.

As a second example of (12.25) consider a simple antiferromagnet, in which (a) all the spins in a unit cell are either parallel or antiparallel to a single direction $\tilde{\boldsymbol{\eta}}$, and (b) all the sublattices are identical. In this instance (12.25) reduces to

$$\frac{d\sigma}{d\Omega} = r_0^2 N \frac{(2\pi)^3}{v_0} \sum_{\boldsymbol{\tau}} \delta(\boldsymbol{\kappa}-\boldsymbol{\tau})\exp\{-2W(\boldsymbol{\tau})\}\,|F_M(\boldsymbol{\tau})|^2\,\{1-(\tilde{\boldsymbol{\tau}}\cdot\tilde{\boldsymbol{\eta}})^2\}_{av}. \tag{12.30}$$

Here $F_M(\boldsymbol{\tau})$ is the *magnetic unit-cell structure factor* defined by

$$F_M(\boldsymbol{\tau}) = \sum_d \tfrac{1}{2}g_d\langle\hat{S}_d\rangle F_d(\boldsymbol{\tau})\sigma_d \exp(i\boldsymbol{\tau}\cdot\mathbf{d}), \tag{12.31}$$

which in the particular case under discussion simplifies to

$$\tfrac{1}{2}g\langle\hat{S}\rangle F(\boldsymbol{\tau}) \sum_d \sigma_d \exp(i\boldsymbol{\tau}\cdot\mathbf{d}). \tag{12.32}$$

In (12.32) σ_d is either ± 1 according to whether the spin at site $\mathbf{d}$ is on the 'plus' or 'minus' sublattice and the suffix av on the last term of (12.30) indicates an average over the domain orientations of $\tilde{\boldsymbol{\eta}}$.

The cross-section (12.30) will be zero when the temperature of the antiferromagnet is raised above its Néel temperature, for then $\langle\hat{S}\rangle$ will be zero. The high Bragg peaks have a weak intensity because of the atomic form factor term $|F(\boldsymbol{\tau})|^2$. Sometimes the orientation factor $\{1-(\tilde{\boldsymbol{\tau}}\cdot\tilde{\boldsymbol{\eta}})^2\}_{av}$ enables us to deduce the absolute orientation of the spin relative to the crystal axes. It is important to remember the distinction between determining the 'configuration' and 'orientation' of the spins. The 'configuration' is the array of plus and minus spins signifying the spins are parallel or antiparallel to $\tilde{\boldsymbol{\eta}}$ and gives no information on the orientation of $\tilde{\boldsymbol{\eta}}$ relative to the crystal axes. The 'configuration' is usually determined easily by finding out just what reciprocal vectors $\boldsymbol{\tau}$ give a contribution to

(12.30). In many cases the magnetic unit cell is larger than the chemical unit cell because two atoms can be chemically equivalent but have opposite spins and sometimes this means that the unit cell has to be enlarged to get a repeating unit.

The elastic magnetic scattering is generally small in comparison with the nuclear scattering, e.g. for iron and nickel it is about 8 per cent and 0.6 per cent respectively, so that accurate measurements would seem out of the question. However, the cross-section for the elastic scattering of polarized neutrons contains a magnetic–nuclear interference term, which, because it is linear in the magnetic scattering amplitude, makes it possible to measure very small magnetic cross-sections with great accuracy. This fact is the reason why most spin-density determinations have been made with polarized neutrons.

There are, in addition to the high sensitivity of the polarized-beam technique, two other advantages that are of great value. First, the fact that the interference term is linear in the magnetic scattering amplitude means that its absolute sign can be determined for a particular reflection. Second, the change in the polarization of the neutrons (eqn (12.11)) can be used to give additional information on the magnetic structure. This second point is more fully discussed in the chapter on polarization phenomena, Chapter 10.

Consider the cross-section for coherent elastic scattering of polarized neutrons by a ferromagnet. In eqns (10.95) and (12.10) it is shown that this is

$$\frac{\mathrm{d}\sigma}{\mathrm{d}\Omega}=\left|\sum_{l}\exp(\mathrm{i}\boldsymbol{\kappa}\cdot\mathbf{l})\right|^2\left|\sum_{d}\exp(\mathrm{i}\boldsymbol{\kappa}\cdot\mathbf{d})\bar{b}_d\right|^2$$

$$+r_0\left|\sum_{l}\exp(\mathrm{i}\boldsymbol{\kappa}\cdot\mathbf{l})\right|^2\left[\tfrac{1}{2}gF(\boldsymbol{\kappa})\langle\hat{S}\rangle\mathbf{P}\cdot\{\tilde{\boldsymbol{\kappa}}\times(\tilde{\boldsymbol{\eta}}\times\tilde{\boldsymbol{\kappa}})\}\sum_{d,d'}2\bar{b}_d\cos\{\boldsymbol{\kappa}\cdot(\mathbf{d}-\mathbf{d}')\}\right.$$

$$\left.+r_0\left|\tfrac{1}{2}gF(\boldsymbol{\kappa})\right|^2\langle\hat{S}\rangle^2\{1-(\tilde{\boldsymbol{\kappa}}\cdot\tilde{\boldsymbol{\eta}})^2\}\left|\sum_{d}\exp(\mathrm{i}\boldsymbol{\kappa}\cdot\mathbf{d})\right|^2\right] \qquad (12.33)$$

where $\mathbf{P}$ is the polarization of the incident beam and

$$\left|\sum_{l}\exp(\mathrm{i}\boldsymbol{\kappa}\cdot\mathbf{l})\right|^2=N\frac{(2\pi)^3}{v_0}\sum_{\boldsymbol{\tau}}\delta(\boldsymbol{\kappa}-\boldsymbol{\tau}).$$

If the experimental arrangement is such that the direction of the ferromagnetic spins $\tilde{\boldsymbol{\eta}}$ is perpendicular to the scattering vector $\boldsymbol{\kappa}$, then for a Bravais crystal (12.33) reduces to

$$\frac{d\sigma}{d\Omega}=\left|\sum_{l}\exp(i\boldsymbol{\kappa}\cdot\mathbf{l})\right|^{2}\{\bar{b}^{2}+2\bar{b}r_{0}\tfrac{1}{2}gF(\boldsymbol{\kappa})\langle\hat{S}\rangle\mathbf{P}\cdot\tilde{\boldsymbol{\eta}}+r_{0}^{2}\,|\tfrac{1}{2}gF(\boldsymbol{\kappa})|^{2}\,\langle\hat{S}\rangle^{2}\},\tag{12.34}$$

i.e.

$$\frac{d\sigma}{d\Omega}=\left|\sum_{l}\exp(i\boldsymbol{\kappa}\cdot\mathbf{l})\right|^{2}\{\bar{b}+r_{0}\tfrac{1}{2}gF(\boldsymbol{\kappa})\langle\hat{S}\rangle\}^{2},\quad\text{for }\mathbf{P}\cdot\tilde{\boldsymbol{\eta}}=1,$$

$$=\left|\sum_{l}\exp(i\boldsymbol{\kappa}\cdot\mathbf{l})\right|^{2}\{\bar{b}-r_{0}\tfrac{1}{2}gF(\boldsymbol{\kappa})\langle\hat{S}\rangle\}^{2},\quad\text{for }\mathbf{P}\cdot\tilde{\boldsymbol{\eta}}=-1.\tag{12.35}$$

Hence, if the scattering length $\bar{b}$ for the material is known, a measurement of the ratio of the two intensities at a Bragg peak for **P** parallel to the spin direction and for **P** antiparallel to the spin direction gives the magnetic scattering amplitude

$$r_{0}\tfrac{1}{2}gF(\boldsymbol{\kappa})\langle\hat{S}\rangle.$$

The ratio of these two intensities is usually called the flipping ratio and denoted by the letter R, i.e.

$$R=(1+(r_{0}/\bar{b})\tfrac{1}{2}gF(\boldsymbol{\kappa})\langle\hat{S}\rangle)^{2}/(1-(r_{0}/\bar{b})\tfrac{1}{2}gF(\boldsymbol{\kappa})\langle\hat{S}\rangle)^{2}.\tag{12.36}$$

The superiority of the polarized-beam method in sensitivity over the direct measurement of the magnetic scattering with unpolarized neutrons is simply demonstrated. Suppose the magnetic scattering amplitude for a given reflection is equal to $\bar{b}/100$. For this reflection

$$R=\left(\frac{1+0.01}{1-0.01}\right)^{2}\simeq1.04,\tag{12.37}$$

whereas the magnetic contribution to the total intensity for the scattering of unpolarized neutrons would be 1 part in 10^4.

Before passing on to discuss examples of the use of the polarized-beam method in determining spin densities we pause to make a few comments on the production of polarized neutrons. It is clear from the cross-section (12.35) that if, for a given reflection, $\bar{b}$ was close to the magnetic scattering amplitude, then the scattered beam would consist primarily of neutrons in one particular state of polarization. This condition on the coherent nuclear and magnetic scattering amplitudes is closely fulfilled in the (111) reflection from a cobalt–iron crystal consisting of 92 atomic per cent cobalt, and is the system most frequently used at present for both polarizing and analysing crystals. There are, besides the fulfilment of this condition, other requirements demanded of a crystal for it to be an efficient polarizer or analyser.

For instance, extinction and also depolarization of the beam as it passes through the crystal can obviously seriously affect the efficiency of a crystal as a polarizer. A detailed discussion of the properties required for an efficient polarizing or analysing crystal is given in Freund and Forsyth (1979).

12.3. Spin densities in metals (Menzinger and Sacchetti 1979; Moon 1982)

The spin density and atomic magnetic form factor for nickel have been determined with great accuracy by Mook (1966) using the polarized-beam technique described above. The spin density maps for the [100] and [110] planes are shown in Fig. 12.1 (a) and (b). The most striking feature of these maps is that they clearly demonstrate that the Ni spin density is quite asymmetric about the lattice sites. To understand the full implication of this result we ask how 3d atomic wave functions behave in a cubic crystal field (Harrison 1980).

The cubic field partially removes the fivefold orbital degeneracy by splitting the wave functions according to the representations E_g and T_{2g} of the cubic group O. The two orbitals of the E_g representation behave like (Fig. 12.2)

$$(3z^2 - r^2)f(r) \quad \text{and} \quad (x^2 - y^2)f(r)$$

where $f(r)$ is the radial part of the wave function. The three orbitals of the T_{2g} representation are proportional to

$$xyf(r), \quad yzf(r), \quad \text{and} \quad zxf(r).$$

Electrons occupying wave functions of E_g symmetry result in a spin density along the [100] cube edges and electrons occupying wave functions of T_{2g} symmetry in a spin density along the [111] cube diagonals. However, if all these orbitals are equally occupied, the resulting form factor $F(\boldsymbol{\kappa})$ is spherically symmetric. In this instance the occupation of the T_{2g} and E_g orbitals would be in the ratio 60 per cent to 40 per cent. With these facts in mind we infer from Fig. 12.1 (a) and (b) that the occupancy of unpaired electrons in the T_{2g} orbitals in Ni is greater than that required for spherical symmetry; Mook in fact finds that 81 ± 1 per cent of the 3d magnetic electrons occupy T_{2g} orbitals. The difference in the radial distribution of the spin density along the symmetry axes is clearly illustrated in Fig. 12.3.

Because the flipping ratio R cannot be measured for all Bragg reflections, there must be errors in the evaluation of the spin density from eqn (12.15). Thus the radial spin density distributions given in Fig. 12.3

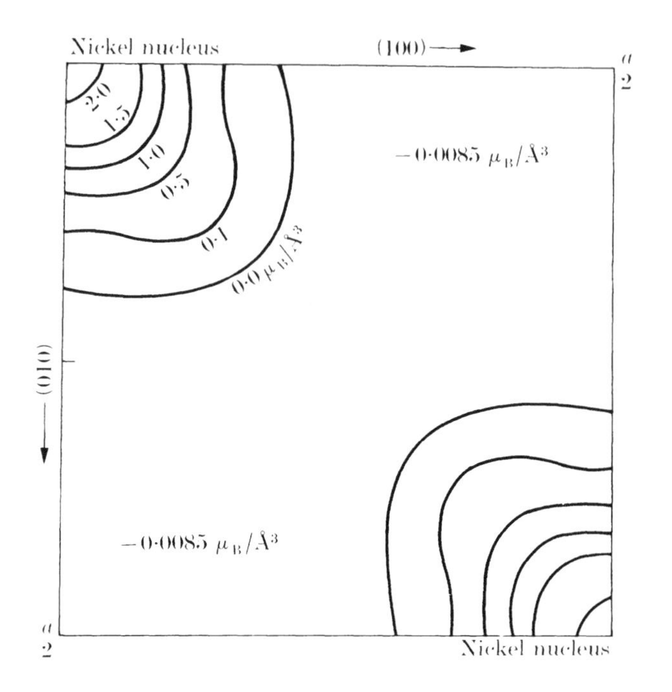

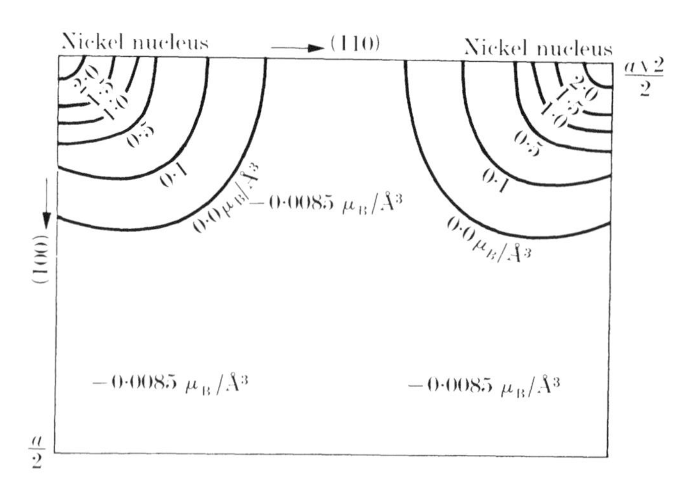

FIG. 12.1. (a) The magnetic moment distribution in the [100] plane for Ni. (b) The magnetic moment distribution in the [110] plane for Ni (Mook 1966).

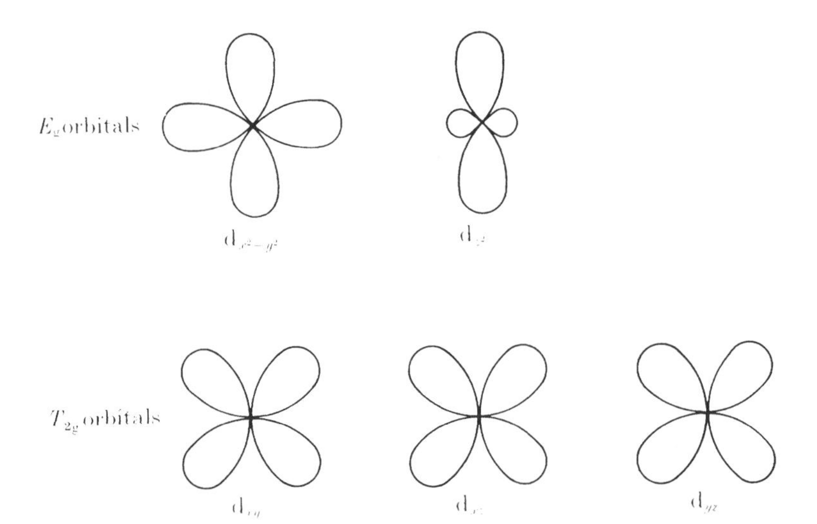

FIG. 12.2. d orbitals in a cubic crystal field.

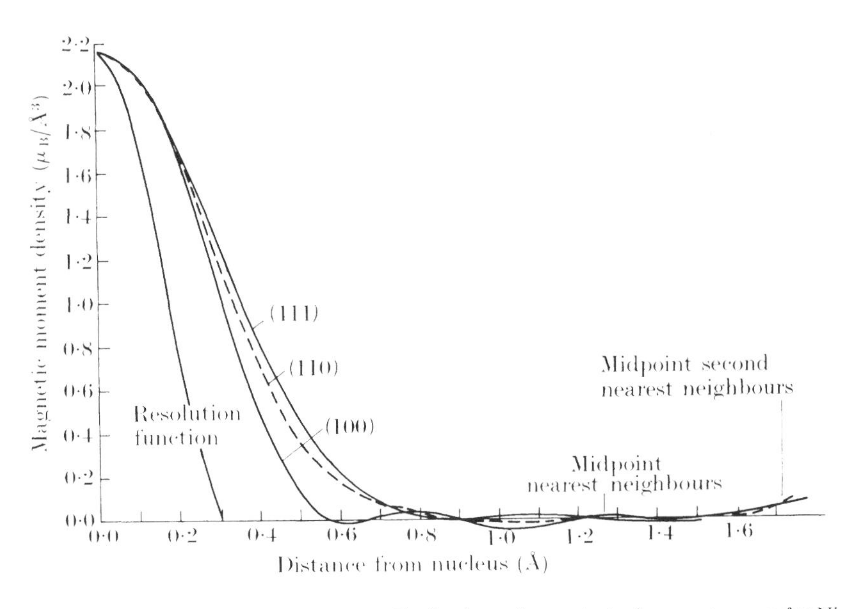

FIG. 12.3. Radial magnetic moment distributions along principal symmetry axes for Ni.

are, more correctly, the true spin density distributions as seen with the finite resolution function arising from measuring, in this case, the first 27 Bragg reflections. Any detail finer than the resolution function given in Fig. 12.3 cannot be resolved.

A further consequence of having R only for a limited number of Bragg reflections is that $\mathscr{s}(\mathbf{r})$ obtained from the series (12.15) will oscillate for large values of $\mathbf{r}$. This would seem to prevent an accurate determination of the spin density between atoms because the oscillations in the calculated $\mathscr{s}(\mathbf{r})$ obscure its true small value in this region. To overcome this difficulty we can ask for the average value of $\mathscr{s}(\mathbf{r})$ over a small volume. The results given in Fig. 12.1 for the spin density far away from the lattice points were derived in this way. If we take the volume to be a cube with edges parallel to the cell axes and of length 2δ, then the spin density averaged over this volume, $\overline{\mathscr{s}(\mathbf{r})}$, is from (12.15)

$$\overline{\mathscr{s}(\mathbf{r})} = \frac{1}{v_0} \sum_{t_1,t_2,t_3} \mathscr{F}(\boldsymbol{\tau}) \left(\frac{1}{2\delta}\right)^3 \int_{x-\delta}^{x+\delta} \mathrm{d}x \int_{y-\delta}^{y+\delta} \mathrm{d}y \int_{z-\delta}^{z+\delta} \mathrm{d}z$$

$$\times \exp\left\{-\frac{2\pi \mathrm{i}}{a}(t_1 x + t_2 y + t_3 z)\right\}$$

$$= \frac{1}{(2\pi\delta/a)^3 v_0} \sum_{t_1,t_2,t_3} \{\mathscr{F}(\boldsymbol{\tau})/t_1 t_2 t_3\} \sin\left(\frac{2\pi t_1 \delta}{a}\right) \sin\left(\frac{2\pi t_2 \delta}{a}\right) \sin\left(\frac{2\pi t_3 \delta}{a}\right). \tag{12.38}$$

The presence of the factor $(t_1 t_2 t_3)^{-1}$ in this series for $\overline{\mathscr{s}(\mathbf{r})}$ makes it converge much more rapidly than the series for $\mathscr{s}(\mathbf{r})$. In calculating $\overline{\mathscr{s}(\mathbf{r})}$ we are asking a less detailed question than when we calculate $\mathscr{s}(\mathbf{r})$ and thus, with the limited data, we obtain a more accurate result. The criterion for $\overline{\mathscr{s}(\mathbf{r})}$ to be a good approximation to the true spin density is that it should be almost independent of the choice of δ for a given value of $\mathbf{r}$. The effect of the averaging procedure can be seen in detail by examining Fig. 12.4, which shows $\mathscr{s}(\mathbf{r})$ and $\overline{\mathscr{s}(\mathbf{r})}$ for Ni at the position $(\frac{1}{2}a, 0, 0)$. The size of the oscillations is a measure of convergence and $\mathscr{s}(\mathbf{r})$ is seen to be oscillating widely at the maximum number of terms included in its corresponding series, whereas $\overline{\mathscr{s}(\mathbf{r})}$ has long settled down to a constant value. Figure 12.5 shows $\overline{\mathscr{s}(\mathbf{r})}$ along the [100] direction.

As a result of his calculations of $\overline{\mathscr{s}(\mathbf{r})}$ Mook was able to conclude that regions of negative (i.e. oppositely directed to the total spin) spin density exist in Ni between neighbouring ions, as shown in Fig. 12.1.

In summary, the careful measurements by Mook of the spin density in Ni have furnished three important facts: (a) the major part of the total spin density is very well localized about the Ni nuclei; (b) there is a

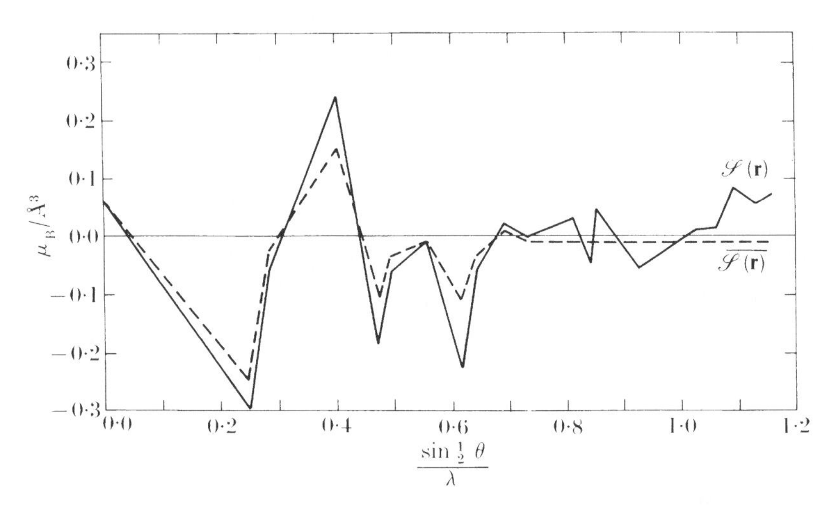

FIG. 12.4. The functions $\mathscr{s}(\mathbf{r})$ and $\overline{\mathscr{s}(\mathbf{r})}$ defined by eqns (12.15) and (12.38) for Ni at the position $(\frac{1}{2}a, 0, 0)$.

distinct asymmetry in the spin density, the occupancy of the T_{2g} orbitals exceeding that required for spherical symmetry by some 20 per cent; and (c) there is a region of almost constant negative spin density between neighbouring Ni atoms.

Before we consider the atomic form factor for Ni we point out an alternative approach to obtaining an average spin density $\overline{\mathscr{s}(\mathbf{r})}$ to that

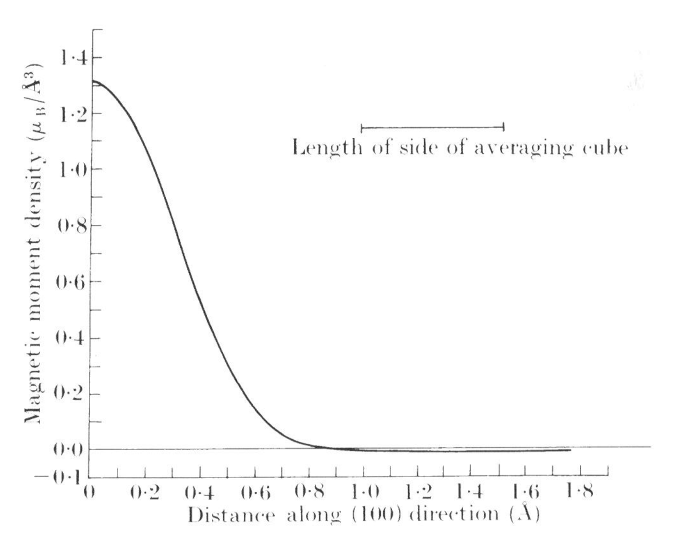

FIG. 12.5. The average magnetic moment density $\overline{\mathscr{s}(\mathbf{r})}$ for Ni for a cubic block 0.15 lattice constants on a side plotted along the [100] direction.

given by (12.38). Instead of averaging $\mathscr{s}(\mathbf{r})$ over a small cube, let us make a Gaussian average about $\mathbf{r}$, i.e.

$$\overline{\mathscr{s}(\mathbf{r})} = \int \mathrm{d}\mathbf{r}' \exp\{-(\mathbf{r}-\mathbf{r}')^2\delta\}\mathscr{s}(\mathbf{r}') \Big/ \int \mathrm{d}\mathbf{r}' \exp(-\delta \mathbf{r}'^2). \tag{12.39}$$

With the defining equation for $\mathscr{s}(\mathbf{r})$ in terms of $\mathscr{F}(\boldsymbol{\tau})$ (eqn (12.15)) we obtain from (12.39) a very rapidly converging series

$$\begin{aligned}\overline{\mathscr{s}(\mathbf{r})} &= \left(\frac{\delta}{\pi}\right)^{3/2} \frac{1}{v_0} \sum_{\boldsymbol{\tau}} \mathscr{F}(\boldsymbol{\tau}) \int \mathrm{d}\mathbf{r}' \exp\{-(\mathbf{r}-\mathbf{r}')^2\delta - \mathrm{i}\boldsymbol{\tau}\cdot\mathbf{r}'\} \\ &= \left(\frac{\delta}{\pi}\right)^{3/2} \frac{1}{v_0} \sum_{\boldsymbol{\tau}} \mathscr{F}(\boldsymbol{\tau})\exp(-\mathrm{i}\boldsymbol{\tau}\cdot\mathbf{r}) \int \mathrm{d}\mathbf{r}' \exp(\mathrm{i}\boldsymbol{\tau}\cdot\mathbf{r}' - \delta r'^2) \\ &= \frac{1}{v_0} \sum_{\boldsymbol{\tau}} \mathscr{F}(\boldsymbol{\tau})\exp(-\mathrm{i}\boldsymbol{\tau}\cdot\mathbf{r})\exp(-\tau^2/4\delta). \end{aligned} \tag{12.40}$$

Mook's atomic form factor $F(\boldsymbol{\kappa})$ for Ni is shown in Fig. 12.6. The experimental points plotted in this figure were obtained by normalizing the magnetic scattering amplitude derived from the measured value of R with the calculated [000] magnetic amplitude, which Mook took to be (at room temperature) 0.155×10^{-12} cm.

The asymmetry in the spin distribution is vividly demonstrated by the difference in the reported measurements for pairs of reflections with the same value of $(\sin\frac{1}{2}\theta)/\lambda$. There are four such pairs included in Fig. 12.6, namely

$$(511) \text{ and } (333)\text{: } (\sin\tfrac{1}{2}\theta)/\lambda = 0.7387$$

$$(600) \text{ and } (442)\text{: } (\sin\tfrac{1}{2}\theta)/\lambda = 0.8530$$

$$(551) \text{ and } (771)\text{: } (\sin\tfrac{1}{2}\theta)/\lambda = 1.0153$$

$$(731) \text{ and } (553)\text{: } (\sin\tfrac{1}{2}\theta)/\lambda = 1.0920.$$

From Chapters 7 and 11 we anticipate that the atomic form factor of Ni can be expressed as

$$F(\boldsymbol{\kappa}) = \frac{2}{g} F_{\mathrm{spin}}(\boldsymbol{\kappa}) + \left(\frac{g-2}{g}\right) F_{\mathrm{orbit}}(\boldsymbol{\kappa}) + F_{\mathrm{core}}(\boldsymbol{\kappa}). \tag{12.41}$$

The value of the gyromagnetic ratio g for nickel is 2.2. The third term in (12.41) represents the almost negligibly small contribution to $F(\boldsymbol{\kappa})$ from the polarization of the argon core due to configuration interaction. By far the largest contribution to $F(\boldsymbol{\kappa})$ comes from the spin, the orbital contribution being only about 10 per cent. Taking note of the constant negative spin density observed in Fig. 12.1. F_{spin} can be written as

$$F_{\mathrm{spin}}(\boldsymbol{\kappa}) = (1+\alpha)F_{3\mathrm{d}}(\boldsymbol{\kappa}) - \alpha\,\delta(\boldsymbol{\kappa}) \tag{12.42}$$

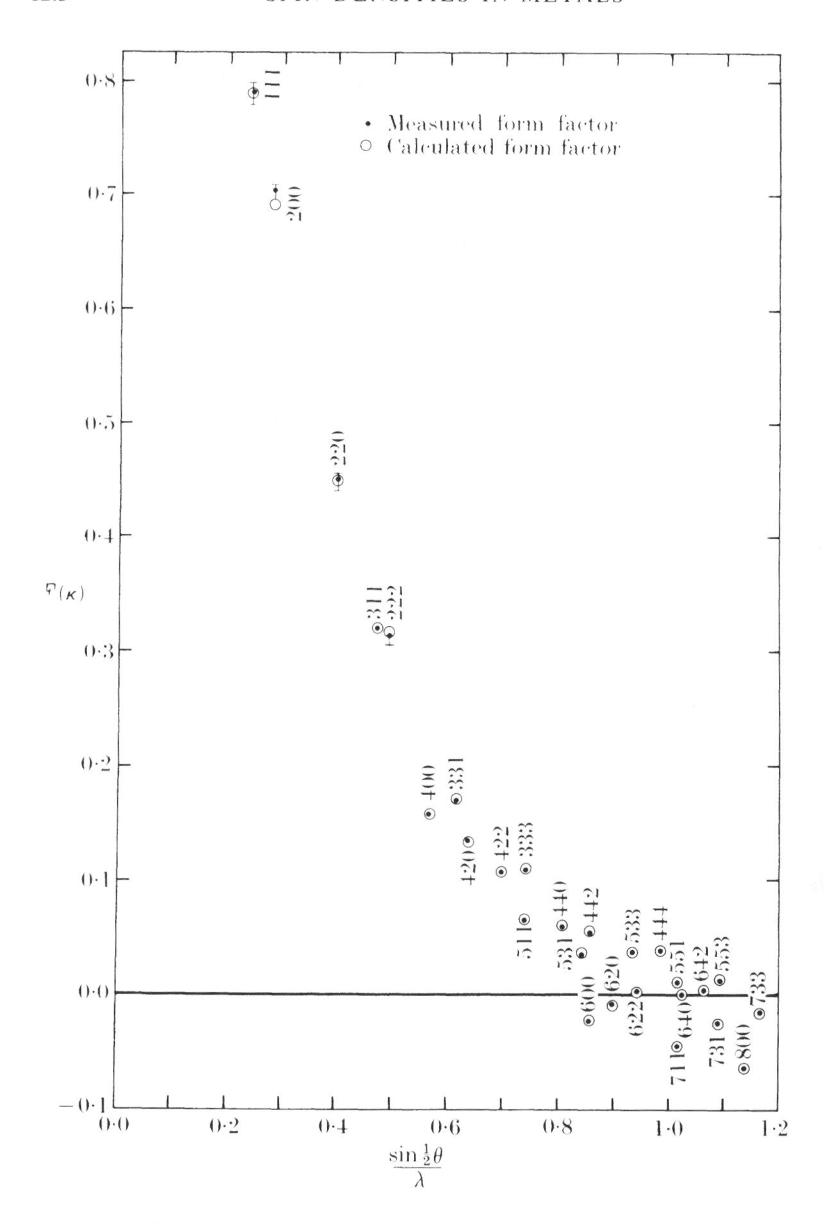

FIG. 12.6. Measured and calculated (eqn (12.41)) atomic form factors for Ni.

where α is a constant and $F_{3d}(\boldsymbol{\kappa})$ is the spin part of the form factor for 3d electrons. Taking $\alpha = 0.19$ and assuming 19 per cent ocupancy of the E_g orbitals, as opposed to 40 per cent for spherical symmetry. Mook achieved the extraordinarily fine agreement between his measured and calculated form factors seen in Fig. 12.6. From this agreement we may conclude that the moment distribution in Ni is almost identical in form to that obtained from the model of free-atom distributions, imposed on a constant, negative background.

The value 0.19 of the parameter α, introduced in (12.42), corresponds to including in the moment distribution a uniform background of $-0.0091\ \mu_B/\text{Å}^3$ so that this value of α is consistent with the negative contribution given in Fig. 12.1.

Taking account of all the facts derived from his experiment, Mook arrived at the following model for the spin-density distribution in Ni

$$\begin{gathered}\text{3d spin } 0.656\ \mu_B/\text{atom} \\ \text{3d orbit } 0.055\ \mu_B/\text{atom} \\ \text{negative contribution } -0.105\ \mu_B.\end{gathered} \tag{12.43}$$

These total to the observed saturation magnetization of $0.606\ \mu_B/\text{atom}$.

The interpretation of the parameter α is open to question (van Laar *et al.* 1979). Indeed, it is not even clear whether it should be regarded merely as a fitting parameter, or whether it admits an interpretation in terms of the basic properties of the electron distribution that can be calculated from a sophisticated band theory. The latter has been used successfully for the interpretation of the spin part of the atomic form factor of nickel (Wang and Callaway 1977). A complete confrontation between experiment and theory must, however, include a calculation of the orbital contribution to the magnetization density.

The occupancy of the E_g orbitals is temperature-dependent. Mook's result of 19 per cent occupancy at room temperature is confirmed by Cable (1981) and the occupancy increases to more 25 per cent at 600 K.

Measurements of the form factor away from Bragg peaks has been accomplished using polarized neutrons to measure phonon scattering (Steinsvoll *et al.* 1981). An expression for the appropriate cross-section is obtained from (10.160) by taking the phonon expansion of the displacement correlation function.

In § 7.5, we noted that the form factor (7.89) is identical with the atomic form factor at a Bragg reflection but it might fall off more rapidly with $\boldsymbol{\kappa}$. Results for iron display this effect, although it has not yet been observed in nickel.

12.4. Spin densities and configurations in antiferromagnetic salts

All the examples we consider in this section are antiferromagnets. The primary reason for this is simply that many magnetic ionic crystals are antiferromagnets.

12.4.1. Manganese ditelluride

As a first example let us discuss manganese ditelluride. Figure 12.7 shows the neutron diffraction pattern found by Hastings *et al.* (1959) for manganese ditelluride ($MnTe_2$) at 4.2 K. This crystal is known to possess a pyrite structure. Hastings *et al.* proved that the magnetic unit cell shown in Fig. 12.8 is consistent with the neutron diffraction pattern. Let us demonstrate that this is so.

The magnetic unit cell in Fig. 12.8 is based on a f.c.c. lattice in which eight of the 12 nearest neighbours of a manganese ion have a spin

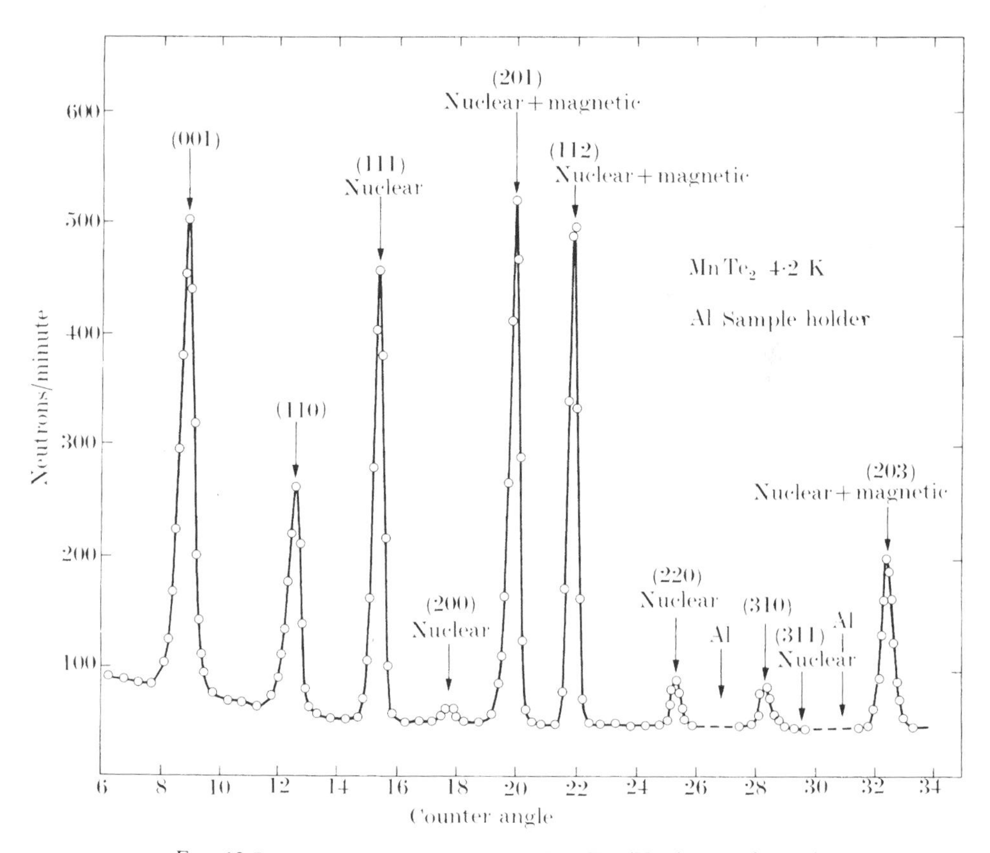

FIG. 12.7. Neutron diffraction pattern of $MnTe_2$ (Hastings *et al.* 1959).

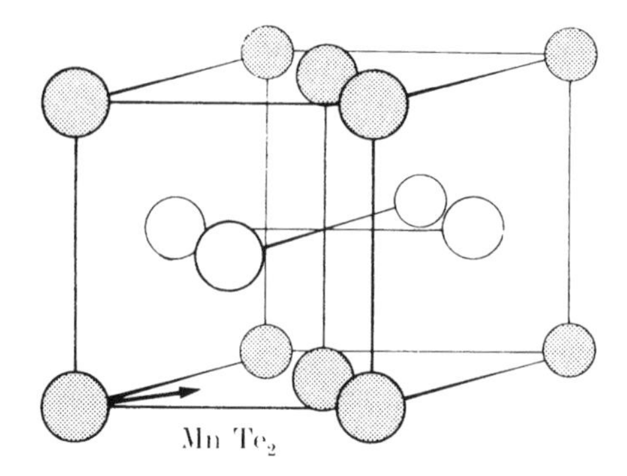

FIG. 12.8. Magnetic structure of $MnTe_2$. Positive and negative orientations of the dipoles are shown by black and white spheres. the direction of the magnetic axis relative to the crystallographic axes is shown by an arrow. For $MnTe_2$, the spin direction lies in the ferromagnetic sheets, but orientation within the plane is unspecified.

oriented antiparallel to it and the remaining four possess spin parallel to it. There are four atoms within the cell, two of 'up' spin and two of 'down' spin, but it is not possible to determine the direction of the spin moments within the plane. The magnetic unit cell structure factor $F_M(\boldsymbol{\tau})$ is therefore simply

$$F_M(\boldsymbol{\tau}) \propto \sum_d \sigma_{\mathbf{d}} \exp(i\boldsymbol{\tau}\cdot\mathbf{d})$$
$$= 1 - \exp i\pi(t_2+t_3) - \exp i\pi(t_1+t_3) + \exp i\pi(t_1+t_2) \qquad (12.44)$$

where the reciprocal lattice vectors $\boldsymbol{\tau}$ are those for simple cubic lattice of cell side a, i.e.

$$\boldsymbol{\tau} = \frac{2\pi}{a}(t_1, t_2, t_3) \quad (t_i \text{ any integers}). \qquad (12.45)$$

Those vectors $\boldsymbol{\tau}$ that coincide with the reciprocal lattice vectors for a f.c.c. latice will give $F_M(\boldsymbol{\tau}) = 0$ because for each of these reciprocal lattice vectors $\exp i\boldsymbol{\tau}\cdot\mathbf{d} = 1$. Hence there is no magnetic scattering at the nuclear reflections arising from the nuclei of the manganese ions. It is easily verified that $F_M(\boldsymbol{\tau})$ is non-zero only for those $\boldsymbol{\tau}$ given in (12.45) for which either

(a) t_1 and t_2 are both even and t_3 odd;
or (b) t_1 and t_2 are both odd and t_3 even;

and that this indexing corresponds to that given in Fig. 12.7.

There are several magnetic reflections shown in Fig. 12.7 that coincide with nuclear reflections. As we have just shown, the nuclear scattering cannot arise from the nuclei of the manganese ions but comes, of course, from the scattering from the nuclei of the anions. The magnetic contributions to the mixed peaks were obtained by subtracting the room temperature pattern with the appropriate Debye–Waller correction.

12.4.2. Manganese fluoride

As a second example let us consider the neutron diffraction pattern for MnF_2 obtained by Erickson (1953). MnF_2 has a lattice such that the Mn atoms form a body-centred tetragonal lattice as shown in Fig. 12.9. The unit cell has a volume a^2c with two Mn and four F in it. The reciprocal

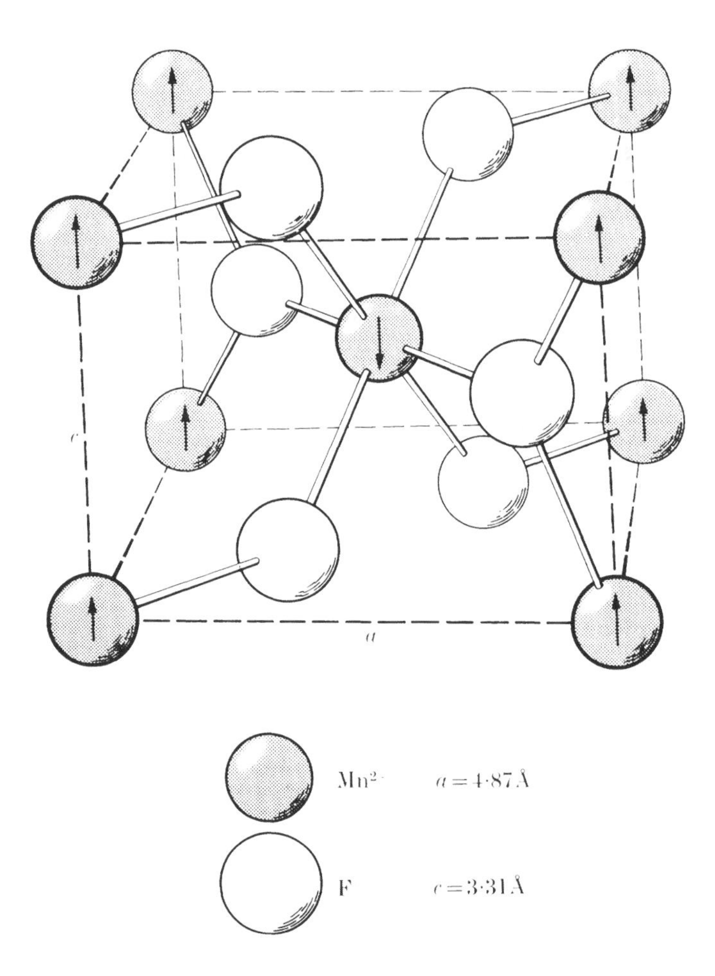

FIG. 12.9. Unit cell of MnF_2.

lattice is given by

$$\boldsymbol{\tau} = \frac{2\pi}{a}\left(t_1, t_2, \frac{a}{c}\, t_3\right) \tag{12.46}$$

and the Mn and F occupy the positions

$$\begin{array}{ll} \text{Mn} & (0, 0, 0) \quad \text{and} \quad \tfrac{1}{2}(a, a, c) \\ \text{F} & a(u, u, 0), a(1-u, 1-u, 0), a(\tfrac{1}{2}+u, \tfrac{1}{2}-u, c/2a), \end{array} \tag{12.47}$$

and

$$a(\tfrac{1}{2}-u, \tfrac{1}{2}+u, c/2a)$$

where $u \simeq 0.31$.

The unit-cell structure factor for nuclear scattering is

$$\begin{aligned} F_N(\boldsymbol{\tau}) &= \bar{b}_{Mn}\{1+\exp i\pi(t_1+t_2+t_3)\} \\ &\quad + \bar{b}_F[\exp 2\pi i u(t_1+t_2) + \exp 2\pi i(1-u)(t_1+t_2) \\ &\quad + \exp i\pi(t_1+t_2+t_3)\{\exp 2\pi i u(t_1-t_2) + \exp -2\pi i u(t_1-t_2)\}] \\ &= \bar{b}_{Mn}\{1+\exp i\pi(t_1+t_2+t_3)\} + 2\bar{b}_F[\cos\{2\pi u(t_1+t_2)\} \\ &\quad + \exp i\pi(t_1+t_2+t_3)\cos\{2\pi u(t_1-t_2)\}]. \end{aligned}$$

Hence

$$\begin{aligned} F_N(\boldsymbol{\tau}) &= 2\bar{b}_{Mn} + 4\bar{b}_F\cos(2\pi u t_1)\cos(2\pi u t_2), \quad &&\text{if } t_1+t_2+t_3 \text{ is an even integer,} \\ &= -4\bar{b}_F\sin(2\pi u t_1)\sin(2\pi u t_2), &&\text{if } t_1+t_2+t_3 \text{ is an odd integer.} \end{aligned} \tag{12.48}$$

If we recall that the reciprocal lattice vectors of a body-centred tetragonal lattice coincide with (12.46) if $t_1+t_2+t_3$ is even, then we understand why in (12.48) the scattering from the nuclei of the manganese ions occurs only when this condition is satisfied. The nuclei of the fluorine ions contribute to all reflections except those of the form $(0, t_2, t_3)$ with t_2+t_3 odd.

The first few reciprocal lattice vectors with their multiplicity and structure factors are

$\boldsymbol{\tau}$	$z(\boldsymbol{\tau})$	$F_N(\boldsymbol{\tau})$
(1, 0, 0)	4	0
(0, 0, 1)	2	0
(1,1,0)	4	$2\bar{b}_{Mn}+4\bar{b}_F\cos^2(2\pi u)$
(1,0,1)	8	$2\bar{b}_{Mn}+4\bar{b}_F\cos(2\pi u)$

Therefore we expect the first nuclear peak to occur at an angle corresponding to the (1, 1, 0) vector because, for this particular symmetry, the first two non-zero reciprocal lattice vectors happen to have zero unit-cell structure factors.

In addition to the expected nuclear peaks at angles corresponding to the (1, 1, 0), (1, 0, 1), and other vectors. Erickson observed peaks at angles corresponding to the (1, 0, 0), (1, 1, 1), (2, 1, 0), and (2, 0, 1) vectors. The first point to note is that the magnetic peaks can be indexed on the same unit cell as the nuclear peaks, i.e. peaks do not appear at points other than those given by (12.46). The magnetic unit cell is therefore of the same size as the chemical unit cell. There are only two Mn atoms in the unit cell; these must have opposite spins. Hence the spin pattern is that with 'up' spins on the corner atoms and 'down' spins on the body centres. The magnetic structure factor is proportional to

$$\begin{aligned} F_{\mathrm{M}} \propto 1 - \exp \mathrm{i}\pi(t_1 + t_2 + t_3) &= 0, \quad \text{if } t_1 + t_2 + t_3 \text{ is an even integer} \\ &= 2, \quad \text{if } t_1 + t_2 + t_3 \text{ is an odd integer.} \end{aligned} \tag{12.49}$$

We now notice that we expect magnetic peaks to appear at both (100) and (001). But actually no (001) peak is observed. It follows that the spins must be oriented in this direction, i.e. $\tilde{\boldsymbol{\eta}} = (0, 0, 1)$, so then the orientation factor $\{1 - (\tilde{\boldsymbol{\tau}} \cdot \tilde{\boldsymbol{\eta}})^2\}$ is zero for this peak. Hence, for MnF_2 we can easily deduce both the spin configuration and orientation. The magnetic peaks disappear at the Néel temperature of 75 K. The intensity of the peaks fits very closely to (12.30) with $\langle \hat{S} \rangle$ equal to 5/2 for very low T and g equal to 2 (no orbital contribution). We conclude that for this model of MnF_2 there are three types of Bragg reflection:

(a) $(0, t_2, t_3)$ reflections with $t_2 + t_3$ odd will be purely magnetic;
(b) all other reflections with $t_1 + t_2 + t_3$ odd will be mixed nuclear and magnetic, the nuclear scattering coming solely from the nuclei of the F ions;
(c) reflections with $t_1 + t_2 + t_3$ even will be purely nuclear.

Let us now seek the conditions required of an antiferromagnet for it to be sensitive to a study with polarized neutrons. If the antiferromagnet has a collinear spin configuration and it is arranged so that $\tilde{\boldsymbol{\kappa}} \cdot \tilde{\boldsymbol{\eta}} = 0$ and that $\mathbf{P}$ coincides with the spin axis, then the flipping ratio R for a particular Bragg reflection will differ from unity provided (i) neither the magnetic nor the nuclear structure factors for the reflection are zero, and (ii) the two structure factors are not out of phase by $\frac{1}{2}\pi$. The MnF_2 structure satisfies both these requirements. The fundamental aspect is a lack of translational symmetry within the unit cell of the environment of the magnetic ions. For MnF_2 the environment of the Mn^{2+} ions differs by a rotation of $\frac{1}{2}\pi$ about the (001) crystal axis and it is the coherence between the nuclear scattering from the non-magnetic F ions and the magnetic scattering from the magnetic Mn^{2+} ions (b type reflections) that produces the polarization dependence of the cross-section.

Even if the above requirements are satisfied it is still necessary to be able to prepare a sample with an overall predominance of one antiferromagnetic domain over the other. This requirement for an almost single antiferromagnetic domain single crystal obviously restricts the usefulness of the polarized-beam technique with antiferromagnets. Nathans *et al.* (1963) showed that a single crystal of MnF_2 cooled to helium temperatures turned out to be predominantly of one domain type. This is determined by a measurement of the polarization ratio of the (210) reflection for which the nuclear and magnetic structure factors are known (eqns (12.48) and (12.49)).

Nathans *et al.* also found that the flipping ratio R differed from unity for some of the supposedly pure nuclear reflections in MnF_2 (c type reflections). The measurement of so-called 'forbidden' magnetic scattering demonstrates the presence of unpaired electron spin on the ligand F ions. The measured structure factors for the 'forbidden' reflections in MnF_2 are shown in Fig. 12.10. The presence of unpaired electron spin on

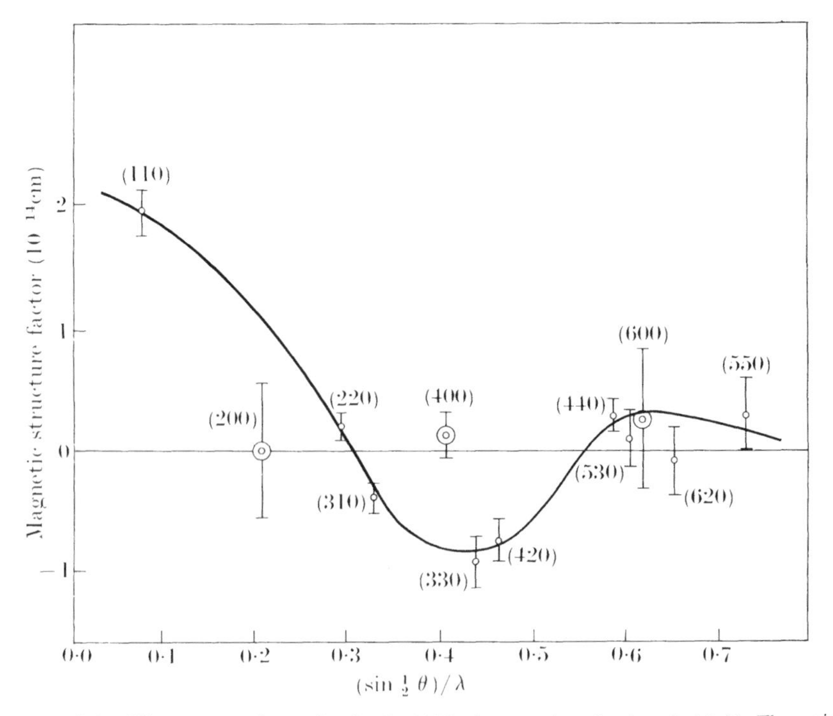

FIG. 12.10. The structure factor for the 'forbidden' magnetic reflections in MnF_2. The points enclosed by a circle are expected to have zero amplitude (Nathans *et al.* 1963).

the ligand ions in MnF_2 has also been demonstrated by measuring the hyperfine field at the F nuclei by nuclear magnetic resonance (n.m.r.). At $T \sim 0$ K the n.m.r. measurements give a hyperfine field of $\sim$40 kG. Now if the magnetic field at a F nucleus was due solely to the dipole field of the Mn^{2+} ions the observed field would be $\sim$12.5 kG, the major part of which is from the nearest-neighbour magnetic ions. The observed additional magnetic field, called the transferred hyperfine field, is believed to be due to the interaction of the unpaired electron spin on the F^- ion with the F nucleus. In essence, what is being observed in both the neutron and n.m.r. experiments is the change in the ions from their state in the truly ionic state, i.e. covalency effects. Clearly an understanding of these experiments is fundamental to our understanding of magnetic ionic crystals and we therefore consider in more detail in the following section the role of covalency in elastic magnetic scattering.

12.5. Covalency effects in ferromagnetic and antiferromagnetic salts (Harrison 1980; Hubbard and Marshall 1965; Tofield 1975, 1976)

We confine our attention in a study of covalency to transition-metal ions, which possess unpaired electrons in d orbits, and we shall for the most part consider the ligand ions to be either F^- or O^{2-}, each of which has the configuration $1s^22s^22p^6$. In a purely ionic crystal the atomic form factor $F(\boldsymbol{\kappa})$ would closely resemble the Fourier transform of a 3d orbital. In a real crystal the atomic form factor will deviate from this because of covalency, and it is our task to investigate just how the form factor is modified. We can readily see what the major modifications will be without resorting to any particular model calculation.

First, we can neglect in discussing the magnetic state of a transition-metal ion in a crystal both its 3s and 3p orbitals, because of their low energy. We also neglect the 4s and 4p orbitals. As for the ligand ions, we can surely neglect their 1s orbitals.

In a complex consisting of a single magnetic ion and its neighbouring ligand ions there is a transfer of electrons from the ligands to the magnetic ion, mainly p-electron transfer to the 3d orbitals, creating an unpaired spin density on the ligand ions. Because of Pauli's principle, the spin of the transferred electron must be opposite to that of the single electron which occupies the 3d state. Hence, for a single 3d orbital, the (back) transfer results in unpaired spin in a 2p state of the ligand that is parallel to the unpaired spin of the magnetic ion. In this instance the moment of the magnetic ion is decreased as a consequence of covalency. Neighbouring magnetic ions in a simple antiferromagnet have opposite spin moments and thus the net unpaired spin density created on the intervening ligand ions can be cancelled, depending on the symmetry of the crystal structure. Since the cross-section for magnetic Bragg scattering is proportional to the square of the magnetic moment associated with the magnetic ions, covalency reduces the intensity of the peaks relative to those from a purely ionic crystal. Now the spatial distribution of unpaired spin is contained in the form factor. If the structure possesses the symmetry necessary to effect cancellation of the unpaired spin on the ligand ions, then it follows that this unpaired spin is 'lost' to the neutrons and the form factor is uniformly diminished with respect to its free ion value. However, if a net spin density resides on the ligands, as it does for MnF_2, then because the distance between the magnetic and ligand ions is large compared to the mean radius of a 3d orbit and, because the 2p orbitals themselves are diffuse, the unpaired spin will manifest itself primarily in a peak about $\boldsymbol{\kappa} = 0$ in the atomic form factor. Whether or not the bulk of this peak lies within the first Bragg reflection for an antiferromagnet depends on the size of the unit cell relative to the distance between the magnetic and ligand ions. For ferromagnets, of course, there is no question of cancellation of unpaired spin created on the ligand ions.

To justify these physical arguments and obtain a more detailed understanding of covalency effects in neutron diffraction, and also to enable us to tie up the results obtained from neutron experiments with those from resonance experiments, we must consider a model in some detail. We choose to base our model calculations on the molecular orbital method. To see how this works we start by looking at a linear chain consisting of alternate single magnetic and ligand orbitals as shown in Fig. 12.11. Each d orbit contains just one electron, and these are taken to be antiferromagnetically ordered in the chain, while each p orbit contains two electrons. In the approximation of taking a crystal to be ionic, electrons leave the magnetic ion to fill the p orbital of the ligands, so the latter are lower in energy than the d orbital, typically by about 10 eV.

Figure 12.12 shows the overlap between neighbouring p and d wave functions for Mn^{2+}–F^- bond. From this diagram it is clear that a wave function of the form (Harrison 1980)

$$\psi_B(\mathbf{r}) = N_B\{p(\mathbf{r}) + Bd(\mathbf{r})\} : \text{bonding} \tag{12.50}$$

piles up electronic charge midway between the nuclei, whereas a wave

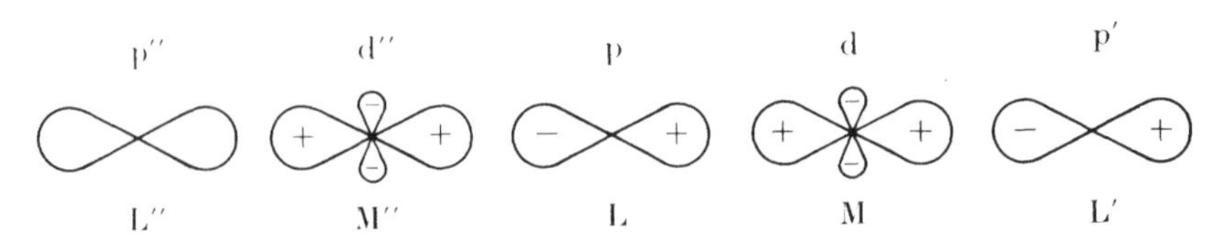

FIG. 12.11. Orbitals of a linear chain antiferromagnet.

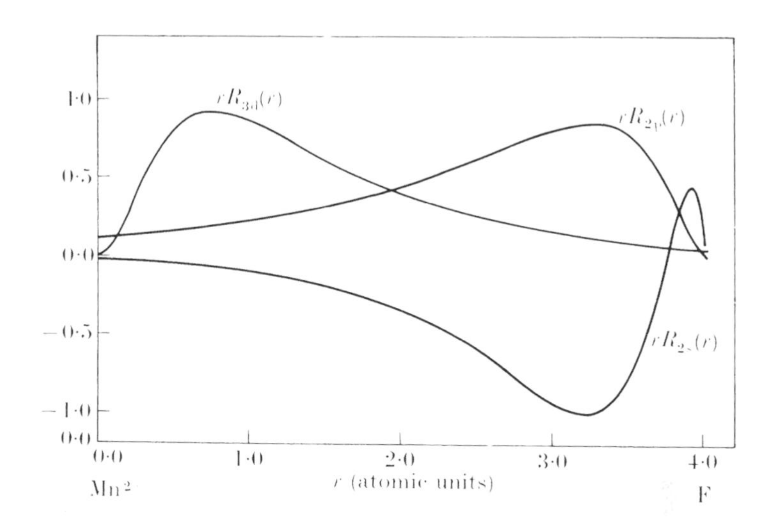

FIG. 12.12. The radial part of the Hartree–Fock free-ion wave functions of Mn^{2+} and F^-. The ions are placed at the interatomic spacing in MnF_2.

function of the form

$$\psi_A(\mathbf{r}) = N_A\{d(\mathbf{r}) - Ap(\mathbf{r})\} : \text{antibonding} \tag{12.51}$$

reduces the charge between the nuclei. In fact the wave function (12.51), the anti-bonding wave function, possesses a node between the two nuclei at which it has zero amplitude. The normalization constants N_A and N_B are given by

$$N_A^{-2} = 1 + A^2 - 2AS,$$
$$N_B^{-2} = 1 + B^2 + 2BS,$$

where S is the overlap integral

$$S = \langle d \mid p \rangle.$$

Because the antibonding and bonding orbitals must be orthogonal, we have the relationship

$$A = B + S - ABS \simeq B + S.$$

Note that, if $B = 0$ which means that there is no *covalent transfer* of spin onto the ligand ions,

$$\psi_A = \frac{1}{\sqrt{(1 - S^2)}}\{d(\mathbf{r}) - Sp(\mathbf{r})\}$$

and

$$\psi_B = p(\mathbf{r}).$$

Thus there remains an *overlap transfer* of spin, which is here seen to be a consequence of orthogonalization of the two states ψ_A and ψ_B. The magnitude of the overlap transfer is given by the square of the overlap integral S. If, for any reason, it was considered necessary in a calculation to include the core states of the magnetic ion, then, because they have a very low energy, it would be sufficient to describe their presence simply in terms of overlap transfer.

The bonding orbital (12.50) must lie lower in energy than the antibonding orbital (the bonding orbital is predominantly p-like and the p orbit lies lower in energy than the d) and consequently it can be assumed that the spin-up and spin-down bonding orbitals are completely full. Thus the bonding orbitals have no unpaired spin associated with them and we therefore need consider only the antibonding orbitals in connection with neutron scattering and resonance experiments. In this way we describe covalency as the spreading of the magnetic orbital on to the ligands, the actual process being a transfer of down spin holes to the ligands.

For the linear chain given in Fig. 12.11, the antibonding orbital to be associated with the ion M is

$$\psi_M(\mathbf{r}) = N_\sigma\{d(\mathbf{r}) - A_\sigma p(\mathbf{r}) + A_\sigma p'(\mathbf{r})\}. \tag{12.52}$$

The letters σ and π denote zero and unit angular momentum about the bond axis, respectively. In (12.52), N_σ is determined by the normalization condition

$$N_\sigma^{-2} = 1 - 4A_\sigma S_\sigma + 2A_\sigma^2 \tag{12.53}$$

where S_σ is the overlap integral between the p and d orbits. The spin density associated with M is

$$D(\mathbf{r}) = \psi_M^2(\mathbf{r}) \simeq N_\sigma^2 d^2(\mathbf{r}) - 2A_\sigma N_\sigma^2 d(\mathbf{r})\{p(\mathbf{r}) - p'(\mathbf{r})\} + A_\sigma^2 N_\sigma^2\{p^2(\mathbf{r}) + p'^2(\mathbf{r})\} \tag{12.54}$$

where we have neglected the overlap between p and p'.

For convenience of notation in this section we use unnormalized form factors defined by analogy with

$$F(\boldsymbol{\kappa}) = \int \mathrm{d}\mathbf{r}\, D(\mathbf{r}) \exp(\mathrm{i}\boldsymbol{\kappa} \cdot \mathbf{r}). \tag{12.55}$$

Correct to second order we have from (12.53) and (12.54)

$$D(\mathbf{r}) = d^2(\mathbf{r})(1 + 4A_\sigma S_\sigma - 2A_\sigma^2) - 2A_\sigma d(\mathbf{r})\{p(\mathbf{r}) - p'(\mathbf{r})\} + A_\sigma^2\{p^2(\mathbf{r}) + p'^2(\mathbf{r})\}. \tag{12.56}$$

The first term in (12.56) involves $d^2(\mathbf{r})$ and is confined to the parent magnetic ion M; the second term is an overlap density and is also confined to the immediate vicinity of M; but the third term gives a density entirely on the ligands and is equally distant from other ions as from the parent ion. The spin density given by M'' is similar to that by M, but in an antiferromagnet is of opposite sign. In particular, we notice that the moment density $A_\sigma^2 p^2(\mathbf{r})$ due to M is exactly cancelled by an equal and opposite contribution from M'. It follows that Bragg scattering can detect only the first two terms of (12.56) and, in effect, in the neighbourhood of M, defined as extending from the nucleus L, through M, to the nucleus L', we may take

$$\mathscr{D}(\mathbf{r}) = \sum_{\mathbf{d}} \sigma_d D(\mathbf{r}-\mathbf{d}) \simeq d^2(\mathbf{r})(1+4A_\sigma S_\sigma - 2A_\sigma^2) - 2A_\sigma d(\mathbf{r})\{p(\mathbf{r}) - p'(\mathbf{r})\} \tag{12.57}$$

and use a similar expression with opposite sign in the neighbourhood of M''. The expressions (12.56) and (12.57) should be contrasted with one another and with the conventional result that ignores covalency and takes $\mathscr{D}(\mathbf{r})$ as simply $d^2(\mathbf{r})$.

We notice that the total density in (12.56), and also the conventional result, is unity, whereas the total density in (12.57) is $1-2A_\sigma^2$. The 'lost' density of $2A_\sigma^2$ has been exactly cancelled off with similar contributions of opposite sign coming from other magnetic ions. Thus, the Bragg scattering from such an antiferromagnet is that appropriate to a smaller spin density per ion than we would naïvely have supposed.

Furthermore, the form factor is not that appropriate to d functions alone but involves the overlap density, as indicated in (12.57). This overlap density makes the form factor appear flatter than would have been expected. To see this it is useful to rearrange (12.57) in the form

$$\mathscr{D}(\mathbf{r}) = d^2(\mathbf{r})(1-2A_\sigma^2) + 2A_\sigma\{2S_\sigma d^2(\mathbf{r}) - d(\mathbf{r})p(\mathbf{r}) + d(\mathbf{r})p'(\mathbf{r})\}.$$

The first term corresponds to a moment of $(1-2A_\sigma^2)$ distributed in the d orbital and the second term represents a correction to the moment that is positive near the centre of M and negative near the boundaries of M, and overall has zero moment. This second term has a form factor that is zero at $\boldsymbol{\kappa}=0$ and passes through a maximum as $\boldsymbol{\kappa}$ increases. The total form factor is, therefore, slightly larger than $(1-2A_\sigma^2)d^2(\mathbf{r})$, except at $\boldsymbol{\kappa}=0$, and this makes the form factor appear flatter.

A third consequence follows from (12.57). Because $d^2(\mathbf{r})$ falls off rapidly with $\mathbf{r}$, the spin density near the ligand nuclei must be dominated by the second term of (12.57) and this is negative. Hence, the spin density, which is positive in the immediate neighbourhood of M, must pass through zero and become negative at some point intermediate between M and the neighbouring ligand L. By symmetry the spin density is zero at the ligand nucleus itself. Hence, each magnetic ion of positive spin is surrounded by a region of negative spin and vice versa. This last conclusion is most unexpected because naïvely we would have argued that as we moved along the chain from the M'' nucleus to the M nucleus we would have only one node in the moment density and that this would be at nucleus L. Instead we conclude there are three nodes, one at nucleus L and one each side of it.

This last conclusion, and also the conclusion that the form factor should appear flatter than expected, depends on the sign of A_{σ}. We now turn to a discussion of three-dimensional systems and verify the choice of sign for A_{σ}.

The three-dimensional system we consider consists of a transition-metal ion surrounded by a regular octahedron of ligand ions. This is the magnetic complex most commonly found. Earlier in this chapter we noted that the fivefold degenerate 3d wave functions split in a cubic environment into two groups, the orbitals of one transforming as the E_g representation of the cubic group O and the orbitals of the other like T_{2g}. In the present case it is necessary to find also the combinations of the ligand orbitals that transform according to these representations. The required combinations are shown schematically in Figs. 12.13 and 12.14, and the

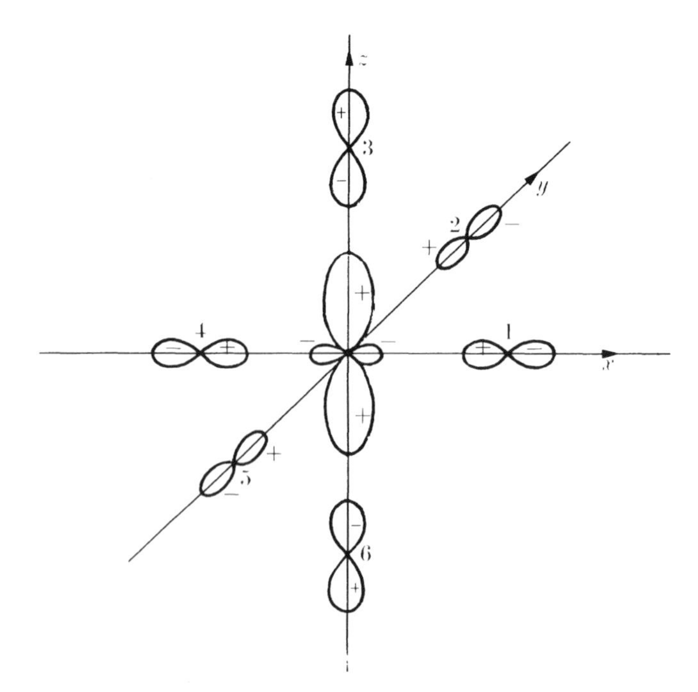

FIG. 12.13. Schematic diagram of the antibonding wave function ψ_z^2 (eqn (12.58)).

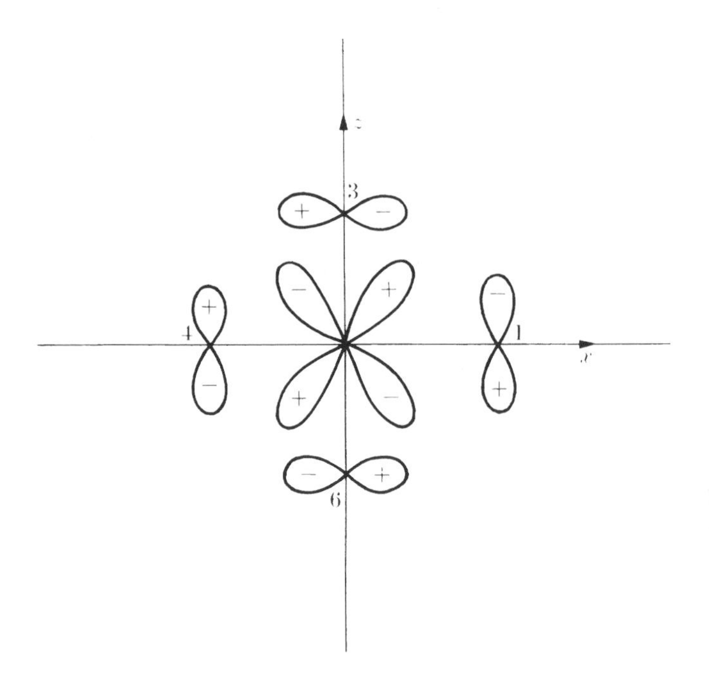

FIG. 12.14. Schematic diagram of the antibonding wave function ψ_{zx} (eqn (12.59)).

corresponding orbitals are

$$\psi_{z^2} = N_\sigma[d_{z^2} - A_\sigma\{p_3^\sigma + p_6^\sigma - \tfrac{1}{2}p_1^\sigma - \tfrac{1}{2}p_2^\sigma - \tfrac{1}{2}p_4^\sigma - \tfrac{1}{2}p_5^\sigma\} - A_s\{s_3 + s_6 - \tfrac{1}{2}s_1 - \tfrac{1}{2}s_2 - \tfrac{1}{2}s_4 - \tfrac{1}{2}s_5\}] \quad (12.58)$$

and

$$\psi_{zx} = N_\pi[d_{zx} - A_\pi\{p_1^z + p_3^x - p_4^z - p_6^x\}], \quad (12.59)$$

where, for example, p_3^σ stands for that p orbital of ligand number 3 and has a positive lobe pointing towards the magnetic ion. These equations define the parameters A_σ, A_π, and A_s. The normalization factors N_σ and N_π are given by

$$N_\sigma^{-2} = 1 + 3A_\sigma^2 - 6A_\sigma S_\sigma + 3A_s^2 - 6A_s S_s,$$
$$N_\pi^{-2} = 1 + 4A_\pi^2 - 8A_\pi S_\pi \quad (12.60)$$

where S_σ, S_π, and S_s are the overlap integrals, of positive sign, defined by

$$S_\sigma = \langle d_{z^2} \mid p_3^\sigma \rangle, \qquad S_\pi = \langle d_{zx} \mid p_3^x \rangle, \qquad S_s = \langle d_{z^2} \mid s_3 \rangle. \quad (12.61)$$

In those complexes with ligands, such as F^-, O^{2-}, Cl^-, the parameters A_σ, A_s, A_π must be positive by the following argument. Because the p orbitals are lower in energy than the d orbitals, the bonding orbitals must be predominantly ligand p orbitals but will have some admixture of d orbitals; in order to get bonding the electron charge must be increased in the region between the nuclei, hence the admixture must be of a sign to give addition of d and p orbitals in the overlap region. Because the antibonding orbital must be orthogonal to the bonding orbital it follows that, for such cases, the antibonding orbital must give subtraction of d and p orbitals in the overlap region and, furthermore, that the antibonding orbital must be predominantly a d orbital; hence A_σ, A_s, A_π are positive and small with the sign choices indicated by (12.58) and (12.59).

The wave functions (12.58) and (12.59) are two of the five antibonding orbitals; the other three can be written down in a similar way. The distribution $D(\mathbf{r})$ associated with orbital ψ_{z^2} is simply $|\psi_{z^2}|^2$ if it is singly occupied, and zero if it is empty or doubly occupied. The total distribution $D(\mathbf{r})$ can be obtained by placing electrons one by one in successive energy levels and summing up the densities due to the unpaired electrons.

Antiferromagnets and ferromagnets are made up of many such complexes and to a good approximation we can take the total spin density as the sum of that due to each individual complex.

For three-dimensional antiferromagnets of the rock-salt type (MnO, NiO), the modification of the spin density by covalency follows exactly the argument for the linear chain. The formulae required follow from (12.58) and (12.59). We shall not write them all down because they are exactly analogous to the linear chain formulae given above. Three conclusions, analogous to those we derived for the linear chain, are as follows.

(1) The absolute intensities are weaker than a simple model neglecting covalency would predict. The effective moment that can be seen is written down by analogy with the linear chain discussion. The moment associated with Bragg scattering from an E_g orbital is

$$1-3A_\sigma^2-3A_s^2 \tag{12.62}$$

where A_σ and A_s are defined by (12.58). The moment associated with Bragg scattering from a T_{2g} orbital is

$$1-4A_\pi^2 \tag{12.63}$$

where A_π is defined by (12.59).

(2) Although the absolute intensities of the Bragg peaks are low, as $\boldsymbol{\tau}$ increases the form factor falls off more slowly than would be expected. The 'overlap density' that causes this is exactly analogous to that discussed for the linear-chain model. This conclusion depends on the sign of A_σ, A_s, and A_π.

(3) Moving along any of the three cube directions, the spin density does not change sign simply according to the sign of the moment on the

nearest ion: there are nodes at points other than those dictated by symmetry and the variation of the spin density along and near to these cube directions is as described in the linear-chain model. This result concerning nodes of the spin density may be sensitive to other effects, and is also dependent on the sign of A_{σ}, A_s, A_{π}.

So far we have discussed in detail only very simple antiferromagnets where the effects of covalency could be seen quite easily. In the more general case the exact cancellation of that spin density entirely on the ligands does not take place: for example, in the ferromagnet no cancellation at all can take place and, in an antiferromagnet such as MnF_2, the environment of each F^- ligand is asymmetrical, so exact cancellation would not be expected. To discuss the more general case we must, therefore, use the full form for $D(\mathbf{r})$, i.e. the three-dimensional analogue of (12.56).

In the full form for $D(\mathbf{r})$, the terms additional to those we considered previously are those entirely on the ligands. We notice that this additional spin density is at such a large distance from the central ion that the form factor for it must have a strong and sharp peak in the forward direction. There are other peaks in the form factor of this ligand spin density but because the ligand wave functions are themselves widespread these peaks are much weaker than the forward peak. In effect, therefore, except near the forward direction, the intensities and form factors should behave like those of the very simple antiferromagnets. In particular the absolute intensities measured for all Bragg peaks except those very close to the forward direction should appear low compared to that expected on a model that neglected covalency. The form factors measured will appear to extrapolate to the values (12.62) or (12.63), or a suitably weighted mean, as appropriate.

However, Bragg scattering can, in principle, show up two other effects in the general case, which were absent for the very simple antiferromagnets. First, the spin density on the ligands gives a sharp forward peak to the form factor and for some crystals the unit cell may be large enough to give Bragg peaks even at angles close to the forward direction: this would make the forward peak directly observable by Bragg scattering. Second, the form factor of the ligand density is not entirely concentrated near the forward direction; it has a small value elsewhere and therefore, in principle, can be detected at these larger angles. In practice this detection would be done most easily by observing the asymmetry in the form factor produced by the departure from spherical symmetry of the ligand spin density or by observing Bragg scattering from peaks which, in the absence of covalency, would be forbidden.

In the discussion given above, we ignored another factor that makes antiferromagnetic diffraction peaks low, namely, the zero-point motion. The deviation of the mean spin from the fully aligned value is small and spin-wave theory does not give a thoroughly good estimate of it (see Chapter 9).

Wedgwood (1976) has made a detailed study of covalent effects in K_2NaCrF_6 which, he points out, is an almost ideal compound to exploit in a confrontation between experiment and theory. Because the unit cell is quite large, the $(CrF^6)^{3-}$ groups are isolated from each other, so that a theory of covalent effects can be based on an isolated $(CrF^6)^{3-}$ cluster, to a good approximation. Unfortunately, fluorine is the most electronegative ligand possible, and covalent effects are small.

Notwithstanding the small nominal covalency, significant deviations are observed at small κ between the experimental results, obtained with polarized neutrons, and calculations for a free Cr^{3+} ion. Data collected out to a maximum scattering vector $\kappa = 9.12$ Å^{-1} permits the construction of an accurate spin density map, shown in Fig. 12.15. The spin density on the Cr^{3+} has T_{2g} symmetry, as expected. The covalent effect is apparent in the small islands of spin approximately centred on the fluorine sites. These islands are not centred exactly on the fluorine ions since the unpaired spin is associated with antibonding wave functions that have a

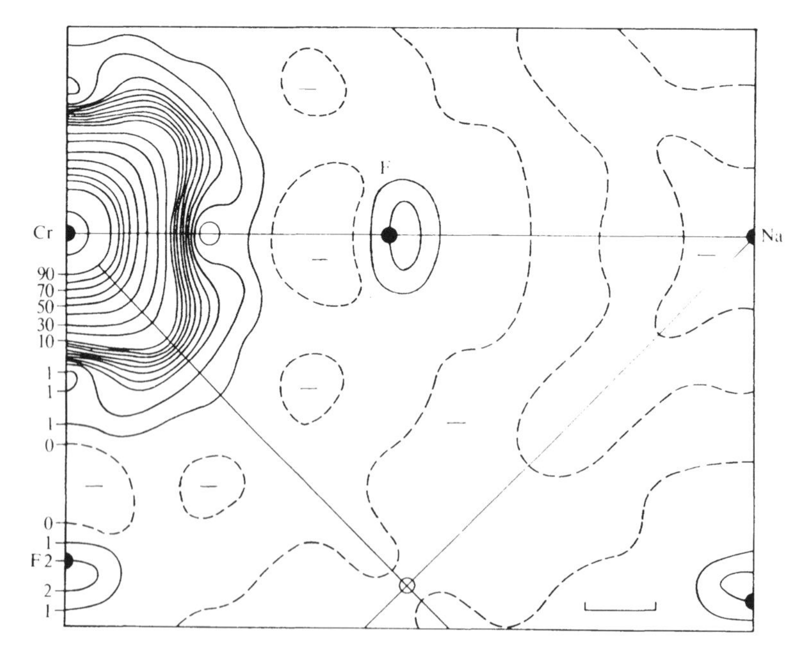

FIG. 12.15. Magnetization density profile in the (001) plane through the chromium site of K_2NaCrF_6. Contour units are 0.01 μ_B/Å^3 (Wedgwood 1976).

negative overlap between the metal and ligand ions, as shown in Fig. 12.15.

In summary it must be emphasized that the modifications introduced in elastic magnetic scattering by covalency are, by their very nature, very small and in systems more complex than those discussed in this section it is difficult to unravel contributions from various effects additional to covalency. The other possible effects include the following.

(a) If the orbital moment of the magnetic ions is not zero, its presence will tend to expand the form factor as compared to the form factor for the spin density alone.

(b) Magnetic ordering in the crystal can establish different radial wave functions for the spin up and spin down states. This is a very complicated effect but it will probably lead to an expansion in the form factor over the free-ion value.

(c) The radial wave functions in a crystal are likely to differ from those in the free ion, mainly because the electrons of the ligand ions help shield the 3d electrons of the magnetic ion from the charge of the nucleus of the magnetic ion. This will clearly result in an expansion of the radial 3d functions from their free-ion state and thereby cause a contraction in the atomic form factor.

(d) For antiferromagnets there is a reduction in the magnitude of the magnetic moment because of zero-point motion.

12.6. Helical spin ordering

In 1959 Yoshimori proposed that the spins in antiferromagnetic MnO_2, which possesses the same crystal structure as MnF_2, i.e. body-centred orthorhombic, form a helical configuration with a pitch that is not commensurate with the lattice. His arguments were based on considerations of the classical energy of the spin system.

The helical configuration for MnO_2 is shown in Fig. 12.16 and Yoshimori demonstrated that such a configuration was consistent with the diffraction pattern obtained for MnO_2. Since Yoshimori's paper, many examples of helical and more complex spin configurations have been found, particularly in the rare earths, (Coqblin 1977) and some ternary rare-earth superconductors that display a propensity for ferromagnetism (for a review see, for example, Fischer and Maple 1982; Sinha 1984) and neutron diffraction has been the principal tool in the investigations. We begin our discussion of helical configurations with an examination of the underlying features of Yoshimori's paper on MnO_2 and then consider the form of the cross-section in a more general case.

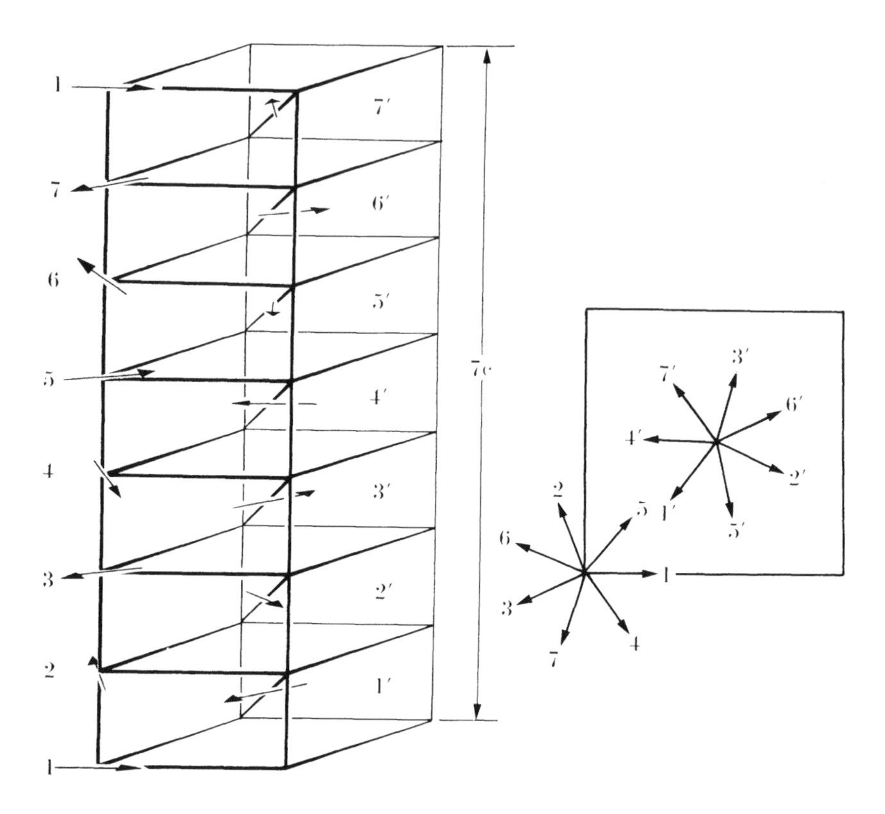

FIG. 12.16. Helical arrangement of magnetic moments in MnO_2.

We consider a simple lattice with just one magnetic atom per unit cell and inquire, by considering just the classical exchange forces (i.e. no anisotropy terms), what spin configurations are stable. The total exchange energy is

$$-\sum_{m,n} J(\mathbf{m}-\mathbf{n})\mathbf{S}_m \cdot \mathbf{S}_n \tag{12.64}$$

where $\mathbf{S}_m$ is the classical spin vector for the ion at the lattice site $\mathbf{m}$ and $J(\mathbf{m}-\mathbf{n})$ is the exchange parameter between spins at sites $\mathbf{m}$ and $\mathbf{n}$. Writing

$$J(\mathbf{m}) = \frac{1}{N}\sum_{\mathbf{q}} \mathscr{J}(\mathbf{q})\exp(i\mathbf{q}\cdot\mathbf{m}), \qquad \mathscr{J}(\mathbf{q}) = \mathscr{J}(-\mathbf{q}) \tag{12.65}$$

and

$$\mathbf{S}_m = \frac{1}{\sqrt{N}}\sum_{\mathbf{q}} \mathbf{S}_{\mathbf{q}} \exp(i\mathbf{q}\cdot\mathbf{m}), \qquad \mathbf{S}_{-\mathbf{q}} = \mathbf{S}_{\mathbf{q}}^* \tag{12.66}$$

where N is the number of unit cells in the crystal (= the number of magnetic ions), we obtain for (12.64)

$$-\sum_{\mathbf{q}} \mathscr{J}(\mathbf{q})\mathbf{S}_{\mathbf{q}} \cdot \mathbf{S}_{-\mathbf{q}}. \tag{12.67}$$

We now seek the lowest minimum of (12.67) subject to the condition that

$$\sum_m \mathbf{S}_m^2 = \text{constant}.$$

From (12.66) this condition is equivalent to

$$\sum_{\mathbf{q}} \mathbf{S}_{\mathbf{q}} \cdot \mathbf{S}_{-\mathbf{q}} = \text{constant}, \tag{12.68}$$

whence the lowest minimum of (12.67) is achieved for that $\mathbf{q}$ for which $\mathscr{J}(\mathbf{q})$ is the highest maximum. We denote this $\mathbf{q}$ by $\mathbf{Q}$ so that, since $-\mathbf{Q}$ is also allowed, we have for the minimum of (12.67) the value

$$-\mathscr{J}(\mathbf{Q})(\mathbf{S}_{\mathbf{Q}} \cdot \mathbf{S}_{-\mathbf{Q}} + \mathbf{S}_{-\mathbf{Q}} \cdot \mathbf{S}_{\mathbf{Q}}). \tag{12.69}$$

The corresponding $\mathbf{S}_m$ is, from (12.66)

$$\mathbf{S}_m = \frac{1}{N^{1/2}}\{\mathbf{S}_{\mathbf{Q}} \exp(\mathrm{i}\mathbf{Q} \cdot \mathbf{m}) + \mathbf{S}_{-\mathbf{Q}} \exp(-\mathrm{i}\mathbf{Q} \cdot \mathbf{m})\}. \tag{12.70}$$

In component form eqn (12.70) reads

$$S_m^{\alpha} = A_{\alpha} \cos(\mathbf{Q} \cdot \mathbf{m} + a_{\alpha}) \tag{12.71}$$

where A_{α} and a_{α} are constants. These equations show that as $\mathbf{m}$ advances in the direction of $\mathbf{Q}$ the spin vector $\mathbf{S}_m$ describes an ellipse. Since, however, the condition $\mathbf{S}_m^2$ = constant must be satisfied for all $\mathbf{m}$, it follows that the ellipse must be a circle. Taking the z-axis perpendicular to the plane of the circle, (12.71) reduces to

$$\begin{aligned} S_m^x &= S \cos(\mathbf{Q} \cdot \mathbf{m} + a) \\ S_m^y &= S \sin(\mathbf{Q} \cdot \mathbf{m} + a) \\ S_m^z &= 0. \end{aligned} \tag{12.72}$$

In general the direction of $\mathbf{Q}$ is not perpendicular to the plane of rotation of the spin vectors, the relative orientation being fixed by anisotropy energy. The magnitude of $\mathbf{Q}$ is determined solely by the exchange parameters and thus, in general, it is incommensurate with the lattice period.

The structure of MnO_2 is shown in Fig. 12.17. Three exchange parameters J_1, J_2, and J_3 are believed to be important. Yoshimori showed that the helical spin configuration illustrated in Fig. 12.16 was favoured in

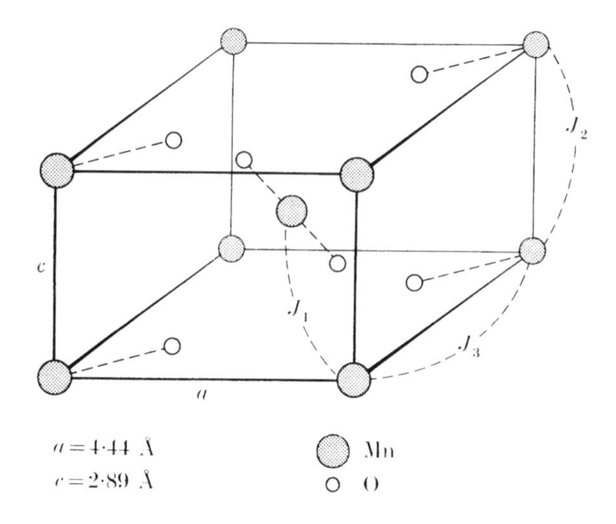

FIG. 12.17. Structure of MnO_2.

MnO_2 if these parameters satisfy

$$J_2 > J_1 \quad \text{with} \quad J_1 > 0, \tag{12.73}$$

the boundary conditions for stability being

$$J_2 = J_1 \quad \text{and} \quad 2J_2J_3 = J_1^2.$$

The spin configuration of Fig. 12.16 is described by

$$\mathbf{S}_l = S \exp(\mathrm{i}\mathbf{w} \cdot \mathbf{l})\{\tilde{\mathbf{x}} \cos(\mathbf{Q} \cdot \mathbf{l}) + \tilde{\mathbf{y}} \sin(\mathbf{Q} \cdot \mathbf{l})\}. \tag{12.74}$$

In (12.74) the reciprocal vector $\mathbf{w}$ is defined by

$$\mathbf{w} = \frac{2\pi}{a}(1, 1, a/c), \tag{12.75}$$

so that

$$\begin{aligned} \exp(\mathrm{i}\mathbf{w} \cdot \mathbf{l}) &= +1 \quad \text{for Mn ions at the corner sites,} \\ &= -1 \quad \text{for Mn ions at the body centre sites.} \end{aligned}$$

The spins screw along the direction of $\mathbf{Q}$ (which coincides with the z-axis) with a pitch $2\pi/|\mathbf{Q}|$.

To calculate the cross-section for the spin configuration described by

(12.74) we need to evaluate

$$\sum_{\alpha,\beta} (\delta_{\alpha\beta} - \tilde{\kappa}_\alpha \tilde{\kappa}_\beta)(S_l^\alpha)^* S_{l'}^\beta$$
$$= S^2 \exp\{i\mathbf{w}\cdot(\mathbf{l}'-\mathbf{l})\}[(1-\tilde{\kappa}_x^2)\cos(\mathbf{Q}\cdot\mathbf{l})\cos(\mathbf{Q}\cdot\mathbf{l}')$$
$$+(1-\tilde{\kappa}_y^2)\sin(\mathbf{Q}\cdot\mathbf{l})\sin(\mathbf{Q}\cdot\mathbf{l}') - \tilde{\kappa}_x\tilde{\kappa}_y \sin\{\mathbf{Q}\cdot(\mathbf{l}+\mathbf{l}')\}]$$
$$= S^2 \exp\{i\mathbf{w}\cdot(\mathbf{l}'-\mathbf{l})\}[\tfrac{1}{2}(1+\tilde{\kappa}_z^2)\cos\{\mathbf{Q}\cdot(\mathbf{l}-\mathbf{l}')\}$$
$$+\tfrac{1}{4}\{(\tilde{\kappa}_y + i\tilde{\kappa}_x)^2 \exp\{i\mathbf{Q}\cdot(\mathbf{l}+\mathbf{l}')\} + \text{c.c.}\}]. \tag{12.76}$$

The cross-section involves (12.76) multiplied by $\exp\{i\boldsymbol{\kappa}\cdot(\mathbf{l}'-\mathbf{l})\}$ and summed over all lattice sites $\mathbf{l}$ and $\mathbf{l}'$. If $2\mathbf{Q}$ does not coincide with a reciprocal lattice vector

$$\boldsymbol{\tau} = \frac{2\pi}{a}\left(t_1, t_2, \frac{a}{c}t_3\right) \quad (t_1+t_2+t_3 = \text{even integer}),$$

there is no contribution to the cross-section from the second term within the square bracket in (12.76); whence

$$\frac{d\sigma}{d\Omega} = r_0^2(\tfrac{1}{2}S)^2 |\tfrac{1}{2}gF(\boldsymbol{\kappa})|^2 N\frac{(2\pi)^3}{v_0}(1+\tilde{\kappa}_z^2)\sum_{\boldsymbol{\tau}}\{\delta(\boldsymbol{\kappa}+\mathbf{w}-\mathbf{Q}-\boldsymbol{\tau})+\delta(\boldsymbol{\kappa}+\mathbf{w}+\mathbf{Q}-\boldsymbol{\tau})\}. \tag{12.77}$$

Hence we see that there are Bragg reflections when either of the two conditions

$$\boldsymbol{\kappa} = \boldsymbol{\tau} - \mathbf{w} \pm \mathbf{Q} \tag{12.78}$$

are satisfied.

If we take

$$\mathbf{Q} = \frac{2\pi}{c}(0, 0, 2/7) \tag{12.79}$$

so that the pitch of the helical structure $2\pi/|\mathbf{Q}| = 7c/2$, and make our unit cell seven times the length of the normal unit cell in the (001)-direction, Bragg reflections occur at

$$(t_1-1, t_2-1, 7(t_3-1)\pm 2), \quad \text{with } t_1+t_2+t_3 = \text{even integer}. \tag{12.80}$$

These coincide exactly with the reflections reported for MnO_2.

Some of the spin configurations that have been found since Yoshimori's analysis for MnO_2 are very complicated. Besides yielding the spin configuration, neutron diffraction has also enabled the turn angle between spins in adjacent planes to be measured as a function of temperature (Coqblin 1977).

In general, for one atom per unit cell the spin vectors obey

$$\begin{aligned}\mathbf{S}_n &= \sum_{\lambda=-\infty}^{\infty} \mathbf{A}_{\lambda\mathbf{Q}} \exp(\mathrm{i}\lambda\mathbf{Q}\cdot\mathbf{n}) \\ &= \mathbf{A}_0 + \mathbf{A}_{\mathbf{Q}} \exp(\mathrm{i}\mathbf{Q}\cdot\mathbf{n}) + \mathbf{A}_{-\mathbf{Q}} \exp(-\mathrm{i}\mathbf{Q}\cdot\mathbf{n}) \\ &\quad + \mathbf{A}_{2\mathbf{Q}} \exp(\mathrm{i}2\mathbf{Q}\cdot\mathbf{n}) + \mathbf{A}_{-2\mathbf{Q}} \exp(-\mathrm{i}2\mathbf{Q}\cdot\mathbf{n}) \\ &\quad + \cdots. \end{aligned} \tag{12.81}$$

Because $\mathbf{S}_n$ is real,

$$\mathbf{A}_{-\lambda\mathbf{Q}} = \mathbf{A}^*_{\lambda\mathbf{Q}}. \tag{12.82}$$

(12.81) is a reasonable description of the spin pattern provided none of the $\lambda\mathbf{Q}$ is equal to $\boldsymbol{\tau}$—because if, say, $\lambda_1\mathbf{Q}$ were equal to $\boldsymbol{\tau}_1$, then $\mathrm{A}_{\lambda_1\mathbf{Q}} \exp(\mathrm{i}\lambda_1\mathbf{Q}\cdot\mathbf{n})$ would be just $\mathbf{A}_{\lambda_1\mathbf{Q}}$ for all $\mathbf{n}$ and could equally well be absorbed in $\mathbf{A}_0$. With this proviso it is easy to show that the elastic magnetic neutron cross-section becomes

$$\frac{\mathrm{d}\sigma}{\mathrm{d}\Omega} = r_0^2 \frac{g^2 N(2\pi)^3}{4v_0} \sum_{\boldsymbol{\tau}} \sum_{\lambda} \delta(\lambda\mathbf{Q} + \boldsymbol{\kappa} - \boldsymbol{\tau})\, |F(\boldsymbol{\kappa})|^2\, \{\mathbf{A}^*_{\lambda\mathbf{Q}} \cdot \mathbf{A}_{\lambda\mathbf{Q}} - |\tilde{\boldsymbol{\kappa}} \cdot \mathbf{A}_{\lambda\mathbf{Q}}|^2\}. \tag{12.83}$$

If the sample has more than one crystallographically distinct atom per unit cell and (12.81) is taken to describe the spin variation from one cell to another, then each term of (12.83) is to be multiplied by the square of the modulus of the unit cell form factor and N is to be taken as the number of unit cells each of volume v_0.

Notice from (12.83) that the uniform component, $\lambda = 0$, produces scattering at the reciprocal lattice vectors only. The other components produce satellites at scattering vectors $\boldsymbol{\tau} - \lambda\mathbf{Q}$. In practice it usually happens that $\mathbf{Q}$ itself is small or the deviation of $\mathbf{Q}$ from a vector that would describe a simple antiferromagnet is small. In these cases the satellites appear clustered around the Bragg peaks or the superlattice peaks respectively. These satellites can only be observed clearly in single-crystal experiments; they have not been properly resolved in powder-diffraction experiments.

The orientation factor in (12.83) is a great help in determining the details of the spin pattern. We shall not at this time give a discussion of particular cases, but it is worth noting that the $\lambda\mathbf{Q}$ satellite of $\boldsymbol{\tau} = 0$ appears with an orientation factor

$$|\mathbf{A}_{\lambda\mathbf{Q}}|^2 - |\tilde{\mathbf{Q}} \cdot \mathbf{A}_{\lambda\mathbf{Q}}|^2,$$

which is zero for a longitudinal wave, i.e. $\mathbf{A}_{\lambda\mathbf{Q}}$ parallel to $\mathbf{Q}$. Hence such longitudinal waves can only be observed as satellites to $\boldsymbol{\tau}$ not zero.

Before leaving this subject it is worth noting that (12.81) describes the spin pattern on a 'rigid ion' model. We could also have spin density waves in the conduction-electron gas of a type described by Overhauser (Currat and Pynn 1979). For such spin-density waves one would, as a first approximation, describe the spin density at $\mathbf{r}$ by

$$\mathcal{S}(\mathbf{r}) = \sum_{\lambda} \mathbf{A}_{\lambda\mathbf{Q}} \exp(\mathrm{i}\lambda\mathbf{Q}\cdot\mathbf{r}) \sum_{n} |\phi(\mathbf{r}-\mathbf{n})|^2, \tag{12.84}$$

which leads to the result

$$\frac{\mathrm{d}\sigma}{\mathrm{d}\Omega} = r_0^2 \frac{g^2 N(2\pi)^3}{4v_0} \sum_{\boldsymbol{\tau}} \sum_{\lambda} \delta(\lambda\mathbf{Q}+\boldsymbol{\kappa}-\boldsymbol{\tau})\, |F(\boldsymbol{\tau})|^2 \{|\mathbf{A}_{\lambda\mathbf{Q}}|^2 - |\tilde{\boldsymbol{\kappa}}\cdot\mathbf{A}_{\lambda\mathbf{Q}}|^2\}. \tag{12.85}$$

The form factor appearing in (12.85) is $F(\boldsymbol{\tau})$, not $F(\boldsymbol{\kappa})$ as in (12.83). The difference comes about because of the different physical descriptions given by (12.81) and (12.84). In (12.81) the spin of an atom is supposed to rotate as a rigid unit, but in (12.84) the spin density is modulated within the atom itself.

12.7. Diffuse scattering (Hicks 1979; Rainford 1982)

Diffuse magnetic scattering from alloys has been used successfully to measure spatial fluctuations of magnetic moments. The cross-section is proportional to the Fourier transform of the magnetic-defect structure, in dilute alloys, which can be inverted to give the moment distribution in the vicinity of an impurity.

The diffuse cross-section is derived from (12.10) by a few simple steps. First, we introduce a convenient notation. We assume that the moments can be ascribed to lattice sites defined by the vectors $\mathbf{R}_i = \mathbf{l}+\mathbf{d}$; thus, following (12.22),

$$\langle\hat{\boldsymbol{\alpha}}\rangle = \tfrac{1}{2} r_0 \sum_{i} \exp(\mathrm{i}\boldsymbol{\kappa}\cdot\mathbf{R}_i)\boldsymbol{\mu}_i^{(\perp)} F_i(\boldsymbol{\kappa}). \tag{12.86}$$

Here,

$$\boldsymbol{\mu}_i^{(\perp)} = \tilde{\boldsymbol{\kappa}} \times (\boldsymbol{\mu}_i \times \tilde{\boldsymbol{\kappa}}); \qquad \tilde{\boldsymbol{\kappa}} = \boldsymbol{\kappa}/|\boldsymbol{\kappa}|, \tag{12.87}$$

where $\boldsymbol{\mu}_i$ is the magnetic moment, which is independent of $\boldsymbol{\kappa}$ for most magnetic structures.

If we allow for the use of a polarized beam, the cross-section contains an interference between magnetic and nuclear scattering amplitudes. The nuclear amplitude involved

$$\overline{\langle\hat{\beta}\rangle} = \sum_{i} \exp(\mathrm{i}\boldsymbol{\kappa}\cdot\mathbf{R}_i)\bar{b}_i \tag{12.88}$$

where $\bar{b}_j$ is the bound, coherent scattering length that is assumed here to be real.

Using the preceding notation, the magnetic contribution to the differential cross-section (12.10) is

$$\left(\frac{d\sigma}{d\Omega}\right)^{el}_{coh} = \tfrac{1}{2}r_0 \sum_{jj'} \exp\{i\boldsymbol{\kappa}\cdot(\mathbf{R}_{j'}-\mathbf{R}_j)\}$$
$$\times[\tfrac{1}{2}r_0 F_j(\boldsymbol{\kappa})F_{j'}(\boldsymbol{\kappa})\{\boldsymbol{\mu}_j^{(\perp)}\cdot\boldsymbol{\mu}_{j'}^{(\perp)} + i\mathbf{P}\cdot(\boldsymbol{\mu}_j^{(\perp)}\times\boldsymbol{\mu}_{j'}^{(\perp)})\} + 2F_j(\boldsymbol{\kappa})\bar{b}_{j'}\boldsymbol{\mu}_j\cdot\mathbf{P}_\perp]. \quad (12.89)$$

In the interference term we have used the relation

$$\boldsymbol{\mu}^{(\perp)}\cdot\mathbf{P} = \boldsymbol{\mu}\cdot\{\tilde{\boldsymbol{\kappa}}\times(\mathbf{P}\times\tilde{\boldsymbol{\kappa}})\} = \boldsymbol{\mu}\cdot\mathbf{P}_\perp. \quad (12.90)$$

We have not included the Debye–Waller factors explicitly in (12.89), but in order to do so we have only to make the replacements (cf. § 8.1)

$$\bar{b}_j \Rightarrow \bar{b}_j \exp\{-W_j(\boldsymbol{\kappa})\}$$

and

$$F_j(\boldsymbol{\kappa}) \Rightarrow F_j(\boldsymbol{\kappa})\exp\{-W_j(\boldsymbol{\kappa})\}.$$

The cross-section (12.89) depends on the precise distribution of moments. Because these are randomly distributed, to a large extent, it follows that the cross-section is not exactly the same from sample to sample. However, by analogy to similar problems in statistical physics, we know that the variation in the cross-section is negligible for bulk samples. It then follows that we really need the average of (12.89) over the configurations of the constituent atoms; we denote the ensemble average by a horizontal bar.

The coherent elastic scattering is given by the square of the average scattering length. The difference between the averaged and coherent cross-sections is the diffuse cross-section. For an unpolarized beam, $\mathbf{P}=0$, the diffuse cross-section is

$$\left(\frac{d\sigma}{d\Omega}\right)_{diff} = (\tfrac{1}{2}r_0)^2 \sum_{jj'} \exp\{i\boldsymbol{\kappa}\cdot(\mathbf{R}_{j'}-\mathbf{R}_j)\}$$
$$\times\left\{\overline{F_j(\boldsymbol{\kappa})F_{j'}(\boldsymbol{\kappa})\boldsymbol{\mu}_j^{(\perp)}\cdot\boldsymbol{\mu}_{j'}^{(\perp)}} - (\overline{F_j(\boldsymbol{\kappa})\boldsymbol{\mu}_j^{(\perp)}})\cdot(\overline{F_{j'}(\boldsymbol{\kappa})\boldsymbol{\mu}_{j'}^{(\perp)}})\right\}. \quad (12.91)$$

Consider a simple ferromagnetic alloy in which the moments are aligned parallel with the z-axis, and are of magnitude μ. The coherent, Bragg scattering is then

$$\left(\frac{d\sigma}{d\Omega}\right)_{Bragg} = \frac{N}{4} r_0^2(1-\tilde{\kappa}_z^2)\{\overline{\mu F(\boldsymbol{\kappa})}\}^2 \frac{(2\pi)^3}{v_0} \sum_{\boldsymbol{\tau}} \delta(\boldsymbol{\kappa}-\boldsymbol{\tau}) \quad (12.92)$$

where v_0 is the volume of a unit cell, $\boldsymbol{\tau}$ a reciprocal lattice vector and

$$\overline{\mu F(\boldsymbol{\kappa})} = \frac{1}{N}\sum_m \mu_m F_m(\boldsymbol{\kappa}). \quad (12.93)$$

The remaining diffuse scattering is given by (12.91)

$$\left(\frac{\mathrm{d}\sigma}{\mathrm{d}\Omega}\right)_{\text{diff}} = \frac{N}{4} r_0^2(1-\tilde{\kappa}_z^2)\Xi(\boldsymbol{\kappa}). \tag{12.94}$$

In (12.94)

$$\begin{aligned}\Xi(\boldsymbol{\kappa}) &= \frac{1}{N}\sum_{m,n} \exp\{\mathrm{i}\boldsymbol{\kappa}\cdot(\mathbf{m}-\mathbf{n})\}\left\{\overline{F_n\mu_n F_m\mu_m} - (\overline{F_n\mu_n})\,(\overline{F_m\mu_m})\right\} \\ &\equiv \frac{1}{N}\sum_{m,n} \exp\{\mathrm{i}\boldsymbol{\kappa}\cdot(\mathbf{m}-\mathbf{n})\}\overline{(F_n\mu_n - \overline{F_n\mu_n})(F_m\mu_m - \overline{F_m\mu_m})},\end{aligned} \tag{12.95}$$

and from the second line it is clear that the diffuse magnetic scattering involves the *spatial* fluctuations of the moments about their mean values.

In an experiment the diffuse magnetic scattering can be resolved from the total scattering by making use of the facts that (a) it is strictly elastic; (b) it is proportional to $1-\tilde{\kappa}_z^2$; and (c) it is distinct from Bragg scattering, which appears only in very sharp peaks when rigid geometrical conditions are satisfied.

The total magnetic moment of the target system is†

$$M = \sum_m \mu_m, \tag{12.96}$$

so that, since the form factors are, by definition, unity in the forward direction $\boldsymbol{\kappa} = 0$,

$$\Xi(0) = \frac{1}{N}\{\overline{M^2} - (\bar{M})^2\} = \frac{1}{N}\overline{(\delta M)^2} \tag{12.97}$$

where

$$\delta M = M - \bar{M}.$$

In other words, $\Xi(0)$ is proportional to the mean-square spatial fluctuation of the total magnetic moment M.

The total magnetic moment varies primarily because of concentration fluctuations, and we shall consider these as follows. We suppose that the binary sample contains $N_1 = cN$ atoms of type 1 and $N_2 = (1-c)N$ atoms of type 2. Each type of atom has a mean magnetic moment, μ_1 and μ_2, respectively. The mean moment μ, as observed in a macroscopic determination of the magnetization, is, clearly,

$$\mu = c\mu_1 + (1-c)\mu_2, \tag{12.98}$$

and we can assume μ to be known for all concentrations. In principle both μ_1 and μ_2 also vary with concentration and therefore cannot be determined from magnetization measurements alone. For the variation of

† Strictly speaking, M is a reduced moment since it is dimensionless (cf. eqn (11.7)).

the total magnetic moment with concentration we note that

$$\delta M = \frac{\partial M}{\partial N_1}\,\delta N_1 = \frac{\partial \mu}{\partial c}\,\delta N_1, \tag{12.99}$$

so, from (12.97),

$$\Xi(0) = \left(\frac{\partial \mu}{\partial c}\right)^2 \frac{1}{N}\,\overline{\delta N_1^2}. \tag{12.100}$$

Let the function p_n be defined such that

$$p_n = 1, \text{ if an atom of type 1 is on the site } n,$$

$$p_n = 0, \text{ if an atom of type 2 is on the site } n.$$

Then

$$\overline{p_m} = c \tag{12.101}$$

and

$$N_1 = \sum_m p_m. \tag{12.102}$$

Hence

$$\overline{N_1^2} = \sum_{m,n} \overline{p_m p_n} = c^2 N^2 + c(1-c)N, \tag{12.103}$$

since

$$\overline{p_m p_n} = c^2 + c(1-c)\,\delta_{m,n}. \tag{12.104}$$

From (12.103) we now have

$$\overline{\delta N_1^2} = \overline{N_1^2} - (\bar{N}_1)^2 = c(1-c)N, \tag{12.105}$$

which together with (12.99) gives the result

$$\Xi(0) = c(1-c)\left(\frac{\partial \mu}{\partial c}\right)^2. \tag{12.106}$$

This is an important relationship because it tells us that, whatever the concentration, the scattering in the forward direction is related to the rate of change of magnetization with concentration and the latter can be determined by straightforward magnetization measurements to give a useful check on the neutron results.

In most alloys (12.106) is completely satisfactory, and it is rigorously valid for small concentrations, but in general the magnetization can vary through parameters other than the concentration. It is therefore worthwhile to consider the effect of short-range order. We suppose that a sample, with a fixed concentration, has a magnetization that varies with the heat treatment it has received in the past. It follows that the magnetization depends upon the short-range order which is established by the particular heat treatment. To discuss this we must describe short-range order parameters as follows.

Let the number of atoms, each of type 1, separated by a distance **R** be $N_{11}(\mathbf{R})$; then by definition

$$N_{11}(\mathbf{R}) = \sum_m p_m p_{m+R}. \tag{12.107}$$

which we write

$$N_{11}(\mathbf{R}) = \{c^2 + s(\mathbf{R})\}N \qquad (\mathbf{R} \neq 0). \tag{12.108}$$

Eqn (12.108) defines the short-range order parameter s(**R**). We now examine the effect of fluctuations of s(**R**) about the mean value of zero; thus

$$\overline{N_{11}(\mathbf{R})} = c^2 N$$

and

$$\{\delta N_{11}(\mathbf{R})\}_c = N\,\delta s(\mathbf{R}) = \delta N_{11}(\mathbf{R}) - 2Nc\,\delta c, \tag{12.109}$$

where the subscript c means that the variation is at constant concentration. In place of the relationship given by eqn (12.99) we now have

$$\begin{aligned}\delta M &= \frac{\partial M}{\partial N_1}\,\delta N_1 + \sum_R \frac{\partial M}{\partial N_{11}(\mathbf{R})}\{\delta N_{11}(\mathbf{R})\}_c \\ &= \frac{\partial \mu}{\partial c}\,\delta N_1 + \sum_R \frac{\partial \mu}{\partial s(\mathbf{R})}\{\delta N_{11}(\mathbf{R})\}_c,\end{aligned} \tag{12.110}$$

where in the second term it is understood that the partial derivatives are taken at constant concentration.

Hence

$$\begin{aligned}\overline{\delta M^2} &= \left(\frac{\partial \mu}{\partial c}\right)^2 \overline{\delta N_1^2} + 2\left(\frac{\partial \mu}{\partial c}\right)\sum_R \frac{\partial \mu}{\partial s(\mathbf{R})}\,\overline{\delta N_1\{\delta N_{11}(\mathbf{R})\}_c} \\ &\quad + \sum_{R,R'} \frac{\partial \mu}{\partial s(\mathbf{R})}\frac{\partial \mu}{\partial s(\mathbf{R}')}\overline{\{\delta N_{11}(\mathbf{R})\}_c\{\delta N_{11}(\mathbf{R}')\}_c}.\end{aligned} \tag{12.111}$$

To reduce this expression to a more convenient form we make use of the relationships

$$\delta N_1 = \sum_m p_m - cN$$

and

$$\delta N_{11}(\mathbf{R}) = \sum_m p_m p_{m+R} - c^2 N, \tag{12.112}$$

which follow directly from (12.102) and (12.107), respectively, to write, following eqn (12.109),

$$\begin{aligned}\{\delta N_{11}(\mathbf{R})\}_c &= \delta N_{11}(\mathbf{R}) - 2c\,\delta N_1 \\ &= \sum_m p_m p_{m+R} - 2c\sum_m p_m + c^2 N.\end{aligned} \tag{12.113}$$

From the product of (12.112) and (12.113),

$$\overline{\delta N_1\{\delta N_{11}(\mathbf{R})\}_c} = \sum_{m,n} \overline{p_n p_m p_{m+R}} - 2c \sum_{m,n} \overline{p_n p_m} + c^2 N \sum_n \overline{p_n} - cN\left\{\sum_m \overline{p_m p_{m+R}} - 2c \sum_m \overline{p_m} + c^2 N\right\} = 0. \quad (12.114)$$

In obtaining this result the identity

$$\overline{p_l p_m p_n} = c^3 + c^2(1-c)(\delta_{l,m} + \delta_{m,n} + \delta_{l,n}) + c(1-c)(1-2c)\,\delta_{l,m}\,\delta_{m,n} \quad (12.115)$$

has been used in conjunction with the condition $\mathbf{R} \neq 0$. Also, from (12.113),

$$\overline{\{\delta N_{11}(\mathbf{R})\}_c \{\delta N_{11}(\mathbf{R}')\}_c}$$

$$= \overline{\left(\sum_m p_m p_{m+R} - 2c \sum_m p_m + c^2 N\right)\left(\sum_n p_n p_{n+R'} - 2c \sum_n p_n + c^2 N\right)}$$

$$= \sum_{m,n} \overline{p_m p_{m+R} p_n p_{n+R'}} - 2c \sum_{m,n} \overline{p_m p_n p_{n+R'}} - 2c \sum_{mn} \overline{p_m p_{m+R} p_n} + 4c^2 \sum_{mn} \overline{p_m p_n} - N^2 c^4.$$

Using (12.115) and also the identity, for $\mathbf{R} \neq 0$ and $\mathbf{R}' \neq 0$,

$$\overline{p_m p_{m+R} p_n p_{n+R'}} = c^4 + c^3(1-c)\{\delta_{m+R-n} + \delta_{m-n-R'} + \delta_{m+R-n-R'} + \delta_{m-n}\} + c^2(1-c)^2\{\delta_{m+R-n}\,\delta_{R+R'} + \delta_{m-n}\,\delta_{R-R'}\},$$

this can be evaluated to give

$$\overline{\{\delta N_{11}(\mathbf{R})\}_c \{\delta N_{11}(\mathbf{R}')\}_c} = c^2(1-c)^2 N(\delta_{R,R'} + \delta_{R,-R'}). \quad (12.116)$$

If we now combine results for $\overline{\delta M^2}$, then from the general relationship between $\Xi(0)$ and $\overline{\delta M^2}$ given by eqn (12.97) we obtain finally

$$\Xi(0) = c(1-c)\left(\frac{\partial \mu}{\partial c}\right)^2 + 2c^2(1-c)^2 \sum_R \left\{\frac{\partial \mu}{\partial s(\mathbf{R})}\right\}^2. \quad (12.117)$$

This formula is a generalization of eqn (12.96). The correction term is negligible at either end of the concentration range because of the factor $c^2(1-c)^2$ and is, therefore, unimportant for dilute alloys. In general, there are an infinite number of short-range order parameters $s(\mathbf{R})$. It follows, therefore, that in order to estimate the second term in (12.117) it is necessary to subject the sample to an infinite set of heat treatments, in each case measuring the total magnetization and short-range order parameters. Fortunately in many cases it ought to be a reasonable approximation to assume that the magnetization changes only with a single parameter, the nearest-neighbour parameter, for instance. We define this parameter as just s, and the number of nearest-neighbour pairs in the sample is $\frac{1}{2}Nr$, where r is the number of nearest neighbours. The number of nearest-neighbour pairs which happen to be both of type 1 is $\frac{1}{2}Nr(c^2+s)$ and then (12.117) becomes

$$\Xi(0) \simeq c(1-c)\left(\frac{\partial\mu}{\partial c}\right)^2 + \frac{2c^2}{r}(1-c)^2\left(\frac{\partial\mu}{\partial s}\right)^2. \tag{12.118}$$

This formula is more tractable than (12.117). To estimate the second term in (12.118) we need measurements only on two samples: the first quenched rapidly to ensure it has no short-range order; the second cooled to have a small residual short-range order s, measurable through X-ray determinations of the chemical order. Magnetization measurements on both then give an estimate of $\partial\mu/\partial s$.

The second term of (12.118) is usually small for the following reasons.

(a) At either end of the concentration range the factor $c^2(1-c)^2$ is small.
(b) The factor r^{-1} is of the order 0.1.
(c) In most cases the dependence of the magnetization on short-range order is small.

For these reasons it is usually satisfactory to retain only the first term in (12.118). But in special situations the second term in (12.118) can be important.

12.7.1. *Shull-Wilkinson model*

The simplest model of an alloy is that in which any fluctuation in the moments of either type 1 or type 2 atoms is completely neglected. In other words, all atoms of type 1 are taken to have a moment μ_1 independent of what their environment might be; similarly all atoms of type 2 are taken to have a moment μ_2. Hence, in general,

$$\mu_m = \mu_2 + (\mu_1 - \mu_2)p_m, \tag{12.119}$$

from which (12.98) immediately follows.

Furthermore,

$$F_m\mu_m = F_2\mu_2 + (F_1\mu_1 - F_2\mu_2)p_m, \tag{12.120}$$

and hence

$$\begin{aligned}\overline{F_m\mu_m F_n\mu_n} &= (F_2\mu_2)^2 + F_2\mu_2(F_1\mu_1 - F_2\mu_2)(\bar{p}_m + \bar{p}_n) + (F_1\mu_1 - F_2\mu_2)^2\overline{p_m p_n} \\ &= (F_2\mu_2)^2 + F_2\mu_2(F_1\mu_1 - F_2\mu_2)2c + (F_1\mu_1 - F_2\mu_2)^2c^2 \\ &\quad + (F_1\mu_1 - F_2\mu_2)^2c(1-c)\,\delta_{m,n} \\ &= \{cF_1\mu_1 + (1-c)F_2\mu_2\}^2 + (F_1\mu_1 - F_2\mu_2)^2c(1-c)\,\delta_{m,n},\end{aligned} \tag{12.120b}$$

where use has been made of the identity (12.104). From the definition of $\Xi(\boldsymbol{\kappa})$, eqn (12.95), we obtain,

$$\Xi(\boldsymbol{\kappa}) = c(1-c)\{F_1(\boldsymbol{\kappa})\mu_1 - F_2(\boldsymbol{\kappa})\mu_2\}^2. \tag{12.120c}$$

In the forward direction, and within the context of the simple model, (12.119) is consistent with (12.106).

In experiments of the Shull–Wilkinson type the major interest is in the extrapolation of the observed cross-sections to the forward direction. Equation (12.120c) shows that in this limit we have a result for $(\mu_1 - \mu_2)^2$. This, coupled with a magnetization measurement of μ, serves to give μ_1 and μ_2 separately.

There are two possible signs of $\mu_1 - \mu_2$ as determined from the neutron cross-section, as is clear from eqn (12.120c). Usually this ambiguity is removed by taking into account some other independent experimental information. Alternatively, we can use polarized neutrons to determine the sign of $\mu_1 - \mu_2$. From eqn (12.89) it is clear that there is an additional term in the diffuse cross-section

$$r_0 \mathbf{P}_{\perp,z} c(1-c)(\bar{b}_1 - \bar{b}_2)\{F_1(\boldsymbol{\kappa})\mu_1 - F_2(\boldsymbol{\kappa})\mu_2\}. \tag{12.120d}$$

Here $\bar{b}_1$ and $\bar{b}_2$ are the coherent bound scattering lengths of the type 1 and type 2 atoms respectively. Implicit in this expression is the assumption that the different atoms have distinct scattering lengths, as we assume them to have distinct magnetic moments. If the experimental arrangement is that the neutron beam is polarized parallel or antiparallel to the magnetization of the sample ($\mathbf{P} \cdot \mathbf{z} = \pm 1$) and $\tilde{\boldsymbol{\kappa}}$ lies in a plane normal to this direction, then the above term has the two values

$$\pm r_0 c(1-c)(\bar{b}_1 - \bar{b}_2)(\mu_1 - \mu_2) \tag{12.121}$$

in the forward direction, according to the two directions of $\mathbf{P}$. The magnitude of this term may be increased by the use of a particular isotope of either type of atom (or both) chosen to give the maximum value of $\bar{b}_1 - \bar{b}_2$.

12.7.2. Dilute alloys

When the concentration c is small, the immediate environment of each type 1 atom is the same and, therefore, one can assume all atoms of type 1 to have the same moment μ_1. However, each type 1 atom will affect the magnitude of the moment of the type 2 atoms that surround it. To allow for this, we define $\phi(\mathbf{R})$ to be the increase in the z-component of the magnetic moment of a type 2 atom at a distance $\mathbf{R}$ from a given type 1 atom, due to the presence of this atom. Thus, we now have

$$\mu_m = \mu_2 + (\mu_1 - \mu_2)p_m + \sum_R \phi(\mathbf{R})p_{m+R}$$

and

$$F_m\mu_m = F_2\mu_2 + (F_1\mu_1 - F_2\mu_2)p_m + F_2\sum_R \phi(\mathbf{R})p_{m+R}. \qquad (12.122)$$

By definition $\phi(0)$ is zero, and because any disturbance must vanish at large enough distances from the type 1 atom, $\phi(\mathbf{R})$ approaches zero as $|\mathbf{R}|$ becomes large. It is convenient to define the spatial Fourier transform

$$\Phi(\boldsymbol{\kappa}) = \sum_R \exp(i\boldsymbol{\kappa}\cdot\mathbf{R})\phi(\mathbf{R}). \qquad (12.123)$$

We expect $\Phi(\boldsymbol{\kappa})$ to be proportional to the isothermal susceptibility of the host material for $\phi(\mathbf{R})$ is, by definition, the response of the host to the type 1 atom situated at $\mathbf{R}=0$. The factor of proportionality will depend on the form of the host–impurity interaction and the type of impurity. However, near a second-order phase transition the temperature dependence of $\Phi(\boldsymbol{\kappa})$ is likely to be dominated by that of the susceptibility.

We shall assume the lattice to be such that every atom is a centre of inversion symmetry so that $\phi(\mathbf{R}) = \phi(-\mathbf{R})$. $\Phi(\boldsymbol{\kappa})$ is then real and $\Phi(\boldsymbol{\kappa}) = \Phi(-\boldsymbol{\kappa})$. From (12.122)

$$\mu = \mu_2 + (\mu_1 - \mu_2)c + \Phi(0)c,$$

and hence

$$\frac{\partial\mu}{\partial c} = \mu_1 - \mu_2 + \Phi(0). \qquad (12.124)$$

From (12.122) we calculate

$$\begin{aligned}\overline{F_m\mu_m F_n\mu_n} &= [F_2\mu_2 + c\{F_1\mu_1 - F_2\mu_2 + F_2\Phi(0)\}]^2 \\ &\quad + c(1-c)\{(F_1\mu_1 - F_2\mu_2)^2\,\delta_{m,n} + 2F_2(F_1\mu_1 - F_2\mu_2)\phi(\mathbf{m}-\mathbf{n}) \\ &\qquad + F_2^2\sum_R \phi(\mathbf{R})\phi(\mathbf{m}+\mathbf{R}-\mathbf{n})\} \\ &= (\overline{F\mu})^2 + c(1-c)\frac{1}{N}\sum_{\boldsymbol{\kappa}} \exp\{-i\boldsymbol{\kappa}\cdot(\mathbf{m}-\mathbf{n})\}\,\{F_1\mu_1 - F_2\mu_2 + F_2\Phi(\boldsymbol{\kappa})\}^2.\end{aligned} \qquad (12.125)$$

If we substitute (12.125) into eqn (12.95), then $\Xi(\boldsymbol{\kappa})$ for a dilute ferromagnetic alloy is given by

$$\Xi(\boldsymbol{\kappa}) = c(1-c)\{F_1(\boldsymbol{\kappa})\mu_1 - F_2(\boldsymbol{\kappa})\mu_2 + F_2(\boldsymbol{\kappa})\Phi(\boldsymbol{\kappa})\}^2. \qquad (12.126)$$

This formula is consistent with (12.106) in the forward direction. A similar calculation shows that the interference term (12.120) becomes

$$r_0 P_{\perp,z}c(1-c)(\bar{b}_1 - \bar{b}_2)\{F_1(\boldsymbol{\kappa})\mu_1 - F_2(\boldsymbol{\kappa})\mu_2 + F_2(\boldsymbol{\kappa})\Phi(\boldsymbol{\kappa})\}. \quad (12.127)$$

To illustrate the behaviour of the diffuse scattering from a dilute ferromagnetic alloy let us assume that the magnetic-moment disturbance is significant only for the type 2 atoms in the first-neighbour shell of a given type 1 atom and that at this distance ($R=\rho$) it has the value ϕ_1. If in addition $F_1 \simeq F_2 = F$ for all values of $\boldsymbol{\kappa}$, then for a polycrystalline sample we can expect

$$\Xi(\boldsymbol{\kappa}) \simeq c(1-c)F^2(\boldsymbol{\kappa})\left\{\mu_1 - \mu_2 + r\phi_1 \frac{\sin(\kappa\rho)}{(\kappa\rho)}\right\}^2, \qquad (12.128)$$

where r is the number of nearest-neighbour atoms. If both $\mu_1 - \mu_2$ and ϕ_1 are positive then the quantity in brackets has the value $\mu_1 - \mu_2 + r\phi_1$, i.e. $\partial\mu/\partial c$, at $\boldsymbol{\kappa} = 0$ and falls rapidly to a value oscillating about $\mu_1 - \mu_2$ as $\boldsymbol{\kappa}$ increases. The forward peak, as a function of $\boldsymbol{\kappa}$, has a width of about π/ρ and thus is well inside the Bragg peaks of most simple crystals. If on the other hand $\mu_1 - \mu_2$ and ϕ_1 have opposite signs then there is a sharp dip in the forward direction.

In general we can draw the following conclusions about the Ξ for a dilute ferromagnetic alloy.

(a) Near the forward direction there will be either a peak or a dip of width about π/R_0, where R_0 characterizes the range of the disturbance of the magnetic moments of the type 2 atoms.

(b) If $\Xi(\boldsymbol{\kappa})$ displays a maximum other than at $\boldsymbol{\kappa} = 0$ then the disturbance $\phi(\mathbf{R})$, taken together with that on the central atom, $\mu_1 - \mu_2$, cannot be of a uniform sign.

(c) At very large values of $\boldsymbol{\kappa}$ the oscillations of $\Phi(\boldsymbol{\kappa})$ are relatively unimportant and, to a good approximation, we obtain the Shull–Wilkinson result of § 12.7.1.

(d) The behaviour of $\Phi(\boldsymbol{\kappa})$ is repeated in every reciprocal lattice cell. Therefore, around each reciprocal lattice vector the same behaviour is observed, apart from the changes in the form factors F_1 and F_2. Moreover, for polycrystalline samples only the behaviour near the forward direction is easy to observe.

(e) In principle, it is possible to make a complete measurement of $\Phi(\boldsymbol{\kappa})$ and hence obtain $\phi(\mathbf{R})$ exactly, but there are limitations set on such a programme by the experimental techniques. Given adequate nuclear contrast, a polarized beam affords great sensitivity (Medina and Cable 1977).

(f) As the temperature of the ferromagnetic alloy is raised towards its Curie point, the elastic diffuse scattering increases rapidly in magnitude because there is an increase in the spatial range of the disturbance on the host. In fact there are two competing processes involved: first, the increase in the range of the disturbance, and second, the decrease in the magnitude of individual changes in moment of the type 2 atoms, but the former process dominates as the Curie temperature is approached so that in this limit $\sum \phi(\mathbf{R})$ tends to infinity (Lovesey 1967; Lovesey and Marshall 1967).

12.7.3. Concentrated alloys (Medina and Cable 1977)

In order to extend the above theory to this, more general, case we make two assumptions. First, we assume that the mean short-range order parameter is zero; second, we assume that, within the range of environments allowed by statistical fluctuations, the disturbance in the magnetic moments is both linear and additive. This second assumption is best explained through the following simplified example.

Let us assume that the lattice is face-centred cubic and that each atom influences only its nearest neighbours. We further suppose that the concentration is $\frac{1}{3}$. Then, on average, each atom will have four nearest neighbours of type 1 and eight nearest neighbours of type 2. However, environments differing from this one have a quite reasonable probability; indeed the probability of n neighbours of type 1 and $12-n$ neighbours of type 2 is

$$P(n)=\frac{12!}{n!\,(12-n)!}\,c^{n}(1-c)^{12-n}.$$

Roughly speaking this means that only values of n ranging from 1 to 7 have appreciable probabilities distributed as follows:

$$\begin{aligned} &P(1)=0.05 \qquad P(5)=0.20\\ &P(2)=0.13 \qquad P(6)=0.12\\ &P(3)=0.22 \qquad P(7)=0.04.\\ &P(4)=0.24 \end{aligned} \tag{12.129}$$

If an atom has, say, three neighbours of type 1, we assume it has the same magnetic moment no matter which particular positions those three might take in the 12 possible positions in the nearest-neighbour shell. This means that the magnetic moment is only a function of n, and we further assume that, within the width of a probability distribution like (12.129), this variation with n is linear. In our particular example, because the total probability is dominated by the values for $n=3$, 4, and 5, producing magnetic moments $\mu(3)$, $\mu(4)$, and $\mu(5)$, respectively, this assumption can roughly be summarized as $\mu(5)-\mu(4)=\mu(4)-\mu(3)$.

The assumption we actually make is more general than illustrated in the foregoing example in the sense that, in principle, each atom is allowed to be influenced by more than just the nearest neighbours (although the disturbance must, certainly, fall off at large distances). The mathematical expression of these assumptions is as follows. If an atom of type 2 is at **m**, it has a moment

$$\mu_m = \mu_2 + \sum_R \phi(\mathbf{R})(p_{m+R} - c), \tag{12.130}$$

whereas if an atom of type 1 is at **m**, it has a moment

$$\mu_m = \mu_1 + \sum_R \chi(\mathbf{R})(p_{m+R} - c) \tag{12.131}$$

where $\phi(\mathbf{R})$ and $\chi(\mathbf{R})$ are the disturbances produced by fluctuations in the number of type 1 atoms at distance **R**. By definition,

$$\phi(0) = \chi(0) = 0 \tag{12.132}$$

and μ_1 and μ_2 are the mean moments for atoms of type 1 and 2, respectively. The two formulae (12.130) and (12.131) can be combined to give for the moment at a general site **m**

$$\mu_m = \mu_2 + (\mu_1 - \mu_2)p_m + \sum_R \phi(\mathbf{R})(p_{m+R} - c) + \sum_R p_m(p_{m+R} - c)\{\chi(\mathbf{R}) - \phi(\mathbf{R})\} \tag{12.133}$$

and

$$F_m\mu_m = F_2\mu_2 + (F_1\mu_1 - F_2\mu_2)p_m + F_2 \sum_R \phi(\mathbf{R})(p_{m+R} - c)$$
$$+ \sum_R p_m(p_{m+R} - c)\{F_1\chi(\mathbf{R}) - F_2\phi(\mathbf{R})\}. \tag{12.134}$$

We define $\Phi(\boldsymbol{\kappa})$ by (12.123) and, in addition,

$$X(\boldsymbol{\kappa}) = \sum_R \exp(i\boldsymbol{\kappa} \cdot \mathbf{R})\chi(\mathbf{R}), \tag{12.135}$$

$$\psi(\mathbf{R}) = \chi(\mathbf{R}) - \phi(\mathbf{R}), \tag{12.136}$$

$$\Psi(\boldsymbol{\kappa}) = \sum_R \exp(i\boldsymbol{\kappa} \cdot \mathbf{R})\psi(\mathbf{R}) = X(\boldsymbol{\kappa}) - \Phi(\boldsymbol{\kappa}), \tag{12.137}$$

$$Y(\boldsymbol{\kappa}) = \sum_R \exp(i\boldsymbol{\kappa} \cdot \mathbf{R})\psi^2(\mathbf{R}), \tag{12.138}$$

$$\Psi(\boldsymbol{\kappa}, \mathbf{Q}) = F_1(\boldsymbol{\kappa})\Psi(\mathbf{Q}) - F_2(\boldsymbol{\kappa})\Phi(\mathbf{Q})$$
$$= \sum_R \exp(i\mathbf{Q} \cdot \mathbf{R})\{F_1(\boldsymbol{\kappa})\chi(\mathbf{R}) - F_2(\boldsymbol{\kappa})\phi(\mathbf{R})\}, \tag{12.139}$$

and finally

$$Y(\boldsymbol{\kappa}, \mathbf{Q}) = \sum_R \exp(i\mathbf{Q} \cdot \mathbf{R})\{F_1(\boldsymbol{\kappa})\chi(\mathbf{R}) - F_2(\boldsymbol{\kappa})\phi(\mathbf{R})\}^2. \tag{12.140}$$

For the model of the concentrated binary alloy defined above, (12.130)–(12.133) represent the full variation of the moments with concentration fluctuations about the mean value of c and with short-range order fluctuations about the mean value of zero. It follows that

$$\frac{\partial\mu_2}{\partial c}=\Phi(0),\qquad \frac{\partial\mu_1}{\partial c}=\mathrm{X}(0), \tag{12.141}$$

$$\begin{aligned}\frac{\partial\mu}{\partial c}&=\mu_1-\mu_2+(1-c)\Phi(0)+c\mathrm{X}(0) \\ &=\mu_1-\mu_2+\Phi(0)+c\Psi(0),\end{aligned} \tag{12.142}$$

$$\frac{\partial^2\mu}{\partial c^2}=2\Psi(0), \tag{12.143}$$

$$\frac{\partial\mu}{\partial s(\mathbf{R})}=\chi(\mathbf{R})-\phi(\mathbf{R})=\psi(\mathbf{R}). \tag{12.144}$$

Using the general formula (12.117) we then have

$$\Xi(0)=c(1-c)\{\mu_1-\mu_2+\Phi(0)+c\Psi(0)\}^2+2c^2(1-c)^2\Upsilon(0). \tag{12.145}$$

From (12.95) and (12.134) it can be shown that, for a random alloy, $\Xi(\boldsymbol{\kappa})$ reduces in this case to

$$\begin{aligned}\Xi(\boldsymbol{\kappa})=c(1-c)\{F_1(\boldsymbol{\kappa})\mu_1-F_2(\boldsymbol{\kappa})\mu_2+F_2(\boldsymbol{\kappa})\Phi(\boldsymbol{\kappa})+c\Psi(\boldsymbol{\kappa},\boldsymbol{\kappa})\}^2 \\ +c^2(1-c)^2\{\Upsilon(\boldsymbol{\kappa},0)+\Upsilon(\boldsymbol{\kappa},\boldsymbol{\kappa})\}.\end{aligned} \tag{12.146}$$

This expression reduces to (12.145) in the limit $\boldsymbol{\kappa}=0$, as it should, and is remarkably simple in form. In most cases the second term will be an order of magnitude smaller than the first and can be neglected.

A case of special importance is that in which the form factors are sufficiently alike to be replaced by a single factor $F(\boldsymbol{\kappa})$; this will usually be a good approximation for alloys within a given period. For this case

$$\left.\begin{aligned}\Psi(\boldsymbol{\kappa},\mathbf{Q})&=F(\boldsymbol{\kappa})\Psi(\mathbf{Q}) \\ \Upsilon(\boldsymbol{\kappa},\mathbf{Q})&=\{F(\boldsymbol{\kappa})\}^2\Upsilon(\mathbf{Q})\end{aligned}\right\}\quad \text{when } F_1\simeq F_2=F(\boldsymbol{\kappa}), \tag{12.147}$$

and the expression (12.146) for $\Xi(\boldsymbol{\kappa})$ simplifies to

$$\Xi(\boldsymbol{\kappa})=c(1-c)F^2(\boldsymbol{\kappa})\{\mu_1-\mu_2+\Phi(\boldsymbol{\kappa})+c\Psi(\boldsymbol{\kappa})\}^2+c^2(1-c)^2F^2(\boldsymbol{\kappa})\{\Upsilon(0)+\Upsilon(\boldsymbol{\kappa})\}. \tag{12.148}$$

A measurement of $\Xi(\boldsymbol{\kappa})$, assuming $F(\boldsymbol{\kappa})$ is known, then gives μ_1 and μ_2 and the combination $\Phi(\boldsymbol{\kappa})+c\Psi(\boldsymbol{\kappa})$. There is no unambiguous method of determining $\Phi(\boldsymbol{\kappa})$ and $\Psi(\boldsymbol{\kappa})$ separately but, because both should vary smoothly with concentration, measurements at slightly different concentrations should be sufficient to determine them with fair accuracy.

It is to be noted that, for this model, (12.143) and (12.144) give some relation between the variation with short-range order and the second derivative of the total magnetization with respect to concentration. This relationship can be made precise only with some further assumption, such as that used in deriving (12.118). In the notation of the present model, this assumption is that $\psi(\mathbf{R})$ is zero beyond the nearest neighbours, i.e.

$$\psi(\mathbf{R}) \simeq \psi \sum_{\rho} \delta(\mathbf{R}-\boldsymbol{\rho}), \tag{12.149}$$

where the sum is over the r nearest neighbours to the site $\mathbf{R}$. Then

$$\Upsilon(0) \simeq r\psi^2, \tag{12.150}$$

$$\frac{1}{2}\frac{\partial^2 \mu}{\partial c^2} = \Psi(0) \simeq r\psi. \tag{12.151}$$

Hence, the second term of (12.145) becomes

$$\frac{1}{2r}c^2(1-c)^2\left(\frac{\partial^2 \mu}{\partial c^2}\right)^2, \tag{12.152}$$

which is to be compared to the first term, which is

$$c(1-c)\left(\frac{\partial u}{\partial c}\right)^2. \tag{12.153}$$

Except for concentrations where $\partial\mu/\partial c$ happens to be near zero, it is usually satisfactory to neglect (12.152) compared with (12.153).

So far we have always assumed the alloy to be completely random but we now relax this condition and examine the effect of chemical order. We shall assume that the short-range order is small and described by parameters $s(\mathbf{R})$. We can retain both (12.133) and (12.134) and note that we need to evaluate products like $\overline{p_m p_n}$, $\overline{p_l p_m p_n}$, and $\overline{p_k p_l p_m p_n}$. Whereas the evaluation of these products was straightforward for the random alloy, much more care is required when short-range order is present. For, by definition, we now have

$$\overline{p_m} = c$$

and

$$\overline{p_m p_n} = c^2 + c(1-c)\,\delta_{m,n} + s(\mathbf{m}-\mathbf{n}),$$

and to calculate the average of any function we have to average over a canonical ensemble subject to these restrictions. Since the algebra is rather tedious, we shall content ourselves only with the main results, namely,

$$\bar{\mu} = \overline{\mu_m} = (1-c)\mu_2 + c\mu_1 + \sum_R s(\mathbf{R})\psi(\mathbf{R}), \tag{12.154}$$

$$c\overline{\mu_1} = \overline{p_m\mu_m} = c\mu_1 + \sum_R s(\mathbf{R})\chi(\mathbf{R}), \tag{12.155}$$

$$(1-c)\overline{\mu_2} = \overline{(1-p_m)\mu_m} = (1-c)\mu_2 - \sum_R s(\mathbf{R})\phi(\mathbf{R}), \tag{12.156}$$

and

$$\overline{F_m\mu_m} = (1-c)F_2\mu_2 + cF_1\mu_1 + \sum_R s(\mathbf{R})\{F_1\chi(\mathbf{R}) - F_2\phi(\mathbf{R})\}, \tag{12.157}$$

where μ_1 and μ_2 are the mean moments in the absence of short-range order. Furthermore, when $F_1 = F_2 = F(\boldsymbol{\kappa})$ it is found that

$$\begin{aligned}\Xi(\boldsymbol{\kappa})/F^2(\boldsymbol{\kappa}) &= \left\{c(1-c) + \sum_R \exp(\mathrm{i}\boldsymbol{\kappa}\cdot\mathbf{R})s(\mathbf{R})\right\}\\ &\times\left[\mu_1 - \mu_2 + \Phi(\boldsymbol{\kappa}) + c\Psi(\boldsymbol{\kappa}) + \frac{1-2c}{c(1-c)}\left\{\sum_R s(\mathbf{R})\psi(\mathbf{R})\{1+\exp(\mathrm{i}\boldsymbol{\kappa}\cdot\mathbf{R})\}\right\}\right]^2\\ &+ c^2(1-c)^2\{\Upsilon(0) + \Upsilon(\boldsymbol{\kappa})\} + (1-2c)^2\sum_R s(\mathbf{R})\psi^2(\mathbf{R})\{1+\exp(\mathrm{i}\boldsymbol{\kappa}\cdot\mathbf{R})\}\\ &+ c(1-c)\frac{1}{N}\sum_Q\sum_R s(\mathbf{R})\exp(\mathrm{i}\boldsymbol{\kappa}\cdot\mathbf{R})\{\Psi(\mathbf{Q}) + \Psi(\boldsymbol{\kappa}+\mathbf{Q})\}^2.\end{aligned} \tag{12.158}$$

In many alloys only the first term is of importance. This formula allows, in principle, corrections to be made for small known amounts of short-range order but it is, clearly, preferable to use a random alloy wheever possible because then the relevant formulae (eqns (12.146) and (12.148)) are relatively simple.

12.7.4. Antiferromagnets (Rainford 1982; Moze and Hicks 1982)

Much of the theory given above can be extended to a simple antiferromagnetic alloy, i.e. an alloy in which the magnetic structure can be divided into two identical sublattices, the magnetic moments associated with one sublattice being aligned antiparallel to those on the other. Experimentally the difficulty involved in observing the magnetic elastic diffuse scattering from such systems is, however, much greater than for a ferromagnetic alloy. This stems from the fact that one cannot, as in the ferromagnetic case, apply an external magnetic field to orientate the magnetization of the sample parallel and then perpendicular to $\boldsymbol{\kappa}$, which enables the diffuse magnetic scattering to be identified. The diffuse magnetic scattering can be isolated by using polarization analysis, or a pure sample as a reference system, although the preparation of two samples identical in all respects except for a concentration c of impurities in one is difficult.

If $\mathbf{w}$ is the smallest reciprocal lattice vector of the superlattice, $\exp(\mathrm{i}\mathbf{w}\cdot\mathbf{m})$ is $+1$ or -1 depending upon whether $\mathbf{m}$ belongs to one sublattice or the other. We define

$$\nu_m = \mu_m \exp(\mathrm{i}\mathbf{w}\cdot\mathbf{m}), \tag{12.159}$$

where

$$\bar{\mu}_m = 0$$

but

$$\bar{\nu}_m = \nu,$$

ν being non-zero at temperatures below the Néel point of the alloy.

Substituting (12.159) into (12.91) gives

$$\frac{\mathrm{d}\sigma}{\mathrm{d}\Omega} = \frac{N}{4} r_0^2(1-\tilde{\kappa}_z^2)\frac{1}{N}\sum_{m,n}\exp\{\mathrm{i}(\boldsymbol{\kappa}-\mathbf{w})\cdot(\mathbf{m}-\mathbf{n})\}\overline{F_n\nu_n F_m\nu_m}.$$

We again separate off the Bragg scattering

$$\left(\frac{\mathrm{d}\sigma}{\mathrm{d}\Omega}\right)_{\text{Bragg}} = \frac{N}{4} r_0^2(1-\tilde{\kappa}_z^2)\{\overline{\nu F(\boldsymbol{\kappa})}\}^2\frac{(2\pi)^3}{v_0}\sum_{\boldsymbol{\tau}}\delta(\boldsymbol{\kappa}-\mathbf{w}-\boldsymbol{\tau}), \tag{12.160}$$

where

$$\overline{\nu F(\boldsymbol{\kappa})} = \frac{1}{N}\sum_m \mu_m F_m(\boldsymbol{\kappa})\exp(\mathrm{i}\mathbf{w}\cdot\mathbf{m}).$$

The elastic diffuse scattering is then given by (12.94) but with

$$\Xi(\boldsymbol{\kappa}) = \frac{1}{N}\sum_{m,n}\exp\{\mathrm{i}(\boldsymbol{\kappa}-\mathbf{w})\cdot(\mathbf{m}-\mathbf{n})\}\{\overline{F_n\nu_n F_m\nu_m} - (\overline{F_n\nu_n})(\overline{F_m\nu_m})\}. \tag{12.161}$$

We now seek general formulae for Ξ analogous to those derived for $\Xi(0)$ for the ferromagnetic alloy. However, when the form factors $F_1(\boldsymbol{\kappa})$ and $F_2(\boldsymbol{\kappa})$ are not equal, it is not possible to derive formulae of equal generality. Thus throughout this section we shall assume that the difference between the form factors is negligible and write $F_1 = F_2 = F(\boldsymbol{\kappa})$. Equation (12.161) then simplifies to

$$\Xi(\boldsymbol{\kappa}) = |F(\boldsymbol{\kappa})|^2\frac{1}{N}\sum_{m,n}\exp\{\mathrm{i}(\boldsymbol{\kappa}-\mathbf{w})\cdot(\mathbf{m}-\mathbf{n})\}\{\overline{\nu_n\nu_m} - \overline{\nu_n}\,\overline{\nu_m}\}. \tag{12.162}$$

The staggered magnetization M_s is defined by

$$M_s = \sum_m \nu_m = \sum_m \mu_m \exp(i\mathbf{w} \cdot \mathbf{m}), \tag{12.163}$$

and we notice the exact analogy to the result for a ferromagnet. For instance, the formula analogous to (12.106) is

$$\Xi(\mathbf{w}) = |F(\mathbf{w})|^2 \, c(1-c)\left(\frac{\partial \nu}{\partial c}\right)^2, \tag{12.164}$$

and the more general formula (12.117) is replaced by

$$\Xi(\mathbf{w}) = |F(\mathbf{w})|^2 \left[c(1-c)\left(\frac{\partial \nu}{\partial c}\right)^2 + 2c^2(1-c)^2 \sum_R \left\{\frac{\partial \nu}{\partial s(\mathbf{R})}\right\}^2\right]. \tag{12.165}$$

The Shull–Wilkinson model of an antiferromagnet, with both atoms of type 1 and type 2 described randomly, would be represented by the formula

$$\nu_m = \nu_2 + (\nu_1 - \nu_2)p_m, \tag{12.166}$$

where it is assumed that all atoms of type 1 have a moment of magnitude ν_1 and of a sign determined by the sublattice they belong to, and a similar assumption is made for atoms of type 2. Then

$$\nu = c\nu_1 + (1-c)\nu_2 \tag{12.167}$$

can be measured as a function of concentration from the intensity of the Bragg superlattice peaks. (12.162) and (12.166) then give the Shull–Wilkinson result analogous to (12.119), namely

$$\Xi(\boldsymbol{\kappa}) = |F(\boldsymbol{\kappa})|^2 \, c(1-c)(\nu_1 - \nu_2)^2. \tag{12.168}$$

We notice that this has a dependence on $\boldsymbol{\kappa}$ only through the form factor $F(\boldsymbol{\kappa})$ and must therefore display a maximum about the forward direction.

For a dilute antiferromagnetic alloy we must replace (12.166) by

$$\nu_m = \nu_2 + (\nu_1 - \nu_2)p_m + \sum_R \phi(\mathbf{R})p_{m+R}, \tag{12.169}$$

from which

$$\nu = c\nu_1 + (1-c)\nu_2 + c\Phi(0), \tag{12.170}$$

$$\frac{\partial \nu}{\partial c} = \nu_1 - \nu_2 + \Phi(0), \tag{12.171}$$

and, finally,

$$\Xi(\boldsymbol{\kappa}) = c(1-c)\,|F(\boldsymbol{\kappa})|^2 \{\nu_1 - \nu_2 + \Phi(\boldsymbol{\kappa} - \mathbf{w})\}^2. \tag{12.172}$$

When the effect of an atom of type 1 is to produce a considerable disturbance in the magnetic order of its environment, then $\Phi(0)$ is large and $\Xi(\boldsymbol{\kappa})$ peaks near the *superlattice* positions given by $\mathbf{w}+\boldsymbol{\tau}$.

Magnetic correlations in diluted antiferromagnets subject to an external magnetic field should reveal the effect of a site-random field on long-range order in ferromagnets (Fishman and Aharony 1979; Grinstein 1984). For local variations in the magnetization of a mixed antiferromagnet mean that an external field couples with varying sign to the local antiferromagnetic order parameter. Neutron diffraction data (Birgeneau *et al.* 1983; Hagen *et al.* 1983) is well described by a structure factor that is the sum of the usual Ornstein–Zernike (Lorentzian) term, and a squared Lorentzian. This finding is consistent with the form of (12.172) since $\Phi(\boldsymbol{\kappa})$ is proportional to the isothermal susceptibility, as noted earlier in this section (Lovesey 1984).

The polarization of the scattered beam follows from (10.109). To isolate the diffuse magnetic from the nuclear disorder scattering, a simple geometry can be used in which $\mathbf{P}\cdot\tilde{\boldsymbol{\kappa}}=\pm P$, i.e. the polarization is parallel or antiparallel to the scattering vector. Discrimination at the detector is set in favour of one direction of the incident polarization, so that spin-flip and non-spin-flip events are distinguished by its orientation relative to $\boldsymbol{\kappa}$. Magnetic scattering contributes to the spin-flip scattering only, together with nuclear spin incoherent scattering described in § 10.4. The latter must be determined separately, in general, although it is negligible for some elements.

REFERENCES

Birgeneau, R. J., Yoshizawa, H., Cowley, R. A., Shirane, G., and Ikeda, H. (1983). *Phys. Rev.* **B28,** 1438.
Cable, J. W. (1981). *Phys. Rev.* **B23,** 6168.
Coqblin, B. (1977). *The electronic structure of rare-earths metals and alloys: the magnetic heavy rare-earths.* Academic Press, New York.
Currat, R. and Pynn, R. (1979). *Treatise on materials science and technology*, Vol. 15 (ed. G. Kostorz). Academic Press, New York.
Erickson, R. A. (1953). *Phys. Rev.* **90,** 779.
Fischer, O. and Maple, M. B. (1982). *Topics in Current Physics*, vols. 32 and 34. Springer-Verlag, Heidelberg.
Fishman, S. and Aharony, A. (1979). *J. Phys.* **C12,** L729.
Freund, A. and Forsyth, J. B. (1979). *Treatise on materials science and technology*, Vol. 15 (ed. G. Kostorz). Academic Press, New York.
Grinstein, G. (1984). *J. Appl. Phys.* **55** (6), 2371.
Hagen, M., Cowley, R. A., Satija, S. K., Yoshizawa, H., Shirane, G., Birgeneau, R. J., and Guggenheim, H. J. (1983). *Phys. Rev.* **B28,** 2602.
Harrison, W. A. (1980). *Electronic structure and the properties of solids.* W. H. Freeman.

Hastings, J. M., Elliott, N., and Corliss, L. M. (1959). *Phys. Rev.* **115,** 13.
Hicks, T. J. (1979). *Nukleonika* **24,** 795.
Hubbard, J. and Marshall, W. (1965). *Proc. Phys. Soc.* **86,** 561.
Lovesey, S. W. (1967). *Proc. Phys. Soc.* **89,** 625, 893.
—— (1984). *J. Phys.* **C17,** L213.
—— and Marshall, W. (1966). *Proc. Phys. Soc.* **89,** 613.
Medina, R. A. and Cable, J. W. (1977). *Phys. Rev.* **B15,** 1539.
Menzinger, F. and Sacchetti, F. (1979). *Nukleonika* **24,** 737.
Mook, H. A. (1966). *Phys. Rev.* **148,** 495.
Moon, R. M. (1982). *J. Physique Coll.* **C7,** 187.
Moze, O. and Hicks, T. J. (1982). *J. Phys.* **F12,** 1.
Nathans, R., Alperin, H. A., Pickart, S. J., and Brown, P. J. (1963). *J. appl. Phys.* **34,** 1182.
Rainford, B. D. (1982). *J. Physique Coll.* **C7,** 33.
Sinha, S. K. (1984). *Condensed matter research using neutrons, today and tomorrow.* (Eds. S. W. Lovesey and R. Scherm) Plenum Press, New York.
Steinsvoll, O., Moon, R. M., Koehler, W. C., and Winsdor, C. G. (1981). *Phys. Rev.* **B24,** 4031.
Tofield, B. C. (1975). *Structure and bonding*, Vol. 21. Springer-Verlag, Heidelberg.
—— (1976). *J. Physique Coll.* **C6,** 539.
van Laar, B., Maniawski, F., Kaprzyk, S., and Dobrzynski, L. (1979). *J. Mag. Mag. Mat.* **14,** 94.
Wang, C. S. and Callaway, J. (1977). *Phys. Rev.* **B15,** 298.
Wedgwood, F. A. (1976). *Proc. R. Soc.* **A349,** 447.
Yoshimori, A. (1959). *J. Phys. Soc. Japan* **14,** 807.

13
PARAMAGNETIC AND CRITICAL SCATTERING

For a perfect paramagnet, in which there are no interactions and no correlations between the ions, the magnetic scattering is strictly elastic, spatially isotropic, and independent of the temperature, as we demonstrated in § 7.4. Scattering from exchanged coupled (Heisenberg) paramagnets is distinctly different, even in the limit of infinite temperatures when the short-range spin correlation vanishes. The difference is revealed in the frequency moments of the spin response function, calculated in § 13.3.2. For the mean-square energy width is found to be proportional to the square of the exchange parameter and the wave-vector dependence reflects local symmetry, is κ^2 for $\boldsymbol{\kappa} \rightarrow 0$, and maximum widths are achieved at the zone boundary. Higher-order frequency moments imply that the response function approximates to a Lorentz curve near the zone centre. For high temperatures and large wave vectors, theory and computer simulations of the spin response function show a peak at finite frequencies which is evidence of a spin oscillation mode.

The gross features of the temperature dependence of the paramagnetic scattering are obtained by noting that the normalized frequency moments are inversely proportional to the wave-vector dependent susceptibility. A phase transition is heralded by a strong increase in the susceptibility at the wave vector that characterizes the ultimate, ordered state. It follows that, for such wave vectors, there is a pronounced decrease in the inelasticity of the scattering as the phase transition is approached by decreasing the temperature. The type of magnetic structure, and hence the value of the ordering wave vector, depends on the sign of the dominant exchange parameter, and the crystal structure (Stanley 1971). For a simple ferromagnet, the ordering wave vector corresponds to a nuclear Bragg reflection, while the ordering vector of a simple antiferromagnet corresponds to a Bragg reflection for a lattice whose cell dimensions are twice those of the real lattice. Consequently, the paramagnetic response of ferro- and antiferromagnetically coupled magnets is quite different, except at infinite temperature.

The nature of the strong critical scattering observed at a second-order, or continuous, phase transition depends on the spatial dimensionality of the magnetic system and the spin dimensionality (Kociński 1983; Kociński and Wojtczak 1978; Hohenberg and Halperin 1977; Patashinskii and Pokrovskii 1979; Wallace and Zia 1978; Lindgård 1978). Indeed, the existence of a phase transition depends on these parameters; for example, a two-dimensional magnet with three-dimensional spins (planar Heisenberg) does now show long-range order at any finite temperature, in contrast to a two-dimensional Ising-spin model which orders at a temperature whose magnitude is of the order of the exchange parameter. The behaviour of the isothermal wave-vector dependent susceptibility, and other static response functions, at the phase transition can be classified according to the space and spin dimensionality, and this feature of static critical phenomena is called universality. The latter is more complicated for dynamic critical phenomena (Hohenberg and Halperin 1977; Patashinskii and Pokrovskii 1979). The term universality is justified by the fact that different physical systems with the same space and spin—or the variable(s) that corresponds to spin-dimensionality—display similar critical properties.

We will not discuss the current theory of static and dynamic critical phenomena, apart from quoting results of immediate interest. Nor will we discuss first-order phase transitions, multicritical points, at which several phases coexist, and tricritical points at which a second-order transition changes continuously to a first-order transition (Lindgård 1978).

Many quasi-one- and two-dimensional magnetic materials are known to exist (Steiner, Villain, and Windsor 1976; Lovesey and Loveluck 1977). Such materials display features expected of one- and two-dimensional magnetic models when the sample temperature exceeds a critical temperature below which three-dimensional magnetic order occurs. Interest in these materials stems, in part, from the fact that theory can often be pushed further in low space dimensions, e.g. the static properties of the one-dimensional, classical and Ising-spin models and the two-dimensional Ising model are known exactly. Low-dimensional systems are, to some extent, a test-bed for theory, and they also display effects that are marginal in three dimensions.

We calculate the wave-vector dependent susceptibility using the mean-field, or Landau, approximation (Patashinskii and Pokrovskii 1979). The approximation has the merit of simplicity, and thus ease of use for relatively complicated systems, and the results are often a useful guide to the expected behaviour. However, the approximation is not consistent with universality, and predicts phase transitions for both one- and two-dimensional Heisenberg magnets, for example. The dynamic response of paramagnets at intermediate and large wave vectors shows evidence of a damped, collective spin oscillation (Hubbard 1971; Lynn 1975, 1981; Mook 1981; Lynn and Mook 1981). The occurrence of a spin oscillation can be understood from a study of frequency moments of the response function calculated with a refined mean-field approximation.

13.1. Response functions

Throughout this chapter we are concerned with Bravais lattices and Heisenberg magnets. For such systems the partial differential cross-section is (cf. Chapter 8)

$$\frac{\mathrm{d}^2\sigma}{\mathrm{d}\Omega\,\mathrm{d}E'} = r_0^2 \frac{k'}{k}\{\tfrac{1}{2}gF(\boldsymbol{\kappa})\}^2 \sum_{\alpha}(1-\tilde{\kappa}_{\alpha}^2)S^{\alpha}(\boldsymbol{\kappa},\omega). \tag{13.1}$$

Here the response function

$$\begin{aligned} S^{\alpha}(\boldsymbol{\kappa},\omega) &= \frac{1}{2\pi\hbar}\int_{-\infty}^{\infty} \mathrm{d}t\, \exp(-\mathrm{i}\omega t) \sum_{ll'} \exp\{\mathrm{i}\boldsymbol{\kappa}\cdot(\mathbf{l}'-\mathbf{l})\}\langle \hat{S}_l^{\alpha}\hat{S}_{l'}^{\alpha}(t)\rangle \\ &\equiv \frac{1}{2\pi\hbar}\int_{-\infty}^{\infty} \mathrm{d}t\, \exp(-\mathrm{i}\omega t)\langle \hat{S}_{\boldsymbol{\kappa}}^{\alpha}\hat{S}_{-\boldsymbol{\kappa}}^{\alpha}(t)\rangle \end{aligned} \tag{13.2}$$

where, as usual,

$$\hat{S}_{\mathbf{q}}^{\alpha} = \sum_{l} \exp(-\mathrm{i}\mathbf{q}\cdot\mathbf{l})\hat{S}_l^{\alpha} \tag{13.3}$$

and $\alpha = x$, y, or z. In applications to real crystals the Debye–Waller factor must be included in (13.1).

It is a simple matter to demonstrate that near the critical point there is a pronounced increase in the paramagnetic scattering.

If we denote

$$\Delta\hat{S}_{\mathbf{q}}^{\alpha} = \sum_{l} \exp(-\mathrm{i}\mathbf{q}\cdot\mathbf{l})\{\hat{S}_l^{\alpha} - \langle \hat{S}_l^{\alpha}\rangle\}, \tag{13.4}$$

then, for a ferromagnet ($\langle \hat{S}_l^{\alpha}\rangle \equiv \langle \hat{S}^{\alpha}\rangle$),

$$S^{\alpha}(\boldsymbol{\kappa},\omega) = N^2\delta_{\boldsymbol{\kappa},\boldsymbol{\tau}}\langle \hat{S}^{\alpha}\rangle^2\delta(\hbar\omega) + \frac{1}{2\pi\hbar}\int_{-\infty}^{\infty} \mathrm{d}t\, \exp(-\mathrm{i}\omega t)\langle \Delta\hat{S}_{\boldsymbol{\kappa}}^{\alpha}\,\Delta\hat{S}_{-\boldsymbol{\kappa}}^{\alpha}(t)\rangle. \tag{13.5}$$

The first term on the right-hand side of (13.5) corresponds to magnetic Bragg scattering and is very small near the critical temperature T_c because $\langle \hat{S}^{\alpha}\rangle \to 0$ at T_c in the absence of a magnetic field. If the spin fluctuations appear almost static to the incident neutrons, the inelasticity of the scattering is small.

Assuming for the present that this is a valid assumption, the cross-section is determined by

$$\langle \Delta\hat{S}_{\boldsymbol{\kappa}}^{\alpha}\,\Delta\hat{S}_{-\boldsymbol{\kappa}}^{\alpha}\rangle. \tag{13.6}$$

Now the static susceptibility χ^{α} is related to the fluctuations in the magnetization $\hat{M}^{\alpha} = -g\mu_{\mathrm{B}} \sum \hat{S}_{l}^{\alpha}$ through (Stanley 1971)

$$\chi^{\alpha} = \frac{\beta}{N} \langle (\Delta \hat{M}^{\alpha})^2 \rangle. \tag{13.7}$$

Thus it follows that, at $\boldsymbol{\kappa} = 0$, (13.6) becomes

$$\frac{N}{\beta} (g\mu_{\mathrm{B}})^{-2} \chi^{\alpha}. \tag{13.8}$$

Near the critical point the fluctuations in the magnetization become very large and in the limit $T \to T_{\mathrm{c}}$, χ^{α} actually diverges. The corresponding critical scattering is, of course, seen at all Bragg peaks, $\boldsymbol{\kappa} = \boldsymbol{\tau}$.

We look at the wave-vector dependence of the diffuse scattering about the Bragg positions and also its energy dependence. To pursue both these aspects it is convenient to express the cross-section in terms of the relaxation function $\mathcal{R}_{\boldsymbol{\kappa}}^{\alpha}(t)$, introduced in Chapter 8 (cf. Table 8.1)

$$S^{\alpha}(\boldsymbol{\kappa}, \omega) = \omega\{1 + n(\omega)\} \mathcal{R}_{\boldsymbol{\kappa}}^{\alpha}(\omega) + \delta(\hbar\omega) \langle \hat{S}_{\boldsymbol{\kappa}}^{\alpha} \rangle \langle \hat{S}_{-\boldsymbol{\kappa}}^{\alpha} \rangle \tag{13.9}$$

where

$$\mathcal{R}_{\boldsymbol{\kappa}}^{\alpha}(\omega) = \frac{1}{2\pi} \int_{-\infty}^{\infty} \mathrm{d}t \exp(-\mathrm{i}\omega t) \mathcal{R}_{\boldsymbol{\kappa}}^{\alpha}(t) \tag{13.10}$$

and $\{1 + n(\omega)\}$ is the detailed balance factor,

$$\{1 + n(\omega)\} = \{1 - \exp(-\hbar\omega\beta)\}^{-1}.$$

From (13.9) and (13.10) it follows that

$$\int_{-\infty}^{\infty} \mathrm{d}\omega \left\{ \frac{1 - \exp(-\hbar\omega\beta)}{\omega} \right\} \{S^{\alpha}(\boldsymbol{\kappa}, \omega) - S_{\mathrm{Bragg}}^{\alpha}(\boldsymbol{\kappa}, \omega)\} = \mathcal{R}_{\boldsymbol{\kappa}}^{\alpha}(t = 0). \tag{13.11}$$

But, from § 8.3,

$$\mathcal{R}_{\boldsymbol{\kappa}}^{\alpha}(t = 0) = \int_0^{\beta} \mathrm{d}\mu \, \langle \hat{S}_{\boldsymbol{\kappa}}^{\alpha}(-\mathrm{i}\hbar\mu) \hat{S}_{-\boldsymbol{\kappa}}^{\alpha} \rangle - \beta \langle \hat{S}_{\boldsymbol{\kappa}}^{\alpha} \rangle \langle \hat{S}_{-\boldsymbol{\kappa}}^{\alpha} \rangle = \frac{N}{(g\mu_{\mathrm{B}})^2} \chi_{\boldsymbol{\kappa}}^{\alpha}, \tag{13.12}$$

where $\chi_{\boldsymbol{\kappa}}^{\alpha}$ is the *isothermal wave-vector dependent susceptibility*; $\chi_{\boldsymbol{\kappa}}^{\alpha}$ gives the response of a magnetic system to an external magnetic field applied in the direction α with a spatial dependence $\cos(\boldsymbol{\kappa} \cdot \mathbf{l})$. Note that in the limit $\boldsymbol{\kappa} = 0$ when the total spin is a constant of motion

$$\hat{S}_{\boldsymbol{\kappa}}^{\alpha}(t) \to \hat{S}_{\mathbf{0}}^{\alpha}(t) = \hat{S}_{\mathbf{0}}^{\alpha}(0),$$

(13.12) becomes

$$\int_0^\beta \mathrm{d}\mu\,\langle \hat{S}_{\mathbf{0}}^\alpha \hat{S}_{\mathbf{0}}^\alpha\rangle - \beta\langle \hat{S}_{\mathbf{0}}^\alpha\rangle\langle \hat{S}_{\mathbf{0}}^\alpha\rangle = \beta\left\{\left\langle\left(\sum_l \hat{S}_l^\alpha\right)^2\right\rangle - \left\langle\sum_l \hat{S}_l^\alpha\right\rangle^2\right\} = \frac{N}{(g\mu_\mathrm{B})^2}\chi^\alpha,$$

i.e. $\chi_{\boldsymbol{\kappa}=0}^\alpha \equiv \chi^\alpha$, as required.

Using (13.12) in (13.11) and recalling that the susceptibility of an isolated ion is

$$\chi_0 = \tfrac{1}{3}(g\mu_\mathrm{B})^2 S(S+1)\beta, \tag{13.13}$$

we have the result

$$\hbar\int_{-\infty}^{\infty} \mathrm{d}\omega \left\{\frac{1-\exp(-\hbar\omega\beta)}{\hbar\omega\beta}\right\}\{S^\alpha(\boldsymbol{\kappa},\omega) - S_{\mathrm{Bragg}}^\alpha(\boldsymbol{\kappa},\omega)\} = N\tfrac{1}{3}S(S+1)(\chi_{\boldsymbol{\kappa}}^\alpha/\chi_0). \tag{13.14}$$

If the inelasticity of the scattering is small, we can replace

$$\{1-\exp(-\hbar\omega\beta)\}/\hbar\omega\beta$$

in the integrand by unity and use the static approximation (Als-Nielsen 1976). This result represents the generalization of the previous discusssion to non-zero $\boldsymbol{\kappa}$, the corresponding cross-section now being proportional to $\chi_{\boldsymbol{\kappa}}^\alpha$ in place of χ^α.

In discussing the energy dependence of the scattering we choose to define the *spectral weight function*

$$F^\alpha(\boldsymbol{\kappa},\omega) = \frac{1}{2\pi}\int_{-\infty}^{\infty} \mathrm{d}t\,\exp(-\mathrm{i}\omega t)\{\mathcal{R}_{\boldsymbol{\kappa}}^\alpha(t)/\mathcal{R}_{\boldsymbol{\kappa}}^\alpha(t=0)\}, \tag{13.15}$$

which possesses the obvious property

$$\int_{-\infty}^{\infty} \mathrm{d}\omega\, F^\alpha(\boldsymbol{\kappa},\omega) = 1. \tag{13.16}$$

On combining (13.12) and (13.15) we obtain

$$S^\alpha(\boldsymbol{\kappa},\omega) - S_{\mathrm{Bragg}}^\alpha(\boldsymbol{\kappa},\omega) = \frac{N}{(g\mu_\mathrm{B})^2}\chi_{\boldsymbol{\kappa}}^\alpha\,\omega\{1+n(\omega)\}F^\alpha(\boldsymbol{\kappa},\omega). \tag{13.17}$$

Having decomposed $S^\alpha(\boldsymbol{\kappa},\omega)$ in this form it is convenient to discuss separately the structure of $\chi_{\boldsymbol{\kappa}}^\alpha$ and $F^\alpha(\boldsymbol{\kappa},\omega)$.

13.2. Wave-vector dependent susceptibility $\chi_{\boldsymbol{\kappa}}^\alpha$

The general features of $\chi_{\boldsymbol{\kappa}}^\alpha$ in the vicinity of the critical temperature T_c can be derived from molecular-field, or Landau, theory. This approximation cannot be expected to yield detailed information but is adequate to give an overall picture. To usefully go beyond this level of discussion requires a sophisticated theory (Hohenberg and Halperin 1977; Patashinskii and Potrovskii 1979; Wallace and Zia 1978) and we shall do no more than indicate results that have been derived. First we discuss the molecular-field approximation as applied to the Heisenberg ferromagnet (cf. § 9.2.1).

If the axis of quantization is taken to coincide with the z-axis, the molecular-field approximation consists of replacing the Hamiltonian

$$\hat{\mathcal{H}} = -\sum_{m,n} J(\mathbf{n})\hat{\mathbf{S}}_m \cdot \hat{\mathbf{S}}_{m+n} \tag{13.18}$$

by a set of single-ion Hamiltonians

$$\hat{\mathcal{H}}_m = -2\sum_{n} J(\mathbf{n})\hat{S}^z_m \langle \hat{S}^z_{m+n} \rangle. \tag{13.19}$$

It follows that the effective magnetic field on $\hat{S}_m$ is $2\sum_n J(\mathbf{n})\langle \hat{S}^z_{m+n}\rangle / g\mu_B$ and a simple calculation (compare, for example, the derivation of (7.39)) gives

$$\langle \hat{S}^z_m \rangle = SB_S\left[\beta \sum_n J(\mathbf{n})\langle \hat{S}^z_{m+n}\rangle\right] \tag{13.20}$$

or

$$\sigma = \frac{\langle \hat{S}^z \rangle}{S} = B_S\left[S\beta\sigma \sum_n J(\mathbf{n})\right] \tag{13.21}$$

where $B_S[x]$ is the Brillouin function defined by

$$B_S[x] = \left(\frac{2S+1}{2S}\right)\coth\{x(2S+1)\} - \frac{1}{2S}\coth x. \tag{13.22}$$

Let us determine the behaviour of σ near the critical temperature. The small-argument expansion of $B_S[x]$ is

$$B_S[x] = \tfrac{2}{3}(S+1)\left\{x - \frac{2x^3}{15}(2S^2+2S+1)\right\} \quad ; \quad x \to 0. \tag{13.23}$$

Hence in (13.21) either $\sigma = 0$ or

$$1 = \tfrac{2}{3}(S+1)[S\beta\mathcal{J}(0) - \sigma^2\tfrac{2}{15}(2S^2+2S+1)\{2\beta\mathcal{J}(0)\}^3]$$

where

$$\mathcal{J}(\mathbf{q}) = \sum_n J(\mathbf{n})\exp(i\mathbf{q}\cdot\mathbf{n}). \tag{13.24}$$

Defining

$$\beta_c = \frac{3}{2S(S+1)\mathcal{J}(0)} = (k_B T_c)^{-1}, \tag{13.25}$$

we find, near T_c,

$$\sigma^2 = \frac{10(S+1)^2}{3(2S^2+2S+1)}\left(\frac{T_c - T}{T_c}\right). \tag{13.26}$$

Thus in the molecular-field approximation σ tends to zero as T tends to T_c like $(T_c - T)^\beta$ where $\beta = \frac{1}{2}$. Both experiment and more sophisticated theories indicate $\beta \simeq 0.32$–0.39 (Patashinskii and Pokrovskii 1979). The molecular-field result (13.26) predicts that magnetic Bragg peaks decrease linearly with temperature as $T \to T_c$.

We turn now to the calculation of $\chi_{\boldsymbol{\kappa}}^{\alpha}$. Above the critical temperature the system is isotropic and $\langle S^\alpha \rangle = 0$. If we impose a magnetic field H_m^α on the system the Hamiltonian (13.18) becomes

$$\begin{aligned}\hat{\mathcal{H}} &= -2\sum_{m,n} J(\mathbf{n})\hat{S}_m^\alpha \langle \hat{S}_{m+n}^\alpha \rangle + g\mu_B \sum_m \hat{S}_m^\alpha H_m^\alpha \\ &\equiv \sum_m \left\{ g\mu_B H_m^\alpha - 2\sum_n J(\mathbf{n})\langle \hat{S}_{m+n}^\alpha \rangle \right\} \hat{S}_m^\alpha. \end{aligned} \tag{13.27}$$

Thus, for $T > T_c$,

$$-g\mu_B \langle \hat{S}_m^\alpha \rangle = \chi_0 \left\{ H_m^\alpha - \left(\frac{2}{g\mu_B}\right) \sum_n J(\mathbf{n}) \langle \hat{S}_{m+n}^\alpha \rangle \right\},$$

and hence

$$M_{\boldsymbol{\kappa}}^\alpha = \chi_0 \left\{ H_{\boldsymbol{\kappa}}^\alpha + \frac{2}{(g\mu_B)^2} \mathcal{J}(\boldsymbol{\kappa}) M_{\boldsymbol{\kappa}}^\alpha \right\},$$

or

$$\chi_{\boldsymbol{\kappa}}^\alpha = \frac{\chi_0}{1 - \{2/(g\mu_B)^2\}\chi_0 \mathcal{J}(\boldsymbol{\kappa})}. \tag{13.28}$$

Since

$$\frac{2}{(g\mu_B)^2}\chi_0 \mathcal{J}(0) = \frac{T_c}{T},$$

(13.28) can be written in the form

$$\chi_{\boldsymbol{\kappa}}^\alpha = \frac{\chi_c}{(T - T_c)/T_c + \{1 - \mathcal{J}(\boldsymbol{\kappa})/\mathcal{J}(0)\}} \tag{13.29}$$

where

$$\chi_c = \frac{T}{T_c}\chi_0.$$

From (13.29) we easily recover the well-known result for the molecular-field susceptibility above the critical temperature. Other theories indicate that $\chi^{\alpha}_{\boldsymbol{\kappa}=0}$ diverges at T_c like $(T_c - T)^{-\gamma}$, $\gamma \sim 1.3$ to 1.4, (Patashinskii and Pokrovskii 1979), whereas molecular-field theory gives $\gamma = 1$.

Further, on writing $\boldsymbol{\kappa} = \boldsymbol{\tau} + \mathbf{q}$ and making use of the property $J(\mathbf{n}) = J(-\mathbf{n})$,

$$\mathcal{J}(\boldsymbol{\kappa}) = \sum_n \exp(i\boldsymbol{\kappa}\cdot\mathbf{n})J(\mathbf{n}) = \sum_n \exp(i\mathbf{q}\cdot\mathbf{n})J(\mathbf{n}) \simeq \sum_n \{1 + \tfrac{1}{2}(i\mathbf{q}\cdot\mathbf{n})^2\}J(\mathbf{n})$$

for small $|\mathbf{q}|$. For cubic crystals we therefore have

$$\mathcal{J}(\boldsymbol{\kappa}) \simeq \mathcal{J}(0) - \tfrac{1}{6}q^2 J^{(2)} \tag{13.30}$$

where

$$J^{(2)} = \sum_n n^2 J(\mathbf{n}). \tag{13.31}$$

Combining (13.29) and (13.30) $\chi^{\alpha}_{\boldsymbol{\kappa}}$ in the vicinity of a Bragg setting has the Ornstein–Zernike form

$$\chi^{\alpha}_{\boldsymbol{\kappa}} = \frac{\chi_c}{r_1^2(q^2 + \kappa_1^2)} \quad (\boldsymbol{\kappa} = \boldsymbol{\tau} + \mathbf{q}) \tag{13.32}$$

where

$$r_1^2 = \tfrac{1}{6}\{J^{(2)}/\mathcal{J}(0)\}, \tag{13.33}$$

$$\kappa_1^2 = \left(\frac{T - T_c}{T_c}\right) r_1^{-2}. \tag{13.34}$$

To see the significance of κ_1, note for the static approximation

$$\begin{aligned}\langle \Delta\hat{S}^{\alpha}_m\, \Delta\hat{S}^{\alpha}_n\rangle &= \frac{1}{N^2}\sum_{\mathbf{q}} \exp\{i\mathbf{q}\cdot(\mathbf{m}-\mathbf{n})\}\langle \Delta\hat{S}^{\alpha}_{\mathbf{q}}\, \Delta\hat{S}^{\alpha}_{-\mathbf{q}}\rangle \\ &\simeq \frac{1}{N^2}\sum_{\mathbf{q}} \exp\{i\mathbf{q}\cdot(\mathbf{m}-\mathbf{n})\} N\tfrac{1}{3}S(S+1)(\chi^{\alpha}_{\mathbf{q}}/\chi_0).\end{aligned} \tag{13.35}$$

Using the result (13.32) for $\chi^{\alpha}_{\mathbf{q}}$ and replacing the summation over $\mathbf{q}$ by an integration (13.35) gives

$$\langle \Delta\hat{S}^{\alpha}_m\, \Delta\hat{S}^{\alpha}_n\rangle = \frac{v_0}{12\pi} S(S+1)\frac{T}{T_c}\frac{1}{r_1^2}\frac{\exp(-\kappa_1 r)}{r} \tag{13.36}$$

where $r = |\mathbf{m} - \mathbf{n}|$. In view of the fact that (13.32) was derived for small $|\mathbf{q}|$, (13.36) gives the asymptotic form of $\langle \Delta\hat{S}^{\alpha}_{m}\, \Delta\hat{S}^{\alpha}_{n} \rangle$, as $|\mathbf{m} - \mathbf{n}| \to \infty$. The result tell us that κ_1 is an inverse correlation range; as $T \to T_c$, $\kappa_1 \to 0$ and the correlation between the spins is extremely long-ranged, falling off as the inverse of their separation.

For $T < T_c$ we take the z-axis to coincide with the axis of quantization. The calculation of $\chi^z_{\boldsymbol{\kappa}}$ proceeds as follows. From (13.27)

$$\sigma_m = B_S\left[\tfrac{1}{2}\beta\left\{2S \sum_n J(\mathbf{n})\sigma_{m+n} - g\mu_B H^z_m\right\}\right]. \tag{13.37}$$

We write

$$\sigma_m = \sigma - \lambda_m \tag{13.38}$$

where λ_m is the change in σ_m from its equilibrium value σ due to H_m, and satisfies $\sigma \gg \lambda_m$. By expanding the Brillouin function about σ, which satisfies eqn (13.21), the λ_m are found to be given by

$$2S\lambda_m = \tfrac{1}{2}\beta\left\{2S \sum_n J(\mathbf{n})\lambda_{m+n} + g\mu_B H^z_m\right\}$$
$$\times\left[\operatorname{cosech}^2\{\beta S\sigma\mathcal{J}(0)\} - (2S+1)^2 \operatorname{cosech}^2\{(2S+1)\beta S\sigma\mathcal{J}(0)\}\right]. \tag{13.39}$$

Because we are particularly interested in temperatures just below T_c the function in the second bracket on the right-hand side of (13.39) can be expanded in σ; to order σ^2 it has the value

$$\tfrac{4}{3}S(S+1) - \tfrac{8}{15}S(S+1)(2S^2+2S+1)\{\beta S\sigma\mathcal{J}(0)\}^2. \tag{13.40}$$

Now

$$\{\beta S\sigma\mathcal{J}(0)\}^2 = \tfrac{15}{2}(2S^2+2S+1)^{-1}\left(\frac{T_c - T}{T_c}\right), \tag{13.41}$$

so (13.40) reduces to

$$\tfrac{4}{3}S(S+1)\left(\frac{3T}{T_c} - 2\right). \tag{13.42}$$

Thus

$$S\sum_m \exp(-i\mathbf{q}\cdot\mathbf{m})\lambda_m = \frac{(g\mu_B)^{-1}\chi_c H^z_{\mathbf{q}}}{(T/T_c) - \{\mathcal{J}(\mathbf{q})/\mathcal{J}(0)\}(3T/T_c - 2)}, \tag{13.43}$$

which on making use of (13.30) yields the result

$$\chi^z_{\boldsymbol{\kappa}} = \frac{g\mu_B S}{H^z_{\boldsymbol{\kappa}}}\sum_m \exp(-i\boldsymbol{\kappa}\cdot\mathbf{m})\lambda_m \simeq \frac{\chi_c}{r_1^2\{(\kappa^z_1)^2 + q^2\}} \quad (\boldsymbol{\kappa} = \boldsymbol{\tau} + \mathbf{q}), \tag{13.44}$$

with

$$(\kappa_1^z)^2 = \frac{12\mathcal{J}(0)}{J^{(2)}}\left(\frac{T_c - T}{T_c}\right) \quad (T < T_c), \tag{13.45}$$

and r_1^2 is given by eqn (13.33). On comparing the expression (13.45) with (13.34) we see that, both above and below T_c the inverse correlation range κ_1 vanishes at the critical point as $|T_c - T|^{1/2}$, their absolute magnitudes for a given $|T_c - T|$ differing only by a factor of 2. Notice that $\chi_{\boldsymbol{\kappa}}^{\alpha}$ given by (13.32) and (13.44) diverge at $T = T_c$ only when $\mathbf{q} = 0$.

Finally, we require to calculate $\chi_{\boldsymbol{\kappa}}^x = \chi_{\boldsymbol{\kappa}}^y$ below the critical temperature. The magnetic field is now applied parallel to the x-axis, say, so that the operative effective fields seen by the atomic spin at $\mathbf{m}$ are

$$-\frac{2}{g\mu_B}\sum_n J(\mathbf{m}-\mathbf{n})\langle \hat{S}_n^z\rangle \quad \text{along the } z\text{-axis}$$

and

$$-\frac{2}{g\mu_B}\sum_n J(\mathbf{m}-\mathbf{n})\langle \hat{S}_n^x\rangle + H_m^x \quad \text{along the } x\text{-axis.}$$

The value of $\langle \hat{S}_m\rangle$ is parallel to the resultant of these two fields so that

$$\langle \hat{S}_m^x\rangle = \langle \hat{S}_m^z\rangle \frac{\left\{g\mu_B H_m^x - 2\sum\limits_n J(\mathbf{m}-\mathbf{n})\langle \hat{S}_n^x\rangle\right\}}{-2\sum\limits_n J(\mathbf{m}-\mathbf{n})\langle \hat{S}_n^z\rangle}.$$

Proceeding as above, we find

$$\chi_{\boldsymbol{\kappa}}^x = \chi_{\boldsymbol{\kappa}}^y = \frac{(g\mu_B)^2}{2\{\mathcal{J}(0) - \mathcal{J}(\boldsymbol{\kappa})\}}. \tag{13.46}$$

Again expanding about $\boldsymbol{\tau} = \boldsymbol{\kappa} - \mathbf{q}$,

$$\chi_{\boldsymbol{\kappa}}^x \simeq \frac{\chi_c}{r_1^2 q^2} \quad (T < T_c) \tag{13.47}$$

with r_1^2 given by (13.33). From (13.46) we conclude that for $T < T_c$, κ_1^x is zero. This molecular-field theory is easily extended to antiferromagnets.

For a simple two-sublattice antiferromagnet we can always find a vector $\mathbf{w}$ such that $\exp(i\mathbf{l}\cdot\mathbf{w})$ has the value $+1$ on one sublattice and -1 on the other:

$$\text{simple cubic antiferromagnet of side } a,\ \mathbf{w} = \frac{\pi}{a}(1, 1, 1)$$

$$\text{body-centred antiferromagnet of side } a,\ \mathbf{w} = \frac{2\pi}{a}(1, 0, 0). \tag{13.48}$$

Hence the magnetic moments (in the molecular-field approximation) are simply

$$\langle \hat{S}_l^z \rangle = \exp(\mathrm{i}\mathbf{w}\cdot\mathbf{l})\langle \hat{S}^z \rangle \tag{13.49}$$

with $\langle \hat{S}^z \rangle$ given by eqn (13.21). This means that the wavelength-dependent susceptibility of an antiferromagnet above the critical temperature is of the same form as eqn (13.29) for the ferromagnet except that $\mathscr{J}(0)$ is replaced by $\mathscr{J}(\mathbf{w})$, the maximum value of $\mathscr{J}(\mathbf{q})$, and the critical temperature T_N is defined as

$$\beta_\mathrm{N} = \frac{3}{2S(S+1)\mathscr{J}(\mathbf{w})} = (k_\mathrm{B}T_\mathrm{N})^{-1}, \tag{13.50}$$

i.e.

$$\chi_{\boldsymbol{\kappa}}^{\alpha} = \frac{\chi_\mathrm{N}}{\{(T-T_\mathrm{N})/T_\mathrm{N}\} + \{1 - \mathscr{J}(\boldsymbol{\kappa})/\mathscr{J}(\mathbf{w})\}} \quad (T > T_\mathrm{N}). \tag{13.51}$$

Below T_N,

$$\chi_{\boldsymbol{\kappa}}^{z} = \frac{\chi_\mathrm{N}}{(T/T_\mathrm{N}) - \{\mathscr{J}(\boldsymbol{\kappa})/\mathscr{J}(\mathbf{w})\}(3T/T_\mathrm{N}-2)} \simeq \frac{\chi_\mathrm{N}}{r_1^2\{(\kappa_1^z)^2 + q^2\}} \quad (\boldsymbol{\kappa} = \boldsymbol{\tau} + \mathbf{q}) \tag{13.52}$$

where

$$(\kappa_1^z)^2 = \frac{12\mathscr{J}(\mathbf{w})}{J^{(2)}}\left(\frac{T_\mathrm{N}-T}{T_\mathrm{N}}\right) \tag{13.53}$$

and

$$r_1^2 = \frac{1}{6}\frac{J^{(2)}}{\mathscr{J}(\mathbf{w})}. \tag{13.54}$$

In many instances the exchange parameter is not isotropic and the appropriate exchange Hamiltonian becomes

$$-\sum_{m,n} \{J(\mathbf{n})\hat{\mathbf{S}}_m \cdot \hat{\mathbf{S}}_{m+n} + \hat{\mathbf{S}}_m \cdot \mathbf{D}(\mathbf{m}-\mathbf{n}) \cdot \hat{\mathbf{S}}_n\} \tag{13.55}$$

where, for example, $\mathbf{D}(\mathbf{m}-\mathbf{n})$ might represent dipole-dipole forces between the atoms.

Define

$$\mathscr{D}(\mathbf{q}) = \sum_n \exp(\mathrm{i}\mathbf{q}\cdot\mathbf{n})\mathbf{D}(\mathbf{n}); \tag{13.56}$$

then the spin ordering is determined by the tensor quantities

$$\mathscr{J}(\mathbf{q})\delta_{\alpha\beta} + \mathscr{D}_{\alpha\beta}(\mathbf{q}).$$

We assume the principal axes of this tensor coincide with the crystal axes. Then for a ferromagnetic system one of these principal values, say along the α-axis, has a maximum at $\mathbf{q}=0$. The critical temperature becomes

$$T_\alpha = \frac{2S(S+1)}{3k_B}\{\mathscr{J}(0)+\mathscr{D}_{\alpha\alpha}(0)\}. \tag{13.57}$$

Further, (13.29) is replaced by

$$\chi_{\boldsymbol{\kappa}}^{\alpha} = \frac{\chi_\alpha}{\{(T-T_\alpha)/T_\alpha\}+1-[\{\mathscr{J}(\boldsymbol{\kappa})+\mathscr{D}_{\alpha\alpha}(\boldsymbol{\kappa})\}/\{\mathscr{J}(0)+\mathscr{D}_{\alpha\alpha}(0)\}]} \tag{13.58}$$

for $T > T_\alpha$. The corresponding expressions for an antiferromagnet are obtained from (13.58) by replacing $\mathscr{J}(0)$ and $\mathscr{D}_{\alpha\alpha}(0)$ by $\mathscr{J}(\mathbf{w})$ and $\mathscr{D}_{\alpha\alpha}(\mathbf{w})$, respectively.

Below the critical temperature the wavelength-dependent susceptibilities are

$$\chi_{\boldsymbol{\kappa}}^{z} = \frac{\chi_c}{2\{(T-T_c)/T_c\}+1-[\{\mathscr{J}(\boldsymbol{\kappa})+\mathscr{D}_{zz}(\boldsymbol{\kappa})\}/\{\mathscr{J}(0)+\mathscr{D}_{zz}(0)\}]} \tag{13.59}$$

and

$$\chi_{\boldsymbol{\kappa}}^{x} = \frac{\chi_c}{\{(T-T_x)/T_x\}+1-[\{\mathscr{J}(\boldsymbol{\kappa})+\mathscr{D}_{xx}(\boldsymbol{\kappa})\}/\{\mathscr{J}(0)+\mathscr{D}_{xx}(0)\}]}, \tag{13.60}$$

etc. Again, the corresponding expressions for an antiferromagnet are obtained from (13.59) and (13.60) by replacing $\mathscr{J}(0)$ and $\mathscr{D}_{\alpha\alpha}(0)$ by $\mathscr{J}(\mathbf{w})$ and $\mathscr{D}_{\alpha\alpha}(\mathbf{w})$, respectively. The range of validity of mean-field theory in the vicinity of the critical point is discussed in Patashinskii and Pokrovskii (1979).

13.3. Spectral weight function

13.3.1. General features

There is not at the present time a single unified theory of the dynamics of Heisenberg magnets that adequately covers the entire temperature range. However, we can gain considerable insight into the essential features of the response function in various ranges of temperature and wave vector by evaluating its moments, namely, for $n \geqslant 1$,

$$\overline{\omega^{2n}} = \int_{-\infty}^{\infty} d\omega\, \omega^{2n} F^\alpha(\boldsymbol{\kappa},\omega) = \frac{i(-1)^{n+1}}{\hbar \mathscr{R}_{\boldsymbol{\kappa}}^{\alpha}(t=0)} \frac{\partial^{2n-1}}{\partial t^{2n-1}} \langle[\hat{S}_{\boldsymbol{\kappa}}^{\alpha}(t), \hat{S}_{-\boldsymbol{\kappa}}^{\alpha}]\rangle\bigg|_{t=0}. \tag{13.61}$$

The proof of (13.61) is as follows. From the definition of $F^\alpha(\boldsymbol{\kappa},\omega)$, eqn (13.15), it follows that

$$\int_{-\infty}^{\infty} \mathrm{d}\omega\, \omega^n F^\alpha(\boldsymbol{\kappa}, \omega) = \mathrm{i}^{-n} \frac{\partial^n}{\partial t^n} \{\mathcal{R}^\alpha_{\boldsymbol{\kappa}}(t)/\mathcal{R}^\alpha_{\boldsymbol{\kappa}}(t=0)\}\bigg|_{t=0}. \tag{13.62}$$

Since $\mathcal{R}^\alpha_{\boldsymbol{\kappa}}(t)$ is an even function of t, the right-hand side of (13.62) is non-zero only if n is an even integer. Further (cf. Chapter 8)

$$\frac{\partial}{\partial t} \mathcal{R}^\alpha_{\boldsymbol{\kappa}}(t) = -\frac{N}{(g\mu_\mathrm{B})^2} \phi^\alpha_{\boldsymbol{\kappa}}(t) = -\frac{\mathrm{i}}{\hbar} \langle [\hat{S}^\alpha_{\boldsymbol{\kappa}}(t), \hat{S}^\alpha_{-\boldsymbol{\kappa}}] \rangle \tag{13.63}$$

from which (13.62) immediately follows.

With regard to eqn (13.62) for the moments $\overline{\omega^{2n}}$, recall that

$$\mathcal{R}^\alpha_{\boldsymbol{\kappa}}(t=0) \propto \chi^\alpha_{\boldsymbol{\kappa}}$$

and, for magnets with an isotropic exchange, the numerator on the right-hand side is proportional to κ^2 for small scattering vectors. In general we shall find that $\overline{\omega^4} \simeq 3(\overline{\omega^2})^2$, which implies that the spectral weight function resembles a Gaussian (Hubbard 1971). This point is further amplified below. Two exceptions to the Gaussian form occur for (a) very small values of $|\boldsymbol{\kappa}|$ at any value of T and (b) at $T \simeq T_\mathrm{c}$ with $\boldsymbol{\kappa}$ near to an ordering vector. In both cases $\overline{\omega^4} \gg (\overline{\omega^2})^2$ and hence $F^\alpha(\boldsymbol{\kappa}, \omega)$ more closely resembles a Lorentzian function than a Gaussian. Note that the spectral weight function is periodic in reciprocal space. Both $\overline{\omega^2}$ and $\overline{\omega^4}$ are proportional to κ^2 and hence in case (a) $\overline{\omega^4} \gg (\overline{\omega^2})^2$, while for (b) $\overline{\omega^4} \sim \chi_{\boldsymbol{\kappa}}^{-1}$ and $\chi_{\boldsymbol{\kappa}}$ is close to a maximum.

It is useful to note the high-temperature limit of eqn (13.61). For small β

$$\mathcal{R}^\alpha_{\boldsymbol{\kappa}}(t) \simeq \beta\{\langle \hat{S}^\alpha_{\boldsymbol{\kappa}} \hat{S}^\alpha_{-\boldsymbol{\kappa}}(t)\rangle - \langle \hat{S}^\alpha_{\boldsymbol{\kappa}}\rangle\langle \hat{S}^\alpha_{-\boldsymbol{\kappa}}\rangle\} = \beta \langle \hat{S}^\alpha_{\boldsymbol{\kappa}} \hat{S}^\alpha_{-\boldsymbol{\kappa}}(t)\rangle \tag{13.64}$$

where the last line follows because in the paramagnetic region $\langle \hat{S}^\alpha_{\boldsymbol{\kappa}}\rangle = 0$. Utilizing (13.64) we obtain

$$\lim_{\beta\to 0} \int_{-\infty}^{\infty} \mathrm{d}\omega\, \omega^{2n} F^\alpha(\boldsymbol{\kappa}, \omega) = \frac{(-1)^n}{\langle \hat{S}^\alpha_{\boldsymbol{\kappa}} \hat{S}^\alpha_{-\boldsymbol{\kappa}}\rangle} \frac{\partial^{2n}}{\partial t^{2n}} \langle \hat{S}^\alpha_{\boldsymbol{\kappa}} \hat{S}^\alpha_{-\boldsymbol{\kappa}}(t)\rangle \bigg|_{t=0} \tag{13.65}$$

and the right-hand side is asymptotically independent of temperature.

13.3.2. Evaluation of moments

We consider the calculation of the first few moments of $F^\alpha(\boldsymbol{\kappa}, \omega)$ for an isotropic ferromagnet and begin by considering $\chi^\alpha_{\boldsymbol{\kappa}}$.

In the limit $T \to \infty$ there is no correlation between spins and hence it follows that

$$\begin{aligned} \lim_{\beta\to 0} \langle \hat{S}^\alpha_l \hat{S}^\beta_{l'}\rangle &= \delta_{\alpha,\beta}\delta_{l,l'}\langle \hat{S}^\alpha \hat{S}^\alpha\rangle \\ &= \delta_{\alpha,\beta}\delta_{l,l'}\tfrac{1}{3}S(S+1). \end{aligned} \tag{13.66}$$

To calculate the first temperature correction to this result we need to evaluate

$$\langle \hat{S}_l^\alpha \hat{S}_{l'}^\beta \rangle = \mathrm{Tr}\{\exp(-\beta \hat{\mathcal{H}}) \hat{S}_l^\alpha \hat{S}_{l'}^\beta\}/\mathrm{Tr}\{\exp(-\beta \hat{\mathcal{H}})\} \tag{13.67}$$

to first order in β where Tr denotes a trace over the spin variables. Clearly

$$\mathrm{Tr} \exp(-\beta \hat{\mathcal{H}}) \simeq \mathrm{Tr}(1 - \beta \hat{\mathcal{H}}) = 1$$

by virtue of the result (13.66) and the fact that $J(0) = 0$. Hence to order β, (13.67) becomes

$$\langle \hat{S}_l^\alpha \hat{S}_{l'}^\beta \rangle = \mathrm{Tr}(\hat{S}_l^\alpha \hat{S}_{l'}^\beta - \beta \hat{\mathcal{H}} \hat{S}_l^\alpha \hat{S}_{l'}^\beta). \tag{13.68}$$

Now

$$\mathrm{Tr}\ \hat{\mathcal{H}} \hat{S}_l^\alpha \hat{S}_{l'}^\beta = -\sum_{m,n} J(\mathbf{m}-\mathbf{n}) \sum_\nu \mathrm{Tr}\ \hat{S}_m^\nu \hat{S}_n^\nu \hat{S}_l^\alpha \hat{S}_{l'}^\beta$$

and

$$\begin{aligned} \mathrm{Tr}\ \hat{S}_m^z \hat{S}_n^z \hat{S}_l^\alpha \hat{S}_{l'}^\beta &= \langle \hat{S}^z \hat{S}^\alpha \rangle \langle \hat{S}^z \hat{S}^\beta \rangle (\delta_{m,l}\delta_{n,l'} + \delta_{m,l'}\delta_{n,l}) \\ &= \delta_{\alpha,z}\delta_{\alpha,\beta}\{\tfrac{1}{3}S(S+1)\}^2(\delta_{m,l}\delta_{n,l'} + \delta_{m,l'}\delta_{n,l}). \end{aligned} \tag{13.69}$$

Hence (13.68) reads

$$\delta_{\alpha,\beta}\tfrac{1}{3}S(S+1)\left\{\delta_{l,l'} + \beta\tfrac{1}{3}S(S+1)\sum_{m,n} J(\mathbf{m}-\mathbf{n}) \sum_\nu \delta_{\alpha,\nu}(\delta_{m,l}\delta_{n,l'} + \delta_{m,l'}\delta_{n,l})\right\},$$

i.e.

$$\langle \hat{S}_l^\alpha \hat{S}_l^\beta \rangle = \delta_{\alpha,\beta}\tfrac{1}{3}S(S+1). \tag{13.70a}$$

and

$$\langle \hat{S}_l^\alpha \hat{S}_{l'}^\beta \rangle = \delta_{\alpha,\beta} 2\beta\{\tfrac{1}{3}S(S+1)\}^2 J(\mathbf{l}-\mathbf{l}'), \tag{13.70b}$$

to first order in β.

Now from (13.64)

$$\begin{aligned} \mathcal{R}_{\boldsymbol{\kappa}}^\alpha(t=0) &\simeq \beta\langle \hat{S}_{\boldsymbol{\kappa}}^\alpha \hat{S}_{-\boldsymbol{\kappa}}^\alpha \rangle \\ &= \beta \sum_{l,l'} \exp\{i\boldsymbol{\kappa}\cdot(\mathbf{l}'-\mathbf{l})\}\langle \hat{S}_l^\alpha \hat{S}_{l'}^\alpha \rangle \\ &\simeq \beta N \tfrac{1}{3}S(S+1)\{1 + \tfrac{2}{3}\beta S(S+1)\mathcal{J}(\boldsymbol{\kappa})\}. \end{aligned} \tag{13.71}$$

i.e. $\mathcal{R}_{\boldsymbol{\kappa}}^\alpha(t=0)$ tends to zero as $\beta \to 0$ like

$$N\beta\tfrac{1}{3}S(S+1).$$

Further

$$\chi_{\boldsymbol{\kappa}}^\alpha = \frac{(g\mu_B)^2}{N}\mathcal{R}_{\boldsymbol{\kappa}}^\alpha(t=0) \Rightarrow \chi_0.$$

If we retain the term in β^2 in (13.71), then

$$\chi_{\kappa}^{\alpha} = \chi_0\left\{1 + \frac{2}{(g\mu_B)^2}\chi_0\mathscr{J}(\boldsymbol{\kappa})\right\}, \tag{13.72}$$

which are seen to be the first two terms in the expansion of the molecular-field result for χ_{κ}^{α}, eqn (13.28). Thus the molecular-field approximation for the susceptibility becomes exact in the limit $\beta \to 0$; this is not surprising since there is no short-range order between the spins at $T = \infty$.

Turning now to the calculation of $\overline{\omega^2}$, we require

$$\overline{\omega^2} = \frac{\mathrm{i}}{\hbar}\frac{1}{\mathscr{R}_{\kappa}^{\alpha}(t=0)}\frac{\partial}{\partial t}\langle[\hat{S}_{\kappa}^{\alpha}(t), \hat{S}_{-\kappa}^{\alpha}]\rangle\Big|_{t=0}$$

$$= \frac{1}{\hbar^2}\frac{1}{\mathscr{R}_{\kappa}^{\alpha}(t=0)}\sum_{l,l'}\exp\{\mathrm{i}\boldsymbol{\kappa}\cdot(\mathbf{l}'-\mathbf{l})\}\langle[[\hat{S}_l^{\alpha}, \hat{\mathscr{H}}], \hat{S}_{l'}^{\alpha}]\rangle. \tag{13.73}$$

Now

$$[\hat{S}_l^x, \hat{\mathscr{H}}] = -2\mathrm{i}\sum_m J(\mathbf{l}-\mathbf{m})(\hat{S}_l^z\hat{S}_m^y - \hat{S}_l^y\hat{S}_m^z) \tag{13.74}$$

and

$$\sum_{l,l'}\exp\{\mathrm{i}\boldsymbol{\kappa}\cdot(\mathbf{l}'-\mathbf{l})\}[[\hat{S}_l^x, \hat{\mathscr{H}}], \hat{S}_{l'}^x]$$

$$= 2\sum_{l,m} J(\mathbf{l}-\mathbf{m})(\hat{S}_l^z\hat{S}_m^z + \hat{S}_l^y\hat{S}_m^y)[1 - \exp\{\mathrm{i}\boldsymbol{\kappa}\cdot(\mathbf{l}-\mathbf{m})\}]. \tag{13.75}$$

from which we obtain the exact expression,

$$\overline{\omega^2} = \frac{4N}{\hbar^2\mathscr{R}_{\kappa}^{\alpha}(t=0)}\sum_m J(\mathbf{m})(1-\cos\boldsymbol{\kappa}\cdot\mathbf{m})\langle\hat{S}_0^z\hat{S}_m^z\rangle. \tag{13.76}$$

Evaluating this expression in the high-temperature limit merely requires the use of (13.70b) and (13.71)

$$\overline{\omega^2} = \frac{8S(S+1)}{3\hbar^2}\sum_m J^2(\mathbf{m})(1-\cos\boldsymbol{\kappa}\cdot\mathbf{m}). \tag{13.77}$$

For nearest-neighbour exchange J,

$$\overline{\omega^2} = \frac{4J(S+1)}{3\hbar^2}\omega_{\kappa} \tag{13.78}$$

where ω_{κ} is the dispersion relation for linear spin waves cf. eqn (9.16). Also, writing $\boldsymbol{\kappa} = \boldsymbol{\tau} + \mathbf{q}$ we obtain for small $\mathbf{q}$

$$\overline{\omega^2} = \frac{4S(S+1)}{9\hbar^2}J^{(2)}Jq^2. \tag{13.79}$$

If a field is applied the expression for the longitudinal second moment is unchanged, while the transverse moment is explicitly field-dependent.

The evaluation of the fourth moment of $F^{\alpha}(\boldsymbol{\kappa}, \omega)$ requires considerable effort. For a pure exchange Hamiltonian and high temperatures the result is (Malinoski 1973)

$$\hbar^4\overline{\omega^4} = 32\left\{\frac{S(S+1)}{3}\right\}^2\left[-\sum_m J^4(\mathbf{m})\{1-\cos\boldsymbol{\kappa}\cdot\mathbf{m}\}\left\{4+\frac{3}{2S(S+1)}\right\}\right.$$
$$+\left[\sum_m J^2(\mathbf{m})\{1-\cos\boldsymbol{\kappa}\cdot\mathbf{m}\}\right]\left[\sum_m J^2(\mathbf{m})\{7-3\cos\boldsymbol{\kappa}\cdot\mathbf{m}\}\right]$$
$$-2\sum_{m,n} J(\mathbf{m})J(\mathbf{n})J(\mathbf{m}-\mathbf{n})\{1-\exp(\mathrm{i}\boldsymbol{\kappa}\cdot\mathbf{n})\}[J(\mathbf{n})\exp(-\mathrm{i}\boldsymbol{\kappa}\cdot\mathbf{m})$$
$$\left.+J(\mathbf{m}-\mathbf{n})\{1-\exp(-\mathrm{i}\boldsymbol{\kappa}\cdot\mathbf{m})\}]\right]. \tag{13.80}$$

It can be shown from this expression that $\overline{\omega^4}$ is proportional to q^2 for small $\mathbf{q}$, as noted previously.

In experiments with polycrystals the cross-section is averaged over all crystal orientations. The second and fourth moments in this instance are

$$\hbar^2\overline{\omega^2} = \tfrac{8}{3}S(S+1)\sum_m J^2(\mathbf{m})\left\{1-\frac{\sin(\kappa m)}{(\kappa m)}\right\} \tag{13.81}$$

and

$$\hbar^4\overline{\omega^4} = 32\left\{\frac{S(S+1)}{3}\right\}^2\left[\!\!\left[-\sum_m J^4(\mathbf{m})\left\{1-\frac{\sin(\kappa m)}{(\kappa m)}\right\}\left\{4+\frac{3}{2S(S+1)}\right\}\right.\right.$$
$$+\sum_{m,m'} J(\mathbf{m})J(\mathbf{m}')\left(J(\mathbf{m})J(\mathbf{m}')\left\{7-10\frac{\sin(\kappa m')}{(\kappa m')}+3\frac{\sin(\kappa\,|\mathbf{m}-\mathbf{m}'|)}{\kappa\,|\mathbf{m}-\mathbf{m}'|}\right\}\right.$$
$$-2J(\mathbf{m}-\mathbf{m}')\left[J(\mathbf{m}')\left\{\frac{\sin(\kappa m)}{(\kappa m)}-\frac{\sin(\kappa\,|\mathbf{m}-\mathbf{m}'|)}{\kappa\,|\mathbf{m}-\mathbf{m}'|}\right\}\right.$$
$$\left.\left.\left.\left.+J(\mathbf{m}-\mathbf{m}')\left\{1-2\frac{\sin(\kappa m)}{(\kappa m)}+\frac{\sin(\kappa\,|\mathbf{m}-\mathbf{m}'|)}{\kappa\,|\mathbf{m}-\mathbf{m}'|}\right\}\right]\right)\right]\!\!\right]. \tag{13.82}$$

In discussing the implications of the above expressions for the second and fourth moments it is convenient to introduce the parameter

$$\alpha = \frac{\overline{\omega^4}}{3(\overline{\omega^2})^2} - 1. \tag{13.83}$$

If the spectral weight function is a Gaussian function of ω, $\alpha = 0$, while for a Lorentzian function it is infinite. The parameter α is shown in Fig. 13.1 for $\boldsymbol{\kappa}$ along the three symmetry axis of a simple cubic lattice with

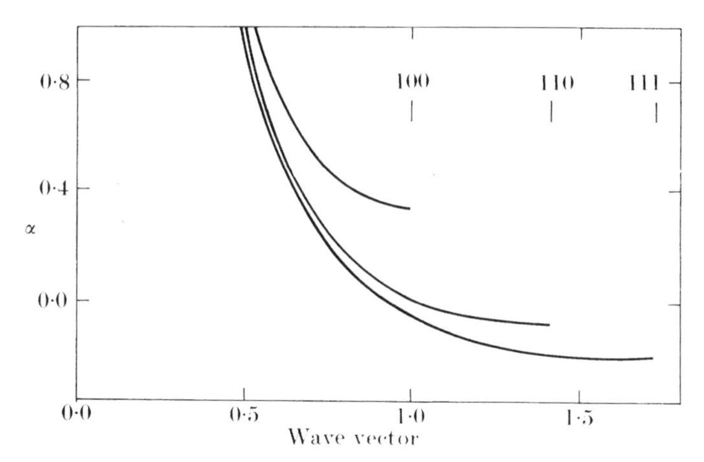

FIG. 13.1. Variation of α (eqn (13.83)) with wave vector along the principal symmetry axes of a simple cubic lattice. α is a measure of the departure of the shape of the distribution from Gaussian. Nearest-neighbour exchange interactions are assumed (Collins and Marshall 1967).

nearest-neighbour exchange coupling and $S = 5/2$. It is evident from the behaviour of α that, at high temperatures, the analysis implies a spectral weight function which is approximately Gaussian for large values of the scattering vector but deviates sharply from this form for small $\boldsymbol{\kappa}$ where it is more closely approximated by a Lorentzian function. This is in accord with our previous general observations regarding the ω-dependence of $F^{\alpha}(\boldsymbol{\kappa}, \omega)$.

13.3.3. *Small* **q** (*Van Hove 1954*; *Patashinskii and Pokrovskii 1979*)

Since both $\overline{\omega^2}$ and $\overline{\omega^4}$ vary like q^2 for small $\mathbf{q}$, the parameter α varies like q^{-2}. Thus $F(\boldsymbol{\kappa}, \omega)$ can in the limit of small $\mathbf{q}$ be represented by a cut-off Lorentzian, viz.

$$\begin{aligned} F(\boldsymbol{\kappa}, \omega) &= \frac{1}{\pi} \frac{\Gamma_{\boldsymbol{\kappa}}}{\omega^2 + \Gamma_{\boldsymbol{\kappa}}^2} \qquad (|\omega| < \omega_c) \\ &= 0 \qquad (|\omega| > \omega_c) \end{aligned} \tag{13.84}$$

where

$$\Gamma_{\boldsymbol{\kappa}} = \frac{\pi}{2\sqrt{3}} \left\{ \frac{(\overline{\omega^2})^3}{\overline{\omega^4}} \right\}^{1/2}$$

and

$$\omega_c^2 = \frac{3\overline{\omega^4}}{\overline{\omega^2}} \tag{13.85}$$

are chosen to give the second and fourth moments correctly. Specializing to high temperatures and a simple magnet with only nearest-neighbour exchange coupling J,

$$\overline{\omega^2} = \left(\frac{2J}{\hbar}\right)^2 \tfrac{2}{3} rS(S+1)(1-\gamma_{\boldsymbol{\kappa}}) \tag{13.86}$$

and

$$\overline{\omega^4} = \left(\frac{2J}{\hbar}\right)^4 \tfrac{4}{9} S^2(S+1)^2 r(1-\gamma_{\boldsymbol{\kappa}})\left\{\frac{7r}{2} - \frac{3r}{2}\gamma_{\boldsymbol{\kappa}} - 2 - \frac{3}{4S(S+1)}\right\}, \tag{13.87}$$

from which we have, to leading order in $|\boldsymbol{\kappa}|^2$,

$$\Gamma_{\boldsymbol{\kappa}} = \frac{J\pi}{3\hbar} r(1-\gamma_{\boldsymbol{\kappa}})\left\{\frac{S(S+1)}{r-1-\{3/8S(S+1)\}}\right\}^{1/2} \tag{13.88}$$

and

$$\omega_c = \frac{4J}{\hbar}\left[S(S+1)\left\{r - 1 - \frac{3}{8S(S+1)}\right\}\right]^{1/2}. \tag{13.89}$$

The geometric factor $\gamma_{\boldsymbol{\kappa}}$ is defined by the relation

$$\mathscr{J}(\boldsymbol{\kappa}) = rJ\gamma_{\boldsymbol{\kappa}}$$

where r is the number of nearest neighbours.

If we go back to the definition of the moments $\overline{\omega^{2n}}$ given by eqn (13.61) we deduce that their temperature dependence close to the critical point is essentially governed by the wave-vector dependent susceptibility that appears in the denominator. The short-range correlation function in the numerator of (13.61) can only have a weak singularity whereas the susceptibility $\chi^{\alpha}_{\boldsymbol{\kappa}}$ contains long-range correlation functions and is strongly temperature-dependent. In view of this we estimate the moments just above T_c by

$$\overline{\omega^2} = \overline{\omega^2_{\infty}}\,\frac{\chi_0}{\chi^{\alpha}_{\boldsymbol{\kappa}}} \quad \text{and} \quad \overline{\omega^4} = \overline{\omega^4_{\infty}}\,\frac{\chi_0}{\chi^{\alpha}_{\boldsymbol{\kappa}}} \tag{13.90}$$

where the suffix ∞ denotes that the moments are evaluated at infinite temperatures. If $\boldsymbol{\kappa}$ is close to an ordering vector, then with $T \simeq T_c$ $1/\chi^{\alpha}_{\boldsymbol{\kappa}}$ is small and, in consequence, $\overline{\omega^4} \gg (\overline{\omega^2})^2$. Hence in this instance the spectral weight function is approximately a (cut-off) Lorentzian function of ω, with a width

$$\Gamma^{\alpha}_{\boldsymbol{\kappa}} = \frac{\pi}{2\sqrt{3}}\left\{\frac{(\overline{\omega^2_{\infty}})^3}{\overline{\omega^4_{\infty}}}\right\}^{1/2}\frac{\chi_0}{\chi^{\alpha}_{\boldsymbol{\kappa}}}. \tag{13.91}$$

For a ferromagnet with small $\mathbf{q}=\boldsymbol{\kappa}-\boldsymbol{\tau}$ we obtain the result

$$\Gamma^{\alpha}_{\boldsymbol{\kappa}}=\Lambda q^2\frac{T_c}{T}r_1^2(\kappa_1^2+q^2) \tag{13.92}$$

where we have made use of the two results that the moments are proportional to q^2 and eqn (13.32) for $\chi^{\alpha}_{\boldsymbol{\kappa}}$.

In the case of a simple antiferromagnet we can replace the ratio of the infinite temperature moments in (13.91) by a constant because, as is illustrated in Fig. 13.1, they are almost independent of the scattering vector for $\boldsymbol{\kappa}$ in the vicinity of $\mathbf{w}$, the superlattice vector. Thus, in this case we write

$$\Gamma^{\alpha}_{\boldsymbol{\kappa}}\propto\kappa_1^2+(q^*)^2. \tag{13.93}$$

where $q^*=\boldsymbol{\kappa}-\mathbf{w}$.

Note that these forms for the width predict that it should decrease as $T\rightarrow T_c$ like $(T-T_c)$, i.e. we should expect to see the inelasticity of the scattering decrease. This latter effect is usually referred to as thermodynamic slowing down of the spin fluctuations. It is emphasized that by small q we mean $q^2\ll\kappa_1^2$, a condition that becomes increasingly difficult to fulfil as $T\rightarrow T_c$.

The wave-vector dependent susceptibility can be measured directly in the static approximation (cf. § 13.1). Schullof *et al.* (1970) verified the validity of the static approximation for the antiferromagnet MnF_2 (cf. § 12.4.2) with $0.04\leqslant T-T_N\lesssim 8$ K and incident energies E in the range

$$56\text{ meV}\leqslant E\leqslant 134\text{ meV};$$

We now consider their results for the wave-vector dependent susceptibility.

Because of the uniaxial anisotropy in MnF_2 it is necessary to assign to it two Néel temperatures, one for the longitudinal properties and another for the transverse. These are determined by (cf. eqn (13.57))

$$T_{\alpha}=\frac{2S(S+1)}{3k_B}\{\mathcal{J}(\mathbf{w})+\mathcal{D}_{\alpha}(\mathbf{w})\}. \tag{13.94}$$

The true Néel temperature is that at which $\chi^{\alpha}_{\boldsymbol{\kappa}}$ for $\boldsymbol{\kappa}=\mathbf{w}$ first diverges as T decreases from infinity. For MnF_2 the longitudinal transition determines the true Néel temperature $T_N=67.3$ K. Because it is the longitudinal component of the susceptibility that diverges at T_N we expect thermodynamic slowing down only in the corresponding $\Gamma_{\boldsymbol{\kappa}}$. The longitudinal and transverse components of the cross-section can be isolated by utilizing the different orientation factors associated with either one. The transverse component has the orientation factor $1+\tilde{\kappa}_z^2$ (assuming the z-axis to coincide with the axis of quantization), the longitudinal the factor $1-\tilde{\kappa}_z^2$; hence for $\boldsymbol{\kappa}$ parallel to the z-axis only the transverse component is observed whereas for $\boldsymbol{\kappa}$ perpendicular to the z axis both are observed.

Above T_N and in the limit $\epsilon=(T-T_N)/T_N \ll 1$, the mean-field form of the spin correlation function (eqn (13.36)) should be replaced by (Patashinskii and Pokrovskii 1979; Als-Nielsen 1976)

$$\langle \Delta\hat{S}_m\, \Delta\hat{S}_n \rangle \propto \exp(-\kappa_1 r)/r^{1+\eta}; \qquad r=|\mathbf{m}-\mathbf{n}|, \tag{13.95}$$

and in consequence ($\boldsymbol{\kappa}=\mathbf{q}^*+\mathbf{w}$)

$$\chi_{\mathbf{q}^*} \propto \left(\frac{1}{\kappa_1^2+q^{*2}}\right)^{1-\eta/2}. \tag{13.96}$$

In view of the above discussion of the properties of MnF_2 and eqn (13.96) for $\chi_{\mathbf{q}^*}$, the differential cross-sections for measurements around (100) and (001) were

$$\frac{d\sigma}{d\Omega} \propto \left(\frac{B_\perp(T)}{\kappa_{1\perp}^2+q^{*2}}\right)^{1-\eta/2}; \qquad \text{(001) reflection} \tag{13.97}$$

and

$$\frac{d\sigma}{d\Omega} \propto \left(\frac{B_L(T)}{\kappa_{1L}^2+q^{*2}}\right)^{1-\eta/2} + \left(\frac{B'_\perp(T)}{\kappa_{1\perp}^2+q^{*2}}\right)^{1-\eta/2}; \qquad \text{(100) reflection} \tag{13.98}$$

Here (Als-Nielsen 1976),

$$q^{*2}=q_a^2+q_{a'}^2+\left(\frac{c}{a}\right)^2 q_c^2. \tag{13.99}$$

Using these forms for $d\sigma/d\Omega$ they obtained the following results for the critical indices (see also Lurie *et al.* 1972)

$$\begin{aligned} &\text{Longitudinal} \begin{cases} \chi_{\mathbf{q}^*=0} \sim \epsilon^{-\gamma},\ \gamma=1.27\pm0.02 \\ \nu=0.634\pm0.02 \end{cases} \\ &\text{Transverse} \quad \begin{cases} \chi_{\mathbf{q}^*=0} \sim \epsilon^{-\gamma},\ \gamma=1.47\pm0.1,\ T_\perp=64.40\pm0.5\ \text{K} \\ \nu=0.63\pm0.08,\ T_\perp=64.9\pm0.7\ \text{K}, \end{cases} \end{aligned} \tag{13.100}$$

where $\kappa_1 \propto \epsilon^\nu$. A best-fit value of $\eta=0.05\pm0.02$ was obtained with the longitudinal measurements closest to T_N.

Turning now to dynamical effects near T_N, several experiments have confirmed the effect of thermodynamic slowing down of the spin fluctuations but disagree quantitatively with the preceding mean-field results. Two other important features revealed by these experiments are: (a) for $T<T_N$ but close to the critical point there is a central peak in the

longitudinal susceptibility, and (b) the spin-wave peaks persist above T_N and finally merge into the central peak as T is further increased.

To illustrate some of these features we again consider the measurements on MnF_2 by Schulhof *et al.* The partial differential cross-section is taken to have the following forms for measurements about (001) and (100)

$$\left(\frac{d^2\sigma}{d\Omega\, dE'}\right)_\perp \propto \left(\frac{1}{\kappa_{1\perp}^2+q^{*2}}\right)^{1-\eta/2}\left\{\frac{\Gamma_\perp}{\Gamma_\perp^2+(\omega-\omega_0)^2}+\frac{\Gamma_\perp}{\Gamma_\perp^2+(\omega+\omega_0)^2}\right\} \tag{13.101}$$

and

$$\left(\frac{d^2\sigma}{d\Omega\, dE'}\right)_L \propto \left(\frac{1}{\kappa_{1L}^2+q^{*2}}\right)^{1-\eta/2}\left\{\frac{\Gamma_L}{\Gamma_L^2+\omega^2}\right\}+\left(\frac{d^2\sigma}{d\Omega\, dE'}\right)_\perp. \tag{13.102}$$

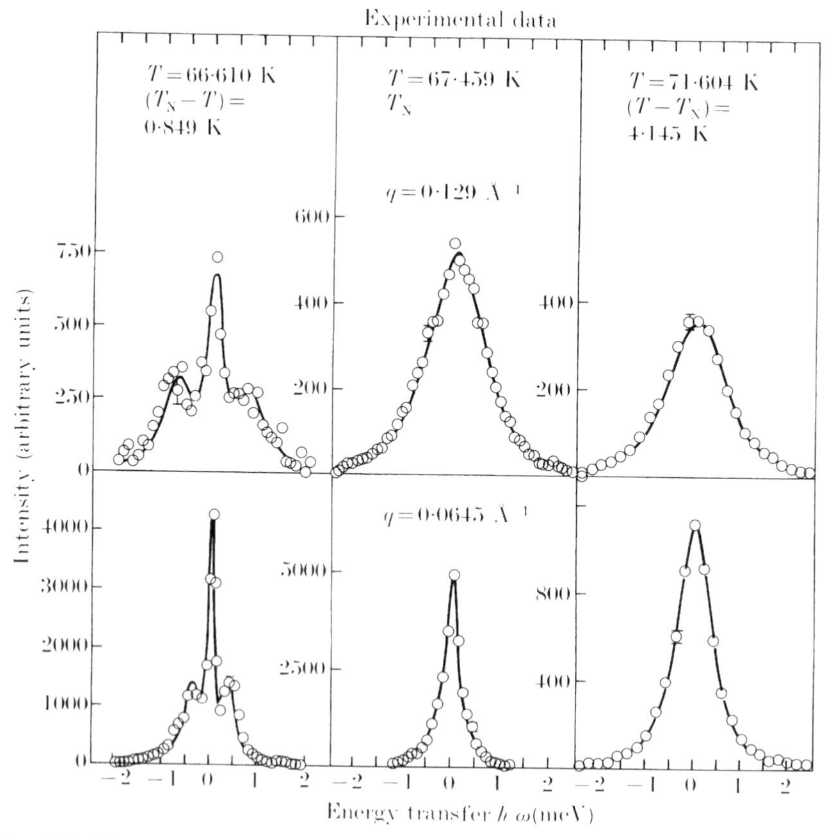

FIG. 13.2. The cross-section measured about (100) is shown as a function of energy transfer and for various values of temperature and wave vector. Both the transverse and longitudinal fluctuations contribute here and the central peak below T_N is to be noted (Schulhof *et al.* 1970).

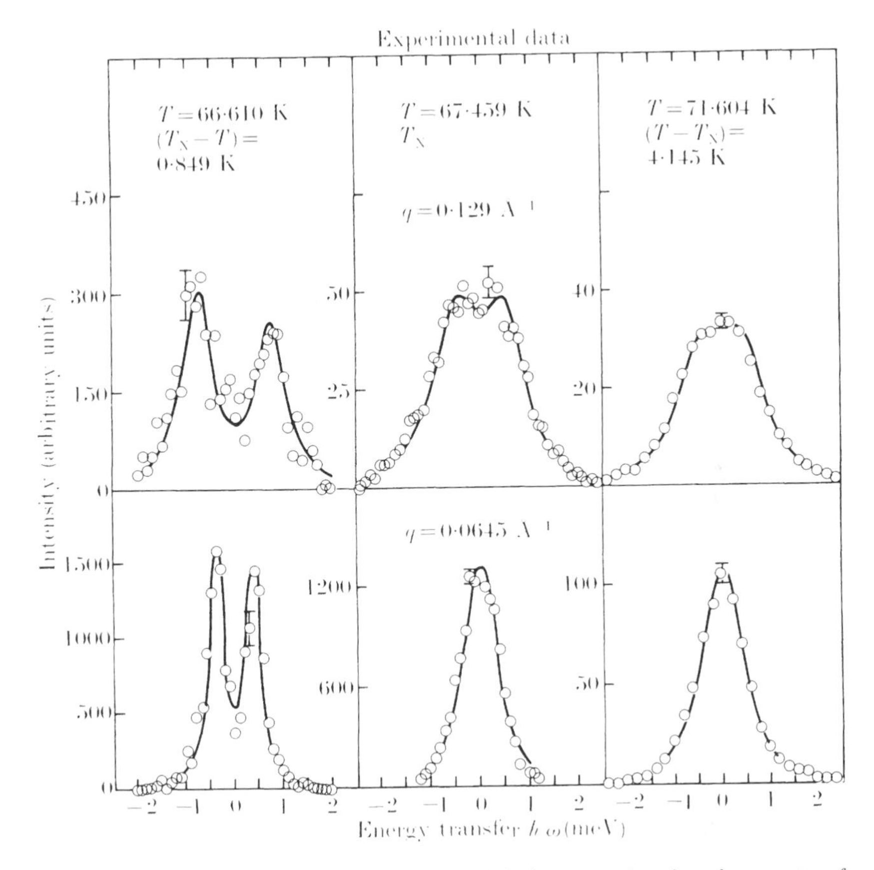

FIG. 13.3. The cross-section measured about (001) is shown as a function of energy transfer and for various values of temperature and wave vector. Only the transverse fluctuations contribute to the cross-section. For the larger of the two values of the wave vector, two distinct spin-wave peaks exist at T_N and are still in evidence at $T_N + 4.145$ K (Schulhof *et al.* 1970).

The measured cross-sections are illustrated in Figs 13.2 and 13.3. These show features (a) and (b) discussed above.

The structure of $RbMnF_3$ is such that dipolar interactions are weak, and it forms an ideal isotropic, cubic Heisenberg antiferromagnet. Experiments by Tucciarone *et al.* (1971) show that, for $T = T_N$, the response function for $q^* = 0.05$ Å^{-1} is a single peaked function centred at $\omega = 0$. On increasing q^* to 0.25 Å^{-1}, the response function develops distinct sidepeaks. Measurements taken in the ordered phase show the existence of a central peak at $\omega = 0$, in addition to the spin-wave scattering. These findings are in accord with the measurements on MnF_2.

Measurements of critical scattering from ferromagnets and antiferromagnets differ in the ordered phase. Whereas a central, diffuse, peak is observed in the antiferromagnets MnF_2 and $RbMnF_3$ for $T < T_N$, no such peak is evident in scattering from Fe, Ni (Minkiewicz 1971) or EuO (Dietrich, Als-Nielsen, and Passell 1976) just below the Curie temperature.

The available experimental results for ferro- and antiferromagnets are largely consistent with so-called dynamic scaling theory (Hohenberg and Halperin 1977; Lovesey and Williams 1986). The latter theory predicts, for example, that at the critical point ($\kappa_1 = 0$) the width of the response functions for ferro- and antiferromagnets should have the following dependence on the wave vector

$$\left.\begin{array}{ll}\Gamma \propto q^{*3/2}; & \text{antiferromagnet}\\ \Gamma \propto q^{5/2}; & \text{ferromagnet}\end{array}\right\}\kappa_1 = 0. \qquad (13.103)$$

The validity of these results is borne out by the experimental findings, together with results for spin diffusion constants (Böni and Shirane 1986, Mezei 1986).

13.3.4. Large **q**

We concluded in § 13.3.2, on the basis of results for the moments of the response function at high temperatures, that for large **q**, near the Brillouin zone boundary, the response resembles a Gaussian function of ω. Measurements on the simple ferromagnets EuO (Mook 1981) and EuS (Bohn *et al.* 1980) reveal distinct structure in the response at finite ω for large **q** and temperatures well above the Curie point. For example, the response of EuO at the zone boundary shows a sharp spin wave at $T = T_c - 10$ which remains a distinct feature at $T = 2T_c$, albeit weaker and more damped. Spin waves at small **q** soften and coalesce, to form a peak at $\omega = 0$, as the temperature is raised through the Curie point. Such behaviour is observed for EuO halfway to the zone boundary for the (111) direction. Theory (Young and Shastry 1982; Lindgård 1983) and computer-simulation data (Takahashi 1983) are in accord with the observation for EuO and EuS. Moreover, at the zone boundary the peak in the response at finite ω persists for $T \to \infty$. This result implies that the spin oscillation responsible for the peak is not a manifestation of short-range order. The result (13.70b) shows that the nearest-neighbour spin correlation vanishes, at high temperature, as $(J/k_B T)$.

Spin waves in the ferromagnetic metals nickel and iron at low temperatures, $T \ll T_c$, are observed for modest qs only, as described in § 9.3. The disappearance of the spin waves with increasing wave vector is ascribed to damping by Stoner, spin-flip, excitations, i.e. a feature of itinerant-electron models. At high temperatures, $T \geqslant T_c$, the observed response for nickel and iron (Lynn 1975, 1981, 1983; Mook and Lynn 1985; Wicksted *et al.* 1984; Martinez *et al.* 1985) is quite different for scans made with constant energy- and wave-vector transfer. Scans of the response function at constant ω as a function $\mathbf{q}$ reveal a well-defined peak whose dispersion is akin to that of the spin wave observed in the ordered phase. No such peak is observed in scans made at constant $\mathbf{q}$ as a function of ω, and the magnetic scattering, separated from nuclear scattering using polarization analysis, is centred around $\omega = 0$ (Steinsvoll *et al.* 1983). The magnetic response of nickel and the ionic ferromagnets EuO and EuS for $T \geqslant T_c$ is, therefore, similar in that a collective oscillation, defined by a peak in the response at constant $\mathbf{q}$ as a function of ω, is not observed at modest *qs* (Böni and Shirane 1986, Mook and Lynn 1986).

The existence of a short-wavelength spin oscillation, in a Heisenbeɪg magnet, at a high temperature can be inferred from a simple calculation which develops the response function as a moment expansion starting with a Gaussian form,

$$F(\boldsymbol{\kappa}, \omega) \simeq F_G(\boldsymbol{\kappa}, \omega) = (2\pi\overline{\omega^2})^{-1/2} \exp\left(-\frac{\omega^2}{2\overline{\omega^2}}\right). \tag{13.104}$$

In order to include information on the fourth moment in the expression for $F(\boldsymbol{\kappa}, \omega)$ we use the Gram–Charlier series (Collins and Marshall 1967)

$$F(\boldsymbol{\kappa}, \omega) = F_G(\boldsymbol{\kappa}, \omega)\{1 + C_2 H_2(x) + C_4 H_4(x) + \cdots\} \tag{13.105}$$

where $H_n(x)$ are Tchebychev–Hermite polynomials and

$$x = \frac{\omega}{\sqrt{(\overline{\omega^2})}}. \tag{13.106}$$

The first two polynomials in (13.105) are

$$H_2(x) = x^2 - 1, \qquad H_4(x) = x^4 - 6x^2 + 3. \tag{13.107}$$

The idea behind the particular expansion (13.105) arises from the orthogonality condition

$$\int_{-\infty}^{\infty} \mathrm{d}\omega\, H_n(x) H_m(x) F_G(\boldsymbol{\kappa}, \omega) = n!\, \delta_{m,n}, \tag{13.108}$$

which means that the addition of a further term, $n+1$ say, leaves the first n moments unchanged. From (13.108),

$$C_n = \frac{1}{n!} \int_{-\infty}^{\infty} \mathrm{d}\omega\, H_n(x) F(\boldsymbol{\kappa}, \omega), \tag{13.109}$$

so that

$$C_2 = \tfrac{1}{2} \int_{-\infty}^{\infty} \mathrm{d}\omega \left(\frac{\omega^2}{\overline{\omega^2}} - 1\right) F(\boldsymbol{\kappa}, \omega) = 0 \tag{13.110}$$

and, using the definition (13.83),

$$C_4=\frac{1}{4!}\int_{-\infty}^{\infty}\mathrm{d}\omega\left\{\frac{\omega^4}{(\overline{\omega^2})^2}-\frac{6\omega^2}{\overline{\omega^2}}+3\right\}F(\boldsymbol{\kappa},\omega)=\tfrac{1}{8}\alpha. \tag{13.111}$$

Hence

$$F(\boldsymbol{\kappa},\omega)\simeq(2\pi\overline{\omega^2})^{-1/2}\exp\left(-\frac{\omega^2}{2\overline{\omega^2}}\right)\left\{1+\frac{\alpha}{8}H_4\left(\frac{\omega}{\sqrt{(\overline{\omega^2})}}\right)\right\}. \tag{13.112}$$

It is a simple matter to show that (13.112) possesses a maximum at finite ω, for fixed $\boldsymbol{\kappa}$, when α is negative. Figure 13.1 shows that, for $T=\infty$, α is negative near the zone boundary for some directions of the wave vector. The condition $\alpha<0$ for the existence of a spin oscillation in a paramagnet is obtained also from a theory based on a generalized Langevin equation for the spin density (Young and Shastry 1982; Lovesey and Meserve 1973) that is much akin to a theory of particle density fluctuations in a monatomic fluid described in § 5.4.3.

We turn now to an approximate calculation of the frequency of the spin oscillation. If the peaks in $F(\boldsymbol{\kappa},\omega)$ arose from an undamped oscillation, then the frequency of oscillation would be given by the square root of (13.76), the normalized second moment of the response. Because the oscillations are damped, (13.76) gives only a first estimate of the observed frequency.

For $T=\infty$, the second moment is given by (13.77), which reduces to (13.78) for a nearest-neighbour exchange coupling. These expressions are independent of the sign of the exchange since ferro- and antiferromagnets have identical properties at infinite temperature. The response function of ferro- and antiferromagnets are distinctly different at lower temperatures, of course, and particularly at large wave vectors.

To estimate the spin-oscillation frequency at a finite temperature we begin with the exact expression for the second frequency moment (13.76). The susceptibility and short-range correlation function can be estimated from (13.29) and (13.35). However, instead of determining the critical temperature from (13.25) we will require the spin correlation function to satisfy the constraint, valid for all $T>T_c$,

$$\langle\hat{S}^{\alpha}_{\mathrm{m}}\hat{S}^{\alpha}_{\mathrm{m}}\rangle=\tfrac{1}{3}S(S+1).$$

This constraint is equivalent to the relation

$$\tfrac{2}{3}S(S+1)\beta\mathscr{J}(0)=\frac{1}{N}\sum_{\mathbf{q}}\left\{\frac{1}{\mu}-\mathscr{J}(\mathbf{q})/\mathscr{J}(0)\right\}^{-1} \tag{13.113}$$

which is used to determine the parameter μ. The physical significance of μ is apparent from the corresponding expression for the wave-vector dependent susceptibility

$$\chi_{\mathbf{q}}^{\alpha} = (g\mu_{\mathrm{B}})^2 \left\{ 2\mathcal{J}(0) \left[\frac{1}{\mu} - \mathcal{J}(\mathbf{q})/\mathcal{J}(0) \right] \right\}^{-1} \tag{13.114}$$

and the isothermal susceptibility

$$\lim_{\mathbf{q}\to 0} \chi_{\mathbf{q}}^{\alpha} = (g\mu_{\mathrm{B}})^2 \left\{ 2\mathcal{J}(0) \left(\frac{1}{\mu} - 1 \right) \right\}^{-1}. \tag{13.115}$$

From (13.115) we deduce that the critical temperature is determined by $\mu = 1$. The relations (13.113) and (13.114) are equivalent to the spherical model (Stanley 1971) and it is known that this model tends to underestimate the critical temperature.

For nearest-neighbour coupling,

$$\mathcal{J}(0) = rJ, \quad \text{and} \quad \mathcal{J}(\mathbf{q}) = \mathcal{J}(0)\gamma_{\mathbf{q}}$$

and (13.113) reduces to

$$\tfrac{2}{3}S(S+1)\beta rJ = \frac{1}{N} \sum_{\mathbf{q}} \left\{ \frac{1}{\mu} - \gamma_{\mathbf{q}} \right\}^{-1} = I(\mu) \tag{13.116}$$

where $I(\mu)$ is an extended Watson integral. The critical temperature for the model is $T_c/I(1)$ with T_c given by (13.25), and

$$\begin{aligned} I(1) &= 1.5164 \quad \text{s.c} \\ &= 1.3932 \quad \text{b.c.c} \\ &= 1.3447 \quad \text{f.c.c} \end{aligned} \tag{13.117}$$

The calculation of the second moment from (13.76) is straightforward for both ferro- and antiferromagnets. We find

$$\overline{\omega^2} = \frac{8r^2J^2S(S+1)}{3\mu^2}(1-\gamma_{\mathbf{q}})(1 \mp \mu\gamma_{\mathbf{q}})\{1 - \mu/I(\mu)\} \tag{13.118}$$

where the upper and lower signs are for ferro- and antiferromagnetic coupling, respectively. The second moment (13.118) increases, from the value (13.77), as the temperature is decreased from $T = \infty$; for a simple cubic lattice and a wave vector at the zone boundary ($\gamma_{\mathbf{q}} = -1$) the frequency of the collective spin oscillation is estimated to increase by a factor two.

Calculations for EuO (Young and Shastry 1982) give values in good agreement with observation (Mook 1981). The second-neighbour exchange in EuO is significant, and so calculations are based on (13.113) and (13.114), and the corresponding expression for $\overline{\omega^2}$. For $T = 1.27T_c$ and a wave vector at the zone boundary, $\hbar(\overline{\omega^2})^{1/2} = 1.9$ meV and the observed energy is 2.2 meV.

Even better agreement between theory and experiment is obtained using an approximation for the generalized Langevin equation which incorporates the fourth frequency moment (Young and Shastry 1982). The latter can be estimated in terms of the spherical model. A collective spin oscillation exists in the theory for $\alpha<0$ only, and this condition is satisfied, for EuO and EuS, for large wave vectors, close to the zone boundary, as in Fig. 13.1. The dynamics of EuS is highly anisotropic, and this is traced, in the theory, to the existence of competing interactions; the second-neighbour exchange in EuS is antiferromagnetic, and approximately half the magnitude of the nearest-neighbour exchange.

13.4. Low-dimensional systems (Steiner *et al.* 1976; Lovesey and Loveluck 1977; Lovesey 1981)

There is now abundant evidence that many anisotropic materials exhibit properties expected of one- and two-dimensional systems, at least in some restricted, but readily accessible, temperature range. The manifestation of one- or two-dimensional magnetic properties reflects a strong intrachain, or intraplane, exchange coupling of magnetic ions, with a relatively weak interchain, or interplane, coupling. The residual interchain, or interplane, interaction (due, for example, to dipolar forces) leads to three-dimensional behaviour at a sufficiently low temperature.

There is no long-range magnetic order in a purely one-dimensional magnet with finite-range forces and $T>0$ (Mermin and Wagner 1966). A definite correlation length exists, which increases as $T\to 0$ and diverges in the limit of absolute zero. The mean-field approximation, considered in § 13.2, fails to reproduce the dependence of the phase transition on spatial dimensionality, since the critical temperature (13.25) is finite for all dimensions. For nearest-neighbour exchange coupling, the spin correlation function in the chain direction,

$$\langle \hat{S}^{\alpha}_{m}\hat{S}^{\alpha}_{m+l}\rangle = \tfrac{1}{3}S(S+1)u^{l}; \qquad l\geqslant 0 \tag{13.119}$$

where $\mathbf{l}=(0,0,la)$, $|u|\leqslant 1$, and the maximum value is achieved for $T=0$ (Stanley 1971; Steiner *et al.* 1976). If the spins are treated as classical variables,

$$u=\coth\{2JS(S+1)\beta\}-\{2JS(S+1)\beta\}^{-1} \tag{13.120}$$

and $-1\leqslant u\leqslant 0$ for antiferromagnetic coupling, $J<0$. The susceptibility and second frequency moment are readily evaluated,

$$\chi_{\mathbf{q}}=(g\mu_{\mathrm{B}})^{2}\tfrac{1}{3}S(S+1)\beta(1-u^{2})/(1+u^{2}-2u\gamma_{\mathbf{q}}) \tag{13.121}$$

and

$$\overline{\omega^{2}}=8J(1-\gamma_{\mathbf{q}})u(1+u^{2}-2u\gamma_{\mathbf{q}})/\{\beta(1-u^{2})\} \tag{13.122}$$

where $\mathbf{l} = (0, 0, la)$, $|u| \leq 1$, and the maximum value is achieved for $T = 0$ and $q = (\pi/a)(J < 0)$ and

$$\overline{\omega^2} = 16J^2S(S+1)(1-\gamma_{\mathbf{q}})(1 \mp \gamma_{\mathbf{q}}); \qquad T \to 0 \tag{13.123}$$

where the upper and lower signs are for $J > 0$ and $J < 0$, respectively. The results (13.123) have the same wave-vector dependence as we find for linear spin waves in a one-dimensional Heisenberg magnet (cf. §§ 9.2.1 and 9.6.1).

In view of the strong short-range order that exists close to the critical temperature, $T = 0$, it is not surprising that one-dimensional magnets support weakly damped spin oscillations at low temperatures. The dispersion of the oscillation frequency is described by (13.123) for $T \ll |J|$.

The damping constant, or inverse life-time, $\Gamma(q)$, for classical spins and ferromagnetic coupling ($J > 0$), is known to leading-order in T/J (Reiter and Sjölander 1980). For $q < (\pi/a)$,

$$\Gamma(q) \propto T \sin(aq)$$

and at the zone boundary, $aq = \pi$,

$$\Gamma(q) \propto T^{3/2}.$$

The damping constant for quantum spins, calculated in § 9.2.4, depends strongly on S (Glaus *et al.* 1983). At the zone boundary, $\Gamma(q) \propto \{T/(S-1)\}^2$, whereas over the rest of the zone the wave vector dependence is the same as for classical spins. For antiferromagnetic coupling ($J < 0$) and classical spins the damping constant is independent of the wave vector, to leading-order in the temperature (Reiter and Sjölander 1980).

The spin fluctuations are suppressed by a magnetic field, as witnessed by the change induced in the inverse correlation length. From (13.121) we obtain, in zero field,

$$a\kappa_1 = (1-u)/u^{1/2} \tag{13.124}$$

and κ_1 is proportional to the temperature in the limit $T \to 0$. In the same temperature limit, but with an external magnetic field, the longitudinal and transverse correlation lengths for $J > 0$ both saturate, and have values proportional to the square root of the field, in the limit of small fields (Lovesey and Loveluck 1976*a*).

The effects of an antiferromagnetically coupled system, $J < 0$, are even more pronounced, since it is forced to display ferromagnetic properties when the field exceeds a critical value. In consequence, the longitudinal inverse correlation length, measured relative to a superlattice peak, (π/a), diverges as the external field approaches the critical value. There are equally pronounced changes in the spin response function, particularly for small wave vectors, near the zone centre, where a weakly damped mode emerges with increasing field (Lovesey and Loveluck 1976*b*; Balcar *et al.* 1984)

The effects of a single-site planar anisotropy, of the form

$$+B\sum_{l}(\hat{S}^{z})^{2} \tag{13.125}$$

have been studied extensively, with a view to interpreting measurements on $CsNiF_3$ (Kakurai *et al.* 1987), which is one of the few known ferromagnetically coupled quasi-one-dimensional magnets. For $T \ll J$, the spins lie in the xy plane, if $B > 0$, to minimize the effect of the anisotropy energy. The latter has the effect of reducing the critical temperature for spin fluctuations in the z-direction, as indicated by (13.57), and since the critical temperature for $B = 0$ is at $T = 0$, the corresponding inverse correlation length for $B > 0$ is finite at $T = 0$. The ordering occurs in the xy plane, as $T \to 0$, so the inverse correlation lengths in the plane, and the allied susceptibility, show critical behaviour. The suppression of spin fluctuations in the z-direction by the anisotropy results, of course, in an increase in the relaxation time of the spin oscillation.

Because the spin-pair correlation function in the cross-section vanishes unless the relative position vector is parallel to the chain axis, the cross-section is independent of the component of the scattering vector perpendicular to the chain. If the chain is in the z-direction, the cross-section is independent of κ_x and κ_y, and the scattering has equi-intensity in planes perpendicular to the z-axis. For ferromagnetic short-range order $J > 0$, the maximum intensity occurs in the planes that pass through magnetic Bragg points, whereas for antiferromagnetic short-range order $J < 0$, the maximum intensity occurs in planes at odd multiples of (π/a) (cf. (13.121)) corresponding to a doubling of the unit cell for an ordered antiferromagnet.

Diffraction from a two-dimensional magnetic system is characterized by equi-intensity along lines in reciprocal lattice space. Of course, the observed intensity will show some variation with wave vector, along the line, which arises from the wave vector dependence of the form factor, for example. A useful feature of both one- and two-dimensional systems is that it is possible to arrange for constant-angle and constant-wave-vector scattering to be identical. Hence, the wave-vector dependent susceptibility is obtained directly from diffraction measurements, using no energy discrimination, without any Placzek corrections. The magnetic anisotropy in low-dimensional systems is very apparent in the excitation spectrum. Spin waves propagating perpendicular to planes containing strongly coupled spins will show negligible dispersion compared with that for spin waves propagating in the planes.

The thermodynamic properties of two-dimensional magnets depend strongly on the dimensionality of the spins. An isotropic two-dimensional Heisenberg magnet does not order at any temperature (Mermin and Wagner 1966) whereas, it is well known that, the two-dimensional Ising spin model shows a phase transition (Stanley 1971). The exact transition temperature of the two-dimensional, nearest-neighbour, Ising model is $(0.567\cdots)T_c$, where T_c is obtained from (13.25) with $r = 4$ and $S = \frac{1}{2}$. At the transition temperature, the susceptibility deviates substantially from the Ornstein–Zernike form, and the exponent $\eta = 0.25$.

REFERENCES

Als-Nielsen, J. (1976). *Phase transitions and critical phenomena*, Vol. 5a (ed. C. Domb and M. S. Green). Academic Press, New York.
Balcar, E., Lovesey, S. W., and Loveluck, J. M. (1984). *Z. Physik* **B54,** 195.
Bohn, H. G., Zinn, W., Dorner, B., and Kollmar, A. (1980). *Phys. Rev.* **B22,** 5447.
Böni, P. and Shirane, G. (1986). *Phys. Rev.* **B33,** 3012.
Collins, M. F. and Marshall, W. (1967). *Proc. Phys. Soc.* **92,** 390.
Dietrich, O. W., Als-Nielsen, J., and Passell, L. (1976). *Phys. Rev.* **B14,** 4932.
Glaus, U., Lovesey, S. W., and Stoll, E. (1983). *Phys. Rev.* **B27,** 4369.
Hohenberg, P. C. and Halperin, B. I. (1977). *Rev. mod. Phys.* **49,** 1.
Hubbard, J. (1971). *J. Phys.* **C4,** 53.
Kakurai, K., Pynn, R., Steiner, M., and Dorner, B. (1987). *Phys. Rev. Lett.* **59,** 708.
Kociński, J. (1983). *Theory of Symmetry Changes at Continuous Phase Transitions.* Elsevier, Amsterdam.
— - and Wojtczak, L. (1978). *Critical scattering theory.* Elsevier, Amsterdam.
Lindgård, P.-A. (1978). *Topics in current physics*, Vol. 6. Springer-Verlag, Heidelberg.
—— (1983). *Phys. Rev.* **B27,** 2980.
Lovesey, S. W. (1981). *Solid-state sciences*, Vol. 23. Springer-Verlag, Heidelberg.
—— and Loveluck, J. M. (1976*a*). *J. Phys.* **C9,** 3639.
—— and —— (1976*b*). *J. Phys.* **C9,** 365.
—— and —— (1977). *Topics in current physics*, Vol. 3. Springer-Verlag, Heidelberg.
—— and Meserve, R. A. (1973). *J. Phys.* **C6,** 79.
—— and Williams, R. D. (1986). *J. Phys.* **C19,** L523.
Lurie, N. A., Shirane, G., Heller, P., and Linz, A. (1972). Magnetism and magnetic materials, 18th Annual Conference, *AIP Conf. Proc.*, Vol. 10.
Lynn, J. W. (1975). *Phys. Rev.* **B11,** 2624.
—— (1981). Physics of transition metals. *Inst. Phys. Conf. Ser.* **55,** 683.
—— (1983). *Phys. Rev.* **B28,** 6550
Malinoski, F. A. (1973). *Int. J. Mag.* **4,** 245.
Martinez, J. L., Böni, P., and Shirane, G. (1985). *Phys. Rev.* **B32,** 7037.
Mermin, N. D. and Wagner, H. (1966). *Phys. Rev. Lett.* **17,** 1133.
Mezei, F. (1986). *Physica* **136B,** 417.
Minkiewicz, V. J. (1971). *Int. J. Mag.* **1,** 149.

Mook, H. A. (1981). *Phys. Rev. Lett.* **46,** 508.
—— Lynn, J. W. (1985). *J. appl. Phys.* **57,** 3006.
—— (1986). *Phys. Rev. Lett.* **57,** 150.
Patashinskii, A. Z. and Pokrovskii, V. L. (1979). *Fluctuation theory of phase transitions.* Pergamon Press, Oxford.
Reiter, G. and Sjölander, A. (1980). *J. Phys.* **C13,** 3027.
Schulhof, M. P., Heller, P., Nathans, R., and Linz, A. (1970). *Phys. Rev.* **B1,** 2304.
Stanley, H. E. (1971). *Phase transitions and critical phenomena.* Oxford University Press, Oxford.
Steiner, M., Villain, J. and Windsor, C. G. (1976). *Adv. Phys.* **25,** 87.
Takahashi, M. (1983). *J. Phys. Soc. Japan* **52,** 3592.
Tucciarone, A., Lau, H. Y., Corliss, L. M., Delapalme, A., and Hastings, J. M. (1971). *Phys. Rev.* **B4,** 3206.
Van Hove, L. (1954). *Phys. Rev.* **95,** 1374.
Wallace, D. J. and Zia, R. K. P. (1978). *Rep. Prog. Phys.* **41,** 1.
Wicksted, J. P., Böni, P., and Shirane, G. (1984). *Phys. Rev.* **B30,** 3655.
Young, A. P. and Shastry, B. S. (1982). *J. Phys.* **C15,** 4547.

INDEX

Made in the USA
San Bernardino, CA
03 March 2014